DICTIONARY

OF MODERN

BIOLOGY

By
Norah Rudin, Ph.D.

BARRON'S

Illustrations and cover art created by the author. In some cases, a textbook illustration was used as a model. A few are modified from original drawings.

All inquiries should be addressed to:
Barron's Educational Series, Inc.
250 Wireless Boulevard
Hauppauge, NY 11788

Library of Congress Catalog Card No. 96-24180

International Standard Book No. 0-8120-9516-2

Library of Congress Cataloging-In-Publication Data
Rudin, Norah.
    Dictionary of Modern Biology / Norah Rudin.
        p.   cm.
    ISBN 0-8120-9516-2
    1. Biology—Dictionaries.   I. Title
QH302.5.R83   1997
574'.03—dc20   96-24180
                                                CIP

PRINTED IN THE UNITED STATES OF AMERICA

9   8   7   6   5   4   3   2   1

CONTENTS

ACKNOWLEDGMENTS

Special thanks to Grace Freedson of Barron's Educational Series, Inc. for giving me the opportunity to author this volume. Pat Wilson got me started, Warren Bratter supervised the work in progress, and Max Reed brought it all home. Myriam Alhadeff provided excellent scientific commentary, and made a number of useful suggestions regarding technical content.

I am indebted to Lynn Margulis for taking time out of her busy schedule to provide me with recent materials and review the relevant sections. (I am, of course, solely responsible for the final content).

I am eternally grateful to Keith Inman, my partner in most other crimes, for constant encouragement in all areas of life.

Last, but not least, my parents, Benjamin and Jenny Rudin, made it possible for me to take the year and a half necessary to complete this project.

KEY TO EFFECTIVE USE OF THIS DICTIONARY

The field of biology is growing and changing at a more explosive rate than ever. Providing a comprehensive, yet clear and usable view of this subject in pocket dictionary form is a formidable challenge. Often several different, yet legitimate views of a particular concept may exist. This is particularly true of biological classification (see Appendix IX). In such cases, rather than attempt to enumerate all opinions, I have presented one view or system, and simply noted that alternate camps exist. Additionally new information is being accumulated at a phenomenal rate, particularly in fields such as immunology and molecular biology applications. Although every attempt has been made to obtain the latest material, some of the definitions contained herein will no doubt already be out of date by the time this volume reaches the shelves. Such is the limitation of printed material.

This said, the following notes will help the reader get started in using this dictionary. My hope is that by using the cross-references as a road map between definitions, the reader will gain an appreciation of biology as an integrated whole, rather than as simply fragments of isolated information.

General Considerations:

Entries: All main entries are given in **boldface type**. In cases where both the singular and plural forms are commonly used, both forms are given and the plural is indicated by the abbreviation "pl." In cases where the acronym for a concept or process is most commonly used, the reader is directed to the entry for it. For instance **polymerase chain reaction** (see *PCR*) directs the reader to the entry for **PCR**.

Alphabetization: All entries have been alphabetized by letter rather than by word; for example, *genetics* precedes *genetic self-incompatibility*, and *polarizing microscope* precedes *polar nuclei*.

Cross-References: There are two types of cross-references, explicit and implicit. Explicit cross-references are *italicized* and enclosed in (parentheses) at the end of a definition. For example:

> **centrolecithal.** Some eggs, typically insect, in which a large yolk occupies the center of the egg. (compare *telolecithal*).

Explicit cross-references are indicated by either a "see" or "compare" qualifier. Although these concepts are not without overlap, "compare" generally refers the reader to a term for the opposite concept or structure, and "see" refers the reader to a more specific or general aspect of the same concept or structure. For example:

> **chemotaxis.** A movement by an organism in response to a chemical concentration gradient (see *taxis*).

Implicit cross-references are indicated by *italic type* within the text of an entry. This calls attention to terms that are defined as separate entries, and that should be understood to assure complete comprehension of the term at hand. Terms given in italics may appear as entries having different forms (*cells* vs. *cell*), as different parts of speech (*mitosis* vs. *mitotic*), or within strings of other words (*mitotic cell division*). In most cases, the reader will be able to readily cross-reference the word or word string; if the word string is not listed as an autonomous term, simply look up the words separately or as pairs (*mitotic, cell* and *cell division*).

The appearance of a term in regular type does not preclude the possibility of its inclusion as a separate entry. Terms that are commonly understood, such as bone or heart, are defined but not italicized with every use.

For most terms, I have tried to orient the reader in the first sentence or two, and provide more details and/or greater clarification in the latter portion of the definition. To get the general idea of a term, without getting bogged down with details, try reading just the beginning of a definition.

Biology-Specific Considerations:

Organismal names: Only the formal Latin binomial names of organisms are italicized. Common names and informal derivatives of the Latin names are not. For example, *Drosophila melanogaster* is italicized, while "fruit fly" is not; *Nematoda* is italicized, while "nematodes" is not. Names indicating hierarchical classifications other than *genus species* (Crustacea) are also not italicized. Although many common and Latin names of organisms are mentioned in the text, neither the listing nor the cross-referencing should be considered comprehensive.

Orientation: Anatomical orientation terms are not cross-referenced, but are defined. They are listed here for easy reference.

> *Dorsal* – Back
> *Ventral* – Front
> *Lateral* – Side

Root Words: Most biological terms have their own origin in Latin or Greek. The most common word roots and elements are defined in the text. Armed with this information, the reader can often decipher an unfamiliar term just from its etymology.

Classification: The following classification concepts may be unfamiliar to some readers with a previous background in biology.
• Traditionally, the hierarchical class below *kingdom* has been termed *phylum* for animals, protoctists (protists), and bacteria and *division* for plants and fungi. There is a trend, at least in some circles, toward using the term *phylum* for this level of classification in all groups. Because this represents a simplification without loss of information, it is followed in this work.
• The term *craniate* is often substituted in this work for the more commonly used term *vertebrate*, because it is more descriptive of the characteristics of organisms belonging to this *phylum*.
• The term *protoctist* is usually substituted for the better-known term *protist*. This work will follow the idea of using *Protoctista* as the kingdom name, while reserving *protist* for those members that are truly single-celled.
• The term *undulipodium* (pl. *undulipodia*) is used to describe *eukaryotic flagella* and *cilia*, while the term *flagellum* (pl. flagella) is reserved for the bacterial organ of quite different structure. This usage is followed because it represents both a simplification and clarification.
The terms and ideas listed above are also cross-referenced and discussed in the text.

◆ A ◆

A (see *adenine*)

aardvark (see *Tubulidentata*)

ABA (see *abscisic acid*)

abaxial The surface of any structure facing away from the main axis. In lateral organs such as a leaf, it refers to the lower surface. (compare *adaxial*)

abdomen
1. The central cavity of vertebrates that contains digestive (*visceral*) organs such as the stomach, intestines, liver, and kidneys. It does not contain the heart or lungs.
2. The posterior segments of an arthropod body.

abdominal Pertaining to the abdomen.

abducens nerves (cranial nerves VI) A pair of nerves originating in the *brainstem* that supply *motor* impulses to the eye muscles.

abiogenesis The creation or genesis of life from inanimate matter. (see *spontaneous generation*)

abiotic The nonliving factors in the environment including climactic, geological, and geographical features that may influence ecological systems. Examples include chemical *pollutants* and seismic activity. (compare *biotic*)

ABO (see *blood groups*)

abomasum The true digestive stomach and the last of four specialized digestive sections of cud-chewing animals (*ruminants*) such as the cow.

aboral Away from the mouth.

abscisic acid (ABA abscisin, dormin) A plant substance produced in mature green leaves and fruits that suppresses bud growth and promotes leaf drop (*senescence*). It also functions in promoting and maintaining *dormancy*, and participates in the opening and closing of *stomata*. It apparently antagonizes growth-promoting substances such as *gibberellins* and *auxins*, possibly by inhibiting *protein synthesis*. (see *growth substance*)

abscisin (see *abscisic acid*)

abscission The normal separation of flowers, fruit, and leaves from plants. The process is promoted by the plant substance *ethylene*. Final separation is accomplished by mechanical forces such as the wind. (see *abscission layer, abscission zone*)

abscission layer In the normal separation of flowers, fruit, and leaves from plants, the layer formed by the breakdown and separation of specialized *parenchyma* cells at the *abscission zone*. (see *abscission*)

abscission zone In the normal separation of flowers, fruit, and leaves from plants, the zone at which the *abscission layer* is formed and where the structure eventually separates. (see *abscission*)

absolute refractory period (see *refractory period*)

absorbtion spectrum The definitive wavelengths (λ) of electromagnetic radiation absorbed by a particular substance. The absorbtion spectrum is depicted as a graph of wavelength variation versus absorbtion of radiation. It may include ultraviolet, visible, or infrared regions of the electromagnetic spectrum. It is often used as an aid in identifying light-absorbing *photosynthetic pigments* such as *chlorophylls*. In a light-induced event, the absorbtion spectrum of a particular *photoreceptor* matches the *action spectrum* of the response. (see *chromophore*)

abyssal The region of the ocean below about 1,000 meters and beyond the continental shelf. The abyssal realm is the largest environment on earth. (compare *benthic, neritic, oceanic, photic zone*)

Acantheria (see *Actinopoda*)

Acanthocephala A phylum of *pseudocoelomate parasitic* animals, the spiny-headed worms. They lack any free-living stage and exist in the gut of a wide range of vertebrates. Mating and *embryonic* development take place within the female worm while still inside the *host*, and *larvae* are released with the feces. It is common for these worms to migrate up the

1

food chain as one organism ingests another.

Acari The order of arachnids that contains the largest number of species, the mites. Most mites are small, less than one millimeter long, have their *cephalothorax* and abdomen fused into an unsegmented body, and as adults have eight legs. Some exhibit *paedomorphosis* in which juvenile stages become reproductive and development then stops at that stage. They inhabit virtually every terrestrial, freshwater, and shallow marine habitat known and are essentially *omnivores*. Some form *mutualistic* relationships with plants that, in turn, protect them from *herbivores*. Others, such as spider mites, cause damage. Both internal and external *parasitic* species are found, including many detrimental to man. These include dust mites, which cause allergy, and chiggers, and ticks, which carry respectively Rocky mountain spotted fever and lyme disease.

accessory nerves (cranial nerves XI) A pair of nerves arising from both the *brainstem* and the *spinal cord*. They supply *motor* impulses to the throat muscles for swallowing and speech.

accommodation The *reflex* process by which the eye focuses an image on the *retina*. At rest, the *lens* and *cornea* of the vertebrate eye are flattened and focused on infinity. To focus on a nearby object, the lens, and to some extent the cornea, must become more curved (convex) or the lens must move in relation to the cornea. In mammals, reptiles, and birds, muscles change the shape of the lens. In fish and amphibians, muscles move the lens in relation to the cornea.

acellular Organisms or tissues in which a cell is subject to the influence of multiple *nuclei*. This occurs when cells are incompletely separated after division or when *nuclear division* occurs in the absence of cytoplasmic division (*cytokinesis*). Examples include *striated muscle*, plasmodial slime molds, and some fungal *mycelia*. (see *syncytium, plasmodium, conidia*)

acentric A *chromosome* lacking a *centromere*. Such a chromosome is rarely inherited, as it contains no attachment point for the *spindle* apparatus that ensures chromosome *segregation* during *nuclear division*. (see *mitosis, meiosis*)

acervulus, acervuli (pl.) A close-packed mat of *vegetative* fungal *hyphae* that gives rise to *conidiophores*. (*asexual spore*-forming structures)

acetic acid (ethanoic acid, CH_3COOH) The acid in vinegar.

acetylcholine One of several *neurotransmitters* that relays the electrical *nerve impulse* in chemical form. Acetylcholine is found at most *excitatory synapses* and *neuromuscular junctions*.

acetylcholinesterase (cholinesterase) An enzyme that removes residual *acetylcholine* from the *synaptic cleft* of *neuromuscular junctions* and *cholinergic nerve synapses*. This prevents any lingering of a single impulse that would interfere with a subsequent initiation. It is one of the fastest-acting enzymes known and permits a rapid succession of *nerve impulses* to be transmitted. Poisons such as nerve gases (tabun, sarin) and some agricultural insecticides (parathion) inhibit cholinesterase, producing continuous neuromuscular stimulation. Such compounds may be lethal to vertebrates, as breathing, for example, requires rhythmical muscular contraction. (see *neurotransmitter*)

acetyl-coenzyme A Better-known as *acetyl-CoA*, it is an important *metabolite* of many *catabolic* processes of the *eukaryotic* cell. In particular, it is formed from *coenzyme A* during the introduction of *pyruvate* into the *Krebs cycle*. It may be oxidized to produce *ATP*, but when levels are high, this pathway is inhibited, and excess acetyl-CoA is channeled into *fatty acid biosynthesis*. Thus, when an organism eats more food than necessary for immediate energy production, the excess is stored as fat. (see *cellular respiration*)

achene A small, dry, single-seeded, *indehiscent* fruit.

achlorophyllous Lacking in *chlorophyll*.

acid Any chemical compound that releases a hydrogen ion (H^+) in aqueous solution, thus raising the relative concentration of protons. It manifests by a decrease in pH of the solution. (compare *base*)

acid mine drainage Acid water that drains from coal mines and some metal mines. It may enter the surface water, making it unfit for life. It is a major cause of environmental damage.

acid rain Natural precipitation that is highly acidic (pH< 5.0). It occurs when gaseous pollutants, such as sulphur dioxide, carbon monoxide, and nitrogen oxide, dissolve in atmospheric moisture, and are subsequently deposited as rain, snow, and fog now containing sulfuric acid, carbonic acid, and nitric acid. Unpolluted rain is slightly acidic (pH 5.0 to 5.6). Damage begins to occur below about pH 4.5. Because pollutants are often introduced in the upper atmosphere in an attempt to disperse them, they often decimate both aquatic and terrestrial environments far from their site of production.

acinus, acini (pl.) The basic functional unit of the lung. Acini occur at the ends of the *bronchioles* and resemble miniature bunches of grapes. Each individual sphere is essentially a membrane-bound air bubble called an *alveolus*. Acini serve to increase the surface area of the lung, enabling efficient exchange of respiratory gases.

acoelomate (aperitoneal) Animals possessing essentially no body cavity other than a digestive tract. These include the solid worms such as the flatworms and ribbon worms. (compare *pseudocoelomate, coelomate*)

acorn worms (see *Hemichordata*)

acoustic nerve (see *auditory nerve*)

acoustico-lateralis system (see *lateral-line system*)

acquired immune deficiency syndrome (see *AIDS*)

acquired immunity Any form of *immunity* that is not innate. It may be acquired actively, passively, or through vaccination. (see *active immunity, passive immunity, vaccine*)

acquired immunological tolerance A learned acceptance of an otherwise foreign *antigen* introduced into an *embryo* before full development of the *immune system*. The mature animal responds to the tissue as "self," and will successfully accept *grafts* of it.

Acrasea (see *Rhizopoda*)

Acrasiomycota (see *Rhizopoda*)

acrocentric A *chromosome* in which the *centromere* is located nearer one end or the other.

acropetal The successive development of plant structures from the base (oldest) to the tip (youngest). It is also used to describe the movement of substances from the base to the apex.

acrosome A specialized vesicle on the tip of a *spermatozoon* head containing *lytic* enzymes that are released on contact with the egg at *fertilization*. This helps dissolve the egg coating, facilitating entry of the sperm head with its *nucleus*. (see *vitelline membrane*)

acrylamide (see *polyacrylamide*)

ACTH (see *corticotrophin*)

actin A family of abundant *structural proteins* present in all cells as a constituent of the *cytoskeleton*. Actin is found in *eukaryotes* but not *prokaryotes*. It is involved in organismal, cellular, and *intracellular* movement. In muscles, subunits of actin can *polymerize* to form *thin filaments*, which with *thick filaments* of the protein *myosin,* form *myofilaments*, or *actomyosin*, the basic contractile unit. (see *globular proteins*)

Actinobacteria A phylum of *Gram-positive aerobic* filamentous bacteria, the actinomycetes. Many species are important in soil fertility and *antibiotic* production (for example, *Streptomyces*), but the actinomycetes also include the agents of leprosy and tuberculosis. *Frankia*, an *actinorrhizae*, forms *symbiotic nitrogen-fixing* nodules on some tree roots, including the alder, similar to the *Rhizobium* nodules on legumes. The odor of fresh earth is imparted mainly by actinomycetes.

actinomorphic The exhibition of *radial symmetry* in flowers.

actinomycetes (see *Actinobacteria*)

Actinopoda A phylum of *unicellular heterotrophic* protoctists, the actinopods. They have been known for centuries as sun *animalcules* because of the rays, or *axopods*, that project from their cells. Also known as radiolarians (marine) or heliozoans (freshwater), they are classified based on the particular arrangement of *microtubules* along the axis of each ray. Most classes of actinopods have a silica skeletal spine, but in one group, the skeleton is made of barium sulfate. Freshwater actinopods are covered with silicon scales.

Actinopterygii The subclass of bony fishes containing the ray-finned fish. They include the vast majority of living fish.

actinorrhizae The symbiotic association of a genus of actinobacteria with certain *temperate forest* tree roots. They form *symbiotic nitrogen-fixing* nodules similar to the *Rhizobium* nodules on legumes.

action potential The transitory reversal of localized electrical potential that occurs across the membrane of a *neuron* or *muscle cell* during the passage of a *nerve impulse*. It is characterized by an *all-or-none* phenomenon (it either happens or it does not), the occurrence of which depends on the magnitude of the stimulus. When a stimulus initiates *depolarization* of the membrane, *voltage-gated sodium channels* open, a process that is self-perpetuating, allowing sodium ions (Na^+) to flow back into the cell from the exterior, where they have been sequestered by the *sodium-potassium pump*. When enough stimuli or a large enough stimulus causes sufficient depolarization of the membrane to open all the channels, the polarity of the cell reverses and an action potential is fired. When it occurs at the *end plate* of a *motor neuron* connecting to a *muscle fiber*, it is called an *end-plate potential* (*EPP*). (see *resting potential, depolarization, facilitation, summation, threshold, active transport, ion channel, EPSP, IPSP*)

action spectrum The definitive wavelengths of electromagnetic radiation that incite a light-induced event. It is depicted as a graph of the intensity of a response plotted against the wavelength (λ) of incident light. The *nerve impulse* in the eye, *phototropism*, and *photosynthesis* are ex-

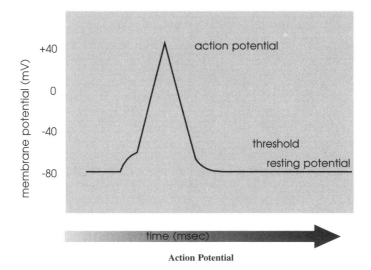

Action Potential

amples of events exhibiting an action spectrum. In any light-induced event, the *absorbtion spectrum* of a particular *photoreceptor* matches the action spectrum of the response.

activating enzymes The enzymes responsible for attaching the correct *amino acid* to each *tRNA* molecule. A specific activating enzyme exists for each of the 20 common amino acids. (see *protein synthesis, translation, genetic code*)

activation The series of biochemical and physiological events that are initiated when a *spermatozoon* head penetrates and fuses with the *plasma membrane* of an egg cell during *fertilization*.

activation energy The heightened state of energy that a molecule must obtain in order to undergo a specific chemical reaction. In biological systems, enzymes lower the *activation energy* for many reactions, enabling them to proceed when they otherwise would not.

activator A *regulatory protein* that binds to a *regulatory site* on *DNA*, permitting *transcription* of the adjacent gene(s). In some cases, this is achieved by unwinding the *DNA duplex*, permitting entry by *RNA polymerase*. (see *CAP site*)

active immunity The secondary immune response, in which residual *B-cells* (*memory cells*) that have previously recognized an *antigen* are primed to mount a swift defense. Many more antigen-sensitized cells are initially available, and their transformation to antibody-secreting *plasma cells* is also hastened. Active immunity may be artificially induced by vaccination with *attenuated* or inactivated microorganisms. (see *immune system, immune response*; compare *passive immunity*)

active site The region of an enzyme that interacts with the *substrate*. The three-dimensional configuration of *amino acids* at the *active site* determines the specificity of substrate binding. Most active site pockets fit the substrate like lock and key, forming an *enzyme-substrate complex*. In a number of enzymes, the configuration of the active site exactly matches the shape of the substrate only after binding has occurred. This process is called *induced fit*. (see *competitive inhibition, cofactor, enzyme inhibition*)

active transport The energy-dependent transport of individual molecules or ions across a *cell membrane* against a concentration gradient (uphill). It is expedited by "pumps" composed of protein molecules embedded in the membrane and fueled by *ATP*. (see *sodium-potassium pump, calcium pump, proton pump*; compare *facilitated diffusion, osmosis*)

actomyosin The protein complex involved in *muscle contraction* and other forms of cellular and organismal movement. It is formed between *polymerized* molecules of *actin* in the form of *thin filaments*, and *myosin* in the form of *thick filaments*, which interdigitate. The actin and myosin molecules in adjacent thick and thin filaments interact to pull the filaments past each other, resulting in a localized contraction. (see *myofilament, myofibril, smooth muscle, striated muscle, cardiac muscle, sarcomere*)

adaption Generally, changes that occur to organisms with regard to their environment and result in increased fitness.
1. In evolutionary adaption, *natural selection* ultimately results in genetic variation.
2. In physiological adaption, changes occur over an individual's lifetime as a result of, for instance environmental conditions that enable it to respond more effectively to them.
3. In *sensory* adaption, a decrease in the excitability of a sense organ that results from repeated stimulation, such that increasingly intense stimuli are required to produce the same response.

adaptive radiation The evolution of divergent species from a single ancestor each adapted to a different *ecological niche*. Darwin's finches are a classic example of this phenomenon. (see *speciation*)

adaptive significance The degree to which any particular behavior may increase survival and reproductive *fitness* as applied to the species. (see *behavioral ecology*)

adaptors Small synthetic *DNA* fragments that exhibit an overhang of a few *nucleotides* at one end, comparable to a *sticky end* generated by a specific *restriction endonuclease*. They may be attached to DNA molecules *in vitro* in order to generate sticky ends complementary to those in, for instance, a preferred *cloning vector*. (compare *linkers*)

adaxial The side of a lateral structure facing towards the main axis. In lateral organs, such as a leaf, it refers to upper surface. (compare *abaxial*)

adenine (A) A nitrogenous base found in *nucleic acids, nucleosides,* and certain *coenzymes* (for example, *NAD, FAD*). It is based on a *purine* ring structure.

adenosine A *nucleoside* of *adenine* linked to D-ribose with a β-glycosidic bond. Adenosine triphosphate (*ATP*), the energy currency of the cell and a building block of *nucleic acids*, is derived from adenosine.

adenosine diphosphate (see *ADP*)

adenosine monophosphate (see *AMP*)

adenosine triphosphate (see *ATP*)

Adenovirus A *DNA* virus causing tumors in animals.

adenyl cyclase A membrane-bound enzyme that converts *ATP* to *cAMP*. (see *second messenger*)

ADH (see *vasopressin, alcohol dehydrogenase*)

adipose tissue A type of *connective tissue* containing adipocytes (fat cells). *White fat* is the most common fat in animal bodies. Hibernating and newborn animals contain deposits of *brown fat* that are well supplied with nerves and blood vessels. It is rich in *unsaturated fatty acids* and *cytochromes*, containing an easily mobilized source of heat energy. Some theories postulate a lack of brown fat in man as one contributor to obesity.

A DNA A form of the *DNA double helix* that contains fewer water molecules (is less hydrated), thus is more stable than *B DNA*. Like the B form, the helix is right handed, but A DNA is found less frequently in living systems. (compare *Z DNA*)

ADP (adenosine diphosphate) The nucleoside *adenosine* with two phosphate groups attached. ADP is important in the energy *metabolism* of the cell. (see *ATP*)

adrenal glands A pair of *endocrine glands* located above each kidney in mammals. Each gland consists of a central *medulla* and an outer *cortex*. The medulla secretes *adrenaline* and *noradrenaline, hormones* involved in the *fight-or-flight* response. The cortex produces some *sex hormones* as well as other hormones (*corticosteroids*) that regulate the salt and water balance of the body and other functions.

adrenaline (epinephrine) The *hormone* produced by the *adrenal medulla* responsible for the *fight-or-flight* response in mammals. It mobilizes *glycogen,* a readily available energy source, from stores in the liver and has a variety of enabling effects on the *cardiovascular* and *muscular systems.*

adrenergic The type of *nerve fiber* that releases *noradrenaline* or, less commonly, *adrenaline* at its terminus when stimulated by a *nerve impulse*. It is characteristic of vertebrate *sympathetic motor nerve fibers*. (compare *cholinergic*)

Adrenocorticotrophic hormone (see *corticotrophin*)

adult A mature individual capable of producing *gametes* (eggs, sperm, *pollen*) that can fuse to form an *embryo*. The adult is typically *diploid* and usually produces *haploid* gametes by *meiosis*. (see *agamont*)

adventitious Any structure arising from an unusual place, such as shoots from roots (adventitious shoots) or roots from stems (adventitious roots).

aer-, aero- A word element derived from Greek denoting a connection with air or oxygen. (for example, *aerobic*)

aerenchyma Nondifferentiated plant tissue (*parenchyma*) containing large air spaces between the cells. It is found primarily in aquatic plants and, to some extent, in partially submerged land plants. It facilitates the diffusion of gases and provides buoyancy.

aerobe An organism that requires free oxygen (O_2) to live and grow. (see *aerobic respiration*; compare *anaerobe*)

aerobic Any biological process utilizing oxygen.

aerobic respiration The main and most efficient source of energy for most organisms. It makes up the portion of *cellular respiration* that requires free oxygen (O_2) as a *terminal electron acceptor* for the *oxidation* of organic substrates and includes *pyruvate* oxidation and the *Krebs cycle*. Aerobic respiration produces a high yield of energy in the form of *ATP* along with carbon dioxide (CO_2) and water (H_2O) as byproducts. The overall reaction for one molecule of glucose is $C_6H_{12}O_6 + 6O_2 = 6CO_2 + 6H_2O + 36ATP$. The reaction occurs in a number of steps. The initial reactions (*glycolysis*) occur in *anaerobic respiration* and in the cell *cytoplasm*. The remaining steps take place in the *mitochondria* in *eukaryotes*. They include the *Krebs cycle* leading to the *electron transport chain*. Some specialized bacteria employ an oxygen-containing compound such as nitrate (NO_3) as the terminal electron acceptor in place of molecular oxygen and produce corresponding byproducts. (see *oxidative phosphorylation*)

Aeroendospora (see *Endospora*)

aerotaxis The movement of an organism or cell in response to an oxygen concentration gradient. The movement may be positively aerotactic (towards) (for example, motile *aerobic* bacteria) or negatively aerotactic (away) (for example, motile *obligate anaerobic* bacteria). (see *taxis*)

afferent The conveyance of impulses or substances from outer regions of the body toward the center. It may apply to *sensory nerve impulses* as well as blood or other fluids. (compare *efferent)*

affinity The strength of the binding reactions between specific biological molecules, for instance, an *antigen* and its *antibody* or an enzyme and its substrate.

affinity chromatography The separation of a specific *macromolecule*, often a protein, from a complex solution based on its strong biological interaction with another molecule called the *ligand*. Typical examples include the noncovalent binding of an *antibody* to a specific *antigen* or *cell surface receptor* to a particular *hormone*. The ligand, in these examples the antibody or hormone, is attached to large beads of an inert column material. The solution containing the molecules to be separated is allowed to flow over the column. The molecules to be isolated, in these examples the antigen or receptor, bind tightly and specifically to their respective ligands and are retained in the column. They are then *eluted* from the column in various ways, often with an excess of free *ligand* or by changing the salt, pH, and/or temperature of the wash *buffer*. The entire reaction may also be carried out in solution with the ligand attached to inert, often magnetic, beads.

aflatoxin A poisonous substance produced by the *Aspergillus* fungus. It is highly toxic, and so mutagenic that it is used experimentally to induce high levels of *mutation* for study. Aflatoxin is sometimes found on foods such as peanuts and almonds that have been stored under conditions where *Aspergillus* is permitted to grow.

AFLP (see *Amp-FLP*)

after-ripening (see *stratification*)

aftershaft (see *contour feathers*)

agamont An adult *life cycle* stage that is capable of *sexual reproduction* but does not produce *gametes*. (compare *gamont*)

agar A solid culture media used in growing microorganisms. Various nutrient combinations are added to the agar base depending on what is required. Agar is a crude extract from red algae, one of the large seaweeds, in particular the genus *Gracilaria*.

agarose A powder that forms a porous gel when reconstituted with water and heated. It is used in the separation of *macromolecules* such as *DNA* and *RNA* by *electrophoresis*. Agarose is a highly purified version of *agar*.

agglutination The clumping together of cells, often red blood cells (*erythrocytes*), bacteria, or *spermatozoa*. It may occur spontaneously, but commonly results from the interaction of free *antibodies* or other linking agents with complementary *antigens* on the cell surface. The result is a network of linked cells. (see *blood groups, lectin*; compare *aggregation*)

aggregation The formation of a cluster of organisms or cells. They are not physically linked. (compare *agglutination*)

aggression A type of behavior involving threats and attacks. It often occurs in response to territorial threats or competition for a mate between individuals of the same species.

Agnatha (see *Cyclostomata*)

agonism The cooperation of two or more processes or systems. For example
  1. Drugs or *hormones* having *synergistic* effects such that their individual activities are complemented and/or multiplied. (compare *antagonism*)
  2. The complementary movements of paired muscles so that contraction of one is accompanied by relaxation of the other. One muscle regains its relaxed shape after contraction when the antagonistic muscle contracts. (see *antagonism*)
  3. (see *agonistic behavior*)

agonistic behavior (agonism) An animal behavior pattern exhibiting features of both *aggression* and *avoidance* and arising from a conflict between aggression and fear. It may apply to rival individuals of the same species, for instance in territorial conflicts. It can be also be ritualized into threat displays, reducing the need for physical violence.

agranulocyte A type of white blood cell (*leukocyte*) lacking any granules in its *cytoplasm*. *Lymphocytes* and *monocytes* are agranulocytes. (compare *granulocyte*)

agroecosystem An *ecosystem* that differs from naturally formed ones in a number of different factors. They include the ar-

rest of natural *succession*, the practice of *monoculture*, the growth of concentrated food sources in geometric patterns (encouraging pests), oversimplified ecosystems, and the continual disruption of the soil by plowing.

AIDS (acquired immune deficiency syndrome) A viral disease of the *immune system* caused by the human immunodeficiency virus (HIV), and characterized by destruction of the *T-cells* responsible for *cell-mediated immunity*. The HIV *retrovirus* is transmitted in bodily fluids but is generally not transferable through casual contact. As a *retrovirus*, it can remain inactive in its *host* for years without causing symptoms. Asymptomatic carriers may still transmit the disease. AIDS is virtually always fatal, although some genetic variants are beginning to emerge that appear to confer immunity at different stages of infection. (see *mucormycosis*)

air bladder (see *swim bladder*)

alanine (ala) An *amino acid.*

albatross (see *Procellariiformes*)

albinism The failure to develop external pigment due to a *recessive genetic mutation*. In humans, an albino has very light skin and hair, and pink eyes due to reflection of the blood vessels behind the noncolored *retina*. The defective *gene* codes for the enzyme tyrosine 3-monooxygenase that is involved in the production of the dark brown pigment *melanin*. Albinism results when defective genes are inherited from both parents.

albumen The egg white of birds and reptiles, comprised mostly of the protein *albumin*. Along with the yolk, it supplies nutrients to the developing *embryo* as well as providing cushioning.

albumin A group of simple, water-soluble proteins found in plants and animals. They are present in large quantities in blood, and involved in salt and water balance as well as the transport of small molecules. (see *albumen*)

alcohol dehydrogenase (ADH) An enzyme found in the liver that initiates the *metabolism* of alcohols in the body into water (H_2O) and carbon dioxide (CO_2).

aldosterone A *steroid hormone* secreted by the *cortex* (outer layer) of the *adrenal gland*. It controls *electrolyte balance* in *mammals*, in particular, the retention of sodium ions (Na^+) and excretion of potassium ions (K^+) in the *renal tubules*. Since all cellular functions, and, in particular, the *nervous* and *muscular systems*, depend on proper ion balance, aldosterone is one of two hormones essential for survival (the other being *parathyroid hormone*).

aleurone grains Membrane-bound granules of storage protein and digestive enzymes occurring in the outermost cell layer (*aleurone layer*) of the *endosperm* (*embryonic* nutritive tissue) in cereal seeds.

aleurone layer The outermost layer of cells in cereal seed *endosperm* (*embryonic* nutritive tissue) containing *aleurone grains*. It is *metabolically* active. At *germination*, it produces digestive enzymes that mobilize starch reserves stored in the rest of the endosperm.

aleuroplast A *plastid* in which *aleurone grains* containing storage protein and digestive enzymes are stored. They are common in cereal seeds. (see *aleurone grain, aleurone layer, proteoplast*; compare *amyloplast, elaioplast*)

alga, algae (pl.) A common term encompassing a large mixed group of *photosynthetic* protoctists. Algae are essentially single-celled plants and found in various freshwater, marine, or damp environments. In fact, this group of organisms is considered ancestral to modern-day green plants. They all contain *chloroplasts*, and are often *unicellular*, but they may also be *colonial* or *multicellular*. The former "blue-green algae" now known as *cyanobacteria* are actually photosynthetic bacteria. (see *phycology*)

alimentary canal A hollow muscular tube running the length of the *digestive system* from mouth to *anus*. Different processes necessary to the digestion of food and absorbtion of nutrients take place along its length. Numerous *glands* introduce digestive enzymes at various points along the canal, and the food is moved along by muscular contractions (*peristalsis*).

alkaline phosphatase Any *phosphatase* enzyme that works optimally at an alkaline pH. Various alkaline phosphatases are widespread in nature and often used as tools in the molecular biology laboratory.

alkaloid A group of complex, organic compounds found in plants, many of which have pharmacological effects. They occur mainly in the poppy, buttercup, and nightshade families of plants, and may be the end products of plant nitrogen *metabolism*. Alkaloids include opium and its derivatives morphine, codeine, and scopolamine, as well as quinine, nicotine, strychnine, and atropine.

alkylating agent A chemical agent that can add alkyl (for example, ethyl, methyl) groups to another molecule. Many *mutagens* act through alkylation, for example, nitrosoguanidine (NG) and ethyl methanosulfonate (EMS).

allantois One of the three *extraembryonic* membranes of *amniotes* (reptiles, birds, and mammals). It is extruded as an outgrowth of the *embryo's* hindgut and functions in waste *metabolism*. In higher *primates* and *rodents*, it persists into later life as the *urinary bladder*.

allele One of two or more variants of a *gene* or *genetic marker*. Alleles occupy the same physical position on *homologous chromosomes* in the same or different individuals. (see *homozygous, heterozygous, dominant, recessive, recombination, meiosis, phenotype, genotype, allele frequency*)

allele frequency The proportion of a particular *allele* (genetic variant) among individuals in a population.

allelochemic Any chemical compound released by one species that affects the behavior or physiology of another. The effect may be negative or positive. (compare *allelopathic*)

allelopathic Any chemical compound released by one plant species that inhibits the *germination*, growth, or reproduction of another. It is widespread as an anticompetition mechanism in the *flora* of particular *ecosystems*. (compare *allelochemic*)

allergic response An inappropriately severe *immune response* to an otherwise relatively harmless *antigen* such as *pollen*, some drugs, minor insect stings, or certain foods. It is mediated by IgE *antibodies*, which are attached to *mast cells*, and produced in abnormally large amounts by people who suffer from allergies. The binding of antigen to these antibodies stimulates the release of *histamines*, the chemical that mediates the symptoms typically associated with allergy. (see *anaphylaxis, immunoglobulin*)

alligator (see *Crocodilia*)

allo- A word element derived from Greek meaning "other." (for example, *allosteric*)

allogamy *Cross-fertilization* in plants. (see *cross-pollination*; compare *autogamy*)

allometric growth The pattern of growth in which components of an organism grow at different rates at different times. For instance, a puppy's paws are proportionally larger than in a full-grown dog, as is the head of a newborn infant. (see *allometry*)

allometry The study of the relative relationship between size and shape during organismal growth. It may also be used on an evolutionary time scale to describe shifts in body makeup of an evolutionary line or to describe shifts in the proportion of *taxonomic* groups. (see *allometric growth*)

allopatric Populations or related species or subspecies that are unable to interbreed because of geographical separation. (compare *sympatric*)

allopolyploid Typically, a *polyploid* organism derived from *chromosome* doubling of a *hybrid* made between distinct *diploid* species. Plant hybrids, in particular,

often show an increase in vigor and fertility and are found in modern agricultural crop strains. Cultivated wheat is an allohexaploid, containing three related but distinct diploid *genomes*. (see *allotetraploid, amphidiploid, homeologous chromosomes*)

allorecruitive A type of *colonial* growth whereby young recruited into the colony come predominantly from other colonies, increasing *genetic diversity*. (compare *autorecruitive*)

all-or-none The type of response pattern exhibited by certain excitable tissues. A stimulus will fail to produce a response until it reaches a certain level of intensity. It then produces a fixed maximum response determined by the properties of a particular tissue. The firing of a *neuron* or *muscle cells* follow an all-or-none response pattern. Amplitude variations may be produced by activating groups of cells with different inherent thresholds to the stimulus. (see *action potential*)

allosteric An alteration of the three-dimensional configuration of a biological molecule, typically protein, in response to its immediate environment. Allosteric enzymes are important in the regulation of biochemical pathways. *Hemoglobin*, which changes structure in response to blood acidity as determined by oxygen saturation, is an example of an allosteric nonenzyme protein. (see *retinal*)

allosteric site An enzyme site at which specific binding produces a modulation of enzyme activity. Interaction at the allosteric site influences, either negatively or positively, activity at the enzyme's *active site*. An enzyme may have more than one allosteric site.

allosteric transition A change from one physical conformation of a protein to another. (see *allosteric*)

allotetraploid (amphidiploid) An *allopolyploid* organism whose *chromosomes* are derived from two different *diploid* species, thus containing four times the *haploid* number of chromosomes. (see *allopolyploid*)

alpha-actinin (α-actinin) The protein in *striated muscle* that serves as an anchor for the *actin* filaments in each *sarcomere*. It forms the *Z-lines* that delineate each contractile unit.

alpha-blocker (α-blocker) Any chemical that blocks the *adrenergic α-receptors* in the *sympathetic nervous system* that preferentially bind the *neurotransmitter noradrenaline*. (see *autonomic nervous system*; compare *beta-blocker*)

alpha helix (α-helix) Generally, any right-handed spiral structure.

1. The specific *secondary structure* in which some *polypeptide* chains are coiled to form protein. Each turn consists of approximately 3.6 *amino acid* residues. Hydrogen bonding between successive coils stabilizes the helix, and the *R-groups* of each amino acid extend away from the structure. Alpha helices are typical of *globular proteins* such as those found in muscle. (compare *beta pleated sheet*)

2. (see *double helix*)

Alpha Helix

alpha receptor (α-receptor) An *adrenergic postsynaptic receptor* found in the *sympathetic nervous system* that preferentially binds the *neurotransmitter noradrenaline*. (see *autonomic nervous system, alpha-blocker;* compare *beta receptor*)

alternation of generations The alternation of *haploid* and *diploid* individuals in the *life cycle* of an organism. Both are capable of reproduction, the haploid by *mitosis (asexual reproduction)*, and the diploid by *meiosis (sexual reproduction)*. Haploid *gametes* generally mate to reform the diploid organism thus completing the cycle. The types may differ greatly in *morphology*. Alternation of generations is found to some degree in all organisms exhibiting a sexual cycle, although the haploid phase in higher organisms is usually reduced to the formation and mating of *gametes*.

altricial Mammals and birds that are blind and helpless at birth. (see *nidicolous*; compare *precocial*)

altruism The behavior of an animal that favors the survival of others of the same species at its own expense. Altruism favors the survival of the genetic contribution of an individual rather than the individual itself. The most common example is parents risking themselves for their offspring. This increases the chance that the parents' *genes* will be passed on, particularly if the reproductive capacity of the parent has been exhausted. (compare *kin selection*)

Alu sequences A family of dispersed *repeated sequences* found in the human genome. Each is about 200 *base pairs* long and usually contains an *Alu restriction endonuclease* site. They seem to have originated and spread as *retroposons*. They are part of the "a" group of short interspersed elements collectively called *SINEs*.

alveolus, alveoli (pl.) In general, any saclike structure.

1. An air sac in the lungs of mammals. Alveoli occur in clusters called *acini* situated at the ends of *bronchioles*. They are surrounded by a network of *capillaries* enabling free exchange of gases between the bloodstream and lungs.

2. A sac forming an internal termination of a glandular duct.

amacrine cell One of the *neuron* types in the vertebrate *retina*. It conducts signals laterally between *ganglion cells* without firing *action potentials*. (see *horizontal cell*)

amastigote A microorganism or *life cycle* stage of an organism that lacks *undulipodia*. (compare *mastigote*)

amber codon The *nucleotide triplet* UAG, which is a *nonsense* or stop *codon*.

amber suppressor A *mutation* causing a *stop codon* to be ignored during *protein synthesis*. It occurs in a *tRNA gene* and causes the *anticodon* to be altered in such a way that an *amino acid* is inserted at an *amber codon* during protein *translation*.

ameba (amoeba)
1. A genus of single-celled *eukaryotic* protists characterized by the ability to continuously change shape due to the formation of *pseudopodia* used for locomotion and food capture.
2. Any *unicellular* protoctist life cycle stage that moves by means of pseudopods and whose shape is therefore subject to constant change.
3. An informal name for the protoctist phylum *Rhizopoda*.

ameboid (amoeboid) Any organism or characteristic that resembles an ameba. It typically refers to the style of movement in which an organism oozes along by extending *pseudopodia*. (see *Rhizopoda*)

American marsupials (see *Didelphimorphia*)

Ames test A biological test used to detect chemical *mutagens* that may be potential *carcinogens*. It is based on the *mutagenic* effect of a substance in a special strain of the bacterium *Salmonella typhimurium*, which was selected to be both hypersensitive to *mutation* and unable to survive outside the laboratory. Any chemical to be tested is first exposed to a human liver extract in order to approximate any processing that might alter its effect. Interpretation of the test depends on extending the *mutation rate* in bacteria to predict a carcinogenic effect in humans, a process that

has proved surprisingly useful.

AMFLP (see *Amp-FLP*)

amino acids The basic building blocks of all *polypeptides* and proteins. They are a group of organic compounds containing an acidic carboxyl (COOH) group, a basic amino (NH_2) group, and a distinctive side group (R-group) that determines the individual chemical properties of each amino acid. They are chemically classified as (a) neutral, basic, or acidic; or (b) nonpolar, polar, or charged. Twenty common amino acids are found in proteins.

R Varies With Each Amino Acid

The Basic Structure of an Amino Acid

amino sugar A *monosaccharide* sugar in which an amino group has been substituted for one or more hydroxyl groups. Examples include *glucosamine*, a major component in *chitin*, and galactosamine, a major component of cartilage and *glycolipids*. Amino sugars are also important components of some bacterial *cell walls*.

amitosis A type of *nuclear division* characterized by the absence of a *nuclear spindle* and the consequent unequal distribution of *nuclear chromosomes*. It is characteristically found in the *macronucleus* of *ciliates* and the *endosperm nucleus* of *angiosperms*. Such species are often *polyploid* (containing extra chromosome sets), reducing the chance of inheriting a *genome* completely lacking in one or more essential *genes*.

amniocentesis. The process whereby a sample of *amniotic fluid* is withdrawn from a pregnant woman in order to aid in the diagnosis of fetal abnormalities. Diseases commonly diagnosed by amniocentesis include chromosomal abnormalities such as *Down's syndrome* and enzyme deficiencies such as *Tay-Sachs.*

amnion (amniotic membrane) One of the three *extraembryonic membranes* of *amniotes* (reptiles, birds, and mammals). It is the layer closest to the *embryo* and contains the *amniotic fluid.*

amniote A nontaxonomic designation for the group of animals distinguished by the presence of *extraembryonic membranes* in fetal development and which develop either inside a protective egg shell or the mother's body. Reptiles, birds, and mammals are amniotes. (compare *anamniote*)

amniotic egg An egg that is protected from the environment by a watertight shell and is completely self-sufficient, requiring only the diffusion of oxygen from the outside during its *embryonic* development. It is typical of reptiles and birds and was important to their adaption to the terrestrial environment.

amniotic fluid The fluid contained within the amniotic membrane (*amnion*) of reptiles, birds, and mammals.

amniotic membrane (see *amnion*)

amoeba (see *ameba*)

amoebocyte A type of cell that moves actively and freely within animal tissues. It is found in the walls of *invertebrates* such as sea sponges and is also characteristic of some *leukocytes* in the vertebrate bloodstream. It is named for its general resemblance to an *ameba* in appearance and movement quality.

amoeboid Any cell resembling an ameba in shape and, in particular, mode of movement. (see *pseudopodium, Rhizopoda*)

Amoebomastigota (see *Discomitochondria*)

amorph A *mutant* organism that completely lacks a particular structure or substance.

AMP (adenosine monophosphate) A *nucleotide* of *adenosine* with one phosphate group attached. AMP is important in the energy *metabolism* of the cell. (see *ATP, cAMP*)

Amp-FLP (amplified fragment length polymorphism, AFLP, AMFLP) A type of *genetic marker, polymorphic* for length, that is typically analyzed using *PCR* (polymerase chain reaction). Amp-FLPs are used primarily in personal identification. (see *length polymorphism, DNA analysis*)

Amphibia The class of vertebrates containing the amphibians, including frogs, toads, newts, and salamanders. They are defined by their dual existence on both land and water, and are the most primitive terrestrial tetrapods. Amphibians have four *pentadactyl* limbs, a moist skin with no scales, and the ability to detect sound, but no external ear. Adults have lungs and live on land. The qualities of the skin that allow it to act as a secondary respiratory organ also confine the animal to damp places. Amphibians must return to the water to breed because *fertiliza-tion* is external. *Larvae* are aquatic and breathe through *gills* until they *metamorphose* to the adult form. Some amphibians, for example the Mexican axolotl, are permanently aquatic (*paedomorphic*), retaining larval gills and containing *atrophied* lungs.

amphibolic A *metabolic* pathway that contains both *anabolic* and *catabolic* reactions. An example is the *Krebs cycle.*

amphidiploid (see *allotetraploid*)

amphids A pair of *sensory* organs on the head of nematodes believed to be *chemoreceptors.* (compare *phasmids*)

amphimixis The fusion of *meiotically* produced *haploid gametes* to form a *diploid* organism. This is the most common form of *sexual reproduction.* (compare *apomixis, parthenogenesis*)

Amphioxus (see *Branchiostoma*)

amphipathic Any compound that contains both strongly nonpolar and strongly polar groups such as *phospholipids.* (see *phospholipid bilayer, plasma membrane*)

amphistylic A specialized jaw suspension found in sharks. The jaws are suspended by ligaments from the skull and by the *hyomandibular bone.* (compare *autostylic, hyostylic*)

amphoteric
1. Any molecule containing both acidic and basic functional groups. When dissolved in aqueous solution it can donate both protons and electrons. *Amino acids* are examples of amphoteric molecules. (see *zwitterion*)
2. A female animal that is capable of producing both *diploid* female and *haploid* male eggs in the same clutch. It is characteristic of some invertebrates, in particular social insects including bees, wasps, and ants. (see *haplodiploidy*)

amplicon The region of *DNA* that is copied in *PCR* (polymerase chain reaction). It is defined at one or both ends by the choice of synthetic *oligonucleotide primers.*

amplification The production of many *DNA* copies from one master region of DNA.
1. When used in reference to *PCR* (polymerase chain reaction), the term is used to describe the faithful *replication* of DNA molecules *in vitro*. Any one DNA segment may be amplified millions or billions of times by the process.
2. (see *gene amplification*)

amplified fragment length polymorphism (see *Amp-FLP*)

ampulla, ampullae (pl.)
1. A muscular sac at the base of each *tube foot* of an echinoderm that serves as a fluid reservoir in the hydrostatically powered movement of the organism.
2. In the vertebrate *inner ear*, structures attached to the ends of the *vestibulocochlear nerves* that function in the detection of motion. They contain the *hair cells* that extend into a gelatinous matrix called the *cupula*. The whole structure protrudes into the openings of the *semicircular canals* and translates movements in the liquid into *sensory* impulses that are transmitted to the brain.

ampullae of Lorenzini Specialized *receptors* in electric fish that detect disruptions in an electric current that they continuously discharge. This enables them to sense objects in murky water.

amygdala Along with the *hippocampus*, a component of the *limbic system* that resides deep in the brain and is responsible for emotional responses. It is also involved in memory recall.

amyl-, amylo- A prefix derived from Greek, meaning starch. (for example, *amylase*)

amylases A group of digestive enzymes that break down (*hydrolyze*) starch or *glycogen* to the simple sugars maltose, glucose, or dextrin. Widely distributed, both α- and β-amylases occur in plants and some microorganisms. β-amylase is found in malt used in the brewing industry. Only α-amylase is found in animals, in the *pancreatic* secretions and saliva. (see *ptyalin*)

amylopectin The water-insoluble fraction of *starch.*

amyloplast A specialized *plastid* that synthesizes and stores starch. It is found in plant storage organs such as the potato tuber. It may play a role in *gravitropism* (gravity-directed plant root growth) by settling towards, thus indicating the direction of, gravity in the cell. (see *tropism, statolith*; compare *elaioplast, aleuroplast*)

amylose The water-soluble fraction of *starch.*

anabolic steroid A class of *steroid hormones* derived from *androgens* (for example, *testosterone*) that promote growth and formation of new lean tissue. It is used to treat some diseases, to boost livestock production, and, controversially and illegally, to build athletes' muscles. Use of such male hormones in women leads to masculinization.

anabolism The enzymatic synthesis of more complex molecules from simple ones. Starch, *glycogen*, fats, and proteins are products of anabolic pathways. Anabolic reactions require *ATP* or an equivalent as an energy source. (see *metabolism*; compare *catabolism, amphibolic*)

anaerobe An organism that can exist in the

15 **androgen**

absence of free oxygen (O_2). *Facultative* anaerobes usually respire *aerobically* (using oxygen) but have the option of *anaerobic respiration* when their oxygen supply is limited. *Obligate* anaerobes have the capacity only to respire anaerobically and may even be poisoned by free oxygen. Anaerobic bacteria often produce their own local anaerobic environment by emitting various gases, such as hydrogen sulfide (H_2S) and ammonia (NH_3), that they form as *metabolites*. (see *glycolysis*; compare *aerobe*)

anaerobic Any biological process that can occur in the absence of gaseous oxygen (O_2).

anaerobic respiration *Cellular respiration* in which free oxygen (O_2) is not involved. Found in yeasts, bacteria, and occasionally in muscle tissue, it is a less efficient process of energy production than *aerobic respiration*, with a relative increase in toxic byproducts. *Fermentation* is an example of anaerobic respiration in which certain yeasts produce ethanol and carbon dioxide as end products. Anaerobic respiration is also seen in muscle cells during periods of increased activity. The end product of *glycolysis* in this instance is *lactic acid*, which is responsible for muscle soreness.

analogous Structures or organs that are apparently similar but have a different *evolutionary* origin and thus a different *embryological* origin and structure. For instance, the wings of birds and insects serve a similar function but have completely different origins. (compare *homologous*)

anamniote The group of vertebrates, including fish and amphibians, whose *embryos* rarely possess *extraembryonic membranes*. Therefore, they must lay their eggs in aquatic or damp environments. (compare *amniote*)

anaphase A stage in *nuclear division* during which either *bivalents* (*meiosis I*) or *sister chromatids* (*mitosis* and *meiosis II*) separate and move to opposite poles of the cell. In mitosis and *anaphase II* of

meiosis, paired chromatids separate and are drawn toward opposite poles. During *anaphase I* of meiosis, a pair of chromatids still connected at their doubled *centromere* (*bivalent*) move together toward the *spindle* poles, away from their *homologues*. (compare *prophase, metaphase, telophase, interphase*)

anaphylaxis A hypersensitive *immune response* to a foreign substance (*antigen*) resulting in the release of massive amounts of *histamine* from *mast cells*. It occurs in reaction to ingestion of a substance to which the organism has previously been *sensitized* such as penicillin, shellfish, or insect venom. (see *allergic response*)

anastomosis The formation of a network by interconnection of branches, filament, or tubes.

anatomy The study of the structure of an organism and its parts. (compare *morphology, physiology*)

anatropous The position of the *ovule* in flowering plants (*angiosperms*) when it is oriented so that the *micropyle* points toward the *placenta*. (compare *campylotropous, orthotropous*)

ancient forest A forest originating prior to an arbitrary date, often set approximately at 1600 A.D.

androdioecious Plant species that bear both male and *bisexual* (*hermaphrodite*) flowers on separate plants. (compare *andromonoecious, gynodioecious*)

androecium A collective term referring to the *stamens* (male parts) of a flower. (compare *gynoecium*)

androgen A general term for *steroids* with male sex *hormone* activity in vertebrates. Androgens are important in the development, function, and maintenance of secondary male characteristics in mammals (for example, facial hair and deepening voice), male accessory sex organs, and *spermatogenesis*. Produced chiefly by the *testes* and, to a significantly lesser extent, by the adrenal *cortex* and *ovaries*, androgens have *anabolic* activity, promoting growth of lean tissue. *Tes-*

tosterone is the most prevalent example.

andromonoecious Plant species that bear both male and *bisexual* (*hermaphrodite*) flowers on the same plant. (compare *androdioecious, gynodioecious*)

androsterone A *steroid hormone* formed in the liver from the *metabolism* of *testosterone*. It has only weak *androgenic* activity.

anemophilic Plants that are *pollinated* by wind. In contrast to those pollinated by insects and other animals, anemophilic plants have only insignificant, unscented flowers with large, feathery *stigmata*.

aneuploid A condition in which an abnormal number of *chromosomes* is present in a cell. It is often caused by *nondisjunction* of *homologous* chromosomes at *meiosis*, resulting in unequal distribution to the progeny. For instance, an offspring may receive one or three copies of a particular chromosome instead of the two normally found in a *diploid* organism. (see *monosomy, trisomy, polysomy*)

angiosperm A plant in which the seeds develop enclosed in flowers. (see *Anthophyta*; compare *gymnosperm*)

angiotensins A group of substances responsible for raising *blood pressure*. Angiotensin I is a *peptide* produced in the *kidney* in response to a drop in *blood pressure*. In the lung, it is converted to Angiotensin II, which raises *blood pressure* by causing constriction of blood vessels and promoting sodium (Na^+) retention by the kidney. Angiotensin III stimulates *aldosterone* secretion, which decreases sodium loss in urine. They are all converted from the same *plasma* protein by the enzyme *renin*, secreted by the kidney. (see *osmoregulation*)

angstrom (Å) A unit of length equal to 10^{-10} (1/10,000,000,000) meters. It is used to describe interatomic distances and wavelengths (λ).

anhinga (see *Pelecaniformes*)

animal An organism typically distinguished by motility, quick reaction to stimuli, and the need to obtain complex nutrients from external plant or animal material (*heterotrophic*). All animals develop from a *blastula*. Distinctions between organisms in the animal kingdom (*Animalia*) and other kingdoms are not always discrete. Exceptions can always be found, particularly in various *life cycle* stages.

animalcules A historical name for the minute animal-like organisms, in particular radiolarians and heliozoan protists, first observed under magnification. They are informally referred to as *protozoa*.

Animalia One of the main groups or *phylogenetic* kingdoms into which all life forms are divided for academic discussion. It contains those organisms generally thought of as animals and that exhibit motility and quick reactions to stimuli. They are *heterotrophic* (obtain their food by digestion of compounds synthesized by other organisms), *multicellular*, and have cells that are generally enclosed by a *plasma membrane* rather than a *cellulose wall* (as in the case of most plants). Motility of *gametes* and specialized single cells is accomplished by 9 + 2 *undulipodia*. Exchange of genetic material takes place by *fertilization* and *meiosis,* and the *embryo* develops from a *blastula*.

animal pole The pole of the *blastula* in which the yolk of an egg is less concentrated. It is found in the characteristically *telolecithal* (asymmetric yolk distribution) eggs of most fish, amphibians, reptiles, and birds. The animal pole is situated opposite the *vegetal* pole, toward which the yolk is more concentrated.

anisogomous (heterogamy) The *sexual* fusion of *gametes* differing in size and/or motility. An example of this is the human egg and sperm. (see *oogamous*; compare *isogamous*)

annealing The formation of a *double-stranded nucleic acid* molecule (*DNA* or *RNA*) from two *complementary* single strands.

Annelida A phylum of *coelomate* worms, the annelids, typified by *bilateral symmetry* (mirror image) and *segmentation* (similar repeating body units). Annelids contain well-developed *circulatory* and *nervous systems*, and *nephridia* for excretion. They have a long, soft, cylindrical body covered by a thin *cuticle* and contain *chitinous* bristles that, in conjunction with segmentation, provide a means of locomotion. Many are *hermaphroditic* (contain male and female reproductive structures in one individual), although *fertilization* usually takes place between two individuals. Many are important in soil turnover and aeration.

annelids (see *Annelida*)

annelid worm (see *Annelida*)

annual A plant that completes its *life cycle* within a single growing season. (compare *biennial, ephemeral, perennial*)

annual ring (growth ring, tree ring) An annual increment of growth in the stems or roots of woody plants. Concentric rings are produced when a new, woody, outer layer (*secondary xylem*) is produced from the middle, *vascular* layer (*cambium*). A lighter-colored ring containing larger *cambium* vessels is produced in spring, followed by a darker layer containing smaller vessels laid down in summer. The process is repeated annually, leading to a succession of alternating light and dark-colored concentric rings. Variations in the climactic conditions that regulate this process make dating by annual rings somewhat unreliable. (compare *annulus*)

annular thickening In plant stems, the thickening of *vessel elements* or *tracheid* walls, laid down in rings along their length. It provides mechanical strength, allowing extension of the *xylem* between the rings so that the xylem is not ruptured as the surrounding tissues grow. (see *annual ring*)

annulus Generally, a ring of cells that may be adapted to specialized uses in different organisms. Examples include

1. A ring of tissue surrounding the stalk of mature *fruiting bodies* (mushrooms) of certain fungi.
2. A line of specialized cells involved in the process leading to violent rupture

of moss capsules and fern *sporangia*, facilitating *spore* dispersal.

3. One of a series of concentric rings formed in the scales of bony fish. Winter rings may be narrower and denser than summer rings. (compare *annual ring*)
4. The external segment of an *annelid* worm.

Anoplura An order of *endopterygote* insects containing the sucking lice. They are characterized by a lack of wings and eyes, the presence of piercing mouth parts, prehensile clawed legs, and fusion of all three *thorax* parts. All species are obligate *parasites* of mammals and feed on blood, which is digested with the aid of *symbiotic* bacteria. Eggs are attached to the *host's* hair and develop into blood-sucking *nymphs*. Some species are *vectors* of diseases such as typhus in humans.

anoxia A lack of dissolved oxygen.

anoxic A habitat that is devoid of molecular oxygen (O_2), such as the gut of an animal.

anoxygenic Nonoxygen producing.

ANS (see *autonomic nervous system*)

Anseriformes The order of birds comprising the waterfowl, including swans, ducks, and geese.

antagonism The opposition of two or more processes or systems. For example

1. Drugs or *hormones* having opposing effects such that their individual activities are partially or completely nullified. (compare *agonism*)
2. Any organism interfering with the growth or presence of another.
3. The complementary movements of paired muscles so that contraction of one is accompanied by relaxation of the other. A muscle regains its relaxed shape after contraction when the antagonistic muscle contracts. (see *agonism*)

ant bear (*see Tubulidentata*)

anteaters (see *Xenarthra*)

antelope (see *Ruminantia*)

antenna, antennae (pl.) A pair of mobile

appendages on the heads of many arthropods. They may be wirelike, feathery, or club shaped. They generally have a *sensory* function, but some crustaceans use them for swimming or attachment.

antennules A pair of small *sensory* appendages that occur in front of the *antennae* on the heads of crustaceans.

anterior The front side of an animal. In *bilaterally symmetrical* (mirror image) animals, this is the end that leads in locomotion. In *bipedal* (two-legged) animals such as man, the anterior side corresponds to the *ventral* side of animals that walk on all fours. (compare *posterior*)

anther In plants, the portion of the *stamen* (male reproductive structure) that produces *pollen* (*haploid germ cells*).

antheridium, antheridia (pl.) The male reproductive structure in seedless plants and some fungi. It is a *multicellular* structure consisting of an outer layer surrounding specialized cells that develop into *spermatozoa*. (compare *archegonium*)

antherzoid (see *spermatozoid*)

Anthocerophyta A phylum of non*vascular plants*, the hornworts. The *gametophytes* generally resemble the liverworts, but the *sporophyte* generation sends up distinctive elongated *sporangia* that resemble horns, thus the name.

anthocyanins One of a group of water-soluble pigments found in higher plant cell *vacuoles*. They are brightly colored and present in flowers and fruits, where they are important in attracting insects and birds. Anthocyanins sometimes contribute to the tint in spring buds and to the autumn colors of leaves. They are natural pH *indicators*, as color may be modified by acidity level; color may also be altered by the presence of trace metals and organic substances.

Anthophyta A phylum of *vascular* seed plants, the *angiosperms*, that flower and produce seeds fully enclosed by fruits (which develop from *ovaries*). Flowers contain the male and female reproductive structures, and *double fertilization* provides both an *embryo* and nutritive *endosperm* tissue. The *alternation of generations* that is so prominent in more primitive plants is reduced to a microscopic scale. The male *microgametophyte* or *pollen sac* consists of just three cells, and the female *megagametophyte* or *embryo sac* contains only seven cells. *Spermatozoa* are immotile and carried to the egg cell by a *pollen tube*, facilitating a terrestrial living habit. They are the most highly evolved plants and dominate modern *flora*. (see *Monocotyledon, Dicotyledon*; compare *gymnosperm*)

Anthozoa A class of cnidarians, the sea anemones and corals, in which only the *polyp* form has survived evolution. The sea anemone has numerous feathery *tentacles,* while the corals are *colonial. Polyps* are contained in a gelatinous matrix (soft corals), horny skeleton (horny corals), or a calcium carbonate skeleton (stony or true corals), Accumulations of these corals in warm shallow seas form *coral reefs.*

anthropoid The first mammals that began to exhibit monkeylike features.

Anthropoidea The suborder of primates containing monkeys, apes, and man.

antibiotic A general term for any member of a group of organic compounds produced by microorganisms that are lethal or inhibitory to other microorganisms. They generally act by inhibiting the growth of *cell walls* (for example, penicillin) or the synthesis of *nucleic acids* and/or proteins (for example, streptomycin). They are effective clinical drugs due to their specificity and lack of side effects, but widespread use has generated new selective pressures on their target organisms, resulting in the spread of antibiotic-resistant strains. In 1928, Sir Alexander Fleming discovered (quite by accident) the first known antibiotic, penicillin, from the fungus, *Penicilum*. He shared the Nobel prize in physiology with Chain and Florey in 1945. (see *resistance transfer factor, gene mobilization*)

antibodies Protein molecules produced by vertebrate organisms that fight disease and confer *immunity*. Specifically, they are *immunoglobulins* produced and released by *B-cells*. Antibodies seek out and bind foreign entities (*antigens*) such as those introduced by bacteria, tissue grafts, or blood transfusions. Each B-cell produces a different antibody that fits the structure of a particular invading antigen like a lock and key. Antibodies do not destroy a virus or bacterium directly but, rather, mark it for destruction by one of three mechanisms: 1) osmotic *lysis* of the cell by *complement*; 2) ingestion and consumption by *phagocytes*; 3) binding and killing by *killer cells*. The type of antibody produced by a specific B-cell is determined by *somatic recombination* that occurs during B-cell production. Proliferation of a specific *clone* of B-cells is induced by the introduction of a complementary antigen. (see *clonal selection, lymphocyte, immunity, immune receptor genes, bone marrow, helper T-cells*)

antibody-binding site The specific site on an *antigen* to which a specific *antibody* binds. (see *antigenic determinant*)

antibody diversity The millions of *antibody genes*, each coding for an antibody specific for a different *antigen*, generated by *site-specific somatic recombination* within the *immune receptor genes*.

antibody specificity The ability of an individual *antibody* molecule to bind one and only one specific *antigenic determinant*.

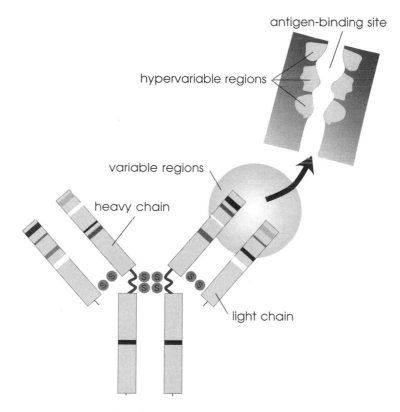

The Structure of an Antibody Molecule
(Immunoglobulin G)

anticlinal In plants, the alignment of the plane of *cell division* at approximately right angles to the outer surface of a structure. (compare *periclinal*)

anticoagulant Any substance preventing *clotting* of blood. Blood itself naturally contains small amounts anticoagulants that prevent clotting under normal circumstances (*fibrin, antithrombin III, heparin*). Bloodsucking animals (leeches, insects, and one species of bat) contain large amounts of anticoagulants in their saliva. Artificial anticoagulants include sodium citrate and EDTA, which are used in blood drawing and transfusion. The rat poison warfarin is also an anticoagulant.

anticodon A *nucleotide triplet* on a *tRNA* molecule that specifically pairs with a corresponding nucleotide triplet (*codon*) on an *mRNA* molecule in the *ribosomal* complex (*polysome*) during *protein synthesis*. (see *translation*)

antidiuretic hormone (ADH, vasopressin) A *peptide hormone* that raises blood pressure by stimulating the contraction of muscles around *capillaries* and *arterioles*. It has a secondary effect on the muscles of the gut, increasing *peristalsis*, and the *uterus*. It also raises blood pressure by increasing water resorbtion in the *renal tubules*, with the side effect of concentrating urine. Its production in the *hypothalamus* and *posterior pituitary* is triggered by low blood pressure. Drinking excessive alcohol increases urine output because alcohol inhibits ADH. (compare *oxytocin*)

antigen Any substance foreign to the body that induces the proliferation of a specific *clone* of *antibody-producing B-cells*. The binding site on the antibody fits the antigen like a lock and key. *Macromolecules* such as proteins, polysaccharides, and nucleic acids are effective antigens and are often presented on the outside of invading cells, such as bacteria. Most small foreign molecules do not elicit antibody formation. However, if the small molecule or *hapten* is attached to a macromolecule called a *carrier*, specific antibodies will be produced against the hapten as well as the carrier. (see *antibody, immunity, antigen-antibody reaction*)

antigen-antibody complex The result of specific binding of an *antibody* to an *antigen*. (see *antigen-antibody reaction*)

antigen-antibody reaction The specific binding between the *antigenic determinant* of an *antigen* and the *antigen-binding site* on an *antibody*. The antibody recognizes the antigen by its *tertiary structure* and charge distribution as

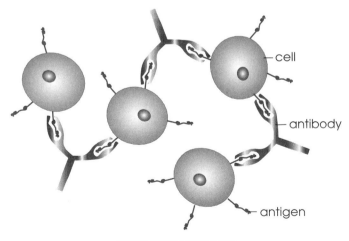

Antigen-Antibody Interactions

determined by *amino acid* sequence. *Antibodies* do not destroy a virus or bacterium directly, but rather mark it for destruction by one of several other mechanisms. (see *complement, macrophage, K-cells, T-cells*)

antigen-binding site One of two identical sites found on each arm of an individual *antibody* molecule. Each site consists of a complex of both *heavy* and *light chains*, each containing *hypervariable regions* that determine *antibody specificity*. (see *immunoglobulin*)

antigenic determinant (see *epitope*)

antigen-shifting The process used by some microorganisms to effectively defeat the vertebrate *immune system*. For instance, the flu virus undergoes periodic *recombination* within the *gene* encoding its major surface *antigen*, periodically provoking new outbreaks. Trypanosomes, the protists responsible for sleeping sickness, possess a *promoter* located within a *transposable genetic element*. This enables the promoter to jump at random to any of the thousands of different versions of the *gene* encoding the surface *glycoproteins* and express a different version. (see *double-strand break*)

antimorph A *mutant allele* whose *expression* antagonizes a normal *gene product*.

antioxidant Any compound, usually organic, that prevents or retards the oxidation by molecular oxygen (O_2) of other materials. It acts by scavenging the *free radicals* generated in auto-oxidation chain reactions. It is destroyed in the process and thus is not enzymatic. Antioxidants are important in removing free radicals in the body, which are thought to be a contributing factor to the initiation of some *cancers*.

antiparallel A term used to describe the opposite orientation of the two strands of a *DNA double helix*.

antipodal cells The three *haploid* cells found opposite the *micropyle* in the *ovule* of seed-bearing plants. They arise as byproducts of the three *meiotic* divisions that produce the egg cell, *synergid cells*, and *polar nuclei*, and are eventually absorbed by the developing *embryo*. Their function, if any, is uncertain.

antiport The transport of two different chemical species in opposite directions across a *cell membrane*. (see *coupled channel*; compare *cotransport*)

antisense DNA In *double-stranded DNA*, the strand opposite the one that is *transcribed* into *mRNA* and *translated* into protein. Antisense DNA is beginning to emerge from the research lab as, among other things, a potential anticancer agent. If a particular *mRNA* message, such as that from a viral invader or an out-of-control *gene*, is well characterized, antisense DNA specific to it may be synthesized and artificially introduced into the cell. The idea is that it would *hybridize* with the unwanted message and prevent its translation into protein, thus negating the potentially harmful effect on the cell. Sometimes the antisense molecule is a hybrid of DNA and *RNA*. (see *protein synthesis*)

antisense technology (see *antisense DNA*)

antiseptic Any agent, often chemical, that slows or stops the growth of microorganisms but may not necessarily kill them. (see *bacteriostatic*)

antiserum An *immune serum* containing *antibodies* against a specific *antigen(s)*. It is used clinically to confer *passive immunity* against certain diseases and also in the laboratory to identify unknown *pathogens* (disease-causing organisms). It is obtained from blood of a human or other animal that has been exposed to the specific antigen(s) either naturally (through disease) or artificially (through *immunization*). (compare *vaccine*)

antithrombin III (see *anticoagulant*)

antlions (see *Neuroptera*)

Anura An order of tailless amphibians that contains the frogs and toads. Adults are highly specialized for jumping, with powerful hind limbs, no tail, and a strengthened *pectoral girdle* to absorb shock upon landing. The hind feet are webbed for swimming. Oxygen obtained through the lungs is limited, drawn in only by the pumping action of the floor of the mouth, and is highly supplemented by absorbtion through the moist skin. *Oxygenated* and *deoxygenated* blood are incompletely separated in the heart. Jelly-covered eggs are laid in the water and hatch into aquatic *larvae* that undergo a *metamorphosis* in which the tail is absorbed and the *gill slits* are replaced by lungs. Frogs require a damper environment than toads; toads are better adapted to drier habitats.

anus The specialized *posterior* exit from the *alimentary canal* found in most *mammals*. *Feces* and other semisolid waste are expelled through it, often under muscular control. In some aquatic *invertebrates,* it is also used in respiration. (see *sphincter*)

anvil (see *incus*)

aorta In vertebrates, the main artery conveying freshly *oxygenated* blood to the body. In mammals, it arises from the left *ventricle* of the heart and is divided into an ascending portion, an arch, and a descending portion. (see *dorsal aorta, ventral aorta*)

aortic arches Six pairs of blood vessels (occasionally more) present in all vertebrate *embryos* that link the ventral aorta leaving the heart with the dorsal aorta. In adult tetrapods, arches one, two, and five disappear. Arch three becomes the *carotid arch* supplying the head, arch four becomes the *systemic arch* supplying the body, and arch six becomes the *pulmonary arch* supplying the lungs. In adult fish, the remaining arches lead to the gills. The study of their comparative anatomy in *embryos* and adults provides striking support for *macroevolutionary* change.

aperitoneal (see *acoelomate*)

aperture In gastropods, such as snails and some other mollusks, the opening in the shell from which the foot and body protrude.

apes (see *Primates, Anthropoidea*)

apex The part of any structure opposite the base by which it is attached.

Aphragmabacteria A phylum of extremely small bacteria comprising the mycoplasmas and spiroplasmas. They are the smallest cellular organisms known and are distinctive among bacteria because of their lack of *cell walls.* The mycoplasmas are responsible for diseases such as bovine tuberculosis and some types of pneumonia, and the spiroplasmas cause a number of plant diseases such as the lethal coconut yellowing disease. They are resistant to penicillinlike *antibiotics* specifically because of their lack of cell walls.

apical bud The growing tip at the apex of a plant. Removal of the apical bud results in the development of side branching. (see *apical meristem, apical dominance, auxin*)

apical complex The specialized structure at the apex of an apicomplexan protoctist that facilitates attachment and penetration of the organism into the tissue cell of its *host.*

apical dominance The phenomenon in which the presence of a growing *apical bud* (shoot tip) on a plant inhibits the growth of *lateral buds.* It is controlled by the interactions of *growth substances,* particularly *auxin* (produced by the shoot tip), and *abscisic acid.* Removal of the growing tips of plants permits lateral branching by removing the inhibitory effects of *auxins.* This technique is commonly used in horticulture to produce bushy plants. (see *apical meristem*)

apical meristem A group of actively dividing cells constituting the growing point at the tip of the root or stem in *vascular plants.* The apical meristem in lower plants such as ferns consists of one cell only but contains groups of cells in higher plants. (see *apical dominance, apical bud, auxin, meristem;* compare *lateral meristem, intercalary meristem*)

Apicomplexa A phylum of *unicellular heterotrophic* protoctists, the sporozoan parasites. They were once grouped together with several other phyla as the sporozoans. Apicomplexans derive their name from a cellular structure located at the anterior portion of the cell, the *apical complex.* They are spore-forming *parasites* of animals, notably arthropods and chordates, that have a complex *life cycle* involving both *sexual* and *asexual* phases. The sexual phase alternates between *haploid* and *diploid* generations. The distinctive arrangement of subcellular *organelles* assures their successful forced entry into a *host.* The most infamous apicomplexan is the malarial parasite *Plasmodium sp. Plasmodium* passes through several *life cycle* stages in the gut of the *Anopheles* mosquito and are injected into a human host through the saliva of a mosquito bite. In the blood, they infect red blood cells (*erythrocytes*) and pass through several more stages, including one that is reingested by a mosquito through a blood meal. They undergo their *sexual* phase and *meiosis* in the mosquito. (see *alternation of generations, oocyst, sporozoite, trophozoite, merozoite*)

aplan-, aplano- A word element derived from Greek meaning "unmoving." (for example, *aplanogamete*)

aplanogamete Any nonmotile *gamete* such as that commonly found in seed plants. (see *aplanospore*)

aplanospore A nonmotile, *asexual spore* commonly found in nonseed plants.

apo- A word element derived from Greek meaning "away" or "different from." (for example, *apoenzyme*)

apocarpous A plant *ovary* made up of unfused *carpels* (the female reproductive structure), as in the buttercup. (compare *syncarpous*)

Apoda (Gymnophionia) A small order of legless, tropical amphibians, the caecilians, that have vestigial scales embedded in their skin. *Fertilization* is internal.

Apodiformes The order of birds containing the swifts and hummingbirds.

apoenzyme An enzyme whose *cofactor* has been removed (for example, via *dialysis*), rendering it catalytically inactive. It comprises the protein component of a *holoenzyme* (enzyme-*cofactor* complex). (compare *core enzyme*)

apogamous The development of a *sporophyte* directly from a *gametophyte*, bypassing the *sexual* fusion of *gametes*. It is common in some ferns and mosses. When the process involves an *unfertilized haploid female* gamete, it is called *parthenogenesis*. It is one form of *apomixis*. (compare *aposporous*)

apomixis A type of *asexual* reproduction in which offspring are produced without *fertilization* and *meiosis*. It is common in plants and affords the advantages of seed dispersal and survival without the risks involved achieving *pollination*. *Vegetative apomixis* may involve propagation by *rhizomes*, *stolons*, and *runners*. Apomixis is also seen in invertebrate animals. (see *parthenogenesis*)

apoplast In some plants, a system of *cell walls* forming a continuum throughout a plant body and along which water that contains dissolved salts can move passively. It is an alternative to water movement within the *xylem.* (compare *symplast*)

apoptosis Programmed cell death. It is a mechanism to trigger the demise of a cell when damage to its DNA is beyond the repair of cellular mechanisms. Apoptosis destroys these cells rather than permitting them to become, for instance, *cancerous.*

aposematic Qualities such as color, sound, or behavior that advertise noxious or otherwise potentially harmful qualities of an animal. Aposematic coloration is common in insects and reptiles, warning a predator not to ingest a poisonous animal. (see *mimicry*)

aposporous The development of a *gametophyte* directly from a *sporophyte*, bypassing *meiosis* and *spore* production. This results in a *diploid* rather than a normal *haploid* gametophyte. Apospory is found in some ferns and fungi. (compare *apogamous, apomixis*)

aposymbiotic An organism lacking its normal complement of *symbionts.*

apothecium, apothecia (pl.) A cup-shaped *fruiting body* (*ascocarp*) of certain fungi and *lichens*. The *asci* (*meiotic spore* products) line the interior and are thus exposed to the environment.

appeasement A behavior by an animal that serves to reduce aggression shown toward it by another member of the same species but that does not involve avoidance or escape. In dominance hierarchies, subordinate members of the group will use appeasement behavior to avoid attack by dominant animals.

appendix (vermiform appendix) An apparently functionless appendage attached to the first part of the large intestine. It is evolutionarily *homologous* with the end of the *cecum* in some *herbivores*, which contains *symbiotic* bacteria that aid in food digestion.

appetitive behavior The first phase of a goal-oriented series of actions performed by an animal. For instance, the initial actions of a hungry animal in search of food.

apposition The deposition of successive layers of *cellulose* on the inner wall of a plant cell, resulting in thickening of the wall. (compare *intussusception*)

apposition eyes (see *ommatidium*)

apterous An animal lacking wings.

aquatic Any wet environment or condition. (see *marine*)

aqueous humor The fluid that fills the space between the *lens* and the *cornea* in the vertebrate eye. It links the *circulatory system* to these eye structures, as blood vessels are absent for optical reasons, and maintains constant pressure within the eye. An increase in pressure leads to the medical condition known as glaucoma.

aquifer Permeable underground layers of rock, sand, and gravel in which groundwater accumulates. Aquifers hold more than 95% of the freshwater reservoir in the United States. Because of the large volume of water and slow rate of movement, the removal of steadily increasing levels of *pollutants* from aquifers is virtually impossible. (see *water table*)

Arachnida The class of *chelicerate* arthropods containing the arachnids, including the scorpions, spiders, ticks, mites, and daddy longlegs. All are typically terrestrial and *carnivorous*. Spiders and scorpions have *spinnerets* on the abdomen for web spinning. Although they are often confused with insects, they differ in many features. The arachnid body is divided into two parts. The anterior segment (*cephalothorax*) contains accessory appendages and four pairs of walking legs. *Pedipalps* function as grasping appendages and may have *sensory* functions as well. *Chelicerae* are modified for crushing and chewing in the absence of true jaws and may deliver poison to kill prey. Silk and poison production are common, but the location and method of production varies. Eyes are simple and *antennae* are lacking. Respiration is carried out by *book lungs* and/or *tracheae*.

arachnids (see *Arachnida*)

arachnoid membrane The middle of three membranes (*meninges*) that surround and protect the brain and spinal cord in vertebrates.

Araneae The order of arachnids containing the spiders. They are particularly important as predators of insects and other small animals, which they hunt or catch in webs of silk. Some spiders have poison *glands* leading out through their *chelicerae* that are used to bite and paralyze prey. The black widow and brown recluse have bites that are poisonous to humans.

Arboviruses A group of *enveloped* plus-strand *RNA* viruses that *parasitize* animals and cause diseases including encephalitis and yellow fever. They are distinguished by a lipid-rich outer envelope, which they acquire as they burst through the *cell membrane* of the *host*. They are often transmitted by insects and other arthropods, hence the name. *Retroviruses*, including *HIV*, are part of this group.

Archaea (archaebacteria) A subkingdom of Bacteria comprising the organisms formerly called archaebacteria or ancient bacteria. Like eubacteria, they are *prokaryotes* but have distinctive membranes, unusual *cell walls* lacking muramic acid (which is common in other bacteria), and unusual *metabolic cofactors*. Certain portions of their *genomic* sequences, in particular those for *rRNA genes*, differ markedly from eubacteria as well as from *eukaryotes*. The Archaea may be the most ancient of life forms and the ancestors of all eukaryotes. They typically exist in extreme environments and include the methane-producing bacteria (methanogens), salt-loving halophilic bacteria, and sulfur/acid-tolerant thermoacidophilic bacteria.

Archaebacteria (see *Archaea*)

Archaeprotista The most primitive phylum of protoctist that lack *mitochondria* entirely. They include the giant marine ameba, *Pelomyxa*, tiny *mastigotes* (for example, *Mastigina*), as well as *symbionts* in both the vertebrate and termite guts.

archegonium, archegonia (pl.) The female reproductive structure in seedless *plants* and some fungi. It is a *multicellular,* flask-shaped structure made up of a narrow neck and a swollen base (*venter*) that contains the egg cell. (compare *antheridium*)

archenteron The earliest gut cavity of higher animal (*deuterostome*) *embryos*. It is the result of invagination of the outer surface during the embryonic stage of *gastrulation*. (see *blastula, gastrula*)

archesporium, archesporia (pl.) A cell or cells in the *sporophyte* generation of plants from which *spores* are ultimately derived (for example, in a fern *sporangium).*

arginine (see *amino acids*)

arginine phosphate A high-energy compound important in the energy *metabolism* of invertebrates. Arginine is phosphorylated when *ATP* levels in the cell are high. The phosphate group is transferred from arginine to *ADP* to form *ATP* when needed.

aril In plants, an accessory seed covering often formed from an outgrowth at the base of the *ovule* (preseed). An example is the mace-containing covering around the fruit of the nutmeg. It may be brightly colored, facilitating dispersal by attracting seed-eating animals. (see *caruncle*)

armadillos (see *Xenarthra*)

arousal A general level of alertness in an animal.

arrow worms (see *Chaetognatha*)

ARS (autonomous replication sequence) A region of *DNA* necessary for the *replication* and continued *propagation* of the molecule. A *plasmid* lacking an ARS, for instance, will fail to propagate within its *host*. ARS sequences have been particularly well-characterized in yeast. (see *replicon*)

arteriole The type of blood vessel forming the smallest branch of an *artery*, leading to *capillaries*.

arteriosclerosis A hardening of the arteries, caused when calcium is deposited in arterial walls. It tends to occur when *atherosclerosis* is severe, exacerbating the condition.

arteriovenous A small muscular type of blood vessel that carries blood directly from *arterioles* to *venules*, bypassing the *capillary* network. It is stimulated by *sympathetic nerves* and regulates the amount of blood flowing through a nearby *capillary bed*.

artery The thick-walled, muscular type of blood vessel that carries blood from the heart to the rest of the body. All arteries except the *pulmonary artery,* which leads to the lungs, carry *oxygenated* blood. (see *arteriole*; compare *vein*)

Arthrophyta (see *Sphenophyta*)

Arthropods Historically, a phylum containing the mandibulates, crustaceans, and chelicerates. Since these groups have each been upgraded to autonomous phyla, it is now an informal classification. All of the phyla are characterized by a tough, *chitinous* protective covering (*exoskeleton*), flexible only at the joints. Each segment typically bears a pair of jointed appendages, which are highly adaptable and may account for the evolutionary success of these groups. Many have specialized secretory glands that produce an adhesive used, for instance, in spinning webs to entrap prey.

artificial chromosome A synthetic construct that contains the basic elements necessary for a *DNA* fragment to propagate and *segregate* much like an *endogenous chromosome*. This has been accomplished in yeast by linearizing a circular *plasmid* containing a *centromere* and an *ARS* sequence, and adding *telomeres* to the ends. Artificial chromosomes have been an important innovation in the study of *genes* at close to their normal dosage and are being used as *cloning vectors* for previously intractable large mammalian genes.

artificial insemination The artificial introduction of *semen* into the reproductive tract of a female animal. It is used in breeding animals to select for desirable traits and in humans to correct some causes of infertility.

artificial selection The directed selection of specific genetic traits by humans in wild or domesticated organisms including agricultural crops and livestock. It was the observation of this phenomenon that allowed Charles Darwin to originate an analogy leading to the concept of *natural selection.*

Artiodactyla The order of mammals containing the even-toed, hoofed animals (ungulates), such as sheep, goats, deer, domestic cattle, antelopes, pigs, camels, giraffes, and hippopotamuses. (see *ruminant*)

ascocarp The *fruiting body* of most *ascomycete* fungi (except yeasts) in which the *asci* and *ascospores* are borne. (see *apothecium, cleistothecium, perithecium*)

ascogonium, ascogonia (pl.) The female, *gamete*-producing cell (*gametangium*) of certain *ascomycete* fungi.

ascomycetes (see *Ascomycota*)

Ascomycota A phylum of fungi, the ascomycetes, characterized by their distinctive reproductive structure, the *ascus*, in which hard-walled *ascospores* (*asci*) are formed. It includes common molds, mildews, morels, and truffles as well as some plant *pathogens*. Most ascomycetes have a complex *life cycle* in which *haploid nuclei* of complementary *mating type* are exchanged via a *cytoplasmic* bridge, eventually forming an *ascocarp* in which *zygote* formation, *meiosis,* and *haploid* ascospore formation take place. *Asexual reproduction* may take place via *multinucleate spores* called *conidia.* *Hyphae,* when present, are divided by perforated *septa.* The yeasts form a *unicellular* subset of genera that exhibit their own distinctive characteristics. Some researchers consider deuteromycetes (most which form ascospores) and lichens (which are symbiotic relationships of algae with ascomycetes) in this phylum.

ascorbic acid (vitamin C) A water soluble essential vitamin that occurs in fruits and vegetables, particularly citrus. A deficiency in ascorbic acid causes scurvy.

ascospore The distinctive type of *spore* formed in the *asci* of ascomycete fungi. *Zygote* formation is immediately followed by *meiosis* and results in the production of *haploid* ascospores.

ascus, asci (pl.) A saclike structure found in *ascomycete* fungi in which *ascospores* (*haploid meiotic* products) are generated. After *meiosis*, the *ascus* contains four or eight haploid *ascospores* (depending on the species) that are liberated under optimal nutritive conditions, sometimes explosively. (compare *sterigma*)

-ase An ending added to any word to denote an enzyme. (for example, *lactase*)

asexual Any reproductive process or phase of life in which an organism is not concerned with *gamete* production, *meiosis,* or the exchange of genetic material. (see *asexual reproduction*)

asexual reproduction The formation of one or more new individuals from the genetic material of a single individual. It is characterized by the lack of transfer of genetic material between individuals and by any of the processes normally associated with it. This includes *meiosis*, production of *gametes,* and *fertilization* leading to *genome* union. Asexual reproduction occurs normally in many *unicellular* and *multicellular* organisms by *mitosis, binary fission, budding,* and *fragmentation,* and, in some plants, by *apomixis.* (see *parthenogenesis, apogamous, aposporous, cloning, vegetative propagation;* compare *sexual reproduction*)

asexual spore In lower plants, fungi, and some protoctists, *somatic haploid* cells that are cast off and divide *mitotically* to produce a new *multicellular* individual. (see *spore, sporophyte, alternation of generations;* compare *sexual spore*)

asparagine (see *amino acids*)

aspartic acid (see *amino acids*)

Aspergillus A genus of *ascomycete* fungi, many of which are *pathogens.* It includes species that produce the poison *aflatoxin* present in *Aspergillus*-contaminated nuts and cereals.

assimilation The absorbtion of simple molecules, usually digested nutrients, and their conversion into more complicated molecules that are incorporated into the organism.

association A climax plant community that shows characteristics that are constant over time and geographical location. An example is a coniferous forest association. An association may have several codominant species. (see *ecological community;* compare *consociation*)

associative cortex The portion of the *cerebral cortex* in the brain that appears to be the site of higher mental activities. It occupies a proportionally higher percentage of the total cortex in primates, particularly so in humans.

assortative mating A mating process in which the selection of breeding partner is nonrandom, usually based on visual, physical aspects (*phenotype*). The "choice" (conscious thought is not implied) may be performed by either sex and may be either similar or dissimilar to self. Examples include physical attributes in humans, timing of seasonal *sexual* maturity in some plants and insects (posi-

tive), and plumage color phases in Arctic skuas (negative). Bias toward mating of like with like is called *positive assortative mating* and mating with unlike partners is called *negative assortative mating.*

aster The spherical array of *microtubes* around the *centriole* at each end of the *spindle* during *nuclear division.* Its function may be mechanical support of the *microtubule* array. Both asters and centrioles are absent from higher plant cells. (see *mitosis*)

Asteroidea (see *Stelleroidea*)

astrocyte A type of *glial cell* in the *central nervous system* (*CNS*). They provide mechanical support for the *neurons* by twining around them and attaching them to blood vessels.

Atelocerata (see *Uniramia*)

atherosclerosis A condition in which *cholesterol, fibrin,* cellular debris, and other materials are deposited on the interior walls of the arteries, decreasing blood flow to vital organs including the brain and heart. (compare *arteriosclerosis*)

atlas The ringlike bone in tetrapods that forms the first neck *vertebra* of the *vertebral column.* It meets with the skull in a movable joint, allowing relatively free head movement, particularly in reptiles, birds, mammals, and somewhat less so in amphibians.

atmospheric inversion A condition in which warmer air is found above cooler air, restricting air circulation. In urban areas, it is often associated with a *pollution* crisis.

atomic weight More stringently called atomic mass, it is the sum of the masses of the protons and neutrons in an atom.

ATP (adenosine triphosphate) The molecule that constitutes the energy currency of all cells. It is produced by phosphorylation of *ADP* (adenosine diphosphate) during both *cellular respiration* (*oxidative phosphorylation*) and *photosynthesis* (*photophosphorylation*). When energy is needed for cellular *metabolism,* the terminal *high-energy phosphate bond* is cleaved in an enzyme-mediated reaction, releasing *ADP* and P_i (inorganic phosphate) as by-products. Sometimes the reaction is repeated and *ADP* is hydrolyzed to *AMP* (adenosine monophosphate) and another P_i. ATP is also one of the building blocks used in *nucleic acid* synthesis. (see *ATP synthase*)

ATPase An enzyme that cleaves one, or less commonly two, terminal phosphate groups from *ATP* with the concomitant release of *metabolic* energy. (see *high-energy phosphate bond*)

ATP synthase The enzyme associated with *proton channels* in the *chemiosmotic* generation of *ATP*.

atrioventricular junction The region where the *atria* and *ventricles* meet in the vertebrate heart. (see *AV node*)

atrioventricular node (see *AV node*)

atrium Generally, any cavity or chamber in the body.
  1. (auricle) A chamber in the vertebrate heart. Fish have a single atrium that receives *deoxygenated* blood from the major vein and passes it to the *ventricle*. Most tetrapods that breathe mainly or entirely by lungs (amphibians, reptiles, birds, and mammals) have a two-chambered atrium. The right receives deoxygenated blood from the body, and the left receives *oxygenated* blood from the lungs.
  2. A cavity surrounding the *gills* of primitive chordates.

atrophy A diminution in size of a structure or organ. It may be a normal process, as in *metamorphosis*, or a result of disease or starvation.

atropous (see *orthotropous*)

attenuation
  1. The loss of virulence of a *pathogenic* microorganism. It may be accomplished by heat treatment or by growing continuously *in vitro*. Attenuation is commonly used in the production of vaccines.
  2. see *attenuator*

attenuator A region adjacent to a *structural gene* that, in the presence of large amounts of the *gene product*, reduces the rate of *mRNA transcription* by terminating the transcript prematurely. It is seen in some bacterial *operons*, in particular the *trp* (tryptophan) operon. (see *negative feedback*)

auditory capsule (otic capsule) The *cartilaginous* or bony part of the *vertebrate* skull that encloses the *inner ear*.

auditory cortex A region on the *temporal lobe* of the *cerebral cortex*. Different surface regions of the auditory cortex correspond to different sound frequencies.

auditory nerves (see *vestibulocochlear nerves*)

aufwuchs (periphyton) A German term whose broader North American usage includes all small organisms attached or clinging to plants or other objects projecting above the bottom sediment of freshwater *ecosystems*. They include microinvertebrates, fungi, algae, and bacteria.

auks (see *Charadriiformes*)

auricle (see *atrium*)

Australopithecus (Southern ape) An extinct primate genus whose African fossils show features intermediate between those of apes and man. The braincase resembles the apes, while the jawbone and dentition resemble man.

autecology The *ecology* of a single organism, including the interaction with both living and nonliving components of its environment. (compare *synecology*)

autoclave A laboratory apparatus used in sterilizing materials, solutions, and equipment. It is similar in principle to a pressure cooker, using steam under pressure to kill microorganisms and degrade complex molecules.

autoecious (monoxenous) A *parasitic* organism that can complete its *life cycle* in a single *host* species. An example is the rust fungus. (compare *heteroecious*)

autogamy
  1. A type of *asexual* self-*fertilization* in which *haploid nuclei* are formed and fuse within the same individual.

It is seen in ciliate protoctists. (see *macronucleus, micronucleus*)

2. *Self-fertilization* in plants. (see *self-pollination*; compare *allogamy*)

autoimmunity The production of *antibodies* against a normal component of an animal's own body. It is a contributory factor in autoimmune diseases such as rheumatoid arthritis, lupus, and myasthenia gravis. Autoimmunity is now suspect as an initiation factor in a number of diseases of modern civilization.

autolysis The self-destruction of a cell by its own digestive enzymes, such as seen in *metamorphosis, atrophy*, and *apoptosis*. (see *lysosome, autophagy*)

autonomic Involuntary behavioral responses that are controlled by the *autonomic nervous system*, such as flushing and sweating. (see *autonomic nervous system*)

autonomic nervous system (ANS) The division of the *vertebrate nervous system* that controls involuntary internal body processes such as heart rate, blood pressure, respiration, and digestion. It supplies *motor nerves* to the *smooth muscles* of the gut, internal organs, and heart muscle. The relative activity of each organ is controlled by the *antagonistic* properties of the *sympathetic* and *parasympathetic nervous systems*, which themselves respond to external stimuli. (see *medulla oblongata;* compare *voluntary nervous system*)

autonomous replication sequence (see *ARS*)

autophagy The process by which a cell destroys its own *organelles* or other portions of itself. The organelle or cell portion fuses with a *lysosome* (a cell organelle itself) that releases digestive (*hydrolytic*) enzymes, destroying the engulfed material. It is part of the normal life processes of a cell, particularly as the cell ages. (see *cytolosome, autolysis, apoptosis, metamorphosis, atrophy*)

autopoiesis The self-maintenance of an organism.

autopolyploid A *polyploid* cell or organism in which replicate *chromosomes* have all arisen from one species. It can be the result of chromosome *nondisjunction* during *meiosis* of *diploid gametes* or during the *mitotic* division of a *zygote*. (compare *allopolyploid*)

autorad (see *autoradiograph*)

autoradiogram (see *autoradiograph*)

autoradiograph (autoradiogram, autorad) The visual record of the results of *autoradiography*, generally a piece of X-ray film marked with dark bands or dots. (see *DNA analysis, RFLP analysis, Southern blot, radioactive labeling*)

autoradiography Literally, self-imaging by exposure of *radioactivity* to a photographic emulsion. The term was originally coined to describe the process by which tissues or cells may be labeled with radioactive compounds. The compounds are taken up differentially by various structures, which may then be visualized, typically on X-ray film. The silver compound used in the photographic process is sensitive to radioactive energy as well as light energy. When exposed, the silver compound turns black, marking the spot. The process was co-opted for use in visualizing proteins and *nucleic acids* (*DNA* and *RNA*) by labeling them, with radioactive *amino acids* (proteins) or *nucleotides* (nucleic acids), respectively. Autoradiography is probably the most ubiquitous laboratory tool in modern molecular biology, used in *DNA analysis* and *sequencing* as well as many other research applications. The X-ray film containing the image is variously called an *autoradiogram, autoradiograph*, or *autorad*. Radioactive labeling is slowly being replaced by tags that contain *fluorescent, luminescent*, or colored reporter molecules, obviating the use of radioactivity. The X-ray film itself is being supplanted by various computer recording and storage devices.

autorecruitive A type of *colonial* growth whereby young produced by a colony tend to remain, limiting *genetic diversity*. (compare *allorecruitive*)

autoregulation The regulation of *tran-*

scription of a *gene* by its own *gene product*. (see *attenuation, negative feedback*)

autosome Any *eukaryotic chromosome* not involved in *sex determination*. Autosomes constitute the vast majority of an organism's chromosomal complement. (compare *sex chromosome*)

autostylic The type of jaw suspension found in amphibians, reptiles, and some fish (modern chimeras and lungfishes) whereby the upper jaw is attached directly to the skull. (compare *amphistylic, hyostylic*)

autotoky *Asexual reproduction* in which progeny are produced from a single parent, such as in *apomixis* and *parthenogenesis.*

autotomy An escape mechanism using self-amputation of part of the body. Some lizards, when seized by a predator, can detach part of the tail by snapping off a *vertebra* using a muscular mechanism. Various arthropods and worms can also easily shed and regenerate body parts as a defense mechanism. (see *regeneration*)

autotrophism Literally, self-feeding, a method of obtaining nutrients in which the principle carbon source is inorganic, usually carbon dioxide. Organic materials are then synthesized using light energy (*photoautotrophism, photosynthesis*) or chemical energy (*chemoautotrophism, chemosynthesis*). In the case of chemical energy, it is derived from the oxidation of an inorganic compound. Autotrophs are important ecologically as the *primary producers* of organic carbon for all *heterotrophic* organisms. (see *trophism*; compare *heterotrophism*)

auxins A general classification of plant growth substances. They are synthesized in actively growing plant roots and tips, and have varied effects as they actively move into other regions. Auxins are responsible for growth by longitudinal cell elongation, rather than *cell division*, in the growing region behind the shoot tip (*apical meristem*). They help maintain *apical dominance* by inhibiting *lateral*

bud development and are responsible for *phototropic* (curving toward light) and *gravitropic* (curving toward gravity) responses. They stimulate cell growth by activating a *proton pump* that excretes H^+ ions from the cell. This promotes an acid-induced loosening, thus expanding the *cell wall*. Auxins are also involved in fruit set and growth, flower formation, and delays in leaf fall and fruit drop. They interact synergistically with *gibberellins* and *cytokinins* in stimulating *mitosis* and *differentiation* in the *cambium*. Natural auxins have been isolated from fungi, bacteria, corn *endosperm*, and the urine and saliva of humans. Artificially produced auxins are used in agriculture and horticulture to regulate plant growth and development. Some have differential inhibitory effects on growth and are exploited as weed killers and *herbicides.*

auxotrophic A laboratory strain, usually of a microoorganism, that will proliferate only when the growth medium is supplemented with specific nutrients not required by the *wild-type* organism. It is one way of controlling the escape of such strains from the laboratory environment. (compare *prototrophic*)

Aves The class of vertebrates that contains the birds, characterized by adaptions for flight. The most obvious external adaptions are the modification of forelimbs as wings and the production of feathers, which provide insulation and the large surface area important to flight. The bones are strong and hollow, making the body light enough for flight. The jaws form a horny beak, and teeth are absent. Eggs with a *calcareous* shell are laid and incubated, and the young are cared for until independence. Behaviors are well developed, manifesting in aural (song) and physical displays. Complex migration patterns are often exhibited. Birds probably evolved from Jurassic reptiles and maintain characteristics such as scaly legs and feet, and *nucleated* red blood cells (*erythrocytes*). They have evolved a separation of *oxygenated* and *deoxy-*

genated blood in a four-chambered heart and are *homeothermic.* Some birds have become flightless, and their forelimbs have either been adapted for swimming (for example, penguins) or become essentially useless (for example, ostriches). The subclass Neornithes comprises all living birds.

avian Anything related to birds.

avidin A protein found in raw egg white. It binds extremely tightly to the vitamin *biotin* found in the pith of citrus fruits. This association is exploited in many molecular biology techniques.

AV node The thin connection of *cardiac muscle* cells that passes the wave of *depolarization* from the *atria* to the *ventricles* during a heart contraction. It is propagated relatively slowly, allowing the atria to finish emptying their contents into the corresponding ventricles before those ventricles start to contract. (see *pacemaker, bundle of His*)

avoidance A behavior that tends to protect an animal by reducing its exposure to hazard. It may be learned.

axenic A laboratory culture that contains only one species of microorganism.

axil The angle between the upper side of a leaf, stem, or branch and the shoot bearing it. It is the site of lateral or *axillary* buds.

axillary In plants, refers to any structure located in a leaf *axil.*

axis The second and strongest neck (*cervical*) vertebra in the *vertebral column* of tetrapods. In reptiles, birds, and mammals, it acts in concert with the *atlas* to provide a pivot for rotation of the head.

axon (nerve fiber) The part of a *neuron* that conveys impulses from the *cell body* to a *synapse.* It is an extension of the *cell body* and in most vertebrates, it is surrounded by a fatty (*myelin*) sheath. (compare *dendrite*)

axoneme A *bundle* of *microtubules* extending the length of any of the various *organelles* used for locomotion and movement in single cells. It consists of

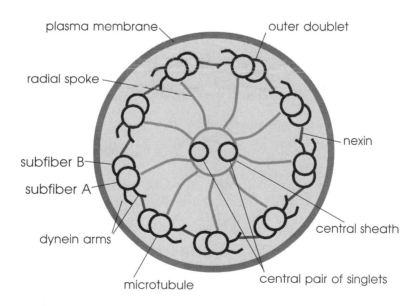

Structure of the Axoneme of an Undulipodium
(Eukaryotic Flagellum or Cilium)

nine pairs of outer microtubules surrounding two single central microtubules, and is enclosed in a membrane continuous with the *plasma membrane* of the cell. Axonemes make up the *undulipodia* used for locomotion in some protoctists, and the tails of *spermatozoa*. (see *flagellum, cilium, pseudopod, axopod*)

axopod A permanent *pseudopod* stiffened by a *microtubular axoneme*.

Azobacter A genus of free-living *aerobic* bacteria found in soil and water. They are characterized by the ability to "fix" atmospheric (inorganic) nitrogen in an organic form that can be used by biological organisms. (see *nitrogen fixation, nitrogen-fixing aerobic bacteria*)

Bacillariophyta A phylum of *unicellular, photosynthetic* protoctists, the diatoms, found in both freshwater and marine environments. They comprise a large proportion of *plankton*, making them an important part of the *food chain.* Diatoms are enclosed in shells, or *tests*, made of opaline silica, structured so that one-half of the rigid shell just fits inside the other like a stack of bowls. They reproduce *asexually* by separating the two halves of a shell. Each half then regenerates another half shell within it. Thus, within a given sequence of *asexual reproduction*, the organisms get progressively smaller. *Sexual reproduction* and fusion of *gametes* regenerates a full-sized individual. Their *chloroplasts* resemble those of brown algae, containing *chlorophylls, carotenes,* and *xanthophylls*. Past deposition of diatoms has formed diatomaceous (siliceous) earth which is harvested for use as an abrasive, asphalt additive, and because of its desiccant qualities, a nontoxic insecticide. The oil stored within fossilized diatoms contributes to petroleum supplies.

bacillus Any rod-shaped *bacterium*. An example is the genus *Bacillus*. (compare *coccus, spirillum*)

backbone (see *vertebral column*)

backcross A mating (crossing) of an offspring back to a parent to ascertain the genetic composition (*genotype*) of the offspring. Typically, a parent bearing a *homozygous dominant* (**AA**) genetic trait is crossed with a homozygous *recessive* parent (**aa**), producing three genetically different kinds of offspring (**AA, aa, Aa**). Two of the offspring (**AA, Aa**) are indistinguishable visually (by *phenotype*) because they both exhibit the dominant **A** trait. These offspring are "backcrossed" to the **aa** parent. If the offspring contain the **AA** genotype, all the progeny of the backcross will be **Aa** and show the **A** phenotype. If the offspring contain the **Aa** genotype, half the progeny of the backcross will be **Aa** and half will be **aa**, thus revealing the recessive **a** phenotype and the **Aa** status of the tested offspring.

back mutation (see *reversion*)

Bacteria One of the five phylogenetic kingdoms into which all living organisms are divided for academic discussion. It comprises all *prokaryotic* microorganisms and specifically excludes all *nucleated* organisms. Bacteria have infiltrated virtually every aspect of the natural world, and in terms of numbers and variety of habitats, they include the most successful life forms. Bacterial *photosynthesis*, in particular from cyanobacteria, accounts for the majority of oxygen production and replenishment in the *biosphere*. Bacteria include the only organisms that have the ability to fix atmospheric nitrogen (*nitrogen fixation*), thus making it available to other organisms, and are also important *decomposers*. They are indispensable in numerous industrial processes, in the food industry (for example, yogurt and vinegar), and as sources of *antibiotics*. They were the first organisms to be used in *genetic engineering* for manipulating and reproducing foreign genes. Bacteria are divided into the subkingdoms of Archaea and Eubacteria.

bacterial photosynthesis (see *photosynthesis*)

bactericidal Any compound that has a lethal effect on bacteria, such as many *antibiotics*.

bacteriochlorophylls The *chromophores* in the *photosynthetic* reaction centers in some bacteria. They differ from the *chlorophylls* found in green plants and algae in that they are activated by longer wavelengths (λ) of light.

bacteriophage (phage) A virus that infects bacteria. It consists of a naked strand of *nucleic acid* surrounded by a complex polyhedral capsid composed mainly of *glycoproteins*. The viral *genome* is injected into a *bacterium* through its helical tail and subsequently subverts the *metabolism* of the *host*. The phage genome is *replicated* and re-encapsulated, eventually being released by cell *lysis*. *Temperate phages* may insert themselves into the *host* genome where they innocuously replicate along with the host until the viral *DNA* is excised and resumes its *lytic* cycle. Bacteriophage DNA has been a common *cloning vector* in *genetic engineering*. (see *lysogeny, virulent phage*; compare *virus*)

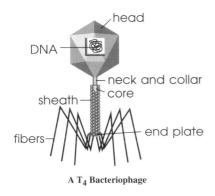

A T$_4$ Bacteriophage

bacteriophage lambda (λ) A *bacteriophage* that infects the bacterium *E. coli*. It was a popular early *cloning vector*.

bacteriorhodopsin A molecular *photoreceptor* that functions as a *light-driven proton pump* in halobacteria. In the absence of oxygen (O_2), the organism switches to *photosynthetic* mode, and bacteriorhodopsin uses light energy via a proton gradient to generate *ATP*. Bacteriorhodopsin uses the same light-absorbing *chromophore, retinal,* found in vertebrate *rod cells*. (compare *sensory rhodopsin*)

bacteriostatic Any compound that prevents the growth and reproduction of bacteria but does not kill them. (see *antiseptic*)

bacterium, bacteria (pl.) Any of the large and diverse group of prokaryotic organisms belonging to the kingdom Bacteria, including Archaea and Eubacteria.

balance of nature A misconception implying that the natural world is or should be in a precarious balance from which it does not or should not budge. In actual fact, both the environment and living organisms in it exist in a continuum of change and evolution. Although human beings have unquestionably interfered in the "normal" progression of evolution, they are also an integral part of the process.

Balbiani ring (see *polytene chromosomes*)

baleen The parallel horny plates made of *keratin* that grow down from the sides of the upper jaw in baleen whales. They are comprised of hollow fibers that form a comblike structure on the inner edge and are used to strain *krill*, a major food source, from seawater.

baleen whales (see *Mysticeti*)

band The visual image representing a particular *nucleic acid* fragment on an *autoradiograph*. (see *DNA analysis, RFLP, Southern blot, Northern blot*)

bandicoots (see *Peramelemorphia*).

barbs The first order of extensions from the *rachis* of a feather, held together by hooked secondary extensions called *barbules.*

barbules The microscopic hooks extending at right angles to the *barbs* of a feather that hook the barbs together. When birds *preen*, they are properly aligning all the barbs and barbules.

bark In woody plants, all tissues external to the *vascular cambium* and comprising the outermost layers. Bark serves a protective function. (see *cambium, cork cambium, rhytidome*)

bark lice (see *Protura*)

barnacles (see *Cirripedia*)

baroreceptor A type of *sensory receptor* in the walls of large blood vessels that functions to maintain steady *blood pressure*. Changes in pressure affect the tension in the vessel wall, stimulating the receptor to fire *nerve impulses* that regulate blood pressure by affecting heart rate and blood vessel diameter. (see *mechanoreceptor*)

Barr body In mammals, the *X chromosome* that is inactivated, equalizing the dosage of genes between the **XX** females and **XY** males. Microscopically, it is seen as a small, dense body. (see *X-inactivation, dosage compensation, heterogametic, sex chromosomes*)

basal The location of a structure at or near the base, or point of attachment, of the main body of an organism.

basal body (see *kinetosome*)

basal cell

1. In flowering plant (*angiosperm*) seeds, one of the two cells resulting from the first *zygotic* division. It goes on to become the *suspensor*, which attaches the *embryo* to the *embryo sac* at the *micropyle* near which the root apex will form. (see *double fertilization*)

2. A cell type found in the lowest layer of the *epidermis* from which tissue is renewed. (see *stratum basal*)

basal ganglia Several large groups of *neurons* in the *corpus striatum* of the vertebrate brain that, together with the *cerebellum* and the *red nucleus* in the *brainstem*, generate complex patterns of activity in *motor neurons*.

basal layer (see *stratum basal*)

basal metabolic rate (BMR) The minimum rate of energy expenditure necessary to maintain vital processes such as circulation, respiration, and involuntary muscle movements. It is measured either directly from heat production or indirectly from oxygen consumption. *Thyroid hormones* in animals are the prime regulators of the BMR.

base

1. A chemical compound that reduces the relative number of hydrogen ions (H^+) when dissolved in aqueous solution. This is often accomplished by releasing hydroxide ions (OH^-) that combine with the hydrogen ions to form water. It is manifest by an increase in the pH of the solution. (compare *acid*)

2. A nitrogen-containing (*nitrogenous*) ring molecule that, in combination with a *pentose* sugar and phosphate group, form a *nucleotide*, the fundamental unit of *nucleic acids* (for example, *DNA, RNA*) and also *ATP*, the energy currency of the cell. Structurally, *cytosine, thymine,* and *uracil* are single ring *pyrimidines*, and *adenine* and *guanine* are double ring *purines*. The terms "base" and "nucleotide" are often used informally and interchangeably in referring to the nucleotide residues that make up a *polynucleotide* chain.

base analogue A chemical whose molecular structure mimics that of a *DNA* or *RNA base*. Many such molecules act as *mutagens* such as 5-bromo-uracil (5-BU) and 2-amino-purine (2AP).

basement membrane A filamentous layer of material in skin that underlies the *epithelial cell* layer and separates it from underlying *connective tissue*. It is composed largely of *collagen* and *glycoproteins*.

base pair (bp) The two *bases* always found chemically bonded together in the *DNA double helix*. *Adenine* always pairs with *thymine* by two hydrogen bonds and *guanine* with *cytosine* by three hydrogen bonds. (see *complementary base pairing*)

base pairing (see *complementary base pairing*)

base pair mismatch (DNA mismatch) A *base pair* in which the *bases* are not *complementary*. Normally, *guanine* pairs with *cytosine* and *adenine* with *thymine*. Occasionally, a mismatched base is inserted during *DNA replication*, a situation that is usually corrected by the cell's *DNA repair* mechanisms. (see *complementary base pairing*)

base pair ratio The ratio of paired *nucleotides* in *double-stranded DNA*, for example, the A-T pairs: G-C pairs. Since the total must be 100%, only one pair need be measured, and the base ratio is often expressed as *GC content*. The measurement varies between *species* and is sometimes used for categorization. A triple hydrogen-bonded G-C pair is stronger than an A-T pair with only two hydrogen bonds. Therefore, the *GC content* also determines the overall *melting temperature* of a double-stranded *nucleic acid* molecule, a characteristic that has ramifications for processes from *DNA* replication to laboratory research. (see *complementary base pairing, PCR*)

base pair substitution The replacement of a specific *DNA base pair* by a different pair, resulting in a *mutation.*

base substitution A *DNA mutation* in which a different *base* is substituted for the original. Various *DNA repair* enzymes usually correct any mismatched *base pairs* that occur spontaneously. (see *transition, transversion*)

basic stain (see staining)

basidiocarp The *fruiting body* of *basidiomycete* fungi, excepting the *rusts* and *smuts*. The basidiocarp is the above-ground portion of the *fungus* that is commonly referred to as the toadstool or mushroom. Within its *gills* reside the *basidia* that, after *meiosis*, give rise to *haploid basidiospores.*

basidiomycetes (see *Basidiomycota*)

Basidiomycota A phylum of fungi, the basidiomycetes, containing mushrooms, toadstools, puffballs, jelly fungi, shelf fungi, and the plant *pathogens* called *rusts* and *smuts*. They have a complex *life cycle* that is mostly *sexual* and occurs underground; the short *asexual* stage manifests as the familiar mushroom or toadstool. *Meiosis*, which occurs directly following *zygote* formation, occurs within *basidia*, characteristic clublike structures contained in the *gills* or pores of a *basidiocarp* (*fruiting body*). The *haploid* meiotic products exhibit two complementary *mating types*, plus (+) and minus (–), that are incorporated into *basidiospores*. They grow as a mass of *hyphae* (*mycelium*) and in *heterokaryotic* mycelia, eventually fuse. These *dikaryotic secondary mycela* immediately give rise to the next generation of basidiocarp. *Asexual reproduction* may take place via *multinucleate spores* called *conidia*. *Hyphae* are divided by perforated *septa*, although septae that cut off the young basidia often become blocked. Many basidiomycete species participate in the formation of *ectomycorrhizae*. Some mushrooms are edible, but others are deadly poisonous.

basidiospore The *haploid* spores of *basidiomycete* fungi. They are borne externally on specialized cells termed *basidia*, on the ends of projections called *sterigmata*. They occur within the *fruiting body* (*basidiocarp*) and come in two complementary *mating types*, which eventually fuse to form a *diploid zygote.*

basidium, basidia (pl.) The club-shaped reproductive structure contained within the *basidiocarp* of basidiomycete fungi that produces *haploid basidiospores*. A *basidium* usually produces four *basidiospores*, two of each complementary *mating type.*

basilar membrane A structure involved in the detection of sound in the terrestrial vertebrate *inner ear*. Specifically, it is the membrane on which the auditory receptors, the *hair cells*, are located in the *organ of Corti* in the *cochlea.*

basipetal The successive development of plant structures from the apex downwards so that the youngest structures are farthest from the apex. The term may also be applied to the movement of substances such as *auxin* toward the base. (compare *acropetal*)

basophil A circulating white blood cell (*leukocyte*) containing granules that rupture and release chemicals that enhance the *inflammatory response*. They are similar in structure and function to *mast cells* (noncirculating) and are also important in *allergic responses*. The granules stain with basic dyes, thus the name. (see *immune system, granulocyte*; compare *eosinophil, neutrophil*)

bats (see *Chiroptera*)

B-cell (B-lymphocyte) A type of *lymphocyte* (white blood cell) that produces *antibodies*. When a B-cell encounters a foreign *antigen* to which it is targeted, it enlarges and begins to divide. In the process, it produces a cell line (*clone*) of *plasma cells* that secretes only one type of antigen-specific antibody. B-cells arise from *stem cells* in the *bone marrow*. (see *clonal selection, somatic recombination, antibody, antibody-antigen reaction, immune system, immune response, humoral immune response, hemopoiesis*; compare *T-cell*)

B DNA The form of the *DNA double helix* that is found most frequently in living systems. It is more hydrated, thus less stable, than *A DNA*, and the rungs formed by the *base pairs* are exactly perpendicular to the *sugar-phosphate backbone*. The helix is right handed. (compare *Z DNA*)

beak The hard mouth structure in birds, some reptiles, and a few fish. It functions as a replacement for lips and teeth.

beard worms (see *Pogonophora*)

bears (see *Carnivora*)

beavers (see *Rodentia*)

bee-eaters (see *Coraciiformes*)

bees (see *Hymenoptera*)

beetles (see *Coleoptera*)

behavioral ecology The study of how natural selection shapes behavior.

Benson-Calvin-Bassham cycle (see *photosynthesis*)

benthic The region of the ocean consisting of the sea bed and the organisms that live on or in it.

berry A many-seeded succulent fruit such as the tomato.

beta-2-microglobin (microglobulin, β-2-microglobin) A small protein found associated with the *MHC I protein* in the vertebrate *cell membrane*. Its structure is similar to certain portions of an *antibody* molecule. (see *MHC proteins, MHC complex, HLA complex*)

beta-blocker (β-blocker) Any chemical that selectively blocks the *adrenergic beta receptors* in the *sympathetic nervous system* that preferentially bind the *neurotransmitter adrenaline*. Clinically, drugs such as propanolol are used to slow heart rate and lower blood pressure. (see *autonomic nervous system; compare alpha-blocker*)

beta carotene (β-carotene) A *carotenoid* that is split into two identical parts to yield *vitamin A* during digestion in vertebrates. It is the orange pigment in carrots. (see *carotenes*)

beta-pleated sheet (β-pleated sheet) A *secondary structure* in which *polypeptide chains* are held together by *hydrogen bonds* between them, giving the resulting protein a pleated appearance. The chains may run in the same direction (parallel) or opposite directions (antiparallel). It is common in fibrous, *structural proteins* such as *keratin* in hair and nails. (compare *alpha helix*)

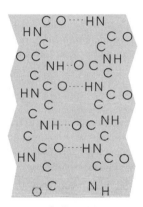

Beta (β)-Pleated Sheet

beta receptor (β receptor) An *adrenergic postsynaptic receptor* type in the *sympathetic nervous system* that preferentially binds the *neurotransmitter adrenaline*. (see *autonomic nervous system, beta-blocker;* compare *alpha receptor*)

bi- A word element derived from Latin meaning "two" or "twice." (for example, *biennial*)

biallelic A *gene* that shows two alternate forms in a population. (see *allele, homozygous, heterozygous*; compare *multi-allelic*)

bicuspid valve (see *mitral valve*)

biennial A plant that requires two growing seasons to complete its *life cycle*. In the first year, it produces only foliage and *photosynthesizes.* The food is stored during the winter in an underground root or stem (*perennating structure*). In the second year, the stored food is used to produce flowers, fruit, and seeds. Some biennials can be induced to act as *annuals* and flower in the first year by appropriate cold or *hormone* treatments. Many root crops, such as carrots and parsnips, are biennials. (compare *annual, ephemeral, perennial*)

bilateral symmetry The mirror image structure of an organism allowing it to be visually divided into two halves along one plane. It is characteristic of most animals where one end leads during locomotion. Some secondary asymmetry of internal organs is common and occurs in vertebrates including man. In plants, bilateral symmetry is seen in some flowers (for example, snapdragon). (see *zygomorphic*; compare *radial symmetry*)

bile A solution of bile salts, bile pigments (for example, bilirubin), cholesterol, lecithin, and traces of other substances secreted by the liver and stored in the *gallbladder.* It enters the *duodenum* via the *bile duct.* Bile aids digestion by facilitating the emulsification of fats in the small intestine. Bile pigments including *bilirubin*, the breakdown product of *hemoglobin*, are merely excretory products.

bile duct A duct in vertebrates that transports bile from the liver to the *duodenum.* (see *gallbladder)*

bile pigments Pigments excreted in *bile* that are the degradation products of *hemoglobin.* The *heme* portion is converted to the green pigment *biliverdin* that is reduced by the enzyme biliverdin reductase to a red-brown pigment, *bilirubin*, that colors the feces. (see *bile*)

bile salts Components of *bile* that aid digestion in the small intestine. Bile salts emulsify fats by lowering their surface tension, causing them to disperse into smaller droplets. This increases their surface area, facilitating the action of the digestive enzyme *lipase.* (see *bile*)

bilirubin (see *bile pigments*)

biliverdin (see *bile pigments*)

binary fission A type of *asexual* reproduction in which a parent cell or simple organism divides into two similar progeny of roughly equal size. In *prokaryotes*, the genetic material, usually a single circular *chromosome*, is replicated, and the cell pinches into two at a region located approximately between the two *genomes.* This ensures that one copy will end up in each progeny cell. Binary fission is occasionally found in some lower eukaryotes as well. It is differentiated from *mitosis* in *eukaryotes* in large part by lack of specific *nuclear division.*

binocular vision A type of vision in which the eyes point forward so that the image of a single object can be focused onto the *fovea* of both eyes at once. This allows perception of depth and distance by interpolation between two points. Binocular vision is found in primates and other vertebrates, particularly predators such as owls. (see *stereoscopic vision*)

binomial nomenclature A system of taxonomic classification introduced by Linnaeus, a Swedish botanist, in which each *species* is defined by two names. The first is the *genus* (generic name) and written with a capital letter. The second is the *species* (specific name) and in lowercase. Both names are in Latin and printed in italic type. For example, man belongs to the species *Homo sapiens.* (see *classification, taxonomy, phylogenetics, cladistics, phenetics*)

binucleate A cell containing two *nuclei.*

bio-, bi- A word element derived from Greek meaning "life." (for example, *biochemical*)

bioactive Any substance that possesses biologically significant properties.

bioassay An experimental technique for quantitating the strength of a biologically active chemical by measuring its effect on a living organism. For instance, the *Ames test* assesses the *mutagenic* potential of various chemicals by measuring the *mutation rate* in bacteria.

biochemical A substance produced by

chemical reaction in a living organism.

biochemical oxygen demand (see *BOD*)

biochemical pathway A sequence of enzymatic reactions in which the product of one reaction becomes the *substrate* for the next. They are the organizational units of *metabolism*. (see *feedback inhibition, feedback regulation, epistatic*)

biochemical taxonomy The use of biochemical characteristics in the classification of organisms. In particular, the study of *serology* and the techniques of *chromatography* and *electrophoresis* have altered the classification of many organisms from the initial visual categorization. Most recently, *DNA analysis* has provided some evolutionary surprises and even more reclassification. (see *biosystematics, cladistics, phenetics, phylogenetics, molecular clock*)

biochemistry The study of chemical reactions occurring in living organisms.

biodegradable Organic compounds that can be decomposed through the action of bacteria and other microorganisms. This action occurs over long periods of time and only under certain conditions. For instance, grass clippings in a plastic bag in a landfill will not biodegrade completely, having access neither to external microorganisms nor to oxygen. Other substances, such as *cellulose* bags, are theoretically biodegradable but only over an extremely extended time scale.

biodiversity (see *biological diversity*)

bioengineering The design, manufacture, and use of replacements or aids for body parts or organs that have been removed or are defective, for example artificial limbs and hearing aids.

biofeedback A technique whereby a person can consciously control physiological responses normally controlled by the *autonomic nervous system*, for example heart rate and *blood pressure*. The technique is reputed to provide relief where traditional medicine sometimes fails, such as asthma, migraine, and muscle cramps.

biogenesis The theory that living organisms originate only from preexisting life forms. The theory of biogenesis became generally accepted as a result of the work of Redi (1688) for macroorganisms, and Spallanzani (1765) and Pasteur (1860), in particular, for microorganisms. It preempted the theory of *spontaneous generation*. Biogenesis does not adequately explain the *origin of life*, a subject that continues to provide a fruitful ground for investigation and discussion.

biogenetic law (see *recapitulation*)

biogenic Any substance produced by living organisms or their remains.

biogeochemical cycle The cycling of a chemical element through the *biosphere*.

biohydrometallurgy The relatively new science of using *genetically engineered* microorganisms to assist with different aspects of mining including purification and waste removal.

biological classification (see *classification*)

biological clock The internal mechanism of an organism that regulates physiological and behavioral rhythms relating to environmental cycles. A few examples include *circadian* rhythms (daily) such as sleep cycles, lunar/monthly cycles such as female fertility, and annual cycles of plant flowering. The term has been co-opted into common usage to describe the longer *life cycle* of a woman's fertility.

biological control The use of natural *predators* or *parasites* to supplant chemical control of pest organisms. Examples abound and include lady beetles and green lacewings that feed on aphids, the wasp *Trichogramma* that parasitizes insect eggs, the bacterium *Bacillus thuriengensis* that kills the *larvae* of moths and butterflies, and, on a larger scale, the attraction of insect-eating birds, bats, reptiles, and amphibians.

biological diversity (biodiversity) Although this term is often keyed to the protection of a single species, it encompasses three concepts. *Genetic diversity* is concerned with the total number of genetic characteristics or *genes* in a species, subspecies, or group of species. *Habitat diversity* is concerned with the diversity of habitats in a given unit area. *Species diversity* itself is concerned with the relative abundance and dominance of particular species as well as the total number of different species.

biological evolution The gradual process of genetic change that occurs in populations of organisms. It manifests as new characteristics in a species and, ultimately, in the formation of new species. (see *Darwinism, Lamarckism, natural selection*)

biology The study of the life sciences, comprising the two major disciplines botany and zoology. The term was coined by Lamarck in 1802.

bioluminescence The production of light in biological organisms by a chemical reaction, usually involving the protein *luciferin* and the enzyme *luciferase*. Bioluminescence is found in some insects (for example, fireflies) and many marine organisms often by virtue of a *symbiotic* relationship with bioluminescent bacteria.

biomagnification (see *food chain concentration*)

biomass The total mass of living organisms of any defined group in a population or per unit area. The measure of biomass ameliorates the distinction of many small organisms versus a few large ones. It also allows a measure of the *net primary productivity* of a particular *ecosystem*, basically as a result of *photosynthetic* activity. A healthier ecosystem can support a larger biomass, be it three elephants or three million fleas. (see *ecology, food chain*)

biomass fuel A new name for the oldest fuel used by humans. It most commonly is wood but also includes *peat* and cattle dung.

biome The largest ecological unit, defined by vegetation and climate and delimited by geography. It defines a particular type of *ecosystem* that might be found in different locations. (see *ecology*)

biomineralization The acquisition, incor-

poration, and/or formation of minerals by a living organism.

biophysics The study of the physical properties and mechanical characteristics of living organisms.

biopoiesis The generation of living from nonliving material. (see *origin of life, spontaneous generation*)

biorhythm A periodic physiological or behavioral change that is controlled by a *biological clock*. Examples include *circadian* rhythms, *estrus* and *menstrual cycles*, *hibernation*, and *migration*.

biosphere The part of the earth and its atmosphere that is inhabited by living organisms.

biosynthesis Enzyme-mediated chemical reactions in a living cell in which complex organic molecules are manufactured from simpler organic or inorganic molecules. (see *catabolism, carbon cycle, nitrogen cycle, photosynthesis, chemosynthesis*)

biosystematics The branch of systematics in which the genetic and evolutionary relationships between various *taxa* are investigated by experimental means. Traditionally based on the results of anatomical and ecological studies, *serology*, *cytology*, and *DNA analysis* have recently contributed a great amount of information. (see *biochemical taxonomy, cladistics, phenetics, phylogenetics, molecular clock*)

biota All living things existing within a given area or on the Earth.

biotechnology The application of biological discoveries to industry, agriculture, and medicine. For instance, bacteria are used to produce various *antibiotics*, and yeasts are used to make beer and bread. *Genetic engineering* has enabled the large-scale production of medically important products such as *insulin*, *hormones*, and other *bioactive* molecules. Genetically engineered products are used in industries from agriculture to waste disposal. Most recently, biotechnology has moved into deciphering and cataloging the fundamental information contained in *DNA*. In the worldwide *Human Genome Project,* information has been and continues to be obtained that impacts diagnostics, *gene therapy*, and personal identification.

biotic All living factors in the *environment* affecting an organism, distinct from physical factors . Examples include care of a child by a parent, predation by another organism, or infection by a microorganism. (compare *abiotic*)

biotic climax A stable *ecological community* arising from an *ecological succession* that has been deflected or arrested, either directly or indirectly, as a result of human activities. Some use the term to describe deflected successions, where a recovery to normal succession is not possible even when the interfering factors are removed. Others use the term *plagioclimax* to describe direct human intervention, reserving *biotic climax* for more indirect effects such as grazing by introduced (but nondomesticated) species. Biotic climax may also be used to describe a natural succession initiated by a biological agent such as *guano* deposits.

biotic potential The rate at which a population of a given species will increase when no limits are set on its growth. It may be depicted as an exponential *growth curve*. In practice, this sort of unrestricted growth is unrealistic but may be approximated by, for instance, a newly inoculated bacterial culture. (compare *carrying capacity*)

biotic province A region inhabited by a characteristic set of life forms and bound by barriers that inhibit both the spread of the distinctive kinds of life to other regions as well as the immigration of foreign species.

biotin (vitamin H) A member of the *vitamin B complex* found in many natural sources including egg yolk, kidney, liver, yeast, and the pith of citrus fruits. It is also produced by normal intestinal bacteria and is required as a *coenzyme* in the *Krebs cycle* for energy production. Biotin

is tightly bound by *avidin*, a component of raw egg white. Overconsumption of raw egg white would result in a biotin deficiency. This extremely strong interaction is exploited in some laboratory processes involving the labeling of various molecules in order to visualize them.

biotype (see *ecotype*)

bipedal Any organism that walks mainly on two legs.

bipolar cell A specialized *neuron* found in vertebrate and cephalopod eyes. It connects the *rod* or *cone cells* (light receptors) to the *ganglion cells* leading to the *optic nerve*. Within the *fovea* (center of the *retina*), the relationship of *receptors* to bipolar cells to ganglion cells is 1:1:1. Toward the periphery, many receptor/bipolar cells may feed into each ganglion cell, increasing sensitivity but reducing acuity.

biramous appendage A two-branched appendage found in crustaceans. It consists of three sections, each of which may have several segments. The outer, shorter branch is called the *exopodite*, and the inner, longer branch is called the *endopodite*. (compare *uniramous appendage*)

birds (see *Neornithes, Paleognathae, Neognathae, Aves*)

bisexual

  1. When referring to a species in general, it describes the differentiation of sexes among individuals, specifically male and female. (see *dioecious*)

  2. (see *hermaphrodite*)

  3. In man, it also describes the sexual attraction of an individual to both sexes.

bison (see *Artiodactyla*)

bitterns (see *Ciconiiformes*)

bivalent A pair of *replicated homologous chromosomes* in association during *meiotic prophase*. Pairing (*synapsis*) of homologous chromosomes is visible at one or several points along the chromosome at which *recombination* may take place. (see *meiosis*)

bivalves (see *Bivalvia*)

Bivalvia A class of marine and freshwater

mollusks, characterized by a flat body in a hinged shell. They have a poorly developed head and large, paired gills used for respiration and often for *filter feeding* as well. Some are anchored to the substratum by tough filaments, while others burrow into sand, rocks, or wood. Some (for example, scallops) swim by clapping their shell valves together.

bladder

1. A general term for any body sac used to hold gaseous, fluid, or solid material.
2. The *urinary bladder* in vertebrates is a thin-walled, muscular sac used as a temporary store for urine in most vertebrates (except birds and some reptiles). In mammals, the urine enters the bladder directly from the *ureters* and is discharged to the *urethra* under the control of a *sphincter* muscle. (see *gallbladder, swim bladder*)

bladderworm (cysticerus) The stage in the *life cycle* of a tapeworm when it encysts in muscle of the *intermediate host* (for example, a pig). It consists of a *bladder* containing the inverted head of the worm. When infected raw meat is ingested, the head everts and attaches itself to the lining of the gut of the *final host* (humans) and becomes an adult worm.

blade The flattened portion of a plant leaf.

-blast- A word element derived from Greek meaning *embryo*. (for example, *blastoderm*)

blastocoel A cavity that appears during early *embryonic* development in animals. It is formed within the mass of cells that make up the *blastula* and is filled with fluid.

blastocyst A mammalian egg in the later stages of *cleavage* but before implantation in the uterus. It consists of a hollow, fluid-filled ball of cells and an inner cell mass from which the *embryo* develops. (see *trophoblast*)

blastoderm The cellular mass that results from *cleavage* of the *cytoplasm* (*blastodisc*) of eggs with a large amount of yolk, such as those of birds, sharks, and

cephalopods (for example, squid, octopus). The term is also used for the cellular coat of cleaved insect eggs.

blastodisc In yolk-rich eggs, the small disk of *cytoplasm* located at one pole at which *cleavage* occurs. This pattern is called *meroblastic cleavage*.

blastomere One of the cells formed from the early division of a *fertilized* animal egg during *cleavage* and prior to *gastrulation*. Blastomeres may differ in size.

blastopore In *embryonic* development, the transitory opening on the surface of the *gastrula* between the internal cavity (*archenteron*) and the exterior. It is formed during *gastrulation* by an inward movement of cells. In *amniotes* (for example, mammals, reptiles, and birds), the invagination is incomplete and an open pore does not develop. This is called the *primitive streak* or virtual blastopore. The blastopore represents the future mouth in *protostomes* and the future anus in *deuterostomes*.

blastula The stage in animal *embryonic* development following *cleavage* of the *zygote*. It is a hollow, fluid-filled ball of cells, one layer thick.

Blattoidea (see *Dictyoptera*)

blind spot A point on the *retina* of the vertebrate eye that is insensitive to light. It is the point at which the *optic nerve* joins the eyeball and contains no light sensitive cells (*rods* and *cones*).

blood The transport medium of an animal's body. It is a fluid tissue circulated by muscular contractions of the heart (vertebrates) or blood vessels (invertebrates). It carries oxygen and nutrients to body tissues and organs, and carbon dioxide and *nitrogenous* waste away to be excreted. It also distributes heat and conveys other substances such as *hormones* throughout the body. Blood consists of liquid *plasma* in which numerous specialized cells and substances are suspended. Red blood cells (*erythrocytes*) contain *hemoglobin* (vertebrates) or *hemocyanin* (some invertebrates) that combine with and convey oxygen. In some invertebrates, the *respiratory pigment* is carried in solution in the plasma, and in most insects, it is completely absent. White blood cells (*leukocytes*) are involved in the *immune system* and include types of cells that produce *antibodies* or engulf and destroy foreign invaders. Also carried in blood are cell fragments called *platelets* that function in *blood clotting*.

blood-brain barrier The barrier between circulating blood and the *cerebrospinal fluid* bathing the spinal cord and brain. It protects the *central nervous system*. The *capillary* walls in this region are impermeable to many, although not all, large particles, such as microorganisms. Although various *macromolecules* are excluded, smaller molecules, such as ethanol, pass the barrier easily.

blood capillary (see *capillary*)

blood cell Any of the cells contained within the fluid *plasma* of blood. In human blood, about 45% of the volume is made up of red cells (*erythrocytes*) and 1% of white cells (*leukocytes*).

blood clotting The conversion of blood from a liquid to a solid state. This occurs when blood is exposed to air, either through injury or after removal from the body. The clot closes wounds and prevents further blood loss. Clotting occurs in a cascade requiring 12 different *clotting factors*. It is initiated when *platelets* encounter a rough surface, such as injured tissue, and release the enzyme *thrombokinase* (*thromoplastin, Factor III*). Thrombokinase changes to enzyme precursor *prothrombin* (*Factor II*) into the active enzyme *thrombin*, requiring the presence of calcium ions. Thrombin then converts *fibrinogen* (*Factor I*), a soluble protein, into the insoluble protein *fibrin*. Fibrin forms a network of fibers in which the blood cells become entangled and form a clot. The clot is stabilized by *Factor XIII* in the presence of calcium ions (*Factor IV*). Factor VIII, required for thrombokinase formation, is absent in *X-linked hemophilia*.

blood groups (blood type) Groups into which individuals are classified based on genetically determined *serological* (blood) factors. The best-known typing system is **ABO** blood-typing, in which the presence or absence of a particular *antigen* or antigens on the surface of red blood cells (*erythrocytes*) determines a type. The type affects transfusion compatibility. The antigenic *glycoproteins* **A** and **B** may be present either singly (genetically *homozygous*), defining a type **A** or type **B** individual, or together (genetically *heterozygous*), defining a type **AB** individual. **AO** and **BO** heterozygotes, show **A** and **B** phenotypes, respectively. The absence of both **A** and **B** antigens defines type **O**. If blood from, for instance, a type **B** donor is transfused into a type **A** recipient, anti-**B** *antibodies* will attack the donor's antigens, causing the cells to clump together (*agglutinate*), an adverse response. Type **O** is considered the "universal donor," having no antigens to provoke an antibody reaction. Type **AB**, producing neither antibody, is considered the "universal recipient," and can receive blood from any individual. Many other serological typing systems are used both to determine serological compatibility and for individual identification, including Rh, Duffy, Kidd, Lewis, Lutheran, MN, and P. Techniques such as *electrophoresis, chromotography*, antibody response, and functional color-change tests are used to define and distinguish different types. Various protein typing systems are also used extensively in the classification of other organisms, although *DNA* identification is beginning to supplant protein typing in general. (see *Rhesus factor (Rh), transplantation, lectins*)

blood plasma The translucent liquid that remains when all cells are removed from blood. It consists of about 91% water, 7% proteins, and dissolved salts. Plasma is slightly alkaline (pH 7.3), maintained by the buffering action of its constituents. It transports digested and dissolved nutrients, excretory products of cellular

Antigen and Antibody Distribution

	Blood Groups			
	A	**B**	**AB**	**O**
Antigen(s) on red blood cells	A	B	AB	—
Antibod(ies) in plasma	anti-**B**	anti-**A**	—	anti-**A** anti-**B**

Donor-Recipient Compatibility

		Donor			
		A	**B**	**AB**	**O**
Recipient	**A**	√	X	X	√
	B	X	√	X	√
	AB	√	√	√	√
	O	X	X	X	√

√- compatible   X- incompatible

Phenotype vs. Genotype

phenotype	possible genotypes
A	$I^A I^A$ or $I^A I^O$
B	$I^B I^O$ or $I^O I^O$
AB	$I^A I^O$
O	$I^O I^O$

The ABO Blood Group System
(see *blood groups*)

metabolism, dissolved gases, *hormones*, and *vitamins*. Much of the body's physiological activities are concerned with maintaining the correct blood values for all these constituents (*homeostasis*) since plasma supplies the fluid that is the environment of all body cells. (compare *blood serum*)

blood platelet (see *platelet*)

blood pressure This usually refers to the pressure in the main arteries. It is measured as a *systolic pressure* (when the heart is contracting) over the *diastolic pressure* (in between heart contractions). A normal human adult blood pressure might be approximately 120/80.

blood serum The translucent fluid that remains after blood has clotted in a test tube. It is basically *blood plasma* minus the factors and substances involved in *blood clotting*, such as *fibrinogen*.

blood sugar The total amount of dissolved blood *glucose*. It is regulated by the *antagonistic* hormones *insulin* and *glucagon* produced by the pancreas and by other *hormonal* systems. The normal values vary between 0.8 mg/L and 1.0 mg/L. Levels above 1.0 mg/L are indicative of diabetes.

blood type (see *blood groups*)

blood vascular system The system of blood vessels and/or spaces through which blood flows in animal bodies. Most animals have a closed system in which the blood is contained in blood vessels and circulated by muscular contractions of the vessels or a heart. Some invertebrates (mollusks, arthropods) have an open system in which blood flows in blood spaces (for example, crustaceans). (see *vascular system; circulatory system*)

blood vessel A tubular structure found in vertebrates and some invertebrates that transports blood throughout the body. Blood vessels vary in diameter and composition in order to distribute blood to all parts of the body efficiently. (see *artery, arteriole, capillary, venule, vein*)

bloom A dense growth, usually short-lived, of a single species, typically algae

in a freshwater pond or lake. (see *eutrophic*)

blubber A thick layer of fatty tissue just below the skin. It is commonly found in aquatic mammals, such as whales, where it serves as a form of thermal insulation.

blue-green algae (see *Cyanobacteria*)

blunt ends The type of cut produced by some *restriction endonucleases* that cleave straight across the two strands of a *DNA* molecule. Since no overhanging *bases* are generated, they are somewhat more difficult to manipulate when creating *recombinant DNA*. (compare *sticky ends*)

B-lymphocyte (see *B-cell*)

BMR (see *basal metabolic rate*)

BOD (biochemical oxygen demand) A measurement used as an indication of organic *pollution*, such as sewage. It measures the amount of dissolved oxygen that disappears from a water sample in a given time at a certain temperature. When dead organic matter enters waterways, bacterial decay begins, a process which uses oxygen. Enough bacterial activity may reduce the available oxygen in the water to such a low level that fish and other organisms die. Sewage effluent must be diluted to comply with the statutory BOD before it can be disposed into the environment. (see *bloom*)

body cavity A cavity in most animals that contains the heart, digestive and other organs, and is surrounded by the body wall. In most animals, the body cavity is the *coelom*, except for arthropods where it is the *hemocoel*. (see *triploblastic*)

Bohr effect The *reduction* in affinity of the *respiratory pigment* in blood (*hemoglobin* in vertebrates) for oxygen as the level of dissolved carbon dioxide (CO_2) rises. This facilitates the delivery of oxygen to tissues where the concentration of (CO_2) is rising due to *metabolic* activity. Conversely, where the (CO_2) concentration is low, in lungs or gills, high oxygen affinity enables it to be collected by the *respiratory pigment* for distribution. Chemically, this is achieved by the decrease in blood pH caused by increased (CO_2) levels and its consequent *allosteric* effect on the hemoglobin (or *hemocyanin*) molecule.

bone A hard *connective tissue* comprising the vertebrate skeleton. Bone cells (*osteoblasts*) are embedded in a matrix that they have secreted consisting of *collagen* fibers coated with a calcium phosphate salt. The cells are connected by channels (*Haversian canals*) carrying blood vessels and nerves that permeate the matrix. Certain large bones contain *bone marrow*, in which both red blood cells (*erythrocytes*) and white blood cells (*leukocytes*) are produced. Most bones are first formed as cartilage, which is replaced by bone during development. A few, called *dermal* or *membrane bones*, are formed by the *ossification* of connective tissue. Bone is a living tissue and subject to change, particularly under *hormonal* influence.

bone marrow A spongy bone tissue at the ends and in the interior of long bones in which both red blood cells (*erythrocytes*) and white blood cells (*leukocytes*) are manufactured. (see *hemopoiesis, myeloid tissue*)

bony fish (see *Osteichthyes*)

boobies (see *Pelecaniformes*)

book gills (gill books) Gills are composed of a series of plates, resembling the pages in a book. They are found in horseshoe crabs and are *analogous* to the *book lungs* of terrestrial *arachnids*.

book lice (see *Psocoptera*)

book lungs (lung books) Respiratory organs found in many spiders and some other chelicerates, consisting of a cavity containing folds of the body wall through which blood circulates and between which air circulates. They may exist alongside or replace *tracheae* and are *analogous* to the *book gills* found in some aquatic organisms.

boreal forest (see *taiga*)

boron (see *trace elements*)

botany The study of plants.

bothrosome (sagenogen) An *organelle* on the membrane of slime net protoctists that produces the *extracellular* matrix within which the single cells of the colony reside.

bottleneck effect The result of extreme forces that drastically reduce the size, thus genetic variability, of a breeding population. It may occur through a *founder effect*, in which a few individuals colonize a new area, or through natural or human-induced disaster. (see *gene pool, inbreeding*)

Bowman's capsule Part of the filtration device of the vertebrate kidney. It is a cup-shaped structure at the end of each *nephron* and surrounds a knot of capillaries (*glomerulus*), forming a *Malpighian corpuscle*. Within this structure, *plasma* is driven by blood pressure from the glomerular capillaries into the Bowman's capsule, after which it passes through the nephron where most water and ions are reabsorbed into the bloodstream. The residue is excreted as urine.

box jellies (see *Cubozoa*)

bp (see *base pair*)

Brachiopoda One of the three phyla of *lophophorates*, the lamp shells. They are *bivalved, coelomate* invertebrates that live attached to the seabed by a muscular stalk or are secondarily cemented, free living. They superficially resemble bivalve *mollusks* in the possession of a bivalve shell, but the valves are placed dorsally and ventrally, as opposed to laterally in the mollusks. They gather food particles with a spiral structure bearing ciliated tentacles, the *lophophore*. Lamp shells are important in dating fossil records, having appeared in Cambrian times and suffering major *extinctions* in other periods. The genus *Lingula* is apparently identical to its Cambrian fossil ancestors.

brachysclereid (see *stone cell*)

brackish Water with a salinity intermediate between seawater and freshwater. It is often found in *wetland* habitats that represent a transition from the river to the sea.

bract A modified leaf that develops below a flower, branch, or an *inflorescence*. It may be highly colored as in the bracts of *Poinsettia* or *Bougainvillea* and is sometimes mistaken for the flower itself.

bracteole A small, sometimes secondary *bract* borne on the flower stalk above the main bract.

bradykinin (see *kinin*)

brain The most highly developed part of the animal *nervous system*. It is located at the anterior end of the body in close association with the major sense organs and is the main site of nervous control within higher animals. Different sections of the brain are specialized for various functions. In humans, the enlarged outer layer of the brain, the *cerebral cortex*, is the site of memory and learning and is considered the seat of consciousness. The *cerebellum* coordinates balance, initiates *reflex* action, and makes adjustments to any voluntary movement initiated by the cerebral cortex. Involuntary muscle actions, such as those involved in breathing and swallowing, are governed by the *medulla oblongata*, located where the *spinal cord* enters the brain. The *hypothalamus* deep in the brain controls various *metabolic* functions and also influences the activity of the *pituitary gland*. In fish and birds, the major *sensory* and *motor* centers occur in the greatly enlarged deeper regions of the cerebral hemispheres, the *corpus striatum*. This may reflect the predominance of *instinct* in bird behavior, as compared with mammals where the cerebral cortex is the dominant region of the brain. The brain has four interconnected internal cavities, the *ventricles*, which are filled with fluid, and is covered by three protective membranes (*meninges*). (see *forebrain, hindbrain, midbrain*)

brainstem The part of the vertebrate brain comprised of the *medulla oblongata*, the *pons*, and the *midbrain*. It links the *cerebrum* to the *pons* and the *spinal cord*, and controls involuntary functions such as breathing, heart rate, and reflex reactions.

In vertebrates, the brainstem is derived from the *neural tube*.

branchial Any structure or organ relating to *gills*.

branchial arches One of a number of skeletal arches in fish that support the *gills*.

Branchiopoda The most primitive class of crustaceans, known as the gill-footed shrimps and including the brine shrimp, fairy shrimp, and water fleas. Most branchiopods live in freshwater (except brine shrimp) and have flat, fringed appendages for *filter feeding*, respiration, and locomotion. Many are pink or red due to the presence of *hemoglobin*. Their use of temporary bodies of water is facilitated by drought resistant eggs. *Asexual reproduction* by *parthenogenesis* is common.

Branchiostoma The genus with the common name amphioxus comprising most of the phylum Cephalochordata. They are small, marine, filter-feeding burrowers with a fish-shaped body including a dorsal and *caudal* fin. The *pharynx* (throat) and *gill slits* are modified for food collection and respiration. Amphioxus has been important in understanding the evolution of vertebrates.

Branchiura A minor class of *parasitic* crustaceans, the fish lice.

breast (see *mammary gland*)

B-receptor A protein found exclusively on the surface of *B-cells*. It is essentially an *antibody* embedded in the *plasma membrane* and functions in identifying foreign *antigens*. When a B-receptor binds such an antigen, it signals the cell to secrete large amounts of the receptor known as circulating antibodies. (see *immunoglobulin, immune system, immune response*; compare *T-receptor, MHC proteins*)

bright field microscope A standard light microscope in which unmodified visible light is used to illuminate the specimen. (compare *polarizing microscope, phase contrast microscope, fluorescent microscope*).

bristle feather Stiff feathers, often found around the nostrils of a bird.

bristle worms (see *Polychaeta*)

brittle stars (see *Ophiuroidea*)

broadleaf A plant having broad leaves as opposed to needles.

bronchioles The branching system of passages carrying inhaled, *oxygenated* air from the *bronchi* to all regions of the lung. At the ends of the bronchioles, gaseous exchange with the bloodstream takes place in the *alveoli* (air sacs). The walls of the bronchioles are lined with mucous-secreting cells that trap dirt and bacteria, and *ciliated* cells that facilitate excretion of foreign matter out of the lungs. The walls of the larger bronchioles are stiffened by incomplete rings of cartilage to prevent kinking, while the walls of the finer bronchioles are thin to allow limited gaseous exchange with *capillaries* surrounding them. (see *respiration*)

bronchus, bronchi (pl.) An air passage leading from the *trachea* and splitting to enter each lung in tetrapods. Each bronchus is a wide tube with walls that are stiffened by thick, incomplete rings of cartilage and contain *mucus*-secreting *glands*. This allows flexibility yet prevents collapse due to excess external pressure and kinking due to bending of the tubes. The main bronchi branch to form a number of smaller bronchi that lead to the *bronchioles*. (see *lungs, respiration*)

brown algae (see *Phaeophyta*)

brown earth A freely draining, slightly acidic type of soil found under *deciduous* forests in temperate climates. When cleared, it provides fertile agricultural soil.

brown fat (see *adipose tissue*)

Brownian motion The random movement of molecules or particles.

brush border A specialized covering found on the outer surface of *epithelial cells* lining the intestine, *renal tubules*, and other organs involved in absorbtion of digested nutrients. Made of hairlike structures called *microvilli*, it greatly increases the surface area of the cell for absorption of dissolved nutrients. There may be as many as 3,000 microvilli on one epithelial cell. (see *digestion*)

Bryophyta A phylum of simple nonvascular, mainly terrestrial plants, the mosses. It once included all non*vascular plants*, but the hornworts and liverworts have been given their own phyla. They are commonly found in moist habitats and show a *heteromorphic* (differently shaped) *alternation of generations,* the *gametophyte* (*sexual* generation) being the dominant generation. The female *gametes* are borne in *archegonia* and the male gametes in *antheridia*. The sperm requires water to travel between the two for *fertilization*. The gametophyte may show some differentiation into stem and leaves, but there are no roots or *vascular tissues*. The *sporophyte* (*asexual* generation), which is wholly dependent on the gametophyte, is simply a stalk (*seta*) on which the *haploid meiotic spore* capsule is borne.

bryophyte An informal, historical name that still encompasses the group of nonvascular plants comprising the mosses, hornworts, and liverworts.

Bryozoa (Ectoprocta) One of the three phyla of aquatic invertebrates characterized by a *lophophore*, a circular or U-shaped ridge surrounding the mouth bearing *ciliated tentacles*. Ectoprocs look like miniature tube worms and are distinguished by an anus located external to the lophophore. Individuals secrete a *chitinous* chamber, the *zoecium*, through which the organism attaches to rocks and other members of a colony. They reproduce *sexually* and also by *budding*. Individuals communicate through pores between the chambers. Ectoprocts include the only species of freshwater lophophorates.

buccal An entry, such as the mouth (buccal cavity), by which food is taken into the *alimentary canal.*

bud
  1. In plants, a compacted underdeveloped

shoot consisting of a shortened stem and immature leaves or floral parts. A bud has the potential to develop into a new shoot or flower. Terminal buds are formed at the stem or branch tip. *Axillary* or lateral buds develop in the leaf *axils*.

2. In some microorganisms, fungi, and lower animals, an outgrowth of the organism that subsequently breaks off and becomes autonomous. Budding is a common form of *asexual reproduction* in organisms such as jellyfish, sponges, and some yeasts It is essentially a specialized form of *mitosis*, in which the division of *cytoplasm* to the progeny is asymmetric. However, the genetic material is distributed equally. (compare *bulbil*)

buffer A chemical solution containing salts that absorb changes in *acidity* or *alkalinity* so that the solution remains at a stable pH value. Since all liquids in living organisms are buffered, any laboratory solutions that seek to replicate biological conditions must be buffered, too. Chemically, buffering is accomplished by exploiting a chemical equilibrium between a weak base and its salt or between a weak acid and its salt.

bug (see *Hemiptera*)

bulb In plants, an underground storage structure providing the means for survival of the plant from one season to the next or for *vegetative propagation*. It is a modified *shoot*, consisting of a short, flattened stem with roots on its lower surface and fleshy leaves or leaf bases above, which are surrounded by protective scale leaves. In spring or summer, one or more buds grow and produce leaves and flowers, exhausting the food supply. Food from *pho-*

tosynthesis in the new leaves replenishes the storage supply in the bulb each year. Multiple buds lead to multiple bulbs, thus leading to the necessity of dividing some plants every few years. (see *vegetative propagation, perennation, perennating structure*; compare *corm, offset, rhizome, runner, stolon, sucker, tuber*)

bulbil A small bulblike structure in plants that is used for *vegetative propagation*. It may form in a leaf *axil, inflorescence*, or at a stem base. The bulbil ultimately separates from the parent plant and becomes an autonomous organism. (compare *bud*)

bundle of His The network of *cardiac muscle* fibers, the *Purkinje* fibers, that conduct the wave of *depolarization* from the *atrioventricular node* at the *atrioventricular junction* through both *ventricles*. It is propagated so rapidly that all the cells of both ventricles contract almost simultaneously. (see *pacemaker*)

bundle-sheath cell In plants, a special cell impermeable to carbon dioxide (CO_2) that is arranged in layers surrounding a *vascular bundle* (leaf vein). It is especially characteristic of plants engaging in C_4 *carbon-fixation*. It serves to concentrate CO_2, directing the enzyme *RuBP* to "fix" carbon rather than to carry out *photorespiration*. (see C_4 *carbon-fixation*)

Burgess Shale A midCambrian sedimentary rock unit in British Columbia that contains extraordinarily well-preserved animal *fossils*.

bustards (see *Gruiformes*)

butterflies (see *Lepidoptera*)

buttress root An *adventitious* root extending from a plant stem that is used for support. They are often seen in plants that grow in bogs and marshes, mostly *monocotyledons*.

C (see *cytosine*)

C$_3$ photosynthesis (see *Calvin-Benson cycle*)

C$_4$ carbon fixation (*C$_4$ photosynthesis*, Hatch-Slack pathway) A form of *carbon fixation* that avoids *photorespiration* in hot climates. This is accomplished by increasing the internal carbon dioxide (CO_2) concentration of cells within which the *dark reactions* occur. In most temperate plants, the first product of *photosynthesis* is a 3-carbon molecule, glyceraldehyde phosphate, fixed by the enzyme *RUBP carboxylase*. *Photorespiration* appears to be an evolutionary relic in which some fixed carbon is oxidized by a secondary activity of *RUBP carboxylase* and lost to the atmosphere without energy production. As atmospheric oxygen (O_2) increased, it was able to compete successfully with CO_2 for the active site of RUBP carboxylase thus skewing the equilibrium more toward the *oxidation* reaction and away from the *carbon fixation* reaction. C$_4$ plants use the more efficient enzyme *PEP carboxylase* to generate a 4-carbon molecule during the initial *carbon fixation* of CO_2, thus avoiding the loss of fixed carbon to the atmosphere. This reaction takes place in the mesophyll cells. The CO_2 is later recycled and concentrated in *bundle-sheath cells* that carry out the *Calvin-Benson Cycle*, facilitating the carbon fixing activity of *RUBP carboxylase* by the high local concentration of CO_2. C$_4$ plants, including many tropical grasses, have up to double the rate of photosynthesis, can use much higher light intensities and temperatures, and lose less water by *transpiration*. (see *CO_2 compensation point, CAM carbon fixation*)

C$_4$ photosynthesis (see *C$_4$ carbon fixation*)

caddis flies (see *Trichoptera*)

caducous In general, a tendency to shed or drop.

1. In plants, the dropping of *sepals* as a flower opens or *stipules* as a leaf unfolds.

2. In animals, the shedding of hair or fur.

caecilians (see *Apoda*)

caimans (see *Crocodilia*)

calcareous Substances containing calcium carbonate ($CaCO_3$).

calciferol (see *vitamin D*)

calcifuge Plant species that prefer soils containing very little calcium carbonate such as loams. (compare *calciole*)

calciole Plant species that thrive on soils containing free calcium carbonate such as alkaline chalk or limestone soils. (compare *calcifuge*)

calcitonin A *hormone* produced by the *thyroid gland* that lowers the calcium concentration in the bloodstream. It acts by preventing calcium and phosphorus withdrawal from bone, thus antagonizing the effect of *parathyroid hormone.*

calcium A mineral essential for both animal and plant functions. It is found in the bones and teeth of animals as calcium phosphate and is used to strengthen *cell walls* in plants. Calcium ions (Ca^{++}) are essential in triggering *muscle contraction*, maintaining the muscles at rest, and preventing spontaneous contraction and spasms. The concentration of calcium ions also influences the breakdown of *glycogen, blood clotting*, and, in mammalian stomachs, the precipitation of *casein* protein from milk.

calcium pump In *striated muscle*, the *active transport* system in the membrane of the *sarcoplasmic reticulum* surrounding each *myofibril*. Calcium pumps serve to concentrate calcium ions (Ca^{++}) within the sarcoplasmic reticulum during rest. The arrival of an *action potential* releases the Ca^{++} into the *sarcoplasm*, initiating a *muscle contraction*. (compare *sodium-potassium pump*)

callose An insoluble carbohydrate that is deposited around the perforations of *sieve plates*, a region of pores in the walls of *sieve cells* in plants. In *angiosperms*, it is also formed in response to injury.

callus A mass of dividing, undifferentiated plant cells growing *in vitro* in *cell culture*.

calmodulin A protein present specifically in *smooth muscle*. It complexes with calcium ions (Ca^{++}) to activate an enzymatic pathway that exposes binding sites for *actin* on the *myosin* heads (*thick filaments*). Smooth muscle contraction, therefore, differs fundamentally from *striated muscle* contraction, which is initiated by uncovering a binding site for myosin on actin (*thin filaments*).

calorimetry The measurement of thermal changes involved in chemical or physical reactions in either an *in vitro* system or an intact organism. For example, a bomb calorimeter is used to determine the caloric value of foods by igniting them in the presence of oxygen and measuring the increase in temperature of water in a jacket surrounding the chamber.

Calvin-Benson cycle The group of light independent, enzyme-catalyzed reactions that follow the *light reactions* in *photosynthesis*. Newly generated *ATP* drives the synthesis of organic storage molecules from atmospheric carbon dioxide (CO_2). Hydrogen atoms from water are used to fix the carbon molecules, releasing gaseous oxygen (O_2) as a result of the overall reaction. The reactions take place in plant *chloroplasts*. (see *photosynthesis, photosystem I, photosystem II, dark reactions*; compare C_4 *carbon fixation*)

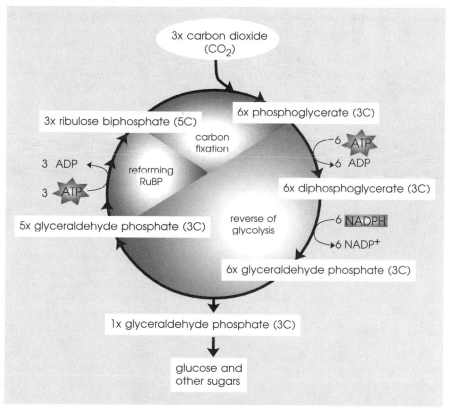

The Calvin-Benson Cycle
(Light-Independent)

calyptra A protective caplike or hoodlike covering on the developing *capsule* in mosses and liverworts, and the *embryo* in ferns. It is derived from the *archegonium* that covers the developing *sphorophyte*.

calyptrogen A layer of actively dividing cells formed over the apex of the growing part of roots in some plants. It gives rise to the *root cap*.

calyx
1. The outermost part of a flower, consisting of green, leaflike *sepals* enclosing the other floral parts during the bud stage.
2. Cup-shaped structures found in some invertebrates.

cambium The ring of cells responsible for lateral growth in *vascular plants*. The cambium layer is found in the stem and root, situated between the *xylem* and *phloem* layers. *Cell division* in the cambium gives rise to the secondary xylem and phloem in woody plants. (see *cork cambium, vascular cambium, procambium, meristem, annual ring*)

Cambrian The first of six geologic periods of the Paleozoic Era, occurring between about 600 to 500 million years ago. It was a time during which many *multicellular* phyla first arose and is characterized by the fossil remains of algae and marine invertebrates, such as trilobites.

CAM carbon fixation (*CAM photosynthesis*) Named for the plants in which it was first discovered (crassulacean acid metabolism), it is a photosynthetic modification that occurs chiefly in plants living in hot, dry environments. The *stomata* of these plants open during the night, releasing carbon dioxide (CO_2) and water vapor, and close during the day. This is the reverse of most plants. It minimizes water loss and reduces *photorespiration*, circumventing the loss of CO_2 without concomitant production of *ATP*. *Carbon fixation* takes place at night by a mechanism closely related to the C_4 pathway. The C_3 cycle (*Calvin-Benson cycle*) takes place in the same cells during the day.

camels (see *Artiodactyla, Tylopoda*)

cAMP (cyclic AMP, adenosine-3′, 5′-monophosphate) A chemical that acts as an *intracellular hormone* mediating many diverse functions, including enzyme activation, genetic regulation, and chemical attraction. It also acts as a *second messenger* in the activation of *peptide hormones*. It is a form of *AMP* derived from *ATP* in a reaction catalyzed by the enzyme *adenyl cyclase*.

CAM photosynthesis (see *CAM carbon-fixation*)

campylotropous The position of the *ovule* in flowering plants (*angiosperms*) when it is oriented so that the *micropyle* points at right angles to the *placenta*. (compare *anatropous, orthotropous*)

cancer A *malignant* tumor or disease caused by it. Cells in malignant tumors are distinguished by uncontrolled division, frequent lack of *differentiation*, and the capability of producing secondary growths (*metastases*) by migration of cells to a part of the body distant from the original tumor. Cancers are classified into two main groups according to the tissue in which they arise: *carcinomas* arise in *epithelial* tissue, *sarcomas* in *connective tissue*. Most cancerous cells in the body are effectively destroyed by the *immune system*. Only when this defense fails do cancerous lesions develop. *AIDS* victims, whose immune systems are severely compromised, often die of cancer.

cane rat (see *Hystricoganathi*)

cane sugar (see *sucrose*)

canine tooth (eye tooth) A mammalian tooth with a single pointed crown occurring on either side of the jaw behind the *incisors*. They are used for piercing and tearing flesh. In *carnivores*, such as the dog, they are long and fanglike. They may be enlarged into tusks (for example, wild boar) and are absent in rodents such as mice, rats, and rabbits.

CAP (catabolite activator protein) A protein whose presence, along with high levels of cAMP, is necessary for the maximal activation of certain *operons*, in particular the *lac operon* in *E. coli.* (see *CAP site, positive control*)

cap A modification of the primary *RNA transcript* that occurs in *eukaryotes* only. During *transcription*, a 7-methylguanosine residue is added to the 5′ end of the molecule. It helps in preventing *degradation* of the molecule during its journey from the nucleus into the *cytoplasm* for *translation*.

capacitation The final stage in *spermatozoa* maturation without which they cannot successfully *fertilize* an *ovum*. It generally occurs within the female reproductive tract and may be initiated by substances secreted from the *ovaries* or uterine lining.

capillary The smallest type of blood vessel in the body. Capillaries branch out from *arterioles* to form a dense network (*capillary bed*) among the tissues and subsequently reunite into a *venule*. Their walls are only one cell thick, allowing exchange of gases, inorganic ions, dissolved nutrients, and excretory products between blood and adjacent tissue cells via *intercellular blood plasma*. (see *circulatory system*)

capillary bed The dense network of *capillaries* found in tissues and organs, where solute exchange takes place between the *circulatory system* and adjacent cells.

capitulum, capitula (pl.) An *inflorescence* (multiple flower) that consists of closely packed flowers or florets that lack stalks and arise on one flattened axis, all at the same level. The capitulum is surrounded or subtended by a ring of *bracts*, giving it the appearance of a single flower. An example is the daisy.

capping A phenomenon in which a particular group of *cell surface antigens* are pulled toward one pole of a cell. Typically, it is seen in response to *antibodies* or a group of plant substances called *lectins*.

Caprimulgiformes The order of birds comprising the nightjars.

capsid The protein coat of a virus surrounding the *nucleic acid*. It is composed of often one, or at most a few, protein species whose molecules (*capsomeres*) are arranged in a highly ordered fashion. A capsid is present only in the inert *extracellular* stage of the virus *life cycle*

CAP site A *DNA regulatory site* located just upstream from the *RNA polymerase* binding site in the *promoter* of certain bacterial *operons* where catabolite activator protein (*CAP*) binds when complexed with *cAMP*. Its binding bends the *DNA*, facilitating the unwinding of the *duplex*, and enables *RNA polymerase* to bind to the nearby promoter.

capsomere The protein building block of a virus *capsid*.

capsule In general, any stiff outer covering.

1. A dry *dehiscent* fruit that is formed from several fused *carpels*. The numerous seeds may be released through pores (for example, snapdragon), a lid (for example, poppy), or by complete splitting of the capsule (for example, iris). The *carcerulus, pyxidium, regma, silicula* and *siliqua*, are all forms of plant capsules.

2. The mucilaginous covering found around the *cell membrane* in many bacteria.

3. The structure within which *spores* of the *sporophyte* generation of mosses and liverworts are formed. It is borne at the end of a long stalk, the *seta*, and ruptures to release the spores.

4. A protective or supportive sheath or envelope that surrounds an organ or body structure. It is usually composed of *connective tissue*, as in the capsule of a joint or kidney.

carapace

1. In reptiles such as turtles and tortoises, the domed top of the shell consisting of platelike bones covered on the outside by horny plates (*scuta*). The *thoracic vertebrae* are attached to the carapace, but the limb attachments are separate and located inside the carapace. The flatter bottom part of the shell is the *plastron*.

2. In some arthropods such as decapod crustaceans, the hard covering of the *cephalothorax*.

carbohydrates One of the four main classes of *macromolecules* (protein, *nucleic acid*, carbohydrate, lipid) found in biological systems and having the general formula type (CH_2O_n). Carbohydrates are used for energy storage and for structural elements in living systems. They may exist as single units (*monosaccharides*), short *polymers* (*disaccharides*), medium length polymers (*oligosaccharides*), or long polymers (*polysaccharides*). Examples include *glucose, fructose, sucrose, glycogen, starch, cellulose,* and *chitin*. They may also occur as components of *glycolipids* and *glycoproteins*.

carbon An essential element in all living systems known on earth today. Molecules in living systems that contain carbon are termed organic. Because carbon can form stable covalent bonds with other carbon atoms and with hydrogen, oxygen, nitrogen, and sulphur atoms, it is an extremely stable and versatile building block. (see *carbon cycle*)

carbon-14 dating (see *radioactive dating*)

carbon cycle The continuous recycling of *carbon* between living organisms and the environment. Atmospheric carbon dioxide (CO_2) is taken up by *autotrophic* organisms (green plants and certain microorganisms) and incorporated into carbohydrates by *photosynthesis*. The carbohydrates, in turn, serve as food for *heterotrophs* (all animals and most other organisms). All organisms return CO_2 to the atmosphere as a product of *respiration* and of decay. The burning of *fossil fuels* also releases CO_2, contributing to abnormally high atmospheric levels of the gas. (see *greenhouse effect*)

carbon fixation The incorporation of the carbon in atmospheric carbon dioxide (CO_2) into organic molecules. It is accomplished most often by *photosynthesis* in green plants, photosynthetic bacteria, and protoctists and, less frequently, by

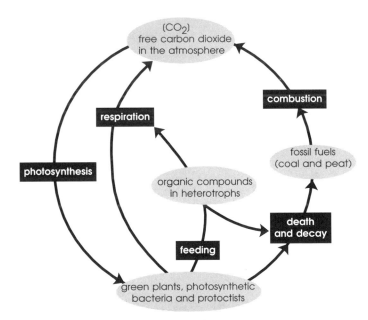

The Carbon Cycle

chemosynthesis in a few bacteria. (see *carbon cycle*)

carbonic anhydrase The enzyme in the *cytoplasm* of red blood cells (*erythrocytes*) that catalyzes the combination of CO_2 with water to form carbonic acid (H_2CO_3). This subsequently dissociates into bicarbonate (HCO_3^-) and hydrogen ions (H^+), lowering the blood concentration of CO_2 and facilitating its continued removal from the surrounding tissue.

Carboniferous The second most recent geologic period of the Paleozoic era, from about 350 to 300 million years ago. It is characterized by the evolution of amphibians, a few primitive reptiles, and giant ferns. Aquatic life included sharks and coelacanths. The period is named after the extensive coal deposits that formed from the remains of vast forests of swamp plants.

carboxylases A group of enzymes that catalyze reactions involving the transfer of carbon dioxide molecules into or out of organic compounds. They are involved in the transfer of carbon dioxide in *respiration*.

carboxylic acid An organic compound of the general formula RCOOH where R is an organic group and -COOH is the carboxylate group. Many carboxylic acids are of biochemical importance, in particular the simpler carboxylic acids (for example, citric, succinic, fumaric, and malic acids) which participate in the *Krebs cycle* and the more complex *fatty acids* which are found in lipids.

carboxysome An *organelle* inside *chloroplasts* that is thought to contain the CO_2-fixing enzyme *ribulose bisphosphate* carboxylase.

carcinogen Any factor resulting in the *transformation* of a normal cell into a cancerous cell. Chemical carcinogens include organic compounds such as hydrocarbons in tobacco smoke and inorganic compounds such as asbestos. Carcinogenic physical agents include ultraviolet (UV) light, X-rays, and other emissions from *radioactive* materials.

Carcinogens are generally *mutagenic*, meaning that they cause changes in *DNA*, and may be detected by biological tests such as the *Ames test*. (compare *mutagen*, *teratogen*)

carcinoma A *cancer* of the *epithelial tissue* (covering and linings of the body). (compare *sarcoma, leukemia, lymphoma*)

cardia (see *cardiac sphincter*)

cardiac muscle The muscle of the vertebrate heart. It is *striated* and contracts *involuntarily*. The rhythmic contractions of the heart arise within the muscle itself at the *pacemaker*. Cardiac muscle consists of short fibers that form a branching network so that the electrical signal can travel through the muscle almost instantaneously and initiate a contraction of all the fibers concurrently. It has a longer *refractory period* than *skeletal muscle* and, consequently, does not fatigue, allowing it to beat continuously. (see *muscle contraction, AV node, bundle of His*)

cardiac sphincter The ring of muscle that regulates passage of material from the esophagus into the stomach.

cardinal veins Two pairs of *veins* found in fish that carry *deoxygenated* blood back to the heart. The *anterior cardinal veins* serve the head while the *posterior cardinal veins* serve the rest of the body. They unite to form the *common cardinal vein* (*Cuvierian duct*), which enters the *sinus venosus* of the heart.

cardiovascular system The closed *circulatory system* of vertebrates. It includes the heart, arteries, *arterioles*, veins, *venules*, and *capillaries*.

carinates An informal classification that contains all modern birds capable of flight, particularly because of an enlarged *sternum* (breastbone). (see *Aves, Neognathae*, compare *Ratite*)

carnassial teeth Specialized teeth found in dogs and other *carnivorous* mammals. They are large and are used to cut meat and shear it from bones.

Carnivora The order that contains the flesh-eating mammals, including wolves, dogs, cats, badgers, otters, weasels, seals, and walruses. The teeth of *carnivores* are specialized for biting and tearing flesh. The claws are well developed and sometimes retractile. Carnivores are typically intelligent animals with keen senses that aid them in capturing prey. Exceptions to the exclusive consumption of flesh include the bear, which is decidedly *omnivorous*, and the panda, which is a *herbivore*. The typical carnivorous teeth have become reduced in those animals that have adopted a herbivorous diet. (see *carnivore, carnassial teeth*)

carnivore An exclusively flesh-eating animal. Carnivores are often but not always a *mammal* of the order *Carnivora*. For instance, several groups of insects and several species of plants are carnivorous.

carnivorous plants A few plants that obtain nitrogen by directly ingesting small insects and other animals. They tend to occur in acidic soils, such as bogs, that do not support the growth of *nitrifying* bacteria. They have developed various adaptions to lure and trap their prey, which they digest with enzymes secreted from various kinds of *glands*. (see *nitrogen cycle*)

carotenes A group of accessory *photosynthetic* pigments, such as *lycopene* (found in tomatoes), α-carotenes, and β-carotenes. β-carotene, the orange pigment in carrots, is split into two identical parts to yield *vitamin A* during digestion in vertebrates. Carotenes also protect the cells, particularly prokaryotic cells, against UV light-induced *DNA* damage. (see *photosynthetic pigments, carotenoids, retinal, rhodopsin*)

carotenoids A group of yellow, orange, or red pigments comprising the *carotenes* and *xanthophylls*. They are found in all *photosynthetic* organisms where they function mainly as accessory pigments in photosynthesis and also protect against the deleterious effects of UV light. An important carotenoid is *β-carotene*, the precursor of *retinal*, the *chromophore* in all known visual pigments. In some animal structures, for example feathers, they simply lend pigment. Along with *anthocyanins*, they contribute to the autumn colors of leaves since the green pigment *chlorophyll* that normally masks the caroteneoids breaks down first. They are also found in some flowers and fruits. (see *absorbtion spectrum, photosynthetic pigments, carotenes, DNA damage*; compare *chlorophylls*)

carotid arch (see *carotid artery*)

carotid artery (common carotid) One of a pair of blood vessels that supplies *oxygenated* blood from the heart to the head and neck. They are derived from the third *aortic arch*. Each common carotid branches into an *internal* and *external* carotid artery in the neck region.

carotid body A small *neurovascular* structure near the branch of the *internal* and *external* carotids (near the *carotid sinus*) containing *chemoreceptors* that monitor carbon dioxide and oxygen concentrations, and blood pH. It responds to a change in any of these factors by sending out *nerve impulses* that influence respiratory and heart rates in order to bring the body back to *homeostasis*. It may also assist *carotid sinus* reflexes.

carotid sinus A small swelling in the *internal carotid artery* near its origin in the neck containing *sensory receptors* that monitor changes in *blood pressure*. An increase in arterial pressure and subsequent sinus stimulation cause a drop in heart rate and *vasodilation*, bringing the body back to *homeostasis*. (see *baroreceptor, carotid body*)

-carp-
  1. A word element derived from Greek meaning "fruit." (for example, *carpal*)
  2. A word element derived from Greek meaning wrist. (see *carpal bones*)

carpal bones Bones in the distal regions of the forelimb of tetrapods. In man, they constitute the wrist (*carpus*). (compare *tarsal bones*)

carpel The female reproductive organ of a flowering plant (*angiosperm*). It consists of an *ovary* containing one or more *ovules* and a stalk, or *style*. At the terminal surface of the style is the *stigma*, the receptive surface for male *pollen grains*, that fertilizes the ovules. Each flower may have one or more carpels that may be borne singly or fused together (*gynoecium*). (see *apocarpous, syncarpous, magasporophyll, ovuliferous scales*; compare *stamen*)

carpus (see *carpal bones*)

carrageenan A complex carbohydrate found in certain red algae (seaweeds). It has a number of commercial uses as a stabilizer of pharmaceuticals and food products.

carrier
1. An individual who is infected with a disease but does not show any symptoms. Such carriers can transmit the infection to others. An infamous historical example is Typhoid Mary. (compare *vector*)
2. An organism that carries a copy of a detrimental *gene* but shows no physical manifestation. Generally, the gene is *recessive* so that in a *diploid* organism, its effects are covered by a second, normal copy of the gene. Such a gene may be *sex linked*, that is carried on the *X chromosome* of a female (*heterozygous*). If sex linked, it will manifest in her **XY** male offspring, where the *Y chromosome* lacks the equivalent gene altogether. For a detrimental recessive gene carried on an *autosome* to manifest as a physical trait, the offspring must receive a faulty copy from each carrier parent, so that the offspring contains two flawed copies (*homozygous*). (see *sex chromosomes*)
3. (see *transmembrane carrier proteins*)
4. A *macromolecule* to which a *hapten* is attached. In this configuration, both elements are capable of eliciting the formation of *antibodies.*

carrying capacity The maximum number of individuals of a species that can be sustained by an environment without decreasing the capacity of the environment to sustain that same amount in the future. Practically, it is the size at which a popu-

lation stabilizes, which is imposed on the *biotic potential* by shortages of important factors such as space, light, water, or nutrients. It is a dynamic characteristic, changing with environmental conditions, and is represented graphically as a *sigmoid growth* curve.

cartilage A *connective tissue* containing *cells* (*chondroblasts*) embedded in a solid fibrous or elastic protein matrix (*chondrin*). In most advanced vertebrates, the skeleton is first formed as cartilage in the *embryo* and then changed into bone. Cartilage persists in a few places, such as the end of the nose, the ends of bones, and in joints. In some fish, such as sharks, cartilage constitutes the entire adult skeleton.

cartilage bone The type of bone that is formed from cartilage in the *embryo*. The cartilage is invaded by bone-forming cells (*osteoblasts*). They convert the cartilage into bone by the deposition of calcium salts (*ossification*). (compare *membrane bone*)

cartilaginous Made of cartilage.

cartilaginous fish (see *Chondrichthyes*)

caruncle A warty outgrowth found on the seeds of a few flowering plants (*angiosperms*), such as castor oil and violet. It is usually brightly colored and may act as an aid in seed dispersal, sometimes by insects. It is similar to but smaller than an *aril*.

Caryoblastea (see *Karyoblastea*)

caryopsis A dry nutlike fruit, typical of the grasses, that is generally known as cereal grain. It is *indehiscent* and similar to an *achene*, except that the *ovary* wall is fused with the *seed coat*.

cascade (enzyme cascade) Any biological mechanism that proceeds via a sequence of biochemical and/or physiological events. A specific signal is usually amplified by the process, and the end result is usually a very localized response. Examples of such processes include *blood clotting, complement* activation, and the action of *peptide hormones*. (see *second messenger, protein kinases*)

casein The protein found in milk as calcium caseinate. It belongs to a group of proteins that provide the *amino acids* necessary as both nutrients and building blocks in early animal development.

Casperian strip A waxy strip that seals the spaces between the *cell walls* surrounding the *endodermis* of plant roots. It regulates the passage of minerals and other dissolved solutes, which must pass through a cell rather than around it to reach the *vascular cylinder* and inner *xylem* layer.

cassowaries (see *Casuariiformes*)

caste One of several specialized groups of individuals that exist in an *ecological community* of social insects such as, ants, bees, wasps, or termites. They are distinguished both by structural and functional differences. For example, honey bees have three castes: the queen (a fertile female) reproduces, workers (sterile females) gather food, and drones (males) mate with the queen. The caste of an individual may be determined by genetic and/or environmental factors.

Casuariiformes The order of birds containing the flightless cassowaries

catabolism The enzyme-mediated breakdown of complex organic substances to simple ones. Catabolic reactions convert energy stored in chemical bonds of foodstuffs to the energy currency of the cell, *ATP*. The ATP is used for work (for example, *muscle contraction* and *nerve impulses*), synthesis of new structures, and general maintenance of *homeostasis*. Digestion and *cellular respiration* are the major catabolic processes. (compare *metabolism, anabolism*)

catabolite activator protein (see *CAP*)

catabolite repression The inhibition of *mRNA transcription* from an *operon* caused by the presence of large amounts of the *metabolic* end product of the operon. For instance, the presence of adequate amounts of *glucose* in a cell tends to repress the *expression* of *genes* encoding *inducible enzymes* that break down alternative forms of fuel. (see *lac operon*)

catalase An enzyme that catalyses the breakdown of hydrogen peroxide, a toxic byproduct of *metabolism*, into water and oxygen. It is present in the *peroxisomes* of animal cells and the *glyoxysomes* of plant cells. Catalase is used commercially in converting latex to foam rubber and in removing hydrogen peroxide from food. (see *superoxide dismutase*)

catalyst A substance that increases the rate of a reversible chemical reaction without being consumed in the reaction or altering its equilibrium point. This is accomplished by decreasing the *activation energy* for a particular reaction. Enzymes are efficient and highly specific biochemical catalysts. (see *active site*)

catecholamines A group of chemicals, many of which act as *neurotransmitters*, including *noradrenaline*, *adrenaline*, and *dopamine*.

cathepsins A specific group of *intracellular* enzymes that degrade proteins (*proteases*) to *amino acids*. They are particularly predominant in the liver, kidney, and spleen where they occur in *lysosomes* and are responsible for *autolysis* after cell death.

catkin A pendulous group of flowers (*inflorescence*) adapted for wind *pollination*. It consists of numerous, usually *unisexual* flowers on a modified *spike*. Male flowers produce very large quantities of *pollen*, as in the hazel tree, and are often protected from dew and rain by rooflike bracts above them. When the pollen or seeds have been shed, the catkin falls as a unit.

cats (see *Carnivora*)

caudal The tail region in those animals that retain one. In primates, caudal refers to the end of the *vertebral column* or the *coccyx*.

caudal vertebrae The bones of the *vertebral column* that protect the *spinal cord* in the tail region. As the tip of the tail is approached, they lose the general features of vertebrae until the terminal ones consist solely of a cylindrical *centrum*. In primates, the caudal vertebrae are much reduced and fuse to form the *coccyx*.

Caudata (see *Urodela*)

caul-, caulo- A word element derived from Latin meaning "stem" or "stalk." (for example, *cauline*)

cauline Any structure or function associated with the stem of a *herbaceous* plant.

CCAAT box An integral *nucleotide* sequence element of *eukaryotic gene promoters*. It is located 40 *base pairs* upstream of the main element, the *TATA box*, and is necessary for efficient gene *transcription in vivo*. (see *GC box*, *UAS*, *enhancer*)

CD marker Cluster of differentiation marker. A type of *cell surface marker* that varies to identify the different groups of T-cells. (see *CD3*, *CD4*, *CD8*)

CD3 A *cell surface marker* that identifies a T-cell.

CD4 A *cell surface marker* that identifies a helper or inducer T-cell (T_4).

CD8 A *cell surface marker* that identifies a

cytotoxic or suppressor T-cell (T_8).

cDNA (complementary DNA) Synthetic *DNA* that is synthesized *in vitro* from an *RNA template* using the enzyme *reverse transcriptase*. A *double-stranded* cDNA molecule is regenerated by synthesis of the *complementary* strand by *DNA polymerase*. It lacks any *introns* or *regulatory* sites that may have been present in the *genomic* DNA sequence. It is used in *cloning* to enrich for functional *genes*, sometimes from a specific tissue. cDNA may also be labeled and used as a genomic DNA *probe*, particularly in a search for *homologous* sequences in different tissues or species.

cDNA library A collection of *cDNAs* reverse transcribed from an *mRNA* fraction of choice and inserted into an appropriate *cloning vector* for propagation, screening, and analysis.

cecum A blind pouch in the *digestive system* usually found at the junction of the

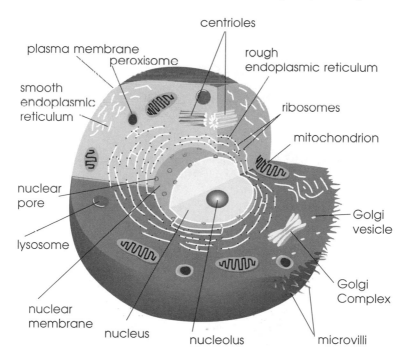

Generalized Animal Cell
(see *cell*)

ileum and *colon*. It ends in the *vermiform appendix* and is *vestigial* in most mammals. In *herbivores* (for example, rabbits and cows), it is large and contains the *symbiotic* bacteria necessary to digest the large amounts of *cellulose* in their diet of plant material.

cell The basic unit of structure and function of all living organisms. *Prokaryotic* cells are typically about 1μm in diameter, as compared with the more complex *eukaryotic* cell which is typically about 20μm in diameter. The largest cells are egg cells (for example, ostrich eggs are 5 cm in diameter), and the smallest are *mycoplasmas* (about 0.1μm in diameter). All cells contain genetic information in the form of *DNA*, which controls cell activities In eukaryotes, the DNA is enclosed in a *nucleus*. All cells contain *cytoplasm* and are surrounded by a *plasma membrane*, which controls the entry and exit of various substances. Plant, most bacterial, fungal, and some protoctist cells are also surrounded by rigid *cell walls*. (see individual cell structures and functions)

cell body The part of a *nerve cell* that contains the *nucleus* and is enlarged compared with the rest of the cell (*axon*). The cell body supplies nutrients to the rest of the *neuron*. (see *dendrites*)

cell capping (see *capping*)

cell constancy A phenomenon seen in a few simple animals where the number of cells or *nuclei* comprising the body is constant over the animal's life. (see *Rotifera*)

cell culture The *in vitro* propagation of cells and tissues.

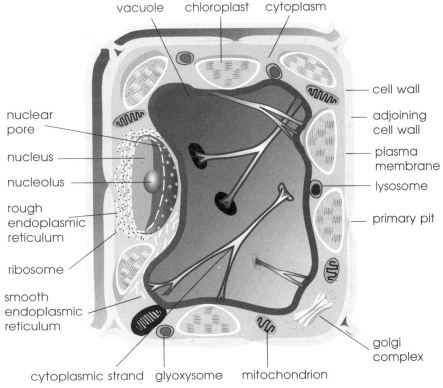

Generalized Plant Cell

cell cycle The ordered sequence of phases through which a cell passes from one *mitotic* (vegetative) *cell division* to the next. It is divided into four phases, G_1, S, G_2, and M. Generation of *protein synthetic* machinery including the *translation* of *mRNA*, *tRNA*, and *ribosomes* occurs in G_1, *replication* of *DNA* happens during S phase, and the formation of materials required for *spindle* formation occurs during G_2. The cell then proceeds through *mitosis*, M.

M - mitosis

G_1 - gap or growth phase 1
   (prior to DNA synthesis)

S - DNA synthesis

G_2 - gap or growth phase 2
   (between DNA synthesis and mitosis)

I - interphase (period between mitoses

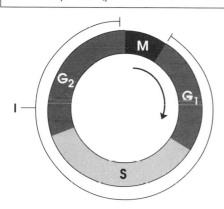

The Cell Cycle

cell division The replication of cell components and distribution into approximately equal parts. In *unicellular* organisms, it is a method of *asexual reproduction*. Its simplest manifestation is *binary fission* in *prokaryotic* organisms. Mitotic cell division in *eukaryotic* cells involves both *nuclear division* and *cytokinesis*. *Multicellular* organisms grow by mitotic cell division, particularly early in development. Mitotic cell division also

contributes to wound healing. Mature tissues may also divide rapidly when continuous replacement of cells is necessary, as in the *epithelial* layer of the intestine. Cell division is a tightly regulated process and disruption of the control mechanisms may lead to *transformation* and *malignancy*. In plants, certain growth regulators (for example, *cytokinins*) stimulate renewed cell division. (see *budding;* compare *meiosis*)

cell fractionation The separation of different cell constituents into homogeneous fractions in the laboratory. Separation is accomplished using techniques such as *centrifugation,* which exploit the inherent differences in the unit *mass* of different cell components.

cell-free translation The synthesis of protein from a particular *mRNA* in an *in vitro* extract. It is a method of producing relatively large amounts of a pure *polypeptide* for identification and analysis. Extracts are typically made from wheat germ or *reticulocytes* (immature red blood cells), as their *translation* machinery is already highly activated.

cell fusion The process of combining the *cytoplasmic* contents of two or more cells within one *cell membrane*. The fate of the *nuclei* depends on the specific cells. Cell fusion occurs naturally, for instance, in the formation of *syncytia* in *striated muscle* and may also be induced in the laboratory in order to study various processes. (see *hybridoma, heterokaryon, somatic cell hybridization*)

cell lineage The cellular ancestry of any cell in a *multicellular* organism. In most cases, it starts with the *zygote* from which all of an organism's cells are derived by *cell division*. It may also imply a more restricted line of cells forming a functional subset of the cell population.

cell-mediated immune response The bodily defense system involving *T-cells* and their products that comes into play after the immediate, nonspecific *immune response*. It targets virus-infected cells, *parasites,* and *cancerous* cells.

Interleukin-1, released by *macrophages* during the initial response, activates *helper T-cells* which release *interleukin-2*, inducing the proliferation of both *inducer* and *suppressor T-cells*. Of particular importance is the activation of cytotoxic T-cells by Interleukin-1. Cytotoxic T-cells recognize and destroy infected body cells. (see *lymphokines, immune system, cell-surface receptor*; compare *humoral immune response*).

spread across the width of the cell, effectively dividing it in two. *Cellulose* is laid down along the new membrane, completing the division. The process begins at the *phragmoplast*, a barrel-shaped structure at the former site of the *spindle* equator. The space between the two cells becomes impregnated with *pectins* and becomes the *middle lamella*. (see *cytokinesis*)

cell sap In many mature plant cells, a solution of sugars and salts contained in a sin-

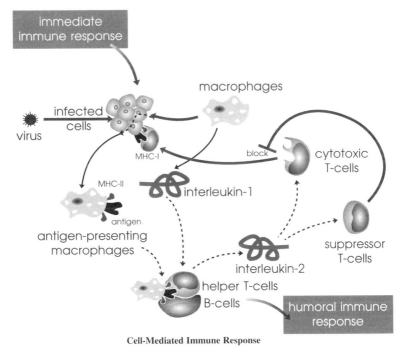

Cell-Mediated Immune Response

cell membrane (see *plasma membrane*)

cell organelles *Subcellular* structures that carry out specialized functions inside each cell. They differ between plant and animal, *prokaryotic* and *eukaryotic* cells. Examples include the *nucleus*, *mitochondrion*, and *chloroplast*. (see listings for individual cell organelles)

cell plate A structure that appears in dividing plant cells and is involved in formation of a new *cell wall*. It is formed by the fusion of membrane components that

gle *vacuole* that occupies most of the cell volume.

cell surface antigen Any *cell surface marker* that is recognized by an *antibody*.

cell surface marker Any of the *transmembrane proteins* involved in the recognition of "self" and "nonself" in the *immune system*. They include the ABO *blood group* markers, the *MHC proteins,* and the *CD3, CD4,* and *CD8* markers.

cell surface receptor (see *transmembrane receptor*)

cell theory The basic theory that all organisms are composed of cells and that growth, development, and reproduction result from the division and *differentiation* of cells. This idea resulted from numerous investigations that started at the beginning of the 19th century and was finally articulated by Schleiden and Schwann in 1839. (see *cell*)

cellular Any structure or function relating to unit *cells*.

cellular respiration The *oxidation* of organic molecules to fuel the chemical reactions in living systems. The final result of respiration is the generation of the high-energy storage compound *ATP*, which is then used to power all *metabolic* processes within the cell. *Cellular respiration* is divided into three stages: *glycolosis,* the *Krebs cycle,* and the *electron transport chain.* Glycolosis may occur

under either *aerobic* (with oxygen) or *anaerobic* (without oxygen) conditions. The Krebs cycle requires free oxygen (O_2) and leads into the electron transport chain. Glycolysis alone results in only partial *oxidation* of the substrate with a net production of just two molecules of ATP for each molecule of glucose and conversion of the *pyruvate* end product into waste (*lactic acid* in animals, *ethanol* in plants). In the presence of free oxygen, the pyruvate is converted into *acetyl CoA*, which permits it to enter the Krebs cycle. In this cyclic series of reactions, high-energy electrons are transferred to *coenzyme* carriers that ferry the electrons into the electron transport chain, resulting in the net production of another 34 molecules of ATP. A total of 36 molecules of ATP results from one molecule of glucose. The enzymes for the Krebs cycle

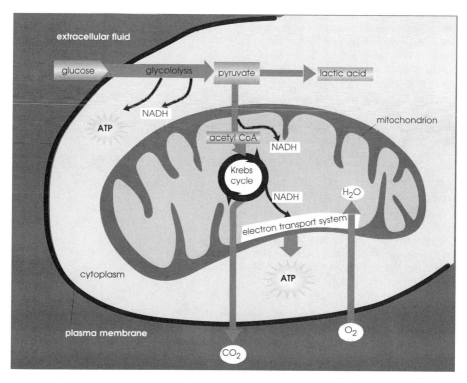

Overview of Cellular Respiration

and the electron transport chain are located within the *mitochondria*. (see *aerobic respiration, chemiosmosis, oxidative phosphorylation, NAD, NADH, FAD, FADH₂*; compare *photosynthesis*)

cellular slime molds (see *Acrasiomycota*)

cellulase An enzyme that breaks down *cellulose*. It is produced by bacteria in the gut of *herbivorous* mammals, such as cows, rabbits, and rodents and in zoomastigote protists in termites and cockroaches, enabling them to utilize cellulose-containing materials as food. (see *ruminant, cecum, rumen*)

cellulose The principal structural material of plant *cell walls* and as such is the most abundant organic compound in the world. It is an insoluble *polysaccharide* composed of chains of *glucose* molecules. (see *cell wall*)

cell wall A rigid wall external to the *plasma membrane*, surrounding the cells of most bacteria, some protoctists, and all fungi and plants. Plant cell walls contain *cellulose* and *pectin*, and those of fungi usually contain *chitin*. The walls of some algae differ, for example, the silica boxes enclosing *diatoms*. Cell walls are freely permeable to gases, water, and solutes. They have a mechanical function, contributing both to the strength of the cell and the organismal support of herbaceous plants. Various modifications of the cell wall include the inclusion of *lignin* or extra cellulose for support, uneven thickening of *guard cells*, and the waterproof covering of *epidermal* and *cork* cells. Bacterial cell walls may contain complex *polysaccharides* cross-linked with *polypeptides*. Bacteria are classified by the differential staining of their cell walls based on their composition into *Gram-negative* and *Gram-positive*. (compare *plasma membrane*)

cement Modified bone that surrounds the root of vertebrate teeth and helps fix them in the socket of the jawbone. It is compact and hard, and has a slightly higher mineral content than bone.

Cenozoic The present geological era, beginning some 65 million years ago, and divided into two periods, the tertiary and the quaternary. It is characterized by the rise of modern organisms, especially mammals and flowering plants (*angiosperms*).

centimorgan (see *cM*)

centipedes (see *Chilopoda*)

central dogma The idea, proposed by Francis Crick in 1958, that molecular information in biological systems flows unidirectionally from *DNA* to *RNA* to protein. Although this is true for the most part, exceptions include *RNA viruses* that *transcribe* DNA from RNA using the enzyme *reverse transcriptase* and some special RNA molecules that are able to act as enzymes and are "self-splicing." (see *ribozyme*)

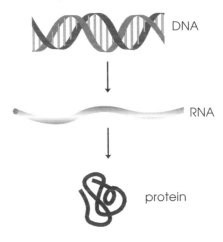

The Central Dogma

central nervous system (CNS) The part of the *nervous system* that integrates *sensory* and *motor* functions in vertebrates and provides direct pathways for *nerve impulses*, usually through the center of the body. In vertebrates, the CNS consists of the brain and *spinal cord*. The CNS of many invertebrates (annelids and arthropods) consists of a connected pair of *ganglia* in each body segment and a pair of ventral *nerve cords* running the length of the body, with a pair of ganglia in the head region serving as the brain. Cnidarians and Ctenophores lack any centralization of the nervous system, possessing a simple *nerve net*. The development of a CNS is associated with the increasing sensory awareness and complex actions that are involved in locomotion, feeding, and reproduction, hence the need for central integration of all nerve activity. (compare *peripheral nervous system*)

centrifuge A laboratory apparatus that separates components suspended in a liquid solution on the basis of mass. This is accomplished by spinning tubes in a circular rotor at the appropriate speed, angle, and length of time for a specific desired separation. The material is spun to the bottom of the tube (the pellet) or suspended at some point within the tube and then collected. (see *ultracentrifuge*)

centriole An *organelle* associated with the assembly and organization of *microtubules* in the cells of animals and protoctists. They possess a distinctive structure, consisting of a cylinder made up of nine triplets of microtubules surrounding two central singlets. Centrioles occur in pairs within the *cytoplasm* near but outside the *nuclear envelope* and are usually arranged at right angles to each other. A pair of centrioles is essential to *spindle* formation during nuclear division. In cells that have *undulipodia*, these structures are anchored to the cell by a structure identical to the centriole, the *basal body (kinetosome)*. Normally, an animal obtains its centrioles from the sperm cell at *fertilization*; an egg may rarely form its own. Higher plants, fungi, and algae lack both centrioles and kinetosomes, and their microtubules are organized by amorphous structures called *spindle plaques*. Centrioles contain their own *DNA* and are self-reproducing; they may well be evolutionarily derived from *endosymbiotic spirochete* bacteria. (see *centrosome, mitosis, meiosis, tubulin*)

centrolecithal Some eggs, typically insect,

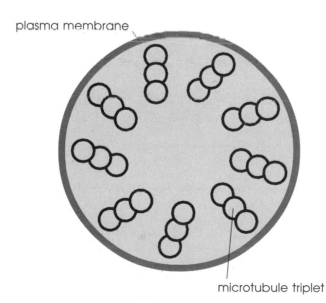

plasma membrane

microtubule triplet

Structure of Basal Body/Centriole

in which a large yolk occupies the center of the egg. (compare *telolecithal*).

centromere (kinetochore) A condensed region on *eukaryotic chromosomes* where *sister chromatids* remain temporarily attached to one another after *replication*. The centromere is a specific genetic *locus* with a defined *DNA sequence* of about 220 *nucleotides* and is usually flanked by *repeated DNA sequences*. Under proper staining, it appears as a constriction, and its exact location along the length is chromosome dependent. (see *satellite DNA, acentric, acrocentric, metacentric, telocentric*)

centrosome One of many sites on the periphery of a *centriole* from which *microtubules* radiate in the formation of a *spindle*. It is an *MTOC* (microtubule organizing center).

centrum, centra (pl.) The main weight-bearing body or center of a *vertebra*, except in the *atlas* and *axis,* and situated ventral to the spinal cord. Each is attached to adjacent centra by *collagen* fibers and also to *intervertebral discs*.

ceph-, cephalo- A word element derived from Greek meaning "head." (for example, *cephalothorax*)

cephalization The evolutionary development of a head in which the brain, sense organs, and feeding organs are concentrated.

Cephalocarida A primitive class of small shrimplike crustaceans, the cephalocarids. They are *hermaphroditic* and *filter feed* from the ocean bottom sediments that they inhabit.

cephalocarids (see *Cephalocarida*)

Cephalochordata A newly autonomous phylum of marine *chordates*, the lancelets, in which the normally juvenile traits of a *notochord, dorsal nerve cord,* and *gill slits* are retained in the adult. They lack a brain and *vertebral column* and so are acraniate chordates. The best known member is *Amphioxus*. (see *paedomorphosis, paedogenesis, neotenic, heterochrony*)

cephalochordates (see *Cephalochordata*)

Cephalopoda The most advanced class of *mollusks*, containing the cuttlefish, squid, octopus, and nautilus. All are aquatic and most are marine. They possess a well-developed head surrounded by a ring of prehensile *tentacles* and a muscular siphon derived from the foot through which water is forced from the *mantle cavity* during locomotion. In nautilus, the animal inhabits the last chamber of a large coiled shell that acts as a buoyancy chamber and has numerous (80 to 90) unsuckered tentacles. In cuttlefish, the shell is internal. In squids it is much reduced. It is absent altogether in the octopus. Squids and cuttlefish have ten tentacles, while octopi have only eight. The complexity of cephalopod eyes rivals that of vertebrates (an example of *convergent evolution*), while the large brain enables powers of learning and shape recognition on par with simple vertebrates. The complex array of color and pattern changes seen in cephalopods may be involved in communication.

cephalopod eye (see *eye*)

cephalopods (see *Cephalopoda*)

cephalothorax In chelicerate arthropods and some crustaceans, the anterior (front) part of the body formed by the fusion of the head and *thorax* that bears the mouthparts and walking legs. In arachnids, the cephalothorax is called the *prosoma*.

Ceratomorpha A suborder of odd-toed ungulates that includes the rhinos and tapirs.

cercaria The last *larval* stage of flukes (*trematode worms*), produced *asexually* by *budding* from preceding larvae (*redia*) inside a secondary *host*. The cercaria is free swimming and emerges from the secondary host, often a snail, to penetrate the skin of its primary host (for example, man in *Schisotosoma*) or to *encyst* as a *metacercaria* awaiting ingestion by the primary host. (see *parasite*)

cerci Appendages, often *sensory*, at the hind end of the abdomen of some insects (for example, mayflies, earwigs, and cockroaches).

cerebellum The part of the brain in higher animals that coordinates involuntary balance, posture, and fine adjustment of locomotion. It monitors the position of limbs, the state of muscle tension, and makes any necessary adjustments to the messages sent out to *voluntary muscles* by the *cerebral cortex*. It is located behind the *cerebral hemispheres*, over the *medulla oblongata*, and originates from the *hindbrain*.

cerebral cortex (pallium) The part of the brain in higher animals that is responsible for the conscious senses of vision, hearing, smell, and touch, for control of voluntary muscle activity, and for language and memory. The cerebral cortex forms the surface layer of the *cerebral hemispheres* of the brain and contains billions of *nerve cell bodies*, collectively called *gray matter*.

cerebral hemispheres (cerebrum) The pair of structures in the brain of higher animals that is considered the seat of consciousness. They contain the centers concerned with the major senses, voluntary muscle activities, language, and memory. In man, each hemisphere has a greatly enlarged and highly folded outer area, the *cerebral cortex*, that contains *gray matter* and overlies other parts of the brain. Beneath the gray matter is the *white matter*, comprising nerve fibers connected to other regions of the brain. Each hemisphere controls the movements of the opposite side of the body. Some functions, such as speech, are confined to one hemisphere or the other. The cerebral hemispheres originate from the *forebrain*.

cerebrospinal fluid The liquid found in the internal cavities and between the surrounding membranes (*meninges*) of the *central nervous system*, including the brain. Cerebrospinal fluid generally lacks blood cells and large molecules, and is filtered from blood by the *choroid plexuses* in the brain. It returns to the bloodstream via either *lymph vessels* or in *venous* blood. It serves to cushion and protect nerve tissues.

cerebrum (see *cerebral hemispheres*)

cervical vertebrae The small *vertebrae* of the neck that form a very flexible portion of the *vertebral column*. Mammals usually have seven: the first and second are modified to form the *atlas* and the *axis*.

cervix The neck of the *uterus* in mammals, a narrow cylindrical passage situated at the end of the uterus leading into the *vagina*. It contains numerous *glands* that secrete mucous into the vagina.

Cestoda A class of *endoparasitic* worms, the tapeworms. They live in the gut of vertebrates and lack a *digestive system*, absorbing predigested food from the *host* through their body wall. Tapeworms have a complex *life cycle* involving one or more *intermediate hosts* (for example, pigs, which become a source of infection when undercooked pork is ingested by humans). The body, up to 10 meters long, is covered by a tough *cuticle* to prevent digestion by the host and has a small head (*scolex*) with hooks and suckers for attachment. The rest of the body consists of a series of repetitive segments (*proglottids*), each of which contains a complete *hermaphroditic reproductive system*. The eggs develop into six-hooked *embryos* that are excreted and eaten by an intermediate host, in which they develop into *cysticercus larva* (bladderworms). They become *sexually* mature in the final host.

Cetacea The order that contains the only completely marine mammals, the whales, dolphins, and porpoises. They have a hairless streamlined body, no hind limbs, forelimbs modified as flippers, and a tail with horizontal *flukes* used for propulsion. An insulating layer of blubber (fat) beneath the skin helps to conserve heat. They have are no external ears, and the respiratory outlet is the dorsal blowhole. The toothed whales (for example, dolphins, porpoises, and some larger whales; *Odontoceti*) feed on fish and other animals and have many peglike teeth. The *baleen* whales (for example, blue whales, humpbacks; *Mysticeti*) feed on *krill*, a type of *plankton*, filtered from the sea by baleen plates in the mouth. Cetaceans are extremely intelligent creatures; the extent of their social structure and means of communication is yet to be completely understood.

CFCs (chlorofluorocarbons) A group of highly stable industrial chemicals (CFCs) once used extensively in cooling systems, aerosols, and styrofoam. They have been shown to have a detrimental effect on the protective *ozone layer* surrounding the earth and are no longer manufactured in the United States. Although worldwide agreements to phase out their production by the year 2000 have been signed, the problem may grow worse before previously manufactured CFCs have been purged from the atmosphere. (compare *hydrocarbons*)

chaeta (see *seta*)

Chaetognatha A minor phylum of marine invertebrates, the arrow worms. They are *coelomate deuterostomes*, and occur mostly in the *plankton* of warm shallow seas, feeding on other *zooplankton* including fish *larvae*.

chalaza
1. In a bird's egg, a twisted strand of fibrous *albumen* that joins the shell membrane to the egg membrane (*vitelline*). A chalaza is at each end of the egg, supporting the yolk sac centrally in the shell on the long axis.
2. The region of an *angiosperm ovule* where the *nucellus* and *integuments* merge.

chamaephyte A *perennial* plant that is able to produce new growth from resting buds near the soil surface. Chamaephytes are usually small bushes (for example, heather). (see *Raunkiaer's plant classification*)

chaparral A unique evergreen scrub *association* found in California. It is historically derived from *temperate deciduous forests* and consists mainly of low bushy shrubs and *annuals*. Its unique *ecology* depends on periodic fires that release seeds from which individual plants re-

sprout. Limiting the fires on chaparral near urban settlements serves only to increase the organic debris and, consequently, the severity of the fire when it does happen. Related associations are found in other areas of the world with dry summers and moist winters ("Mediterranean"), including the Mediterranean itself, central Chile, and the southern tips of South Africa and Australia. (see *biome*)

character An attribute of individuals within a species for which various differences can be defined.

Charadriiformes An order of birds comprising the shorebirds, waders, and seabirds including sandpipers, woodcocks, oystercatchers, plovers, gulls, terns, auks, puffins, murres, jaegers, and skuas.

chazmolithic Any organism living inside the cracks in rocks or microorganism colonizing the surface of rock. They lack the ability to actually penetrate the rock. (compare *endolithic*)

chela The last joint on an *arthropod* limb in, for example, crabs, lobsters, and scorpions. It is characterized by its opposition to the joint preceding it, so that it can be used for grasping.

chelicera One of a pair of appendages on the head of nonmandibulate (without jaws) arthropods. In spiders and scorpions, they are often fed by poison *glands* and are used to inject paralyzing poisons into prey. (see *mouthparts*)

Chelicerata A phylum of arthropods (previously a subphylum), the chelicerates, including spiders, mites, and scorpions. They lack *mandibles* (jaws) and sensory antennae. Chelicerates are characterized by *chelicerae*, mouthparts that often take the form of pincers or fangs. The first two body segments are combined into a *cephalothorax*.

chelicerates (see *Chelicerata*)

Cheloni The order of reptiles containing the turtles and tortoises. The body is enclosed in a shell of bony plates covered by horny scales, with an upper *carapace* and lower *plastron*. The jaws are tooth-less and horny. Turtles are aquatic and different species may inhabit freshwater or marine environments. The giant marine turtle comes ashore only to lay her eggs on sandy beaches. Tortoises are terrestrial and well adapted to the desert areas in which they are often found.

chemically gated ion channel A specialized *ion channel* found, in particular, in the *postsynaptic membranes* of *neurons* and *muscle cells*. It opens and closes in response to the binding of chemicals, usually a *neurotransmitter*, to specialized *receptors* in the *postsynaptic membrane*. Chemically gated ion channels provide for the specific and directional passage of particular ions, usually potassium (K^+), sodium (Na^+), calcium (Ca^+) and, occasionally, chloride (Cl^-). (see *action potential, nerve impulse, neuromuscular junction, acetylcholine, adrenaline, GABA*; compare *voltage-gated ion channel, stimulus-gated ion channel*)

chemiluminescent A term applied to chemical reactions that produce light as a byproduct. Enzymes that can be co-opted to *catalyze* such reactions *in vitro* (for example, *alkaline phosphatase,* horseradish peroxidase) are starting to replace *radioactive* tags in the visualization of biological molecules. (compare *biolumi-nescent*; see *autoradiography, Southern blot, RFLP, DNA analysis*)

chemiosmosis. The generation of *ATP* using energy generated by a *proton pump*. The membranes of *chloroplasts* and *mitochondria* contain protein "pumps" in which the energy from high-energy electrons induces a conforma-tional change, causing protons (H^+) to be exported. This creates a proton gradient. As protons pass backwards through the membrane by diffusion through special channels associated with *ATP synthase*, their release of energy is used to fuel the formation of ATP from *ADP* and P_i. (see *electron transport chain*; compare

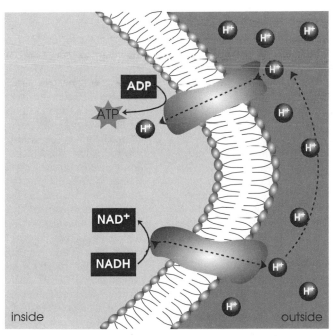

Chemiosmosis

substrate-level phosphorylation)

chemoautotrophism The nutritional mode, utilized only by some bacteria, in which the energy for *biosynthesis* is derived from the *oxidation* of inorganic molecules such as ammonia (NH_3), methane (CH_4), or hydrogen sulfide (H_2S). The energy is then used to fix inorganic carbon, such as that in carbon dioxide, into organic molecules. (see *trophism, autotrophism, chemosynthesis;* compare *chemoheterotrophism, phototrophism*)

chemoheterotrophism The mode of nutrition in which both energy and carbon are obtained by ingesting other organisms. Energy is derived from the splitting of chemical bonds in the ingested foodstuffs. It is used by all animals and fungi, most protoctists, a large portion of prokaryotes, and a few *achlorophyllous* plants. (see *trophism, chemotrophism, heterotrophism;* compare *chemoautotrophism*)

chemoreceptor A specialized *receptor* that responds to chemical compounds. They are elements of the vertebrate *nervous system* that function to collect chemical information about the external or internal environment. All chemoreceptors are *sensory neurons* that *depolarize* in response to the binding of specific molecules, either directly (smell, internal blood chemistry) or indirectly (*taste buds*). (see *olfactory*)

chemosynthesis The conversion of inorganic molecules to organic molecules using chemical energy rather than light energy. Organisms existing in the deep vents of the ocean accomplish primary production in this fashion, often using hydrogen sulfide (H_2S) as an energy source. (see *trophism, chemoautotrophism;* compare *photosynthesis*)

chemotaxis A movement by an organism in response to a chemical concentration gradient. (see *taxis*)

chemotrophism The mechanism by which most organisms obtain energy, that is, from chemical reactions rather than from light energy via *photosynthesis*. Most chemotrophic organisms are *heterotrophic*, using organic compounds that are obtained by ingesting other organisms as their energy source. A few specialized bacteria are *chemoautotrophic*, obtaining energy from reactions of inorganic compounds, such as the *oxidation* of ammonia or hydrogen sulfide. (see *trophism; chemoautotrophism, chemoheterotrophism;* compare *phototrophism*)

chemotropism A leaning or growth by an organism in response to a chemical stimulus. (see *tropism*)

chiasma, chiasmata (pl.) A location at which newly replicated *chromatids* of *homologous chromosomes* are in physical contact during *meiotic prophase I.* Chiasmata represent a mutual exchange of material between homologous, nonsister *chromatids* (*crossing-over*) and provide one mechanism by which the exchange of genetic material (*recombination*) occurs. Each *bivalent* may contain several chiasmata, and their placement may be directed by factors such as *DNA sequence*. As meiosis progresses, the chiasmata slip toward the ends of the paired chromosomes, helping to stabilize the *synaptonemal complex* until they separate at the end of *meiosis I.*

chief cells Specialized cells that secrete *pepsinogen*. They are contained in *exocrine glands* in the *gastric pits* of the stomach lining. (see *gastric juice, pepsin*)

Chilopoda The class of arthropods containing the centipedes, characterized by a flat body divided into numerous segments, each bearing one pair of walking legs. This distinguishes them from millipedes (*Diplopoda*), in which each body segment bears two pairs of legs. They are terrestrial and *carnivorous*, with poison claws on the first body segment. Their poison may be deadly to humans.

chimera An organism that contains tissues of more than one genetic type. In mythology, the Chimera was a creature with the head of a lion, body of a goat, and tail of a serpent. In modern biology, this situation may arise as a result of natural *mutation, grafting* (for example, in agriculture), or the experimental introduction of a cell or cells into a developing *embryo*. Many *variegated* plants are examples of chimeras. *Genetic engineering* is another example of the creation of a chimeric organism in the laboratory. (see *mosaic*)

chimpanzees (see *Primates*)

Chiroptera The order that contains the bats, the only flying mammals. Bats have a thin elastic flight membrane extending from the elongated forearm and four of the elongated fingers to the hind limbs and, usually, the tail. The first finger and the toes are smaller, free, and clawed. Bats are *nocturnal* and have oversized, specialized ears that assist in their use of *echolocation* for navigation, *migration,* and prey capture. Like other mammals, they give birth to helpless young (*altricial*), nurse them from *pectoral* breasts, and care for them until independence. In colder regions, bats migrate or *hibernate*. Any disturbance to a cave full of hibernating bats hanging from the roof, can cause them to use up valuable food stores and reduce their chance of surviving the winter. The approximately 1,000 species of bat constitute about one-fourth of all mammal species. Bats inhabit many diverse *habitats* and have developed extremely specialized adaptions for each. Most bats are *insectivores*, but a signifi cant number are fruit and nectar eaters and are essential to *pollination* and forest ecology. Only a few species are *carnivorous*, only three of which are actually vampire bats.

chitin A nitrogen-containing *polysaccharide* found in some animals and the *cell walls* of most fungi and some protoctists. The outer covering of *arthropods*, the *cuticle*, is impregnated with chitin, which makes the *exoskeleton* more rigid. It is associated with protein to give a uniquely tough yet flexible and light skeleton, which also has the advantage of being waterproof. Chitin is also found sporadically in the hard parts of other animals. It

is a *polymer* of N-acetyl-glucosamine.

chitinous Any structure containing *chitin*.

chitons (see *Polyplacophora*)

chloramphenicol An *antibiotic* that inhibits *translation* of *mRNA* on *prokaryotic ribosomes*, but does not affect *eukaryotic protein synthesis*. It has historically been used to treat bacterial infections. Since the cure often results in a worse situation than the disease, its use has been largely discontinued. It has also been used in the laboratory to inhibit the growth and division of bacterial cells *hosting plasmids* or *cosmids* containing cloned genes. The multiplication of the *episome* is relatively uninhibited and the culture consequently enriched for the *gene* of interest. Improvements in plasmid vectors have obviated the need for this procedure in many cases. (compare *cycloheximide*)

chlorenchyma A subset of the most common plant cell, *parenchyma*, containing many *chloroplasts*.

chlorinated hydrocarbons A group of industrial compounds including DDT, aldrin, and dieldrin as well as other *pesticides* and *herbicides*. Because they are lipid soluble, they accumulate in animal fat. The effect is therefore compounded at each point in the progression of a *food chain*. Although the effects are still poorly understood, among them are reproductive deformities and failures, exemplified by the plight of bald eagles and their broken eggs. DDT was banned for use in the United States in 1971, resulting in a dramatic recovery of affected birds, including the brown pelican and the bald eagle. DDT continues, however, to be produced in the United States and used by other countries, and has accumulated in various deposits from which it continues to seep out. (see *food chain concentration*)

Chlorobia A phylum of *Gram-negative* bacteria containing the *anoxygenic* green sulfur *phototrophs*. They are generally found in mud and pond scum, and typically use the hydrogen atoms from hydrogen sulfide (H_2S) to reduce carbon dioxide (CO_2) to carbohydrates for food. Their photosynthetic byproduct is thus elemental sulfur (S_2) rather than oxygen (O_2).

Chloroflexa The phylum of bacteria containing the green nonsulfur *phototrophic* bacteria. They are somewhat unusual in that they are *photoheterotrophs*. They get their energy from sunlight but obtain their carbon from small organic molecules such as pyruvate and acetate. They are generally found in mud and pond scum.

chlorofluorocarbons (see *CFCs*)

chlorophylls A group of bright green pigments that collect and trap light energy for *photosynthesis*. They are responsible for the green color in plants, as they absorb blue-violet and red light, and hence reflect green light. The molecule consists of a *hydrophilic* head, containing magnesium at the center of a *porphyrin ring*, and a long *hydrophobic* hydrocarbon tail that anchors the molecule in membrane lipids. Different chlorophylls have different chemical groups attached to the head. (see *absorbtion spectrum, photosynthetic pigments, pheophytin*; compare *carotenoids*)

Chlorophyta A phylum of protoctists, the green algae and green seaweeds. They inhabit mainly fresh water but are occasionally found in marine and terrestrial environments. Chlorophytes may be either *unicellular, colonial,* or *multicellular* and contain *chlorophylls, carotenes,* and *xanthophylls*. Green algae store food as starch and fat, have *cell walls* containing *cellulose,* and achieve motility by *undulipodia*. They may propagate *asexually* by *zoospores* or by similar motile cells that function as *gametes* by fusing with a mate and undergoing *sexual reproduction*. This may be followed by the formation of overwintering *zygospores*. Green algae are thought to be the ancestors of modern plants.

chloroplast The cell *organelle* in which *photosynthesis* takes place. It contains *chlorophylls* a and b, as well as other pigments. It is found in all photosynthetic plant cells but not in photosynthetic *prokaryotes*. The typical higher plant chloroplast is lens shaped and about 5μm in length. Various other forms exist in the algae, for instance spiral in *Spirogyra* and cup shaped in *Chlamydomonas*. The number per cell can vary from one to over a hundred, depending on the organ-

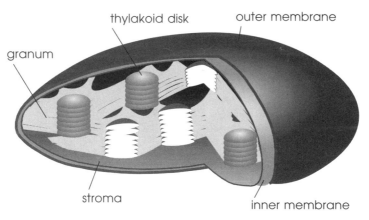

Chloroplast

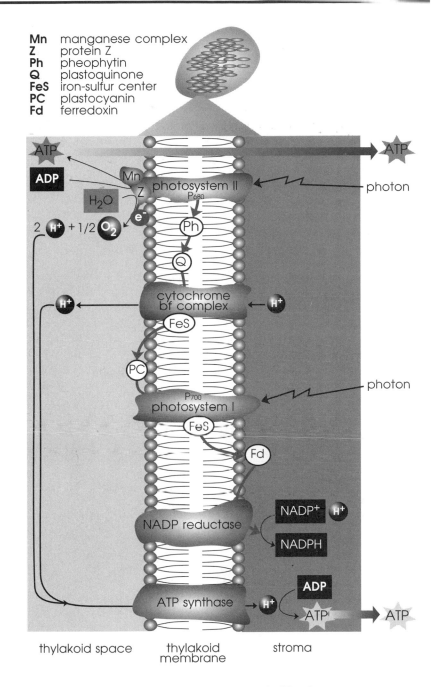

Mn manganese complex
Z protein Z
Ph pheophytin
Q plastoquinone
FeS iron-sulfur center
PC plastocyanin
Fd ferredoxin

Electron Transport and Photophosphorylation in Chloroplasts
(see *chloroplast*)

ism. In a mature chloroplast, two outer membranes typically enclose a *stroma*, where the *dark reactions* occur. In the stroma are embedded a number of *grana*, each consisting of a stack of disk-shaped, membranous *thylakoids* that house the *photosynthetic pigments* and the *electron transport system* involved in the light dependent phase of photosynthesis. The grana are connected by sheet-like layers called *lamellae*. The stroma may contain storage products of photosynthesis, such as starch grains, as well as other substances. In addition to enzymes involved in the dark reactions of photosynthesis, the stroma also contains a circular *DNA* molecule and *protein-synthetic* machinery typical of *prokaryotes*. Because of this, chloroplasts, like *mitochondria* and other organelles are thought to have an *endosymbiotic* origin. Chloroplasts may have originated from cyanobacteria. *Chloroplast DNA, (cpDNA)* codes for some proteins, but chloroplasts are not autonomous and depend on *nuclear DNA* for the synthesis of most other components. (see *serial endosymbiosis theory*)

chloroplast DNA (see *cpDNA*)

chlorosis The loss of *chlorophyll* from plants, resulting in yellow leaves. It may result from the normal process of cell death, a lack of key nutrients necessary for chlorophyll synthesis, or disease. (see *photosynthesis*)

Chloroxybacteria (see *Cyanobacteria*)

choanocyte (collar cell) A type of cell that lines the chambers of a sponge and circulates water by beating its *undulipodia*. They are essentially identical to choanoflagellates, a group of the protist phylum Zoomastigina from which sponges are thought to have evolved.

cholecystokinin A *peptide* digestive *hormone* that is produced upon arrival of food in the small intestine. It is produced in the duodenum and stimulates the gallbladder to release *bile salts* into the intestine and the pancreas to secrete digestive enzymes.

cholesterol A lipid found in animal cells. It is a crucial component of *cell membranes* as well as a precursor of *steroid hormones* (for example, *corticosteroids*, sex hormones). Normal cholesterol synthesis in the liver is suppressed by dietary intake. Improper regulation of cholesterol *metabolism* can result in deposits in the gallbladder (gallstones) and in atrial walls causing *arteriosclerosis*. (see *HDL, LDL*)

choline A member of the *vitamin B complex* that acts to disperse fat from the liver and prevent its excess accumulation. It is an essential nutrient. In man, it is synthesized from *lecithin* as a part of the normal bacterial action in the bowel. As part of *acetylcholine*, it functions as a *neurotransmitter*. (see *nerve impulse*)

cholinergic The type of *neuron* that releases the *neurotransmitter acetylcholine* from its ending when stimulated to transmit a *nerve impulse*. Cholinergic nerves are involved in the *parasympathetic nervous system*. (compare *adrenergic*)

cholinesterase (see *acetylcholinesterase*)

Chondrichthyes The class of vertebrates containing the *cartilaginous* fish, including sharks, skates, rays, and the dogfish. They lack any bone, and their skin is covered with *denticles* (*placoid scales*) that are modified in the mouth to form razor sharp replaceable teeth. Since neither lungs nor a *swim bladder* are present, the fish sinks when it stops swimming.

chondroblasts (see *chondrocytes*)

chondrocranium The first part of the skull to form in vertebrate *embryos*. It consists of *cartilaginous* structures that protect and support the brain, *olfactory* organs, eyes, and the *inner ear*. It usually becomes *ossified* in adults, forming the *membrane bones*. (see *neurocranium*)

chondrocytes (chondroblasts) The cells that extrude and are embedded in the matrix of *cartilage*.

Chordata The animal *phylum* historically containing the chordates, characterized by the presence of a *notochord* in at least some stage of development and *pharyngeal slits* (*gill slits*) at some or all stages of their life history. Urochordates and cephalochordates have been given phylum status. Vertebrates, in which the notochord is replaced by a *vertebral column* (backbone), are all contained in phylum Craniata.

chordates An informal grouping of the newly autonomous phyla Urochordata, Cephalochordata, and Craniata.

chorion One of the three *extraembryonic membranes* of reptiles, birds, and mammals. It encloses the *amnion* (including the *embryo* or *fetus*), *yolk sac*, and *allantois*. In mammals, the outer layer (*trophoblast*) contains *villi* that intercalate with adjacent maternal tissue to form the *placenta*. In reptiles and birds, it lines the shell and in conjunction with the *allantois*, permits gaseous exchange within the egg.

chorionic gonadotrophin A *hormone* secreted in higher mammals by the *chorionic villi* of the *placenta*. It helps maintain the *corpus luteum* (an ovarian structure that functions during pregnancy) in the earlier stages of pregnancy. The detection of human chorionic gonadotrophin (HCG) in the urine is often used as a pregnancy test.

choroid The middle layer of the vertebrate eye, between the *sclera* and the *retina*. It is rich in blood vessels and contains pigment to prevent internal reflection of light. (see *tapetum*)

choroid plexus One of the two projections into the fluid-filled spaces of the brain through which exchange of materials between blood and *cerebrospinal fluid* takes place. The choroid plexus forms the *blood-brain barrier*. It precludes the entry of blood cells and large molecules from the bloodstream into the cerebrospinal fluid. This provides an extra measure of protection for the brain from some potentially harmful substances; ironically, it also complicates the administration of some potentially useful medications.

chrom-, chrome-, chromo- A word element derived from Greek meaning "color." (for example, *chromosome*)

chromatid One of the two daughter strands of a replicated *chromosome*. Chromatids are joined together by a single *centromere* that divides during the last stage (*anaphase*) of *nuclear division* in *mitosis* or *meiosis II*, generating two separate sister chromosomes.

chromatin The entire complex of a *eukaryotic chromosome*, including *DNA*, *chromosomal* proteins, and chromosomal *RNA*. Highly condensed portions, for instance the regions flanking the *centromere*, are seldom or never *transcribed* into *RNA* and are termed *heterochromatin*. The remainder, termed *euchromatin*, is normally maintained in an open configuration from which *genes* can be readily activated during cell growth.

chromatography An analytical technique for separating the components of complex mixtures based on their physical properties. Generally, this involves partitioning between a mobile phase (gas or liquid) and a stationary phase (liquid or solid) based on the charge and/or size of the different molecules. Separation is dependent on the method of chromotography used. (see *affinity chromatography, paper chromatography, thin-layer chromatography, gas chromatography, gel filtration, ion exchange, Rf value*; compare *electrophoresis*)

chromatophore Generally, any pigmented, structure.
1. Membrane bound vesicles (*thylakoids*) bearing pigments in *photosynthetic* bacteria.
2. (see *chromoplast*)
3. A pigment-containing cell in animals used for camouflage. The color in *melanophores*, *lipophores,* and *guanophores* can be changed by concentration or dispersal of the pigments within the cells in response to various stimuli, such as light intensity, temperature, fright, or courtship.

chromomeres Small beadlike structures visible on *chromosomes* during the initial stages (*prophase*) of both *mitosis* and *meiosis.*

chromophore The light absorbing molecule in a *photoreceptor.*

chromoplast A plant cell *plastid* containing non*photosynthetic* pigments. Chromoplasts contribute to the coloration of fruits and flowers, which function as animal and insect attractants. They are often derived from *chloroplasts* in which *carotenoid* pigments accumulate.

chromosome The physical structure into which *DNA* is organized and on which *genes* are carried. In *prokaryotes,* the chromosome consists of a single circle of naked DNA. In *eukaryotes,* DNA is packaged, along with small amounts of *RNA* and *histone* proteins, into linear entities that reside in a membrane bound cell *nucleus.* The number of chromosomes in the nucleus is characteristic of the species; human cells have 23 matched pairs for a total of 46 chromosomes (22 *autosome* pairs plus a pair of *sex chromosomes*). Chromosomal substructures include the *centromere* (center) and *telomeres* (ends). The name comes from the fact that brightly colored stains were originally used to visualize chromosomes as distinct nuclear bodies. (see *cell division, meiosis, mitosis, chromatin*)

chromosome jumping A method of ordering the *DNA* information on a *chromosome* across relatively large areas and across regions that are resistant to *cloning.* Large fragments of the *genome* of interest are generated by *partial digestion* using a *restriction endonuclease* and cloned into a *vector,* generating a jumping library. If two *probes,* one containing a known *gene* and the other containing the new gene of interest, hybridize to the same large clone, their distance on the chromosome can be easily deduced by

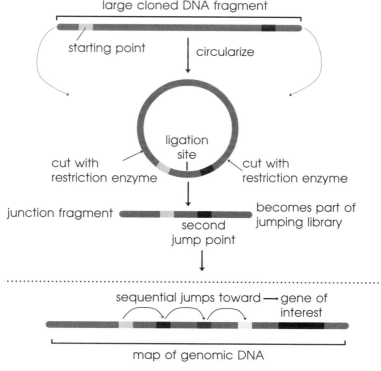

Chromosome Jumping

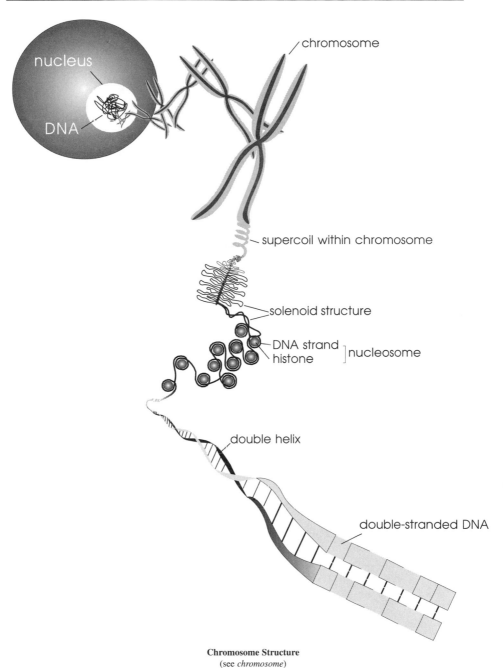

Chromosome Structure
(see *chromosome*)
Modified from an illustration in *An Introduction to Forensic DNA Typing*, ©1997, by Inman and Rudin, CRC Press, Boca Raton, Florida.

intrachromosomal

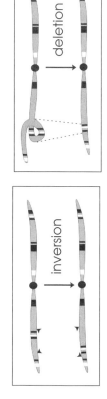

interchromosomal

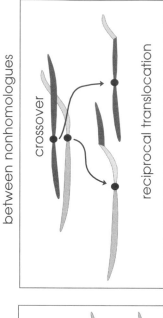

between nonhomologues

crossover

reciprocal translocation

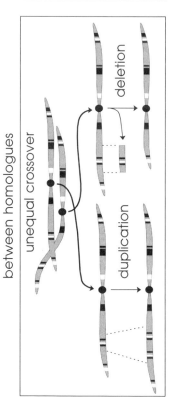

between homologues

unequal crossover

deletion

duplication

Some Types of Chromosomal Rearrangements
(see *chromosome*)

measuring their distance apart on that fragment. The region between them can then can also be analyzed by *chromosome walking*.

chromosome loss The failure of a *chromosome* to become incorporated into an offspring *nucleus* at the end of *mitosis* or *meiosis*.

chromosome map (see *genetic map*)

chromosome mutation A change in the gross structure of a *chromosome* or the number of chromosomes.

chromosome painting (see *FISH*)

chromosome puff (see *polytene chromosome*)

chromosome theory of inheritance The unifying theory stating that inheritance patterns may be generally explained by assuming that *genes* are located at specific sites on *chromosomes*.

chromosome walking A method of ordering the *DNA* information on a *chromosome* by identifying fragments containing overlapping *nucleotide* sequences. Known, *cloned* fragments are used to isolate overlapping, adjacent fragments and the process is continued in a linear fashion along the length of the chromosome. (compare *chromosome jumping*)

chrysalis The *pupae* of butterflies and moths.

Chrysomonada (Chrysophyta) A phylum of *photosynthetic* protoctists, the chrysomonads or golden-yellow algae.

The group includes *unicellular planktonic* organisms as well as *colonial* forms. Freshwater chrysomonads are among the many organisms manifest as pond scum. Photosynthesis in *chrysoplasts* is based on *chlorophylls* a and c, with *fucoxanthin* as an accessory pigment. Reproduction is *asexual*. The marine forms incorporate silica into their *tests*, which are useful as a *paleoecological* indicator as each species grows only under very specific environmental conditions.

chrysoplast A yellow *plastid* that comprises the *photosynthetic organelles* of chrysophytes, diatoms, and haptomonads. It is analogous to a *chloroplast* but characteristically contains *chlorophylls* a and c rather than *chlorophylls* a and b.

chyme Partially digested food that has been processed in the stomach of vertebrates and is ready to enter the small intestine.

chymotrypsin A *proteolytic* (protein cutting) *enzyme* secreted by the pancreas in an inactive form, *chymotrypsinogen*. It is activated by the enzyme *enteropeptidase*, a proteolytic enzyme itself. Chymotrypsin aids in the digestion of proteins in the *small intestine* by catalyzing the partial *hydrolysis* of *polypeptide chains* at the *amino acid* residues phenylalanine, tyrosine, and tryptophan.

chytrid body The structure in *chytridiomycotes* in which *zoospores* form.

Chytridiomycota A phylum of funguslike

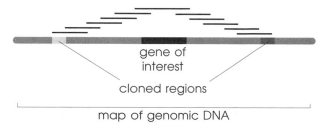

overlapping hybridization fragments

gene of interest

cloned regions

map of genomic DNA

Chromosome Walking

protoctists, the chytrids. Like fungi, they contain *chitin* in their *cell walls* and obtain their nutrition by absorbtion (*osmotrophy*). In most cases, they develop a *chytrid body* in which motile *zoospores* form. Some genera are *obligate anaerobes*, growing inside the *rumen* of animals such as cattle and elephants that have a fibrous cellulose diet. *Aerobic* members of the phylum inhabit and feed on decaying pond vegetation. Chytrids are thought to be the ancestors of modern fungi.

Ciconiiformes An order of birds including the herons, bitterns, storks, ibises, and spoonbills.

-cidal, -cide A word element derived from Latin meaning "killer." (for example, *bactericidal*)

ciliary body A ring of muscular tissue in the front of the *vertebrate* and cephalopod eye, at the junction of the *choroid* with the *iris*. The *ciliary muscles* (circular smooth muscle tissue) permit *accommodation* for near and distant vision by altering the curvature of the lens (*amniotes*) or by moving the lens with respect to the *retina* (cephalopods, sharks, and amphibians). It contains the *ciliary glands*, which secrete the *aqueous humor* that fills the eye.

ciliary feeding (see *filter feeding*)

ciliated epithelium A single layer of *epithelial* cells lined with numerous *cilia* that beat in *metachronal rhythm* (each moves sequentially, a split second after the one before it). (see *ciliary feeding, filter feeding*)

Ciliates (see *Ciliophora*)

Ciliophora A phylum of *unicellular* or *colonial* protoctists containing the *ciliates*, organisms typically covered in *cilia*, including the well-known *Paramecium*. They are among the most complex cells in the phylum and contain a *macronucleus*, *micronucleus*, and a *cytostome* at the end of a depression where food *vacuoles* form. The body wall is a tough but flexible *pellicle*. Most are *heterotrophic*, but some have become secondarily *photosynthetic* through algal *symbionts* or borrowed *chloroplasts*. Waste is expelled through a special region of the pellicle, the *cytoproct*. Reproduction includes both a complex process known as *conjugation* as well as *binary fission*. Most are single celled, but a few others give rise to swimming forms or form *spore* stalks and are genuinely *multicellular*.

cilium, cilia (pl.) An exterior *organelle* of some cells used for locomotion or transport of liquid past a cell. It has a characteristic structure, the *axoneme*, consisting of 9 pairs of *microtubules* surrounding an inner pair and is anchored to the cell by a *basal body* (*kinetosome*). Usually, many cilia occur in an area and work in concert. They are essentially a short version of *undulipodia*. (see *Ciliophora*, *ciliated epithelium*; compare *flagellum*)

circadian rhythm A biorhythm based on a 24-hour cycle controlled by an *endogenous biological clock*. In the absence of environmental cues, the endogenous rhythm will continue but will tend to drift out of synch. An environmental cue, usually light, resets the clock. In humans, the biological clock lies in a region of the *hypothalamus*, which responds to the *amino acid melatonin* secreted by the light sensitive *pineal gland*, located just beneath the skull. In some insects, the clock may be located in the *optic lobes* of the brain. Research continues into the molecular mechanisms underlying biorhythms. (compare *circannual behaviors*)

circannual rhythm Animal behaviors that occur on a yearly cycle, such as breeding, *hibernation,* and *migration*. They seem to be mediated by *hormones* keyed to *exogenous* factors such as day length, but the physiological and molecular mechanisms are largely unknown. (compare *circadian rhythm*)

circulatory system A continuous series of vessels or spaces in animals that transports fluids and dissolved solutes around the body. The system ranges from a series of open spaces (*open system*) in arthropods to a fully enclosed system of vessels (*closed system*) most highly developed in mammals. It enables all parts of the body to receive a constant supply of oxygen and food, and to have waste products removed promptly. (see *vascular system, blood vascular system*)

circumnutation (see *nutation*)

cirripedes (see *Cirripedia*)

Cirripedia A class of crustaceans, the barnacles, that are free-swimming as *larvae* and *sessile* as adults. They ultimately attach themselves to rocks, pilings, or sometimes other animals by their head. They stir food into their mouth with their feathery legs. Their body is covered with *calcareous* plates, which are solidly attached to the substrate. Barnacles are *hermaphroditic* but *cross-fertilize*.

cis-acting A genetic element that is physically on the same *chromosome* and located near the region that it affects. It is usually defined by a specific *nucleotide* sequence. (compare *trans-acting*)

cisterna One of a number of flattened, membrane bound spaces between the membranes of cell *organelles* such as the *endoplasmic reticulum* and the *Golgi apparatus.*

cis-trans test (see *complementation test*)

cistron A term originally defined as a functional genetic unit within which two *mutations* cannot *complement* to generate a *wild-type phenotyp*e. It is now equated with the term *gene*.

citric acid A weak acid, occurring naturally in the juice of citrus fruits. It is important as an intermediate in the *Krebs cycle*.

cladistics A school of thought in which only evolutionary relatedness (*phylogenetics*) is considered when assigning taxonomic grouping, exclusive of physiological *morphological* attributes. It is particularly suited to take advantage of *DNA* sequence data. Cladistics shows the order of evolutionary decent but not the extent of divergence. Practically, phylogenetic data is considered together with physiological morphological information (*phenetics*) when classifying organisms,

although the genetic component is coming to have more weight as information accumulates. (see *biochemical taxonomy, biosystematics, molecular clock*)

cladode (see *phylloclade*).

clams (see *Bivalvia*)

clamworms (see *Polychaeta*)

claspers The *pelvic* fins of male *cartilaginous* fish (for example, sharks) that are modified to serve as *copulatory* organs.

class A *taxonomic* category used in the classification of organisms. It consists of a collection of similar *orders* and is a subdivision of a *phylum.*

classification The grouping, defining and naming of organisms based on comparative relationships. Biological classification systems today use components of both evolutionary relatedness (*cladistics*) and morphological similarity (*phenetics*), along with comparative physiology and biochemistry. With the wealth of DNA information being amassed, the genetic component is beginning to take precedence. (see *binomial nomenclature, biosystematics, phylogeny, phylogenetics*)

clast A rock particle or fragment.

clathcrins A group of proteins found in *eukaryotic cell membranes* that serve to anchor certain proteins to specific sites.

clavicle One of a pair of bones that lie on each side of the base of the neck in some vertebrates. In man, they form the collar bones, which connect the shoulder blade with the breastbone. They are *membrane bones.*

clay Extremely fine-textured sticky soil. It has a tendency to become waterlogged and compacted, reducing the availability of oxygen to plants.

clear-cutting The practice of cutting all trees in a stand at the same time. Clear-cutting over large areas, on steep slopes, and in areas of high rainfall may be extremely detrimental to the local *ecology*, dramatically increasing erosion including runoff of nitrates leading to *pollution* of local water sources. The impact of clear-cutting of small areas in flatter, drier regions may be less. (see *second growth*)

cleavage The rapid divisions of the vertebrate *zygote* that divide the *fertilized* egg into smaller and smaller cells (*blastomeres*) while retaining the same overall size of the *embryo*. The resultant mass of about 32 cells is called a *morula*. Each blastomere inherits a different portion of the egg *cytoplasm* which contains nonhomogeneous components that determine *differentiation* during the developmental process. (see *holoblastic, meroblastic*)

cleidoic egg An egg with a tough shell that permits gaseous exchange but restricts water loss. It is characteristic of terrestrial animals such as reptiles, birds, and insects and usually has a large food store of yolk and/or *albumin.*

cleistogamy *Self-fertilization* within a closed flower. It may occur toward the end of the flowering season in certain plants (for example, wood sorrel, violet) when no seed has been set by *cross-pollination.*

cleistothecium A type of *fruiting body* (*ascocarp*) in some fungi that is completely closed and from which the *spores* are eventually liberated through decay or rupture of the wall.

climacteric The phase of increased *cellular respiration* rate found in some plant species at fruit ripening and *senescence.* Complex carbohydrates are broken down into simple sugars, *cell walls* soften, and volatile compounds associated with scent and flavor are produced. (see *ethylene*)

climax community The *ecological community* resulting from a relatively rapid *ecological succession* under a particular set of conditions. It continues to change slowly in response to environmental conditions, including climactic changes and industrial influences such as *pollution* and grazing. It may also be defined as the stage at which the greatest *biomass* or *biological diversity* is achieved. (see *climax vegetation*)

climax state A hypothetical steady state stage occuring at the end of *ecological succession* in which a *climax community*

is self-perpetuating and in equilibrium with the existing environmental conditions. In reality, this is probably never achieved in nature.

climax vegetation The plant community resulting from the colonization of a bare surface. The vegetation passes through a succession of stages as more complex organisms and *communities* replace earlier, simpler ones. Major influences include climate, soil conditions, and human activity. (see *climax community, ecological succession*)

cline A continuous gradation of a particular characteristic. It may be applied to environmental factors (the temperature gradient in an ocean or lake is called a *thermocline*) or to biological factors, such as the characteristics of a species correlated with changing ecological variables like environment or geography (*ecocline*). The populations at each end of a biological cline may be substantially different from each other.

clisere A succession of *climax communities* that occur in an area as a result of climactic changes. (see *ecological succession*)

clitellum A saddle-like swelling prominent in some *sexually* mature *annelids* (earthworms and leeches). It secretes a mucus tube that binds the two worms together during *copulation*. Afterwards, it secretes a *cocoon* that encloses and protects the *fertilized* eggs during development.

clitoris A small, erectile rod of tissue in female mammals that lies in front of the *vagina* and *urethra*. It is *homologous* with the male *penis.*

cloaca In all vertebrates except *placental* and *marsupial mammals,* the common chamber into which solid and liquid wastes as well as *gametes* are discharged.

clonal selection The method by which one or a few *stem cells* in the *bone marrow* are provoked to proliferate in response to a particular *antigen*. This creates a line of *B-cell clones* all expressing the same *B-receptor*, which is also released into the bloodstream as free *antibody.*

(see *immune receptor genes, somatic recombination immune system, immune response, humoral immune response*)

clone A group of genetically identical cells or individuals derived by *asexual* division from a common ancestor. The term may also describe one cell or organism genetically identical to the one from which it has been produced, usually by some asexual process. The term is also applied to segments of *DNA* when they are isolated and copied *in vitro*. The term "clone" has been adopted in common language to mean anything that is similar or identical to the original, such as "computer clones." (see *gene cloning, parthenogenesis*)

cloning (see *gene cloning*)

cloning vector (see *vector*)

closed-canopy forest A forest in which the leaves of adjacent trees overlap or touch so that the trees essentially form a continuous cover. It is common in *tropical rainforests*, and essential to their specialized *ecosystem*.

clotting (see *blood clotting*)

clotting factors A group of substances, all of which are essential for *blood clotting*. They are normally activated in a cascade when blood exits the *circulatory system*, for example due to injury, and include various proteins, *enzymes, vitamin K, calcium ions*, and *platelets*.

club mosses (see *Lycophyta*)

cluster of differentiation (see *CD marker*)

cM (centimorgan) One *genetic map unit*.

Cnidaria (coelenterates) A phylum of aquatic, mostly marine, invertebrates including jellyfish, sea anemones, and *colonial* corals. Cnidarians are the most primitive *multicellular* animals. They typically exhibit *radial symmetry* and have a body wall of two layers separated by a layer of jelly (*mesoglea*) enclosing the body cavity (*coelenteron*). The mouth is surrounded by a circle of tentacles that are used for food capture and defense and may bear stinging cells (*cnidocytes*). Either the sedentary *polyp* or mobile *medusa* form may be used, with some

species alternating between them in a *dimorphic life cycle* (*alternation of generations*). They reproduce both *sexually* and *asexually*. (see *nematocyst*)

cnidocil The trigger responsible for firing a *nematocyst* (barb) when a *cnidocyte* (stinging cell) is stimulated. (see *Cnidaria*)

cnidocyte (thread cell) A specialized stinging cell found only in *Cnidaria* and a few of their predators (*nudibranchs*) that incorporate them. They contain *nematocysts*, each of which extrudes a barbed thread upon stimulation, that is used for adhesion, penetration, or injection of poison.

CNS (see *central nervous system*)

CoA (see *coenzyme A*)

coacervates Inorganic colloidal particles (for example, clay) onto which organic molecules have been adsorbed. It is thought that this may have been an important concentrating mechanism in *prebiotic* evolution. (see *origin of life*)

coarctate In insects that undergo *metamorphosis*, a *pupae* in which the last *larval cuticle* is retained and forms a hardened shell or *puparium* around the body. (compare *exarate, obtect*)

coatis (see *Carnivora*)

cobalt (see *trace elements*)

coccoid Any spherical or nearly spherical structure, or cluster of such.

coccolith An external, platelike structure on some haptomonad protists that is made of calcium carbonate.

coccolithophorids (see *Haptomonada*)

coccus, cocci (pl.) Any spheroid *bacterium*. Examples include the genera *Diplococcus* and *Streptococcus*. (compare *bacillus, spirillum*)

coccyx The three to five fused *vertebral* bones at the bottom of the *vertebral column* in man and other *primates*.

cochlea A tubular cavity in the *inner ear* in mammals, birds, and crocodilians containing the essential organ of hearing, the *organ of Corti*. The membrane is differentially sensitive to the frequency of sound waves, the apex responding to lower vibrations and the base to higher frequencies, enabling distinction of pitch. In mammals, it is spirally coiled. (see *organ of Corti*)

cockatoos (see *Psittaciformes*)

cockroach (see *Dictyoptera*)

cocoon A protective covering for the eggs and/or *larvae* produced by many invertebrates such as spiders and earthworms. The larvae of many insects also spin cocoons for protection during the *pupal* stage. The cocoon of the silkworm moth is the source of raw silk fibers.

coding region Any portion of the *chromosome* that contains *DNA* sequences directing the synthesis of a *polypeptide*. They are distinguished from *introns* which are removed during *RNA* processing, *regulatory regions* which are not *transcribed*, and other portions that appear to have no function. (see *exon, splicing*)

codominance The situation in which both genetic *alleles* (variants) in a *heterozygous* individual are expressed, producing a third *phenotype* different from either single allele.

codon The basic genetic coding unit. It is a triplet of three consecutive *nucleotides* that defines either a specific *amino acid,* or initiation or termination of a *polypeptide* chain. Codons are arranged consecutively along the *DNA* molecule and must be *transcribed* in the appropriate register (*reading frame*) into *RNA* for *translation* into protein. In the *polysome* array, the *mRNA* molecule directs *tRNA* molecules, which contain a complementary *anticodon* on one end and a specified amino acid on the other, in the synthesis of polypeptides. The nucleotide triplets on the DNA and on the tRNA molecules run in the same direction (have the same sense); the triplets on the mRNA molecule run in the opposite direction. Sixty-four possible triplet combinations of the four DNA or RNA bases are possible, specifying 20 common *amino acids*. A number of amino acids are determined by more than one triplet combination, lead-ing to redundancy in the *genetic code.* (see *protein synthesis, ribosome, genetic code*)

-coel- A word element derived from Greek meaning "cavity." (for example, *Coelom*)

coelacanth (see *Crossopterygii*)

Coelacanthini (see *Crossopterygii*)

Coelenterata A historical animal phylum that contained the lower invertebrates, now separated into the phyla Cnidaria and Ctenophora. The term is still sometimes used as a synonym for Cnidaria.

coelenteron (gastrovascular cavity) The body cavity of cnidarians and ctenophores. It has a single opening, the mouth, through which food is ingested and waste products *egested.* In some groups, *gametes* are also discharged into it.

coelom A fluid-filled cavity arising in the *mesoderm* of the more advanced (*triploblastic*) animals. In annelids, mollusks, echinoderms, and chordates, it is the main body cavity, containing the *viscera*. In mammals, it is divided into separate cavities enclosing the heart, lungs, and gut. In arthropods, the coelom is reduced to cavities surrounding the *gonads* and excretory organs, the main body being a blood-filled cavity, the *hemocoel*. The coelom functions as a hydrostatic skeleton in some worms (for example, earthworm). The evolutionary development of a coelom enabled separation of the gut from the body wall. This facilitated muscle driven body movement, provided an environment for specialized organs, and permitted the development of a closed *circulatory system.*

coelomate (peritoneal) Animals possessing a true body cavity. Evolutionarily, the advent of a *coelom* enabled the development of separate, more efficient body systems. (compare *pseudocoelomate, acoelomate*)

coelomoduct In invertebrates, a *ciliated* duct formed from the lining of the *coelom* and connecting it with the exterior environment. It may provide an exit for waste products and/or *gametes.*

coenocyte A plant cell or organism (typically fungi or algae) containing multiple *nuclei* not separated by *cell walls*. It is the result of repeated division of the nucleus without the formation of additional cell walls separating the *cytoplasm*. (see *acellular, plasmodium, syncytium*)

coenzyme A nonprotein, organic molecule that acts as a *cofactor* for some enzymes. It plays an accessory role, often by acting as an electron donor or acceptor. Coenzymes are frequently derivatives of water soluble *vitamins*. (see *cofactor, apoenzyme, holoenzyme, coenzyme A, coenzyme Q, FAD, FMN, NAD, NADP*)

coenzyme A (CoA) A coenzyme integral to *fatty acid* synthesis and *oxidative respiration*. In general, it functions as a carrier of acyl groups. In respiration in particular, it facilitates the entry of *pyruvate* into the *Krebs cycle* by combining with it in the presence of NAD^+ to form *acetyl-CoA*. This produces *NADH* and results in the loss of a CO_2 molecule. (see *respiration, aerobic respiration*)

coenzyme Q (CoQ) A coenzyme that funnels electrons into the *electron-transport system* in *mitochondria*. (see *respiration, aerobic respiration*)

coevolution The long-term, mutual evolutionary adjustment of the characteristics of the members of a biological community in a reciprocal relation to one another. For instance, certain insects and birds have developed specialized appendages in order to obtain *nectar* from particular flowers, while the flowers have developed specific attractive forms and colorations. The animals aid the plant in *pollen* dispersal. (see *pollination, style*)

cofactor Any nonprotein molecule essential for a particular enzyme reaction. When cofactors are organic molecules, they are known as *coenzymes*. An enzyme-cofactor complex is termed a *holoenzyme*, while the inactive protein portion of such an enzyme alone is termed the *apoenzyme*.

cohesive ends (see *sticky ends*)

cohort A group of individuals in a population all born in the same year.

colchicine An *alkaloid* derived from the roots of the autumn crocus that inhibits *tubulin polymerization*. One consequence is to prevent *spindle* formation during *nuclear division*, resulting in the failure of the replicated *chromosomes* to separate from each other. These cells are arrested in *metaphase*. Colchicine is an important laboratory tool in chromosome analysis.

cold-blooded (see *poikilothermic, ectothermic*)

Coleoptera The largest order of *insects* and possibly the largest order in the animal kingdom. It contains the beetles and weevils. They are found universally in a variety of terrestrial and freshwater habitats and undergo *complete metamorphosis (endopterygote)*. Their forewings are modified to form hard leathery *elytra* that protect the membranous hind wings and soft abdomen when resting. The head contains biting mouth parts. The *larvae* may be legless grubs, caterpillarlike, or active predators. Many larvae and adults are serious pests of crops, stored produce, and timber (woodworm, wireworm). Some borers of live wood may transmit fungal disease (for example, Dutch elm disease). Others are considered beneficial, for instance the lady beetle which eats aphids.

coliform bacteria (see *Proteobacteria*)

Coliiformes The order of birds comprising the mousebirds.

collagen The major *structural protein* present in fibrous tissues such as bone, skin, cartilage, and tendon in higher vertebrates. It also occurs in invertebrates, for example in the *cuticle* of nematodes. The *secondary structure* of collagen is that of a *triple helix* of *polypeptide chains*, three of which are further entwined into a "super helix" that is responsible for its great tensile strength. When boiled, collagen yields gelatin.

collar cell (see *choanocyte*)

Collembola An order of *exopterygote* primitive wingless insects, the springtails, that are abundant in soil, under bark and on pond surfaces. They are important *decomposers*.

collenchyma The first type of tissue providing mechanical support in young growing plant structures. The walls are irregularly strengthened with *cellulose* and *pectin*, leaving the cells still able to expand as the tissue continues to develop. (see *parenchyma*)

colloid Any substance that, when mixed with an appropriate solvent, exhibits properties intermediate between a suspension and a true solution. Small particles remain suspended by random Brownian motion and are often surrounded by a shell of solvent, preventing flocculation. The mix is called a colloidal solution and is important in a number of biological situations, such as carbohydrate storage.

colon The large intestine in vertebrates, located between the *ileum* and the *rectum*. It receives the indigestible residue from food, mostly *cellulose*, and contains beneficial bacteria. In the colon, most of the water from ingested food is absorbed back into the blood, together with *vitamins* (in particular *vitamin K*) manufactured by the bacteria. The leftover waste products are converted into solid masses of feces that are moved by muscular motion into the rectum for excretion. (see *egestion*)

colonial A group of cells or organisms that live in a *colony*. Each individual maintains the capacity for individual growth and reproduction.

colony
1. A group of organisms of the same species, generally dependent on each other to some degree, and containing individuals that may have specialized functions. Colonialism may be found in organisms ranging from bacteria to vertebrates such as birds. Corals are one example of a colonial organism.
2. In microbiology, a visible growth of cells, often *clonal* descendants of one or a few starting cells.

color blindness The imperfect perception of color caused by the malfunction or absence of one of the three *cone* cell pigments in the *retina* of the vertebrate eye. The defect is usually inherited as a *sex-linked recessive gene* located on the *X* *chromosome* and is therefore exhibited mostly in men. The most common form of color blindness is the inability to distinguish between reds and greens.

colostrum The liquid secreted by the *mammary glands* for the first few days after birth, preceding the full-scale secretion of milk. It is rich in protein and *antibodies*, and contains enzymes to clear mucus from the digestive tract of the newborn.

Columbiformes The order of birds containing the pigeons and doves.

columella auris The single *ear ossicle* present in amphibians and primitive reptiles. It is a rod of bone or cartilage that connects the *tympanic membrane* directly with the *inner ear* and transmits sound. Higher vertebrates have evolved two additional ear ossicles, the *incus* and the *malleus*, and the columella auris has become the *stapes*. It is *homologous* with the *hyomandibular* bone of fish. (see *middle ear, ear ossicles*)

columnar epithelium Elongated *epithelial* cells that are often *metabolically* active. They are found in interior linings, including the respiratory tract, testes, and small intestine. (see *villi, microvilli*)

comb jellies (see *Ctenophora*)

commensalism
1. A *symbiotic* association between two organisms in which one benefits and the other remains unaffected. For instance, the barnacles on the back of a whale enjoy mobility and access to fresh food sources while the whale does not appear to either suffer or gain. (see *symbiotroph*; compare *mutualism, parasitism*)
2. A *symbiotic* association between two organisms in which neither species necessarily takes nutrients from the other.

committed A particular *embryonic* tissue

after the point where its developmental fate is fixed and may not be altered. (compare *determined*)

common cardinal vein (see *cardinal vein*)

community (see *ecological community*)

community-level effect An interaction between species in an *ecological community* that results in major changes, including the presence, absence, or abundance of other species in the community.

compact bone The concentric layers of solid bone surrounding the spongy *bone marrow*. A dense deposition of *collagen* provides mechanical strength. (see *Haversion canals, bone*)

companion cell A thin-walled cell in flowering plants (*angiosperms*) that functions to exchange soluble food molecules and waste with the less *metabolically* active, *enucleated sieve elements*. They arise, along with their paired sieve elements, by unequal longitudinal division of a common parental cell and are characterized by dense *cytoplasm* and prominent *nuclei*.

compass plant Any plant with its leaf edges permanently aligned due north and south. Such plants avoid receiving the strong midday rays of the sun directly on the leaf blades but are positioned to use fully the weaker rays of the morning and evening sun from the east and west. The best known example is *Silphium laciniatum.*

compensation point In *photosynthetic* cells, the light intensity at which the rate of *respiration* equals the rate of photosynthesis such that the net exchange of oxygen and carbon dioxide is zero. At normal daylight intensities, the rate of photosynthesis exceeds respiration. Compensation points for most plants occur around dawn and dusk but vary with species. (see *Calvin-Benson cycle, C_4 carbon-fixation, photorespiration*)

compensatory hypertrophy The replacement of a lost or damaged part of an organ by an increase in size of the remainder.

competent bacteria Bacteria that have

been rendered permeable to *exogenous DNA*, usually by exposure to calcium chloride ($CaCl_2$). It is a way of introducing *cloned* DNA into a bacterial *host* for propagation. (see *transformation*)

competition The result of a common demand by two or more organisms for a limited resource. Examples include food, water, minerals, light, mates, and nesting sites. When *intraspecific*, it is a major factor in limiting population size or density; when *interspecific*, one of the competing species may be eliminated.

competitive exclusion The principle stating that two species that have exactly the same requirements cannot coexist in exactly the same *habitat*. One of the species will be able to use a limited resource more efficiently and eventually eliminate the other, at least locally. (see *ecological niche*)

competitive inhibition The reversible inhibition of an enzyme by a molecule structurally similar to the normal *substrate*. The inhibitor molecule fits the *active site* and blocks access to the normal substrate. The equilibrium for the overall reaction is established according to molecular concentrations and affinities.

complement

1. A defensive response in vertebrate systems that is triggered by the *cell walls* of bacteria, fungi or any *antibody*-marked complex. It consists of a group of about 20 proteins, collectively called complement. They assemble into doughnut-shaped channels that pierce the *cell membrane*, causing the cell to take in water and ultimately burst by *osmotic lysis*. Some complement proteins also amplify other body defenses by stimulating *histamine* release and attracting *phagocytes* to the site of infection. (see *complement fixation, immune system, immune response*; compare *killer T-cells*)

2. (see *complementation*)

complementary (see *complementary base pairing, complementation*)

complementary base pairing The chemical attraction between specific *base pairs* in a *nucleic acid* molecule. *Adenine* always pairs with *thymine* by two hydrogen bonds and *guanine* with *cytosine* by three hydrogen bonds. In pairing between *DNA* and *RNA*, the *uracil* of RNA pairs with adenine. Complementary base pairing is responsible for holding *DNA* in the form of a *double helix*, each base pair forming a rung of the ladder. This obligate pairing is what ensures that genetic material is replicated accurately from generation to generation. It is also exploited ubiquitously in DNA research and testing. (see *replication, nucleic acid hybridization, DNA analysis, RFLP, Southern blot, PCR*)

complementary DNA (see *cDNA*)

complementary male Small and usually degenerate males of certain animals (for example, some barnacles and angler fish) that live on or in the body of the female. They are dependent on the female for nutrition and are sometimes reduced to little more than an attached *testis*. Their attachment ensures that *fertilization* occurs.

complementation The restoration of a normal (*wild-type*) *phenotype* when two copies of a *gene* each containing different *mutations* are present, usually on *homologous chromosomes* in a *diploid* organism. Complementation may also occur when a second copy of the gene is introduced into a nominally *haploid* organism under a variety of circumstances. The phenomenon depends on the action of compensatory *gene products* to produce a functional protein *Intragenic* complementation can occasionally occur when the products from the same gene interact to form a multisubunit protein. (see *complementation group, complementation test;* compare *recombination*)

complementation group Different *mutant alleles* (genetic variants) that, when present in a *heterozygous* state, fail to *complement* each other. In other words, the cell or organism continues to show a mutant *phenotype*, thus the mutations reside within the same *gene* or functional genetic unit. (see *complementation, complementation test*)

complementation test (cis-trans test) A test performed to determine whether two or more *mutations* lie in the same *gene* or functional genetic unit. The test is performed by crossing strains or individuals such that *homologous chromosomes*, each bearing a different mutant, are present in the same cell. If the cell or organism is *phenotypically* normal, the mutants are *complementary* and reside in different genes or at least in different protein-coding units. If a *gene* produces a multisubunit protein, *intragenic complementation* may occur. (see *complementation group, complementation*)

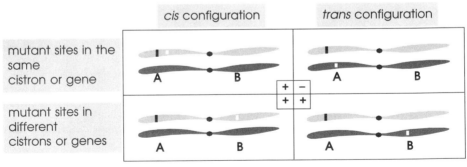

(+) complementation
(-) no complementation

Cis-Trans Test for Complementation

complement fixation An *immune response* in which an *antigen-antibody complex* combines with and "fixes" *complement* proteins, consequently inactivating them. (see *complement fixation test*)

complement fixation test A diagnostic procedure used to determine the presence of antibodies against specific diseases. It is based on the ability of the *antigen-antibody complex* to fix and, consequently, to inactivate *complement* proteins. Briefly, any native complement is removed from the questioned blood sample and *antibodies* or *antiserum* from it are combined with exogenous *antigen* and a known amount of complement. In the following step, an indicator system, usually red blood cells and the corresponding antibodies, is added. If the suspect antibodies are present in the original blood sample, then the complement is fixed and no longer available for *lysis*. If the red blood cells lyse, this indicates that complement is still available, and therefore was not fixed at the first step. Consequently, the suspect antibodies were not present in the blood sample. (see *complement fixation*)

complete metamorphosis (see *endopterygote*)

compound eye The type of eye commonly found in arthropods. It consists of several thousands of units (*ommatidia*). The spots of light focused by the ommatidia form a mosaic image that lacks visual acuity but renders the eye extremely sensitive to movement. (see *eye*)

compound leaf A leaf with a *blade* divided into several distinct *leaflets*.

con A (see *concanavalin A*)

concanavalin A (con A) A *mitogenic* plant *lectin* extracted from pokeweed. It is used in the laboratory to specifically provoke the division and *transformation* of *T-cells* in *culture*. It binds to specific carbohydrate residues on the cell surface, making it useful as a structural probe. It is also used as a *ligand* in *affinity chromatography* of *glycoproteins*.

conceptacle A reproductive cavity that develops on the tips of the *thalli* in some brown algae (for example, bladderwrack).

condensation reaction A chemical reaction in which two molecules are joined together with the concurrent elimination of a water (H_2O) molecule. It occurs, for instance, in the joining of *amino acids* by a *peptide bond* in the formation of a *polypeptide*.

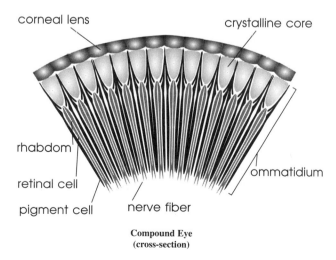

corneal lens

crystalline core

rhabdom

retinal cell

pigment cell

nerve fiber

ommatidium

Compound Eye
(cross-section)

conditional mutation A *DNA* mutation that acts or looks normal (*wild-type*) under certain conditions (*permissive*) and shows a mutant *phenotype* under other (*restrictive*) conditions. (see *temperature-sensitive mutation*)

conditional reflex A reflex action in which the original unconditional stimulus is replaced by a novel conditional stimulus. It is a simple form of *learning* but is not permanent and requires periodic reinforcement. The most famous example is Pavlov's experiment with dogs, in which he was able to evoke salivation just by ringing a bell that they had come to associate with food.

conditioning A form of learning in which external stimuli are used to reinforce behavior. *Classical conditioning* depends on the pairing of two stimuli so that, eventually, the response is produced upon presentation of only the second, novel, stimulus. An example of this is Pavlov's experiment with dogs, in which he was able to evoke salivation just by ringing a bell that they had come to associate with food. In *instrumental conditioning*, reinforcement (or discouragement) occurs only in response to a behavior, without any initial stimulus. A common example is the house-training of pets.

condyle The curved surface at the end of a bone that articulates with the socket of another bone, forming a joint that allows planar movement but no rotation. An example is the condyle at each side of the lower jaw.

cone (see *strobilus*)

cone cell The light sensitive cell type in vertebrate and cephalopod eyes that is responsible for color vision. There are three groups of cones, each type containing a slightly different light sensitive molecule (*opsin*) that is differentially sensitive to red, green, or blue light. The brain interpolates the level of intensity received from each of the three colors, translating the *nerve impulses* into full color vision. Cone cells respond only to a high level of stimulation, explaining why color vision fades in dim light. In the *fovea*, where cone cells are concentrated in some species, each is connected to a *ganglion cell*. This provides an individual nerve impulse for each stimulation and results in high visual acuity. (compare *rods*)

conformational change (see *allosteric*)

congenital A condition present at birth. It generally applies to deformities and does not address the cause.

conidiophore In *ascomycete* fungi, a specialized *hypha* upon which one or more *asexual spores* (*conidia*) are borne.

conidiospore (see *conidium*)

conidium, conidia (pl.) (**conidiospore**) A *multinucleate asexual spore* produced by some ascomycete fungi. They are borne on specialized *hyphae* called *conidiophores*.

Coniferophyta The largest phylum of *gymnosperms*, the conifers, containing the evergreen trees and shrubs. Examples are pine, spruce, yew, and fir as well as the largest living plant, the giant sequoia of California. They generally show a pyramidal growth form, have simple needle-, fan-, or pan-shaped leaves, and bear both large female and smaller male and reproductive structures in cones (*strobili*) on the same plant (*monoecious*). Their *spermatozoa* are immotile and carried to the egg cell by a *pollen tube*. In contrast to *angiosperms*, seeds are not covered by *diploid ovary* tissue but remain embedded in the haploid *megasporangium* from which the *megaspores* are produced.

conifers (see *Coniferophyta*)

conjugated protein A protein to which a nonprotein component (*prosthetic group*) is attached. This may be another organic molecule such as lipid or carbohydrate or an inorganic metal ion.

conjugation A type of *sexual reproduction* employed by some *unicellular* organisms (for example, ciliates, some types of algae, and bacteria) for the purpose of exchanging genetic material. Two non-*gamete* individuals unite temporarily by a tube or bridge and either *gametes*, or more commonly naked *DNA*, may be transferred either unidirectionally or bidirectionally. Details of the process differ greatly between organisms. (see *F-factor*; compare *fertilization, transduction*)

Conjugophyta (see *Gamophyta*)

conjunctiva A layer of transparent tissue covering the *cornea* of the eye and the inner eyelid in vertebrates.

connective tissue A general type of tissue that serves structural, sequestering, and defensive functions. It is derived from *mesoderm*. Connective tissue is typified by living cells embedded in a nonliving matrix that they often secrete, ranging in consistency from fluid *blood plasma* to semicrystalline bone. Defensive connective tissue includes the *macrophages, lymphocytes,* and *mast cells* of the *circulatory system*. Structural connective tissues include cartilage and bone. *Fibroblasts* produce *collagen*, the main structural matrix protein, and are embedded in it. In bone, the collagen fibers are covered with calcium phosphate. Sequestering connective tissue includes the fat cells of *adipose tissue* as well as pigment-containing cells such as red blood cells (*erythrocytes*). (see *immune system, circulatory system*)

connexins A group of proteins that form channels between adjacent cells at *gap junctions*.

consensus sequence The *nucleotide* sequence of a *chromosome* region that tends to be conserved within an organism or particular group of organisms. Regulatory regions such as *promoters* and *enhancers* and specialized regions such as *telomeres* and *centromeres* are examples of elements for which consensus sequences have been obtained.

conservation The adjustment of needs or habits in order to minimize the use of a particular resource, such as energy, land, or water.

conserved protein A protein that is present and has a highly uniform *amino acid* sequence in a particular, often diverse,

group of organisms. (see *conserved sequence, polymorphism*)

conserved sequence A *nucleotide* sequence that is similar or identical in the *genomes* of many, often diverse, organisms. Conserved sequences often code for key functional proteins, whose essential activity would be disrupted by even minor changes. (see *polymorphism*)

consociation A *climax community* that shows characteristics that are constant over time and geographical location. It is usually dominated by one particular plant species. Several consociations together may form an *association*. For example oakwood, beechwood, and ashwood consociations together make up a *deciduous* forest association.

conspecific Any element or idea regarding individuals of the same species.

constant regions In *immune receptor genes*, the "C regions" that remain the same for any particular *gene* (or class of genes). They are incorporated into both *B-receptors* (which are also released as *antibodies*) and *T-receptors*, and provide the means by which the molecules are embedded in the *plasma membrane* of a cell. The T-receptor and B-receptor *light chain* genes contain only one choice of "C region." The B-receptor heavy chain contains 5 choices, each of which determines a different class of released antibody. (compare *variable regions, joining regions, diversity regions*)

constitutive enzyme A type of enzyme that is synthesized regardless of the availability of its *substrate*. Some normally *inducible enzymes* may become constitutive through a *mutation* in a *regulatory* gene.

consumer Any organism that feeds on living or dead organic material. This includes all animals, *parasitic* and *insectivorous* plants, and many bacteria and protoctists. (see *food chain, carnivore, herbivore*; compare *producer*)

consummatory In animals, this term is applied to behavior associated with the achievement of a goal. Examples include, eating as the consummatory act of feeding behavior and copulation as the consummatory act of sexual behavior. (compare *appetitive behavior*)

contact inhibition The phenomenon in which animal cells grown in culture cease to divide wherever they contact another cell, resulting in a confluent layer, one cell thick. It is probably a manifestation of a similar behavior *in vivo* that regulates tissue and organ size. Contact inhibition tends to be lost in *transformed cancer* cells, again mimicking their probable behavior in an organism. (see *cell culture*)

continuous variation (see *quantitative variation*)

contour feathers Bird feathers that give a streamlined shape and provide the flight surfaces of the wings and tail. The longer contour feathers are the flight or *quill* feathers with mostly interlocking *barbs*. The shorter contour feathers have a greater region with separate barbs.

contractile ring A ring of protein filaments that assembles just beneath the *plasma membrane* of animal cells at the end of *mitosis* (*anaphase*) and pinches opposite sides of the *cytoplasm* together in the process of *cytokinesis*, creating two new cells. It is composed largely of *actin* in association with *myosin*.

contractile roots Specialized roots found in certain *bulb-* or *corm*-forming plants that serve to pull the root down deeper in the soil. The large amount of carbohydrate stored in the root *cortex* may be rapidly absorbed by the plant, causing collapse and subsequent contraction of the root. Contractile roots counteract the tendency for each year's new growth to be raised above the growth from the previous year.

contractile vacuole A cell *organelle* found in many protists that functions to remove excess water from the cell (*osmoregulation*). The contractile vacuole expands as it fills with water from the *cytoplasm* and contracts to empty it to the exterior.

control Any sample that is tested in parallel with experimental samples and for which the expected outcome is known. Controls demonstrate that a particular procedure worked correctly and provide a basis of comparison for differences in experimental samples. (see *controlled experiment*)

controlled burning The use of proscribed fire to reduce the risk from wildfires, control tree diseases, increase food and *habitat* for wildlife, and manage forests for the greater production of desirable tree species. Several types of *biome* depend on periodic fires for the health and propagation of *endemic* species.

controlled experiment A type of experiment in which only one *independent variable* is changed at a time in order to determine its effect on a defined *dependent variable*. At least two parallel experiments are run, which are identical except for the factor being tested. Therefore, any change in the outcome must be attributable to the effects of the independent variable. (see *control*)

conus arteriosus The final chamber in the fish heart, located after the single *ventricle*, that distributes *oxygenated* blood to the body.

convergent evolution The development of similar structures in unrelated organisms, presumably as a result of adaption to similar environmental needs and independent of any close genetic association. The development of similarly structured eyes of cephalopods and vertebrates and the development of succulent plants in geographically separated desert environments are examples. (compare *parallel evolution, divergent evolution*)

coots (see *Gruiformes*)

Copepoda A class of minute crustaceans, the copepods, whose members lack a *carapace* and *compound eyes*, and have the first *thoracic* appendages modified for *filter feeding*. The remaining thoracic appendages are used for swimming. Many are important members of the *plankton* and a vital component of the base of the *food web*.

copepods (see *Copepoda*)

copialike elements A family of *mobile genetic elements* found in *Drosophila* (fruit flies). Each is about one *kb* (kilobase) long and is able to mediate its own *transposition* to a distant location in the genome. (see *transposon*; compare *TY element, P element, FB element, LINEs*)

copper (see *trace element*)

copulation The *sexual* union of animals of opposite sexes in which *spermatozoa* are introduced into the body of the female to facilitate *internal fertilization* of *ova* (egg cells).

Coraciiformes The order of birds containing the kingfishers, hornbills, rollers, and bee-eaters.

coral (See *Anthozoa*)

coral reef A complex, highly productive, *symbiotic ecological community* of *colonial* organisms found in tropical and subtropical waters. The reef itself is formed over long periods by the accumulation of carbonates secreted by coral animals. Destruction of even a small portion of coral reef represents the obliteration of dozens or even hundreds of years of accumulation. Like tropical rainforests, the health of the world's coral reef has been recognized as an indicator of the overall state of the environment. (see *Anthozoa*)

core enzyme The smallest functional unit of a particular enzyme. Its activity may be modified by additional protein subunits. (compare *apoenzyme, holoenzyme*)

cork (phellem) In woody plants, a protective layer of dead, impermeable cells formed by the *cork cambium*, which may eventually replace the *epidermis*. Cork is abundant on the trunk of certain trees, for instance the cork oak, from which it is stripped for commercial use.

cork cambium (phellogen) In woody plants, a sub*epidermal* layer of cells that give rise externally to the *cork* and internally to *phelloderm* tissue.

corm In plants, an underground storage structure, providing the means for survival of the plant from one season to the next or for *vegetative propagation*. It is formed from a swollen stem base, bearing *adventitious roots* and scale leaves. Unlike *tubers*, corms lie upright. They are often renewed annually, each new corm forming on top of the preceding one. (see *vegetative propagation, perennation, perennating structure*; compare *bulb, offset, rhizome, runner, stolon, sucker, tuber*)

cormorants (see *Pelecaniformes*)

cornea The firm transparent front part of the outer layer of the vertebrate and cephalopod eye, covering the *iris* and the *pupil*. Its curved surface is responsible for the initial refraction of incident light (coarse focus) which then passes through the *lens*. The cornea is composed of *collagen* and lacks a blood supply, deriving its nutrients via the *aqueous humor* from the *ciliary body*.

cornification (see *keratinization*)

corolla A collective term for all the *petals* of a flower. It is a nonreproductive structure. Often arranged in a whorl, it encloses the reproductive organs and with the *sepals*, when present, protects them. Petals are often brightly colored and attract *pollinating* animals.

corona A series of petal-like structures in a flower, either outgrowths of the petals or modified from the *stamens*. An example is the trumpet of the daffodil flower.

coronary Any structure or function related to the heart.

coronary vessels Either of two pairs of blood vessels present in vertebrates that carry blood to and from the heart. Blockage of these vessels may lead to coronary heart disease.

corpus allatum, corpora allata (pl.) *Glands* near the brain of insects that produce *juvenile hormone*. A gradual decrease in production is responsible for progression through *larval* stages, *pupation*, and final differentiation into the adult stage. (see *metamorphosis*)

corpus callosum A bundle of *nerve fibers* connecting the two *cerebral hemispheres* of the brain in *placental mammals*. It facilitates communication and coordination between the two hemispheres.

corpus luteum An *endocrine gland* formed from a ruptured *Graafian follicle* in the *ovary* after *ovulation*. It is temporary, secreting *progesterone* that maintains the uterine lining until *menstruation* and during pregnancy. (see *chorionic gonadotrophin, luteinizing hormone, menstrual cycle, estrus cycle*)

corpus striatum A portion of the brain involved with complex *motor* activity. It is located deep in each *cerebral hemisphere* and is most highly developed in birds, where it is the site of higher brain functions.

cortex In general, the outer layer of any organ or structure. For example, the surface of the gray matter in the brain is the *cerebral cortex* and the outer layer of the kidney is the *renal cortex*. In the green stem of a plant, it refers to the outer layer of *ground tissue*, situated beneath the *epidermis*.

cortical granules Membrane bound vesicles found in the *cortex* of many animal eggs. The contents are extruded at *fertilization*, changing the *vitelline membrane* into the *fertilization membrane*. These changes serve to prevent multiple *spermatozoa* from penetrating the egg.

corticosteroid Any *steroid hormone* produced by the *adrenal cortex*, including *glucocorticoids* and *mineralocorticoids*. Corticosteroids are important in many body functions, including carbohydrate and protein *metabolism* and maintenance of *osmotic balance*. Synthetic corticosteroids are used in the treatment of adrenal insufficiency as well as inflammatory conditions such as rheumatoid arthritis.

corticosterone A specific *steroid hormone* produced by the *adrenal cortex* that promotes salt retention, thus is antidiuretic.

corticotrophin (ACTH, adrenocorticotrophic hormone) A hormone secreted by the *pituitary gland* that stimulates the secretion of *corticosteroid hormones* from the *adrenal cortex*. Its se-

cretion is stimulated by the *hypothalamus* and by stress, and is modified by the level of circulating *corticosteroids*. (see *feedback inhibition*)

cortisol (see *hydrocortisone*)

cortisone A precursor of the *steroid hormone hydrocortisone*. It is produced in the *adrenal cortex*.

corymb A type of *racemose inflorescence* (multiple-flowering structure) that is flat topped and *indeterminate*.

cosmid A specialized type of *cloning vector* that contains various desired *plasmid* elements that enable it to be propagated in bacteria and those enabling it to be packaged *in vitro* into *bacteriophage* λ. Cosmids allow the cloning of relatively large *genomic DNA* fragments, an important factor when searching for mammalian genes as well as their propagation into *E. coli*.

cosmid library A collection of large fragments of *genomic DNA* contained in a *cosmid cloning vector* and propagated into *E. coli*.

cosmoid scales Spiny scales found in the skin of *cartilaginous* fish. The outer part is formed from a *dentine*like substance (*cosmine*) covered with an enamel-like material (*vitrodentine*). (compare *ganoid scales*)

cosmopolitan species Any species with a broad distribution, occuring all over the world wherever the environment is appropriate. For instance, the moose is found in the northern forests of both North America and Europe.

C_0t curve A graphical depiction of the *renaturation* kinetics of *single-stranded DNA* to *double-stranded DNA*. It stands for the logarithm of the initial concentration (C_0) of DNA strands in solution multiplied by the time (t) renaturation has been allowed to proceed. Because *repeated sequences hybridize* more rapidly, various classes of DNA will introduce variation into the idealized curve. Experiments involving C_0t curves were historically important in determining that *eukaryotic* DNA contains various classes of repeated DNA as well as *unique sequences.*

cotransport The simultaneous transport of two different chemical species across a *cell membrane.* (see *coupled channel*; compare *antiport*)

cotyledon The first leaf that is borne on the *embryo* of seed plants (*gymnosperms* and *angiosperms*). It is usually simpler in structure than those that follow. Cotyledons often act as storage organs in seeds lacking an *endosperm* (for example, legumes) and in others (for example, sunflower) form the first *photosynthetic* organs. *Monocotyledons* and *dicotyledons* are so termed because they respectively develop one or two cotyledons, but exceptions exist.

countercurrent exchange The situation obtained when two liquids with dissolved solutes or gases flow in opposite directions in adjacent semipermeable tubules. Diffusion takes place across the membranes. Instead of stopping when the concentrations have equalized (the situation when two liquids are stagnant or flowing in the same direction at the same rate), the portion carrying the diffused materials moves away and is replaced by new liquid, again at lower concentration. The phenomenon is utilized in both the *gills* of fish and in the vertebrate kidney.

coupled channel A *transmembrane protein channel* that takes advantage of the sodium ion (Na^+) gradient created by the *sodium-potassium pump* in order to cotransport more complex molecules such as sugars and *amino acids* across the *cell membrane.* Sodium ions will reenter the cell through the coupled channel by *facilitated diffusion* but only when accompanied by a sugar or amino acid. The cell is therefore able to concentrate these important *metabolites* against their own concentration gradients. (see *transmembrane ion channel, symport, antiport*)

courtship The specialized pattern of behavior that is preliminary to mating and reproduction.

coxa, coxae (pl.) The first segment of an insect leg, linking the *thorax* with the second leg segment.

coxal gland In some arachnids, a type of excretory organ opening at the base of the legs. (see *Malpighian tubules*)

cpDNA (chloroplast DNA) The *extranuclear DNA* that resides in each *chloroplast* in the *photosynthetic eukaryotes* that contain them. It is a self-replicating circular *chromosome.* cpDNA uses a *genetic code* that contains *prokaryotic* features, supporting the probable origins of chloroplasts as *endosymbionts.* Chloroplast DNA is self-*replicating.* Like the *mitochondrial genome*, it usually exhibits *uniparental inheritance.* (see *maternal inheritance*; compare *nuclear DNA, mtDNA*)

crabs (see *Crustacea, Malacostraca*)

cranes (see *Gruiformes*)

cranial Any structure or function relating to the head.

cranial nerves The paired nerves that originate directly from the brain of vertebrates, bypassing the spinal cord. With the exception of the *vagus nerve*, they supply the sense organs and muscles of the head. The 12 cranial nerves present in mammals are: (I) *olfactory*, (II) *optic*, (III) *oculomotor*, (IV) *trochlear*, (V) *trigeminal*, (VI) *abducens*, (VII) *facial*, (VIII) *vestibulocochlear*, (IX) *glossopharyngeal*, (X) *vagus*, (XI) *accessory*, and (XII) *hypoglossal.*

Craniata (Vertebrata) The chordate phylum containing those organisms in which the *notochrod* has been replaced by *cartilage* or bone, forming a segmented *vertebral column.* Craniates also have a distinct head, containing a brain protected by a skull. This group, which includes all fish, amphibians, reptiles, birds, and mammals, has long been designated as the subphylum Vertebrata. The group of organisms remains the same, but the name has been changed to Craniata and the designation to phylum to represent the distinguishing characteristics more accurately. The names are used interchangeably in this work.

cranium, crania (pl.) The skull of vertebrates.

crassulacean acid metabolism (see *CAM carbon-fixation*)

crayfish (see *Crustacea, Malacostraca*)

creatine phosphate A high-energy compound important in the energy *metabolism* of vertebrate muscle. It helps to maintain a relatively constant level of *ATP* in the muscle during contraction by transferring its phosphate group to *ADP* to form ATP. The energy necessary for the synthesis of creatine phosphate is provided by *glycolysis.*

creatinine A waste product produced by the catabolism of *creatine* in mammals. It is filtered from the blood by the kidneys and disposed of in urine.

creepers (see *Passiformes*)

Crenarchaeota The phylum of archaean prokaryotes (archaebacteria) containing the thermoacidophiles or acid/heat loving bacteria. They live in habitats like the hot springs in Yellowstone National Park, which maintain a temperature of around 60°C and a pH of 1 to 2. At room temperature, these organisms freeze and die immediately. They protect their *DNA* from the extreme conditions with a protein coat, a mechanism similar to that found in *eukaryotes.*

Cretaceous The last of three geological periods of the Mesozoic geological era, about 135 million to 70 million years ago. It is marked by a rapid *extinction* of dinosaurs as well as certain shelled cephalopods (ammonites) and aquatic reptiles toward the end of this period. The first birds were seen during the Cretaceous period. Flowering plants (*angiosperms*) replaced *gymnosperms* as the dominant terrestrial vegetation. The Cretaceous is named for the large amounts of chalk (*fossilized plankton*) found in rocks of the period.

crickets (see *Orthoptera*)

Crinoidea The most primitive class of *echinoderms*, including the sea lilies and feather stars. The mouth is on the upper surface of the body and is surrounded by feathery arms bearing tube feet. Adult forms may be free-living or *sessile*, varying by species, but the *larvae* are always sessile, attached to the substratum by a stalk.

crinoids (see *Crinoidea*)

crista, cristae (pl.) The folds of the inner *mitochondrial* membrane that are the site of *ATP* production. They contain the enzymes and other proteins (for example, *cytochromes*) involved in *aerobic respiration* and the *electron transport chain.*

cRNA Synthetic *RNA* that is *transcribed in vitro* from a *single-stranded DNA template.*

Cro-magnon man A group of early representatives of the species *Homo sapiens* that lived in Europe about 40,000 to 13,000 years ago. They are believed to be direct ancestors of modern man.

crocodiles (see *Crocodilia*)

Crocodilia The order of reptiles containing the crocodiles, alligators, and caimans. Crocodilians are large, tropical, aquatic animals, closely related to dinosaurs. Most have a four-chambered heart, while all other living reptiles have a three-chambered one.

crop

1. A part of the *alimentary canal*, present in such animals as earthworms, insects, and birds, that is modified for the storage and partial digestion of food. Food passes from the *esophagus* into the crop before going a little at a time into the *gizzard*. It is particularly prominent in grain-eating birds.

2. Food plants grown in an organized and controlled fashion.

crop milk A secretion of sloughed off *epithelial* cells from the *crop* in both sexes of pigeons, on which nestlings feed. Its production, like mammalian milk, is influenced by the *hormone prolactin.*

crop rotation The practice of growing different agricultural crops in regular succession to assist in the control of insect pests and disease, increase soil fertility (particularly with *nitrogen-fixing leguminous* plants/bacteria), and decrease erosion. (compare *monoculture*)

cross-fertilization *Fertilization* of *gametes* between two different individuals. It encourages exchange of genetic material, leading to population diversity and evolutionary *fitness*. For this reason, even *hermaphroditic* organisms often cross-fertilize. (see *cross-pollination*; compare *self-fertilization*)

crossing-over The physical exchange of corresponding *chromatid* segments at *chiasmata* between *homologous chromosomes* during *meiosis*. Crossing-over leads to genetic *recombination* and is used to map the physical locations of *genes* or *loci* on chromosomes. *Alleles* (variants) of genes (or loci) that reside closer together on a chromosome will *cross-over* less frequently (*genetic linkage*) than those located farther apart. Genes on opposite ends of a chromosome or on different chromosomes tend to show no genetic linkage, rather they are inherited independently (*independent assortment*). (see *homologous recombination*, *genetic mapping*; compare *gene conversion*)

crown The part of a tooth outside the gum, covered by enamel. (see *teeth*)

cruciferous Plants belonging to the family now known as Brassicaceae, including cabbage, cauliflower, horseradish, and mustard. They are characterized by a group of chemicals known as mustard oils that are toxic to many insects. Along with other compounds present in crucifers, the mustard oils may also play a role in preventing colon cancer prevention.

Crustacea A phylum of *mandibulate arthropods* containing the mostly aquatic, gill-breathing crayfish, crabs, lobsters, barnacles, and water fleas, and the terrestrial woodlice and pillbugs. The body is divided into a head, *thorax*, and abdomen. The head bears *compound eyes* and the mouthparts. The thorax (or sometimes *cephalothorax*) is usually covered with a dorsal *carapace* made up of a *chitinous cuticle* that is often impregnated with calcium carbonate, some of the calcium being reabsorbed from excretory wastes. Most crustaceans have separate sexes, although barnacles are *hermaphroditic*.

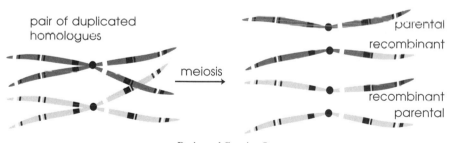

Reciprocal Crossing-Over

Crossopterygii (Coelacanthini) A subclass of bony fish, the coelecanths, containing only one living genus, *Latimeria*. It was thought to be extinct until 1938 when the first live specimen was captured off South Africa.

cross-pollination *Fertilization* of *gametes* (*pollination*) between individuals. (see *cross-fertilization*; compare *self-pollination*)

Copulation may take a number of specialized forms, and the members of some orders carry their eggs with them until they hatch. Crustaceans include many *filter feeders* that are important in marine and freshwater food webs.

crustaceans (see *Crustacea*)

cryo- A word element derived from Greek meaning "icy cold" or "frost." (see *cryobiology*)

cryobiology The study of the effects of very low temperatures on living systems. Some organisms or their parts can be preserved under these conditions and revived with apparently no ill effect.

crypt A pit or depression forming a *gullet* in cryptomonad protoctists.

cryptic coloration A protective coloring that blends in with an animal's usual surroundings, concealing it from predators. Examples are strong contrasting patterns to break the visual outline of the body (for example, angel fish), contershading on the underside of the body to mitigate shadow (for example, caterpillars), and imitation of inanimate objects such as petals, leaves, and twigs (for example, stick bug).

cryptobiosis The dormant state of a *propagule*, such as a *spore*. It is usually a condition in which *metabolism* is suspended due to starvation, desiccation, or extreme temperatures.

Cryptomonada A phylum of protoctists, also known as cryptoprotists or phytoflagllates. They are typically *photosynthetic* and *colonial*, although exceptions exist. Some are *facultative* or *obligate heterotrophs*, and all appear to lack any form of *sexual reproduction*. They have a unique form of *cell division* in which a second *gullet* or *crypt* forms and migrates to the opposite end of the cell. It then simply splits in two (*binary fission*). All are motile and some form colonies of *mucilaginous* sheaths. Many cryptomonads have an unusual feature, a *trichocyst*, that is expelled to sting and capture small protists or bacteria. The reddish color of the *redtide* ciliate *Mesodinium rubrum* is due to hundreds of cryptomonad *endosymbionts*.

cryptophyte (geophyte) A plant in which the resting buds are below the soil surface. (see *Raunkiaer's plant classification*)

Ctenophora A phylum of marine invertebrates, the sea walnuts and comb jellies. The transparent, often globular, saclike body bears eight rows of fused *cilia* used in locomotion. The *archenteron* forms a canal system in the body. Most have *tentacles* armed with adhesive cells for food capture on each side of the body, and many are *bioluminescent*. At one time, they were thought to be closely related to *cnidarians*, but this view has changed.

Cubozoa A class of cnidarians containing the box-jellies and sea wasps.

cuckoos (see *Cuculiformes*.)

Cuculiformes An order of birds including the cuckoos and others.

cud The contents of the *rumen* in animals such as cows that is regurgitated and rechewed. This leads to more efficient digestion of the *cellulose* in their food by the *cellulase* produced by bacteria in the *cecum*. (see *ruminant*)

cultivar Any agricultural or horticultural variety of a plant species. The term is derived from the words *cul*tivated *var*iety

culture A population of microorganisms or dissociated tissue cells that are propagated in liquid or solid medium *in vitro*.

culture medium A nutrient solution, either in liquid form or solidified with *agar*, used in the *in vitro* cultivation of microorganisms such as bacteria or fungi or the dissociated tissue cells from a *multicellular* organism.

cumulus cells Cells derived from the *Graafian follicle* that surround the *ovulated* mammalian egg. They disperse quickly if *spermatozoa* are present, more slowly if they are not.

cupula, cupulae (pl.) Part of the motion detection system in the terrestrial vertebrate *inner ear*. It is a gelatinous material, located at the end of each *ampulla*, into which the *cilia* of *sensory hair cells* extend. Each ampulla protrudes into a semicircular canal. When the cupula is bent by the movement of the liquid, the cilia are bent, and their attached *nerve cells* fire. Complex movements are analyzed and interpreted in the brain by comparing the sensory input from each canal.

curassows (see *Galliformes*)

cusp The conical point on the crowns of *molar* and *premolar* teeth of *carnivorous* and *omnivorous* mammals, such as dog and man. Small premolars have only two cusps, larger molars have three or four. They are used for crushing, cutting, and chewing.

cuticle Any protective covering secreted by the outer layer of cells (*epidermis*). In plants, the cuticle is a waterproof layer of waxy *cutin*, found mainly in aerial portions and some seeds. In animals, protective (though not waterproof) cuticles are found in *endoparasites* such as tapeworms and flukes and the thin *collagen* covering of earthworms. A *chitinous*, waterproof cuticle is found in arthropods and is impregnated with calcium in the shells of marine invertebrates such as crustaceans and mollusks. (see *carapace*)

cutin A group of substances chemically related to *fatty acids* that form a continuous *cuticle* layer on the *epidermis* of plants. Cutin is waterproof, reducing water loss (*transpiration*), and also serves as a protective barrier, preventing invasion by *parasites*. (see *suberin*)

cutis (see *dermis*)

cutting Part of a plant that is removed from the parent and propagated *asexually*. (see *grafting, vegetative propagation*)

Cuvierian duct (see *cardinal vein*)

cyanelle An *intracellular* cyanobacterial symbiont or similarly derived *organelle* that carries out *photosynthesis*. They are found in some protoctists.

Cyanobacteria A phylum of *Gram-negative* bacteria characterized by the possession of a blue-green or cyan *photosynthetic pigment*. They are *photosynthetic aerobes* and found abundantly in a variety of habitats, particularly in freshwater and soil. Cyanobacteria are responsible for generating a large portion of the free oxygen (O_2) in the earth's atmosphere. They produced the accumulation of the massive limestone deposits known as *stromatolites* as well as the bulk of modern petroleum deposits. Some are *nitrogen fixing*, and others are partners in *lichen* associations. They are nutritionally the most independent organisms that exist. The Cyanobacteria have previously been classified as blue-green algae or *Cyanophyta* but have little in common with other algae beyond their photosynthetic pigments. Chloroxybacteria (Prochlorophyta), once thought to comprise a phylum of their own, have recently been shown to comprise a subphylum of cyanobacteria. They lack the characteristic blue-green pigment of other cyanobacteria but do contain the standard green chlorophyll found in most plants. Cyanobacteria are thought to be the evolutionary ancestors of *chloroplasts*. (see *heterocyst*)

cyanocobalamin (vitamin B_{12}) A member of the *vitamin B complex*, it has a complex organic ring structure with a single cobalt atom at the center. Foods of animal origin and some yeasts are the only major dietary sources. In man, a deficiency leads to pernicious anemia because cyanocobalamin is required to develop red blood cells.

Cycadophyta A phylum of seed plants, the cycads. They are palmlike *gymnosperms* with large divided leaves and terminal *cones*. Their *spermatozoa* are *undulipodiated* but carried to the vicinity of the egg by a *pollen tube*. Most cycads are *dioecious*.

cycads (see *Cycadophyta*)

cyclic AMP (see *cAMP*)

cyclic photophosphorylation A simple cyclic process resulting in the conversion of *ADP* to *ATP* using light energy. It stands alone in some primitive *photosynthetic* bacteria. It is an alternative pathway to the more complex systems in other photosynthetic bacteria and in the *chloroplasts* of plants and algae. In cyclic photophosphorylation, ATP is generated without the concomitant generation of *NADPH*, thus no *reducing power* is made available for *carbon fixation*. (see *photosystem I*)

cycloheximide An *antibiotic* that inhibits the *translation* of *eukaryotic nuclear mRNA* on *cytoplasmic ribosomes* but permits the *protein synthesis* on *prokaryotic* ribosomes, including those found in *organelles* such as *mitochondria* and *chloroplasts*. It is used as a laboratory tool to differentiate proteins synthesized by mitochondrial or chloroplast ribosomes and those manufactured by *nuclear genes*. (compare *chloramphenicol*)

cyclosis (see *cytoplasmic streaming*)

Cyclostomata (Agnatha) The class of craniates (vertebrates) containing the earliest and most primitive vertebrates, the jawless fish. The only living orders include the lampreys and hagfish, aquatic vertebrates lacking the paired fins typical of true fish.

cyclostomes (see *Cyclostomata*)

cyme (see *cymose inflorescence*)

cymose inflorescence (cyme, definite inflorescence) An *inflorescence* (multiple flowering structure) in which apical (top) growth is terminated by the formation of a flower at the top of the plant. Subsequent growth continues from lateral buds below the apex, which themselves form flowers and more lateral shoots. (see *sympodial, monochasial cyme, dichasial cyme;* compare *racemose inflorescence*)

cypsela The small, dry fruit characteristic of composite flowers (for example, daisy, sunflower). It is similar to an *achene* but formed from a bicarpellary (two *carpels*) inferior *ovary* in which only one of the *ovules* develops to maturity.

cyst A protective outer covering formed by some microorganisms, often in response to extreme environmental conditions, and enclosing a *dormant life cycle* stage. In particular, it is one mechanism an *endoparasite* uses to protect itself from the *digestive* and/or *immune systems* of its *host*.

-cyst- A word element derived from Greek meaning "bladder" or "sac." (for example, *cysticercus*)

cysteine (cys) An *amino acid*.

cysteine linkage (see *disulfide bond*)

cysticercus (see *bladderworm*)

cystine The chemical compound formed by the linkage of two *cysteine amino acids* through a *disulfide bond* (-S-S-).

Disulfide bonds are important in the structure of certain cysteine-rich proteins, for instance *antibodies*.

cystolith A deposit of calcium carbonate found on a stalk arising internally from the *cell walls* in some flowering plants (*angiosperms*). The stalk is composed of large modified *epidermal* cells. (compare *statolith*)

cytidine The *nucleoside* formed when *cytosine* is linked to D-ribose via a β-glycosidic bond. Cytidine triphosphate (*CTP*) is derived from cytidine.

cyto-, -cyte A word element derived from Greek meaning "cell." (see *leukocyte*)

cytochalasin B A fungal-derived *alkaloid* that interferes with the assembly of *actin* filaments by capping one of the ends so that further growth can take place. It disrupts most aspects of cell motility.

cytochrome Any of several *conjugated proteins* containing *heme* that act as intermediates in the *electron-transport chain* during *aerobic respiration*.

cytochrome *bf* complex A protein complex in green plants and algae that links *photosystem II* to *photosystem I*. It catalyzes the transfer of electrons from *plastoquinol* (QH_2) to *plastocyanin* (*PC*).

cytogenetics The area of study that links the structure and behavior of *chromosomes* with observed genetic phenomena.

cytohet (heteroplasmon) A cell containing a mixture of genetically different *cytoplasms*, including cell *organelles*.

cytokine Any substance that promotes cell growth and division. *Interleukins* are an example.

cytokinesis The division of *cytoplasm* after *nuclear division* in *mitosis* or *meiosis II*. In animal cells, cytokinesis results from an invagination of the *cell membrane* and subsequent division of the cytoplasm between daughter cells. In plant cells, it involves formation of a new *cell wall* that separates the daughter *nuclei*. (see *contractile ring, cell plate, middle lamella*; compare *nuclear division*)

cytokinins A class of plant *growth substances* that stimulate *cell division* and determine the course of *differentiation*. Cytokinins are produced in the roots and distributed to the rest of the plant body. They work synergistically with *auxins*. In contrast, cytokinins stimulate the growth of lateral branches. Naturally occurring cytokinins are derivatives of the *purine base adenine*.

cytology The study of cell structure.

cytolosome Digestive *organelles* in some protists that break down the cell's own organelles in times of starvation.

cytolysis The destruction of cells, usually by the dissolution of their *plasma membranes*.

cytopharynx The throat in *protists* that have a fixed mouth (*cytostome*), such as *Paramecium*.

cytoplasm The living contents of a cell, excluding the *nucleus* and other cellular *organelles*. It lies within the *plasma membrane* and contains all the dissolved and suspended ions and other molecules necessary for life.

cytoplasmic inheritance The determination of heritable characteristics by genetic material residing outside the *nucleus*. This includes the *DNA* contained in *organelles* such as *mitochondria* and *chloroplasts* as well as *plasmids*. Traits determined by *extranuclear genes* are not inherited according to *Mendelian laws* but generally follow a *uniparental* or *maternal inheritance* pattern. In animals, this stems from the fact that only female *gametes* tend to transfer an appreciable amount of *cytoplasm* to the *zygote*. (see *episome*)

cytoplasmic streaming (cyclosis) The movement of *cytoplasm* within the cell. It is a process requiring expenditure of energy by the cell and involves *microfilament* and/or *microtubule* activity. It is particularly noticeable in plasmodial slime molds, plant *sieve tube elements*, and some fungi.

cytoproct In *ciliate* protists such as *Paramecium*, a special region in the outer *pellicle* into which waste is emptied from internal *vacuoles*. It appears periodically when solid particles are ready to be expelled. (compare *cytostome*)

cytosine (C) A *nitrogenous* base found in *nucleic acids*. It is based on a *pyrimidine* ring structure.

cytoskeleton A supporting matrix of protein fibers found in all *eukaryotic* cells. It maintains cell shape and acts as a scaffold on which *organelles*, enzymes, and other *macromolecules* are attached. It is composed mostly of *actin filaments, microtubules*, and *intermediate filaments*. Actin filaments and microtubules are also involved in cell movement.

cytosol The soluble fraction of *cytoplasm* remaining after all particulates have been removed.

cytostome The fixed site in some *ciliate* protists at which food is ingested. It can be complex, as in *Paramecium* where it is lined with *cilia*. From the cytostome, food passes into food *vacuoles* where it is digested. (compare *cytoproct*)

cytotoxic Any substance that is poisonous or lethal to cells.

cytotoxic T-cell (see *T-cell*)

2,4-D A synthetic *auxin* used as a potent, selective weed killer. *Monocotyledenous* species (for example, cereals and grasses) are generally resistant to 2,4-D, while *dicotyledonous* plants are often susceptible, making the compound an effective weed killer in crops and lawns.

daddy longlegs (see *Opiliones*)

daltonism (see *color blindness*)

damselflies (see *Odonata*)

dance of the bees The language of honey bees, communicated by a precise pattern of movements. Performed on the vertical surface of the comb, the "waggle" dance indicates the direction of a food source by transposing the visual angle between the food source and the nest in reference to the sun, to the same angle given by the straight run of the dance relative to gravity on the vertical surface of the comb. Distance is given by the tempo or degree of vigor of the dance. Much of this information was recently elucidated by building a computerized robot bee and performing actual tests with it in the hive.

dark reactions (see *Calvin-Benson cycle*, C_4 *carbon fixation*)

Darwin Charles Darwin, in the 1850s, was the first to propose the theory of *natural selection* as a mechanism for the evolution of a species. His ideas were influenced by the observation of isolated populations that had evolved on the Galapagos Islands, in particular the finches. (see *Darwin's finches*, *Darwinism*)

Darwinian fitness (see *fitness*)

Darwinism The theory proposed by Charles Darwin to explain evolutionary change. It is often summarized as "survival of the fittest," as applied to populations over time. Briefly, it suggests that in any varied population of organisms, those possessing traits best adapted to the environment will tend to survive long enough to reproduce and make a genetic contribution to future generations. (see *Darwin*, *Darwin's finches*, *natural selection*)

Darwin's finches The thirteen species of finches first described by Darwin and unique to the Galapagos Islands. They are related to the finches on the South American continent from which they evolved but have differentiated to fill many of the different *ecological niches* on the islands that were apparently vacant when they arrived. Of particular interest is the breadth of *adaptive radiation* they display, particularly with regard to beak type and the number of different ecological niches that they occupy. For instance, the woodpecker finch occupies a niche that in other parts of the world would be occupied by an entirely different species, the woodpecker. (see *Darwin*, *Darwinisim*, *natural selection*)

Dasyuromorphia The order of marsupials comprising the marsupial carnivores.

daughter cell One of the new cells resulting from various types of *cell division*, such as *mitosis*, *binary fission*, or *meiosis*. They may variously be called offspring or progeny cells.

daughter nucleus One of the new *nuclei* resulting from *nuclear division*, either with or without *cytokinesis* and *cell division*.

day-neutral plant A plant in which flowering is independent of the *photoperiod* (relative periods of dark and light). (compare *long-day plants*, *short-day plants*)

deamination The enzymatic removal of an amino group (NH_2) from an organic compound, frequently an *amino acid*. In mammals, it occurs chiefly in the liver as part of the normal metabolic processes to form *urea* which is excreted. If it occurs at particular *bases* in *DNA*, it can ultimately result in *mutation*.

decapod An informal designation for the group of crustaceans containing the most specialized (and commercialized) members of the class *Malacostraca*: the prawns, shrimps, lobsters, crabs, and crayfish. All except the crabs have a long abdomen ending in a tail used for swimming backwards. The head and *thorax* are characteristically fused into a *cephalothorax* and covered with a *carapace*. Of the five pairs of walking legs, the first and second pair often have pincers (*chelae*) used in feeding and defense.

decarboxylase An enzyme that catalyzes the decarboxylation (removal of CO_2) of organic compounds, often converting *amino acids* to amines.

deciduous Plants, generally woody trees and shrubs that seasonally shed all their leaves, for example before the cold or dry season. In addition to cold survival, it is an adaption to conserve water loss by *transpiration* in regions with severe seasonal drought. (compare *evergreen*)

deciduous teeth (milk teeth) The temporary first set of teeth in a mammal. They fall out and are replaced by a permanent set that are larger and contain additional molars absent in the original set.

decomposer Organisms (mainly bacteria, fungi, and some protists) that feed on dead organic material. They are essential to the *food web* by virtue of their ability to break cells down into simple components and recycle the nutrients making them available again to primary *producer* organisms (typically green plants, algae, and *photosynthetic* bacteria). (see *detrivore*; compare *consumer*)

deer (see *Ruminantia*)

deficiency disease Any disease caused by deficiency of a particular essential nutrient (for example, vitamins or minerals), usually presenting a characteristic set of symptoms. Examples include *Chlorosis* in plants caused by lack of iron and magnesium, and scurvy in humans due to lack of *vitamin C*. (see *trace element, vitamin)*

definite inflorescence (see *cymose inflorescence*)

deforestation The destruction of natural forested regions by the over-harvesting of trees for commercial and other uses, and the burning of forests to convert the land for agricultural purposes. (see *clear-cutting, second-growth, global warming*)

degeneration The *reduction* or loss of all or part of an organ during the course of *evolution,* with the result that it becomes *vestigial.* (compare *atrophy*)

deglutition The process of swallowing.

degradation The breakdown of *macromolecules* into smaller and smaller units. It may occur by the action of *hydrolytic* enzymes, *lipases* (fat), *nucleases* (*DNA*), or *proteases* (protein) inside the cell. Outside the body, it may occur from exposure to various environmental factors such as temperature, sunlight, or chemicals. Degradation is of particular concern when attempting to isolate and analyze biological molecules. (see *DNA analysis, Southern blot, RFLP, PCR*)

dehiscent A fruit or *fruiting body* that opens at maturity to release the seeds or *spores*. The fruit often bursts or pops to aid in seed dispersal, encouraged by desiccation or *programmed cell death.* (compare *indehiscent*)

dehydration The elimination of all traces of water or moisture. It is often used in the preservation and analysis of biological specimens.

dehydrogenase Any enzyme that catalyzes the removal of hydrogen atoms from organic molecules. Often, dehydrogenases are involved in *redox* reactions, where they act as carriers to transfer hydrogen atoms from one *substrate* to another. Examples include *coenzymes* such as *NAD* or *FAD* in the respiratory chain. Specific dehydrogenases are named after their substrate, for instance *alcohol dehydrogenase (ADH).*

Deinococci A phylum of *Gram-positive aerobic* bacteria that possess the unusual property of being particularly resistant to damage from radiation. It comprises the former "radioresistant micrococci." The reclassification has occurred based on DNA *homologies* in the small (16S) *ribosomal RNA* (*rRNA*) subunit.

deletion A *chromosomal mutation* where a segment of *DNA* is missing. When deletions occur in *expressed genes*, they are usually deleterious, as relatively large chunks of genetic information are lost. They also tend to produce at least partial sterility, as *homologous* chromosomes fail to pair correctly at *meiosis* They may occur by various *recombinational* mechanisms including *unequal crossing-over* of *homologous chromatids* during *meiosis*, and the breakage and unequal rejoining of chromosomes. (see *ionizing radiation*)

demography The statistical study of populations, especially their patterns in space and time.

denaturation The dissolution of the *secondary* and *tertiary structure* of proteins or *nucleic acids*. These changes may occur as a result of change in pH, temperature, exposure to air, or mechanical manipulation. Usually, the ionic interactions that keep organic molecules dissolved in an aqueous solution are destroyed, resulting in their precipitation out of solution. An example is the heating of the *albumin* protein in egg white during cooking, which results in its solidification. Another important example is the separation of the normally *double-stranded DNA double helix* into two separate strands (*single-stranded DNA*).

denaturing gel electrophoresis An *electrophoresis* system containing agents to *denature* the *macromolecules* being separated in order to examine characteristics conferred only by their *primary structure*. Detergents (for example, sodium dodecyl sulfate, also known as SDS) are common denaturants for proteins, while formaldehyde is often used for *RNA* and urea for *DNA*.

dendrite One of several branching projections that arise from the *cell body* of a *neuron* to make *synaptic* connections with other nerve cells. (compare *axon*)

dendrochronology A method of dating using the *annual rings* of trees, employed when the life spans of living and *fossil* trees in an area overlap. Although not absolutely precise because of climactic variation from year to year, the method may be more accurate than *radioactive*

dating techniques under certain circumstances. Bristlecone pines, which can live for up to 5,000 years, have been used in such work.

denitrification The process in which fixed nitrogen is converted to nitrogen gas (N_2) and nitrous oxide gas (N_2O), and returned to the atmosphere. It is carried out by several genera of soil bacteria under *anaerobic* conditions, and is an essential part of the *nitrogen cycle*. Historically, attempts to apply highly concentrated nitrogen fertilizers to crops have been frustrated by this process. (compare *nitrogen fixation, nitrification*)

dense bodies In *smooth muscle*, protein structures to which *myosin* molecules are sometimes attached. They are the functional equivalent of *Z-lines* in *striated muscle*.

density dependent population effects Processes regulating the growth of a population that are dependent on the number of organisms per unit area. They include *biotic potential* and *carrying capacity*. They come into play particularly as the population size increases. For instance during a famine, the mortality rate increases and the food supply can be said to have a density dependent population effect. (compare *density independent effects*)

density gradient centrifugation A laboratory procedure whereby cell *organelles* and *macromolecules* can be separated by high-speed centrifugation in an *ultracentrifuge*. The separation is based on their individual buoyant densities. Common materials for forming a gradient of varying density include sucrose and cesium chloride (CsCl). The *organelle* or molecule of interest will come to equilibrium at a characteristic solution density that matches its own. Until recently, density gradient centrifugation using CsCl and *ethidium bromide* was commonly used to isolate circular *plasmids* for use in *genetic engineering*.

density independent population effects Processes regulating the growth of a population that are independent of the number or density of organisms. They include climactic factors and physical disruption of the habitat by either natural or artificial phenomena. (compare *density dependent effects*)

denticle (placoid scale) A fish scale resembling a tooth. It is composed of *dentine* covered with *enamel* and surrounding a pulp cavity. They are found primarily in *cartilaginous* fish (for example, dogfish, shark) both on the skin and as teeth in the mouth. Denticles are *homologous* to vertebrate teeth.

dentine The main constituent of teeth, found between the *enamel* and *pulp cavity*. It is secreted by cells (*odontoblasts*) and consists mainly of calcium phosphate in a fibrous matrix. It is similar to bone but higher in inorganic material. Ivory is made of dentine.

dentition The number, type, arrangement, and physiology of teeth in any given species.

deoxycorticosterone A *steroid* hormone produced by the adrenal *cortex* having *mineralocorticoid* activity. (see *adrenal glands*)

deoxygenated A term usually referring to blood returning to the heart from distant regions of the body and therefore lacking in oxygen. (compare *oxygenated*)

deoxyribonuclease (see *DNase*)

deoxyribonucleic acid (see *DNA*)

deoxyribose The sugar found in *DNA*. It is the *monosaccharide, ribose*, in which the hydroxyl group (OH) is replaced with hydrogen (H).

deoxy sugar A sugar in which a hydroxyl group (OH) is replaced with hydrogen (H). An important example is *deoxyribose*, the sugar component of *DNA*.

dependent variable In scientific experimentation, the factor in question. It may or may not be influenced by the *independent variable*, the one being changed. (see *scientific method*)

depolarization The reduction in electric potential across a *nerve cell* membrane as a result of a stimulus. During depolariza-

tion, the membrane becomes more permeable to ions that have been actively sequestered on either side of the cell membrane. Sodium ions (Na^+) that have been excluded from the cell during rest begin to flow into the cell, and potassium ions (K^+) that have been concentrated in the interior flow out. Depolarization may initiate an *action potential* that travels along a *nerve fiber*, propagating a *nerve impulse*. (see *sodium-potassium pump, synapse, patch clamp, voltage-gated ion channel*)

deposit feeding The eating of material that has settled on the bottom of a body of water.

depurination The removal or loss of a *purine base* (*adenine* or *guanine*) from *DNA*. Depending on how this lesion is repaired, it may ultimately result in a *mutation*.

derm-, derma-, -dermis A word element derived from the Greek meaning "skin." (for example, *dermis*)

dermal bone (see *membrane bone*)

dermal tissue In plants, the outer protective covering of the plant that is often covered with a waxy *cuticle*. It is one of the three major types of tissue. (compare *ground tissue, vascular tissue*)

Dermaptera A small order of *exopterygote nocturnal* insects that includes the earwigs. The body is covered with an *exoskeleton,* and the head contains biting and sucking mouthparts, appropriate for their omnivorous diet. The forcepslike *cerci* at the end of the abdomen aid in a complicated folding of the wings, as well as being used in attack and defense.

dermatome The part of the *mesodermal somites* of vertebrate *embryos* that underlies the *epidermis* and develops into *dermis*. (see *mesoderm*)

dermis (cutis) The inner of two layers of vertebrate skin, found beneath the *epidermis*. It is thicker than the epidermis, and consists of *connective tissue* (mostly *collagen*) in which are embedded small blood vessels, nerve endings, and *sweat glands* (mammals only). The mammalian dermis also contains *hair follicles* associated with *sebaceous glands*.

Dermoptera An order of placental mammals with only one living family, the flying lemurs.

desert A *biome* defined by minimal precipitation, usually less than 50 centimeters per year. Water availability is therefore the predominant controlling variable for most biological processes. Additionally, since vegetation is sparse and skies usually clear, heat is rapidly radiated at night, leading to daily extremes in temperature. Both desert plants and animals have evolved specific strategies to deal with this type of environment. (see *annual, succulent, CAM carbon fixation, estivation, biome*)

desmin A protein involved in the formation of *intermediate filaments*, particularly in muscle cells and at *desmosomes*.

desmosome One kind of *intercellular* junction, found between *epithelial* and smooth muscle cells. Desmosomes are mechanically strong, providing a means of attachment and a mechanism for disbursement of shear forces throughout tissues. *Actin* and/or *keratin* may be found in these types of junctions as well as *desmin*. (see *intercellular junction*)

deterivore A general term for organisms that live on the refuse of an *ecosystem*, including dead animals and the castoff parts of animals. They include crabs, vultures, and jackals as well as the *decomposers*, which tend to be microorganismal. (compare *consumer*)

determinant A spatially localized molecule that causes cells to adopt a particular fate or set of related fates during *embryonic* development. (see *determined*)

determinate growth In plants, growth that stops at a certain point. A good comparative example is found in tomato plants; the tall, leggy varieties that continue to grow are *indeterminate*, the short, bushtype are determinate.

determined A particular *embryonic* tissue after the point where its developmental fate may be predicted. (see

determinant; compare *committed*)

detritus Loose natural material that results from the direct disintegration of rocks or organisms, often a mixture of the two.

Deuteromycota (fungi imperfecti) A phylum of fungi, the deuteromycetes, classified solely on the basis that *sexual* reproductive structures are not known, either due to evolutionary loss or lack of discovery. The majority appear to be ascomycetes, along with some basidiomycetes and zygomycetes that have lost the ability to form sexual structures such as *asci* or *basidia*. They produce *asexual spores* (*conidia*), often in great profusion. Some are clinically and economically important, such as the genus that includes *Penicillium* (makes penicillin) and the species responsible for roquefort and camembert cheese. Others are responsible for skin diseases such as ringworm, athlete's foot, and jock-strap itch. Some researchers consider deuteromycetes to be a subdivision of ascomycete fungi. (see *parasexuality*)

deuterostomes One of the two major groups of *coelomate* animals. It is not a formal taxonomic classification. They show a distinct pattern of *embryological* development in which the anus forms from or near the *blastopore* and the mouth forms subsequently on another part of the *blastula*. Additionally, the embryo shows a pattern of *radial cleavage*. Each cell shows a relatively late *commitment* to any particular developmental pathway. The group includes chordates, echinoderms, and a few other small, related phyla. (compare *protostomes*)

deutoplasm The yolk or yolk laden *cytoplasm* at the *vegetal pole* of many eggs.

development The process whereby a single cell becomes a differentiated organism.

Devonian The geological period lasting from approximately 400 million to 350 million years ago. It is named for Old Red Sandstone deposits found in Devon, England and noted particularly for the variety of *fossil* fish. During the late

Devonian, primitive amphibians were evolving, *vascular* land plants appeared, and terrestrial fauna included insects and spiders.

dextran The storage *polysaccharides* found in yeasts and bacteria. It consists of *glucose monomers* linked into branched molecules.

dextrin Any of a class of *polysaccharides* formed as intermediate products in the enzymatic *hydrolysis* of *starch* by *amlyases.*

dextrose The most common form of *glucose.*

di- A word element derived from Greek meaning "two," "twice," or "double." (see *dicentric*)

dia- A word element derived from Greek meaning "passing through" (for example, *diabetes*) or "opposite movement" (for example, *diakinesis*).

diabetes

1. (diabetes mellitus) A disease caused by a deficiency of the *hormone insulin* (secreted by the pancreas) or a deficiency in insulin *receptors.* Insulin works with the hormone *glucagon* to govern levels of sugar in the blood; insulin acts to remove glucose from the bloodstream for storage as *glycogen* in the liver; glucagon releases glycogen when energy is needed. A lack of insulin activity results in the inability to store *glucose,* leading to abnormally high blood sugar levels, excretion in urine, and other symptoms. Some forms of diabetes can be fatal if left untreated but can be controlled by administering insulin. Insulin was one of the first medically important substances to be produced cheaply and in large quantities by *genetic engineering.* (see *islets of Langerhans, endocrine gland*)

2. (diabetes insipidus) A disease producing similar symptoms to *diabetes mellitus* but caused by either a failure of the *pituitary gland* to produce or secrete *antidiuretic hormone* (ADH, *vasopressin*), or a failure of the kidney receptors to respond to it.

diadelphous Flowers in which the *stamens* are formed into two separate groups around the *style.* Sometimes one stamen will occur singly while the rest are grouped. (compare *monadelphous, polyadelphous*)

diagravitropism A *gravitropic* response in which the direction of plant growth is at right angles to the force of gravity, generally horizontal. It is exhibited by the *rhizomes* of many plants. (see *tropism*)

diakinesis The last stage of *prophase* in *meiosis I.* The duplicated, *homologous chromosomes* appear to be maximally condensed. They move apart, and any *chiasmata* (*cross-overs*) that have formed move toward the ends of the chromosomes until only the ends of the *bivalents* are connected. By the end of diakinesis, the *nuclear* membrane and *nucleoli* have disappeared. (see *sister chromatids, synapsis, synaptonemal complex*)

dialysis A technique for separating molecules by size using selective diffusion through a *semipermeable membrane.* Typically, the complex solution is placed inside a dialysis bag that is suspended in a liquid *buffer* of a lower *osmotic pressure* (fewer ions). As the osmotic pressure tends to equalize, molecules diffuse out, limited by the pore size of the dialysis membrane. At the end of the process, only molecules larger than the selected pore size are contained in the bag. Dialysis is performed naturally by the kidneys to extract wastes from the blood.

diapause A hormonally induced period of *dormancy* in the *life cycle* of some insects, during which growth and development cease and *metabolism* decreases. It is often seasonal, analogous to *hibernation*, and enables the insect to survive unfavorable environmental conditions.

diaphragm The muscular partition that divides the chest from the abdomen and functions in respiration. During inhalation, it moves downward, increasing the volume of the chest cavity and causing air to be drawn into the lungs. During exhalation, it moves upward, decreasing the volume and expelling air. Contractions of the diaphragm are basically *involuntary*. (see *autonomic nervous system, respiration*)

diaphysis (see *epiphysis*)

diastase (see *amylase*)

diastole The phase of the heartbeat when the *ventricles* are relaxed and arterial pressure drops slightly while the *atria* are filling and contracting. The measurement of diastolic pressure is reflected in the bottom, lower number of a blood pressure measurement. (compare *systole*)

diatom (see *Bacillariophyta)*

dicentric chromosome An abnormal situation in which a *chromosome* contains two *centromeres*. Such chromosomes often break during *mitosis or meiosis* if the two centromeres get pulled to opposite *poles*. (see *chromosomal mutation)*

dichasial cyme A *cymose inflorescence* (multiple flowered structure) in which each branch gives rise to two or more branches bearing flowers. (compare *monochasial cyme*)

dichogamous The condition in some flowering plants (*angiosperms*) in which the *anthers* and *stigmas* mature at different times, thus helping prevent *self-pollination*. (see *protandry, protogyny*; *self-incompatibility*; compare *homogamous*)

dichotomous Any branching or bifurcation, for instance the division of a growing point into two equal points such as in leaf veins.

dicot (see *Dicotyledons)*

Dicotyledons A class of *angiosperms* (flowering plants) characterized by having two *cotyledons* (first leaves) in the seed. Dicots generally have stem *vascu-*

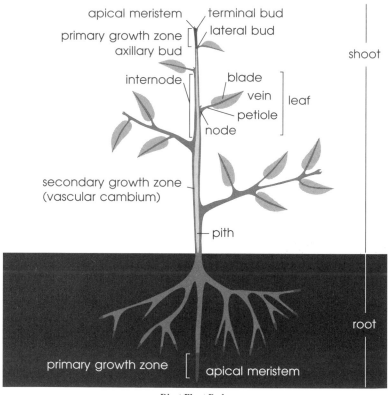

Dicot Plant Body
(see *Dicotyledons*)

lar tissue in the form of a ring of open bundles, flower parts in multiples of four or five, and branched leaf veins. *Pollen* is usually *tricopate* (three furrows or pores), and true secondary growth from a *vascular cambium* is common. There are almost 200,000 dicotyledonous species. (compare *Monocotyledon*)

Dictyoptera (Blattoidea) An order of *exopterygote* insects containing the cockroaches. They are *nocturnal* and *omnivorous*. They are considered pests and may spread certain diseases. They have a flattened body enabling them to hide in crevices, and hardened forewings to protect the larger, delicate hindwings although they seldom fly. Cockroaches are one of the most pervasive and successful organisms on earth.

dictyosome A stack of membrane bound sacs (*cisternae*) that, together with associated vesicles, form the *Golgi apparatus*. The term is usually only applied to plant cells where many such stacks are found. In contrast, the Golgi apparatus of most animal cells is a continuous network of membranes.

dictyostele (see *solenostele*)

Didelphimorphia An order of marsupials comprising the American marsupials, the opossums.

dienchephalon The part of the *forebrain* in higher vertebrates that integrates *sensory* information. It contains the *thalamus*.

differentiation The process during which cells with generalized functions become morphologically and functionally specialized. Differentiated cells make up the different organs and tissues of the body, embodying the concept of division of labor in a *multicellular* organism. This may include the loss of some abilities (including the ability to divide) and the gain of others. All cells (with the few exceptions of those that have lost their *nuclei,* such as *red blood cells*) contain the entire complement of genetic information; differentiation is a process of progressive restriction of *gene expression*. Although the mechanisms of differentiation are not fully understood, developmental fate is determined to a large extent by cell location in the early *embryo* and is affected by chemical gradients of diffusible sub-

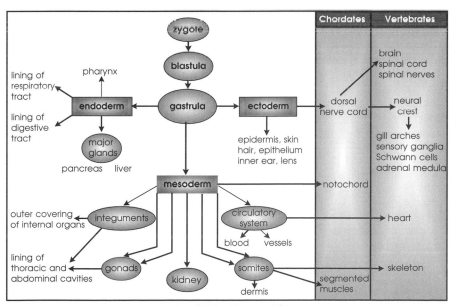

Derivation of Major Tissues and Organs
(see *differentiation*)

stances. Differentiation in plants may be reversible under suitable circumstances, but this is rarely true in animals. (see *determination, reduction, totipotent*)

diffuse-porous Wood in which *vessel elements* of equal diameter are evenly distributed, precluding any obvious *growth rings*. Birch is an example.

diffusion The tendency for gases, liquids, and solutes to disperse randomly and occupy available space. The process is accelerated by raising the temperature, which is translated into increased molecular movement. (see *passive diffusion, facilitated diffusion, osmosis*)

digestion The enzymatic breakdown of complex organic foodstuffs into simpler soluble substances that can be assimilated by cells and tissues. In simple animals (for example, protists, cnidarians, ctenophores, and some other invertebrates) it is *intracellular*, with solid particles being engulfed and digested by individual cells. In higher animals, it is *extracellular*, occurring in an *alimentary canal* or gut into which *digestive enzymes* are secreted. (see *catabolism, digestive enzyme*)

digestive cavity An internal body cavity in which ingested food is stored and digestion begins. It was an evolutionary advancement that permitted *multicellular* organisms to begin *extracellular* digestion of other organisms larger than one cell.

digestive enzyme One of a group of enzymes involved in the breakdown of complex organic foodstuffs into simpler soluble substances that can then pass through *cell membranes* and be assimilated by cells and tissues.

digestive system The body system that breaks down ingested food and assimilates soluble nutrients. The simplest animals (*prokaryotes*, protists, fungi) use only *intracellular* digestion, which limits the size of ingested particles. In cnidarians, food is taken into a *digestive cavity*, but final digestion is still carried out intracellularly. Invertebrates such as roundworms have a true but simple extracellular *digestive system*, in which food is taken in at one end, nutrients extracted, and waste excreted at the other. In vertebrates, the digestive system is complex, the components include the mouth, esophagus, stomach, large and small intestines, liver, and pancreas. (see *duodenum, crop, gizzard, cecum, rumen, digestive enzymes)*

digit A finger or toe at the end of a terrestrial tetrapod limb. (see *pentadactyl limb, hallux, pollex*)

digitigrade The ambulatory mode of some mammals, in which only the *digits* (fingers and toes) are in contact with the ground. It is seen in fast-running animals such as dogs and cats. (see *pentadactyl limb*; compare *plantigrad, unguligrade*)

dihybrid A genetic term for an organism *heterozygous* (different *gene* variants) at two different *loci* (chromosomal locations) and obtained by a cross between parents *homozygous* (same gene variants) for these loci. This can only be accomplished in a *diploid* organism. It is *phenotypically* obvious only when the genetic *alleles* (gene variants) exhibit clear *dominant* and *recessive* characteristics. For example in Mendel's cross between yellow round (YYRR) and green wrinkled (yyrr) peas, all offspring can only have a (YyRr) genotype and show a phenotype of yellow and round. A characteristic ratio of phenotypes (9:3:3:1) is obtained when a dihybrid is *self-crossed*, assuming a simple dominant/recessive relationship between the alleles. (compare *monohybrid*)

	YR	Yr	yR	yr
YR	YRYR	YYRr	YyRR	YyRr
Yr	YYRr	YYrr	YyRr	Yyrr
yR	YyRR	YyRr	yyRR	yyRr
yr	YyRr	Yyrr	yyRr	yyrr

9	yellow round	YYRR, YYRr (2), YyRR (2), YyRr (4)
3	yellow wrinkled	YYrr, Yyrr (2)
3	green round	yyRR, yyRr (2)
1	green wrinkled	yyrr

Dihybrid Cross

dikaryon A cell or fungal *hypha* containing two genetically distinct *haploid nuclei*, each dividing independently at *cell division*. The situation occurs in fungal *mycelia*, in particular *ascomycetes* and *basidiomycetes*, as an intermediate between the haploid and *diploid* phases of the *life cycle*. (see *heterokaryon*; compare *monokaryon*)

dimorphism The existence of clearly separable physical forms of a species. The most common example is *sexual dimorphism* in animals, where the male and female may show different colors and/or sizes. (see *alternation of generations*)

Dinoflagellata (see *Dinomastigota*)

Dinomastigota (Dinoflagellata) A phylum of mostly *photosynthetic* protoctists, the dinomastigotes or dinoflagellates, most of which have two *undulipodia*. The majority are marine, where they are abundant in *plankton*. Some are *bioluminescent*. They are covered with stiff *cellulose* plates, often encrusted in silica. A few species are colorless and *heterotrophic*, and others live as *mutualistic symbionts* inside other marine invertebrates, notably corals. Zooxanthellae, the *symbiotic* form of dinoflagellates, is primarily responsible for the productivity of *coral reefs*, which often occur in nutrient-poor waters. Some contain a trychocyst, an *organelle* that can be expelled to sting prey. The structure of dinomastigote DNA is considered *mesokaryotic*, that is bridging the gap between *prokaryotic* and *eukaryotic*. It condenses into visible chromosomes, but the *chromatin* is organized into fine fibrils, more like bacteria. Some dinoflagellates produce powerful respiratory toxins and are responsible for the destructive "red tides" that poison marine life and contaminate seafood, particularly *filter feeders*.

Dinornithiformes The order of birds containing the flightless moas and kiwis.

dinosaur Any of the large extinct terrestrial reptiles that existed during the Mesozoic era. They included Ornithischia, Stegosaurus, Triceratops, Saurischia (including Tyrannosaurus), and Apatosaurus (Brontosaurs). Dinosaurs were a very successful and diverse group, dominating the terrestrial environments of the earth for 140 million years. Several theories explain their *extinction* at the end of the Cretaceous period, one of which is the major climactic changes induced by continental shift.

dinucleotide A compound of two *nucleotides* linked by their phosphate groups. Important examples are the *coenzymes NAD* and *FAD*.

dinucleotide repeat A *tandemly repeated* sequence in *DNA* that consists of two *nucleotides* linked head to tail many times.

dioecious Plants in which male and female flowers are borne on separate individuals. (compare *monoecious*, *hermaphrodite*)

dioxins A group of compounds composed of oxygen (O), hydrogen (H), carbon (C), and chlorine (Cl). They are produced as a byproduct of manufacturing other compounds, such as *herbicides*, and are highly toxic. Other effects are being studied.

diphosphoglycerate (see *DPG*)

diphycercal tail The type of tail found in cyclostomes, the young stages of all fish, and the *larvae* of amphibians such as frogs and toads. The *vertebral column* extends into the tail, which has *caudal* fins of equal size above and below it. (compare *heterocercal tail*, *homocercal tail*)

diphyodont Any vertebrate in which a set of *deciduous teeth* (milk teeth) is shed and replaced by a second set of permanent teeth.

diploblastic Animals with a body organization derived from two *embryonic* germ layers, *endoderm* and *ectoderm*, that are separated by a jellylike *mesoglea*. This configuration is found only in sponges (Porifera), cnidarians (jellyfish, corals, and sea anemones), and ctenophores (comb jellies and sea walnuts). (compare *triploblastic*)

diploid A cell or organism in which each *chromosome* is present in duplicate. In other words, it contains twice the *haploid* number of chromosomes. *Homologous* pairs of chromosomes separate at *meiosis*, one of each pair being distributed into each *gamete*. In animals, all cells are typically diploid except for *gametes*, which are haploid. The *sporophyte* of plants is diploid, while the *gametophyte* (in higher plants, essentially only the gametes) is haploid. (compare *polyploid*)

diplont An organism representing the *diploid* stage of a *life cycle*. (compare *haplont*)

Diplopoda The class of arthropods containing the millipedes, characterized by a cylindrical body divided into numerous segments each bearing two pairs of walking legs. This distinguishes them from centipedes (*Chilopoda*), in which each body segment bears one pair of legs. Millipedes are *herbivorous* and, although they may produce foul-smelling fluids and small amounts of cyanide gas, are relatively innocuous to humans while centipedes may be deadly.

diplotene In *meiosis*, the stage in late *prophase I* when the pairs of *sister chromatids* begin to separate from each other. *Chiasmata* may be seen at this stage.

Diplura An order of insects, the diplurans, that live in damp soil under stones or logs. They are small, blind, and comprise species that are either *herbivorous* or *carnivorous*.

dipluran (see *Diplura*)

Diprodontia An order of marsupials comprising the diprodont marsupials, the kangaroos.

Diptera The order of insects that contains the flies characterized by only one pair of wings, the hind wings having been modified into *halteres* (balancing organs). The adults have sucking or piercing mouthparts and feed on plant juices, decaying organic matter, or blood. The feeding habits of the legless *larvae* (*maggots*) range from *herbivorous* to *carnivorous*. *Metamorphosis* is complete (*endopterygote*), and the *pupa* is often protected by a barrel-shaped *puparium*.

Many flies are carriers of various diseases, for example the *Anopheles* mosquito that transmits malaria. This order includes the fruit fly, *Drosophila melanogaster*, which has been used extensively in the laboratory to study genetics. (see *polytene*)

directed mutagenesis A specific alteration *in vitro* of a *cloned gene*, preceding its reintroduction back into the same or a different organism.

disaccharide A sugar molecule composed of two *monosaccharide* units. *Sucrose* and *maltose* are examples. (compare *monosaccharide, oligosaccharide, polysaccharide*)

Discomitochondria A phylum of *unicellular* protoctists containing the euglenids, kinetoplastids, amoebomastigotes, and pseudociliates. Euglenids are characterized by the embodiment of classic characteristics of both animals and plants. They are *photosynthetic* and contain *chloroplasts*. They have a different pigment combination than most algae and are surrounded by a *pellicle* rather than a *cell wall*. They are motile due to a single, anteriorly directed *undulipodium*. Euglenids exhibit no *sexual* phase in their *life cycle*, even in *colonial* and *multicellular* species. Some are *facultative heterotrophs* and can survive even if all their *chloroplasts* are lost, making them a valuable laboratory tool for the study of *subcellular organelles*. The other classes of Discomitochondria are highly diverse *heterotrophs*. Each has at least one *undulipodium*, and some have thousands. Several occur as *symbionts* in the guts of termites and other wood-eating insects, excreting the *cellulase* that makes it possible for them to digest hard plant material. *Sexual reproduction* is rare. Included are *parasitic* species of the genus *Trypanosoma*, responsible for sleeping sickness, leishmaniasis, and Chagas' disease. These diseases are difficult to control due to the ability of trypanosomes to rapidly change the nature of the *glycoproteins* making up their *antigenic* coat.

This is accomplished by a unique mechanism at the *DNA* level. (see *antigen-shifting*)

discontinuous variation (see *qualitative variation*)

disinfectant A substance used, particularly on inanimate surfaces, to kill microorganisms.

disomic A *haploid* organism that contains two copies of one *chromosome*. Although the situation may lead to a viable organism, physical abnormalities might be manifest.

displacement activity A type of behavior that appears to be irrelevant to the situation in which it occurs and that may interrupt another activity. It may occur when an animal is presented with equal and opposing situations, such as *fight-or-flight*. For instance, fighting birds may stop and pluck at nest material for a few moments. It may also occur as a consequence of frustration, when an animal is prevented from attaining a goal, and causes the attention to be switched to another stimulus to which it then responds.

display behavior Stereotyped, genetically acquired behaviors that are used by an animal to communicate specific information to another, especially one of the same species. Display behavior is most frequently seen in courtship and aggression. For example, a male bird may puff out its feathers and sing to attract a female during the breeding season and display other actions to establish a territory.

distal Denotes the part of a limb or structure that is farthest from the origin or point of attachment. For example, the fingers are distal to the shoulder. (compare *proximal*)

disulfide bond (cysteine linkage) The covalent cross-linkage between two cysteine *amino acids*. Disulfide bonds play a part in connecting multiple *polypeptide* chains in proteins with quaternary structure.

diuretic Any substance that encourages water excretion from the kidneys. (see *antidiuretic hormone*)

diurnal
1. Daily rhythms, such as normal patterns of waking and sleeping. (see *circadian rhythm*)
2. Animals that are active during daylight hours. (compare *nocturnal*)

divergent evolution The divergence of genetic characteristics in subpopulations of a species that have become separated by physical barriers. Some common characteristics are retained, but others evolve as the organism adapts to the specific environment. For instance, the ostrich, emu, and rhea are thought to have a common ancestor. (compare *convergent evolution*)

diversity regions In *immune receptor genes*, a small region that contributes to *antibody diversity*. It is represented by few or no copies in the *T-receptor* and *B-receptor light-chain* genes, but about twenty different copies are found in the B-receptor *heavy-chain* region. They are incorporated into both B-receptors (which are also released as *antibodies*) and T-receptors. (compare *constant regions, variable regions, joining regions*)

diverticulum A blind tubular or saclike outgrowth from a body tube or cavity. For instance, the *appendix* and *cecum* of a rabbit form a diverticulum.

division One of the major groups into which the plant and fungal kingdoms have historically been divided. Since divisions are analogous to the *phyla* into which the animal and protoctist kingdoms are divided, there is a trend toward using the term phylum for both.

dizygotic twins (fraternal twins) Twins produced from two separate *ova* that have been *fertilized* simultaneously by separate *spermatozoa*. Such twins share no greater percentage of genetic material than do siblings. (compare *monozygotic twins*)

DNA (deoxyribonucleic acid) A *nucleic acid* that contains the genetic information found in most organisms. It is the main component of *eukaryotic* (**H**) *chromosomes*, along with small amounts of *RNA* and *histone* proteins in *eukaryotes*. DNA

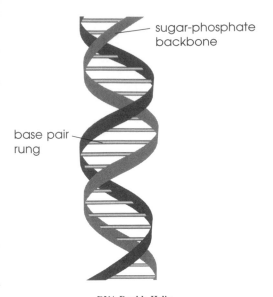

DNA Double Helix
(see *DNA*)

is the sole component of most *prokaryotic* chromosomes as well as *extrachromosomal elements* such as *plasmids*. The DNA molecule is made up of two *polynucleotide* chains that wind around each other to form the distinctive *double helix* structure. Each polynucleotide chain is made up of individual units, each comprised of a *nitrogenous base* (*adenine* (**A**), *guanine* (**G**), *thymine* (**T**), *cytosine* (**C**)) linked via a *deoxyribose* sugar to a phosphate molecule. Linkage via the phosphate molecules (*phosphodiester bonds*) forms a *sugar-phosphate backbone*. The orientation of each chain is designated by its ends. A phosphate group (POy^{2-}) protrudes from the 5′ (*five prime*) carbon of the sugar on one end, and a hydroxyl (OH) group protrudes from the 3′ (*three prime*) carbon of the sugar on the opposite end. The chains are situated *antiparallel* to one another. They are held together by *hydrogen bonding*,

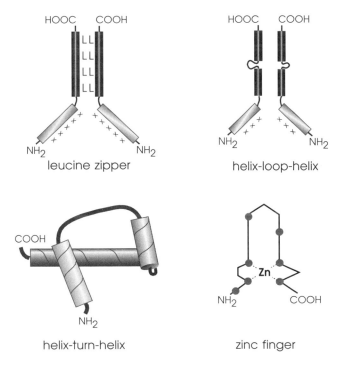

DNA Binding Motifs
(see *DNA binding proteins*)

the nature of which defines the obligate pairing between the bases: a double bond between **A** and **T**, a triple bond between **G** and **C**. (see *base pair, semiconservative replication, transcription, translation, genetic code, gene*; compare *RNA*)

DNA amplification (see *amplification*)

DNA analysis (DNA typing) Any laboratory analysis in which *DNA* is analyzed in order to obtain either specific or comparative information about an organism. Examples include sequence comparison of related organisms to determine their evolutionary relatedness, analysis of specific *loci* to investigate particular *mutations*, or typing at many predetermined *loci* to obtain an individual *DNA profile* used for identification. (see *DNA sequencing, RFLP, PCR, Southern blot, electrophoresis, polymorphism, genetic typing, unique sequence, repeated sequence, tandem repeat, hypervariable, VNTR*)

DNA binding proteins Any of the several classes of proteins that possess specialized structural motifs conducive to DNA binding. (see *leucine zipper, helix-loop-helix, helix-turn-helix, zinc finger*)

DNA clone (see *clone*)

DNA degradation (see *degradation*)

DNA duplex (see *double-stranded DNA*)

DNA fingerprinting (see *DNA profile*)

DNA helicase An enzyme that unwinds the *DNA double helix* by disrupting the hydrogen bonds that hold it together. Helicases probably facilitate DNA *replication*.

DNA hybridization (see *nucleic acid hybridization*)

DNA ligase (ligase) An enzyme that catal-

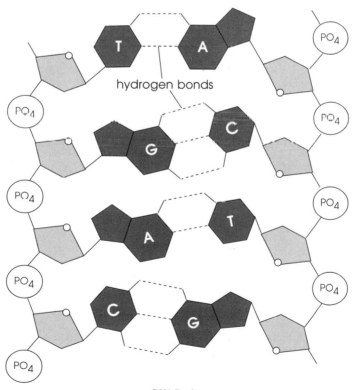

DNA Duplex

yses *phosphodiester bond* formation between two *DNA* molecules. It is used in normal *DNA replication* and also to repair damaged *chromosomes*. DNA ligase is also essential in *cloning* and *genetic engineering* in order to link DNA fragments *in vitro.*

DNA linkers Small, synthetic *DNA* fragments that contain a cut site for a specific *restriction endonuclease*. They may be attached to DNA molecules *in vitro* in order to generate *restriction sites* for further ease of manipulation in *cloning* procedures. (compare *adaptors*)

DNA mismatch (see *base pair mismatch*)

DNA polymerase Any enzyme that specifically assembles *deoxyribonucleoside triphosphates* into *DNA* strands in the order dictated by a *template*. Polymerases are necessary but not sufficient for DNA *replication* and also function in *DNA repair*. Several different DNA polymerases play specific roles in the cellular *metabolism* of DNA, and the particular characteristics of each also vary between organisms. One particular DNA polymerase, *Taq I* (isolated from *Thermus aquaticus*, a bacterium that lives in hot springs) is resistant to heat and is essential to the *polymerase chain reaction (PCR)*. (compare *RNA polymerase, reverse transcriptase*)

DNA probe (see *probe*)

DNA profile The result of *DNA analysis* of an organism at a number of *polymorphic loci* (locations). A DNA profile may be used to classify the organism as belonging to a species, a group of individuals, or ultimately as an individual, depending on the nature of the *genetic markers* employed. A DNA profile compiled from a sufficient number of rare alleles may be considered unique. (see *genetic typing, RFLP, VNTR, PCR*)

DNA repair The mechanisms by which a cell corrects potentially *mutagenic* damage to its *genome*. Numerous mechanisms exist, including the direct repair of the damaged *bases*. Alternatively, a portion of the *DNA* strand containing the damage may be excised and the complementary strand used as a *template* in resynthesis. In postreplication repair, mismatched bases are replaced or a sister molecule is used as a repair template during *recombination*. (see *base pair mismatch, gene conversion, double strand break, thymine dimer, base substitution, transposon, carcinogen*)

DNA replication (see *replication*)

DNase (deoxyribonuclease) Any *nuclease* type enzyme that *hydrolyzes* (breaks) the *phosphodiester bonds* of *DNA*. Two general classes include *endonucleases* that work from the middle of a DNA molecule and *exonucleases* that work from the ends of a molecule. (see *DNA degradation, restriction endonuclease*)

DNA sequencing The determination of the exact *nucleotide* sequence in a defined region of *DNA*. The first method to be used widely was developed by A. Maxam and W. Gilbert in the late 1970s and depended on partial chemical degradation of known DNA fragments. The method of preference during most of the 1980s and 1990s was developed by F. Sanger. It utilized dideoxy nucleotides that can be incorporated into a growing DNA chain but terminate synthesis at that point. If these chemically altered nucleotides are *labeled* and introduced into an *in vitro* synthesis (using a separate tube for each of the four nucleotides), DNA molecules of random lengths are produced, each terminated by a known, labeled nucleotide. This indicates the location of the complementary base on the *template* strand. The fragments are separated using a *polyacrylamide* gel *electrophoresis*, and the DNA *base* sequence is read from the gels. Currently, sequencing methods using *PCR* (polymerase chain reaction) are beginning to dominate the field, and the detection and analysis process is becoming increasingly automated. In automated detection, fluorescent tags are used to label the sequencing products. (see *DNA replication*)

DNA topoisomerase An enzyme that regulates the supercoiling in circular *DNA* molecules, such as bacterial chromosomes and *plasmids*. It facilitates DNA *replication*. (see *tertiary structure*)

DNA typing (see *DNA analysis*)

DNA virus A *virus* that employs *DNA*, either *single-* or *double-stranded*, as its primary genetic material. They include *papilloma viruses, herpesviruses,* and *poxviruses.*

dobsonflies (see *Megaloptera*)

dogfish (see *Chondrichthyes*)

dogs (see *Carnivora*)

dolphins (see *Cetacea*)

domain A physically distinct portion of a protein, determined by the three-dimensional folding in the *tertiary structure*.

dominance hierarchy (pecking order) A strict hierarchy found in many vertebrate social groups, in which each individual occupies a particular position that is recognized by others in the group. The hierarchy is established initially by challenge and fighting, the dominant animal emerging as the one that cannot be dominated by any other. After establishment, a hierarchy is usually stable, as subordinates tend to avoid threatening the dominant animal and perform submissive actions to avoid being threatened themselves.

dominant The *allele* (variant) of a particular *gene* that determines the *phenotype* (physical characteristic) of a *diploid* organism when found in *heterozygous* condition with another, *recessive* allele of the same gene. Often, but not always, it is the "normal" (*wild type*) and most common form, because the dominant allele of the gene generally produces a functional protein. In contrast, the *recessive* allele often produces a nonfunctional protein or no protein.

dominant species The species that are most abundant in a given area, *ecological community,* or *ecosystem*. (see *species diversity*)

dopamine A *neurotransmitter* found in the *corpus striatum* of the mammalian brain. Its action is important in the normal functioning of *motor* systems and also in emotional states. Low levels of dopamine are associated with Parkinson's disease in man.

dormancy A period of minimal metabolic activity in a plant body or reproductive structure. It is a means of surviving a period of adverse environmental conditions, usually cold (when water is not available because it is frozen) or drought. *Seeds, spores, cysts,* and *perennating structures* (*bulbs and corms*) are often used to survive dormancy and *germinate* to produce new plants under appropriate conditions. Adult plants may become dormant by losing their leaves and producing drought resistant buds. The onset and breaking of dormancy are normally controlled environmentally, by factors such as day length (*photoperiod*) and temperature. Dormancy may be promoted by certain *hormones* (for example, *abscisic acid*) and broken by other hormones (for example, *gibberellins*). (see *estivation, diapause, hibernation, stratification, photoperiodism*)

dormant An organism, tissue, or *propagule* in a nongrowing state of suspended activity. (see *dormancy*)

dormin (see *abscisic acid*)

dorsal The side of an organism farthest from the *substrate* upon which it stands, rests, or is attached, usually the upper surface. In four-legged animals, dorsal refers to the spinal region. This is continued when describing bipedal animals, such as man, where the back side would be described as posterior in animals that walk on all fours.

dorsal lip In *embryos* with symmetrical yolk distribution, such as amphibians and aquatic invertebrates, the initial site of invagination in the *blastula* (*gastrulation*) that creates the *archenteron* and the three embryonic *germ layers*. (see *endoderm, mesoderm, ectoderm, blastocoel, blastopore*)

dorsal nerve cord A defining feature in all chordates. It develops from *ectoderm* into a hollow cord containing the central *nerves* of the body. In vertebrates (craniates), it also differentiates into a defined brain. In craniates, it is further protected by the *vertebral column* and is called the *spinal cord.*

dosage compensation In organisms in which *gender* is determined by a pair of *heteromorphic* or *homomorphic sex chromosomes* (for example, **XY** vs. **XX**), the mechanism that allows unrelated *structural genes* on the *sex chromosomes* to be expressed at the same levels in females and males, regardless of the *ploidy* of a particular sex chromosome. In female mammals, one of the two **X** chromosomes in each cell is inactivated (*X-inactivation*), resulting in a *Barr body.* In male *Drosophila*, the non-*sex-linked* genes on the **X** chromosome in each cell are hyperactivated (*X-hyperactivation*).

dot blot A laboratory technique for analyzing nucleic acids (*DNA, RNA*). The sample to be analyzed is applied to a solid support, often a *nylon membrane*, as a drop and challenged with a *labeled probe* for the purpose of identifying particular *nucleotide* sequences that might be present. This technique determines the presence but not the size of a particular fragment. When the sample is applied through a special apparatus designed to conform the sample pattern to lines rather than circle, it is called a "slot blot." (see *reverse dot blot*; *Southern blot*)

double circulation The type of blood *circulatory system* employed by mammals, birds, and many crocodilians. It depends on a four-chambered heart that enables the complete separation of *oxygenated* and *deoxygenated* blood.

double fertilization In flowering plants (*angiosperms*), the union of one *pollen nucleus* (male *gamete*) with the egg nucleus to form the *diploid zygote* and another pollen nucleus from the same grain with the two *polar nuclei* of the egg to form a *triploid endosperm* nucleus.

double helix The specialized, physical structure taken by the two antiparallel *polynucleotide* chains of *DNA*. (see *base pair, complementary base pairing*)

double recessive A *diploid* organism containing both *recessive alleles* (gene variants) of a particular gene and thus exhibiting the corresponding *phenotype*. Double recessives are often used in *test crosses*, since a *dominant gene* contributed by the other parent will always be phenotypically manifest in the offspring.

double strand break A form of *DNA* damage in which the *phosphodiester bonds* on both DNA chains are split. Such a break is often repaired during *synapsis* in *meiosis*, using *a sister chromosome* as a *template*. It is thought that meiosis may have evolved as a mechanism to repair double strand breaks. Double strand breaks also occur normally in some organisms as a method for a directed reshuffling of particular parts of their *genome*. (see *DNA repair, homothallic, immune receptor genes, antigen-shifting*; compare *single strand nick*)

double-stranded DNA (DNA duplex, duplex DNA) *DNA* in its duplex form, with the two *polynucleotide* chains held together by *hydrogen-bonding*. In this form, it is unable to interact with other *nucleic acid* molecules or enzymes. Most DNA *in vivo* is found in the duplex form, except for small regions that temporarily unwind to permit *replication* and access to enzymes. (see *denaturation*, compare *single-stranded DNA*)

doves (see *Columbiformes*)

down feather The first type of *feather* produced by baby birds. They have a short *quill*, noninterlocking *barbules*, and are soft and fluffy, providing insulation. Some *follicles* produce down feathers throughout life.

Down's syndrome A condition seen in humans, characterized by abnormal physical development and mental retardation. It is usually caused by the *nondisjunction* of *chromosome* 21 at *meiosis* It results in an extra copy of the chromosome or a piece of it in the affected individual.

downstream The direction of *mRNA transcription* in a particular region of a *chro-*

mosome. It is used in reference to particular sequence elements, such as *promoters, coding sequences,* and *regulatory sequences.* (compare *upstream*)

DPG (diphosphoglycerate) A small organic molecule found in the bloodstream that binds to *hemoglobin* and alters its shape, facilitating oxygen dissociation in the tissues. It increases in people who stay at high altitudes in response to the lower concentration of oxygen in the atmosphere. (see *allosteric*)

dragonflies (see *Odonata*)

Drosophila melanogaster A species of fruit fly that has been studied extensively with regards to its genetics and development, in part because of the giant *polytene chromosomes* found in its *salivary glands.*

drupe (stone fruit) A fleshy fruit, such as a plum, containing one or a few seeds, each enclosed in a stony layer that is part of the fruit wall. Blackberries and raspberries are collections of small drupes or drupelets. (compare *berry, nut*)

duck-billed platypus (see *Monotremata*)

ducks (see *Anseriformes*)

ductless gland (see *endocrine gland*)

ductus arteriosus A blood vessel present in tetrapod *embryos* that links the *pulmonary artery* with the *aorta.* It enables blood to be shunted from the *right ventricle* of the heart into the systemic circulation, bypassing the lungs. It normally closes at birth and remains as a solid strand.

dugong (see *Sirenia*)

duodenum The first part of the *small intestine* into which the food passes when it leaves the stomach. It forms a loop into which the pancreatic and *bile* ducts open. Alkaline secretions neutralize the acid from the stomach and continue the digestive process.

duplex DNA Double-stranded DNA.

duplication In genetics, the occurrence of extra *genes* or *chromosomal* segments in the *genome.* The duplication may be either tandem (adjacent) or dispersed. (see *repeated sequence, tandem repeat*)

dura mater The outer membrane that surrounds and protects the brain and *spinal cord* in vertebrates. (see *meninges*)

duramen (see *heartwood*)

dyad A pair of *replicated* sister *chromatids* still joined at the *centromere,* as seen in the first *meiotic* division.

dynein A protein found associated with *microtubules* in the *axoneme* of *undulipodia.* It appears to form cross bridges between each microtubule doublet and its neighbor. It is hypothesized that the making and breaking of cross bridges, concomitant with the *ATPase* activity associated with dynein, is responsible for movement in the axoneme. Dynein may also play a part in *chromosome* movement during *anaphase* of *nuclear division.* (see *nexin*)

eagles (see *Falconiformes*)

ear One of a pair of sense organs, located on either side of the head in vertebrates. They function in hearing and balance. In mammals, the ear consists of an *outer ear* separated by the *tympanic membrane* (eardrum) from the *middle ear* that communicates with the *inner ear* through the *oval* and *round windows*.

ear ossicles In mammals, three small bones in the *middle ear*: the *malleus, incus,* and *stapes*. They form a series of levers that transmit vibrations falling on the *eardrum* to the *inner ear*. They are *homologous* with certain jaw bones of lower vertebrates. Amphibians and primitive reptiles have just one ear ossicle, the *columella auris* or stapes.

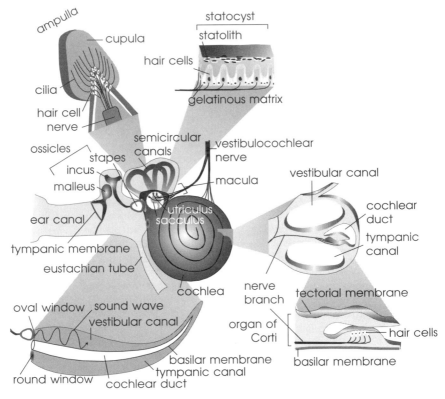

The Vertebrate Ear
(see *ear*)

eardrum (see *tympanic membrane*)

early successional species Species that occur only or primarily during the early stages of *ecological succession*. They are typically rapid growing and short lived, with high reproductive rates. They usually consist of plants and microorganisms. (compare *late successional species*)

earthworms (see *Oligochaeta*)

earwigs (see *Dermaptera*)

ecad A variation within a species that arises in response to a particular set of environmental conditions, and whose characteristics are not heritable. In other words, such changes are entirely *phenotypic* and nongenetic.

ecdysis (molting) The periodic shedding of an outer layer. In arthropods, the *exoskeleton* is shed several times during *metamorphosis* in order to accommodate a larger body. The process is controlled by the *hormone ecdysone*, secreted by the *prothoracic gland*. The stages between arthropod molts are defined as *instars* and numbered accordingly. In reptiles, the *epidermis* is shed, either in a single piece by snakes or in small patches by lizards. The periodic or seasonal loss and replacement of hair or feathers also occurs in mammals and birds, including humans who continuously shed small flakes of *epidermis*.

ecdysone (molting hormone) A *steroid hormone* produced by arthropods that induces *molting* and *metamorphosis*. It is secreted by the *prothoracic gland*.

ecesis The successful *germination* and establishment of migrating plant species colonizing a new area. It represents the third in a series of six phases in *ecological succession*.

Echinodermata A phylum of marine invertebrates including the starfishes, brittle stars, sea urchins, sea cucumbers, and sand dollars. Most echinoderms exhibit *radial symmetry*, typically with five rays extending from a central disc but may have many more. All have calcium-containing skeletal plates (*ossicles*), and most have spines. Part of the *coelom* is modified as the *water vascular system* that extends into hydraulic *tube feet*, used in locomotion. Although *asexual reproduction* is possible in echinoderms, most reproduction is *sexual* and external. The free-swimming *larvae* are *bilaterally symmetrical* and *metamorphose* into adults.

echinoderms (see *Echinodermata*)

Echinoidia The class of echinoderms that contains the sea urchins. They lack distinct arms and are covered by a rigid shell bearing movable spines. They walk by means of *tube feet* or jointed spines, which are also used defensively. The mouth, including a complicated jaw apparatus, is on the ventral surface.

echinoids (see *Echinoida*)

Echiura A phylum of sedentary marine worms, the echiurans or spoon worms. They form burrows in which they spin a mucus-coated net to trap food particles. They pump sea water through their burrows by constricting their trunk muscles and collect the food with a motile *ciliated proboscis*. The water current they create also serves to oxygenate their blood and remove wastes. *Fertilization* produces *trocophore larvae*.

echolocation A method used by some animals, the most familiar being bats and whales, to locate objects, including prey and obstacles, and for general orientation. It is most common in cave-dwelling, *nocturnal* animals or wherever normal eyesight would not be adequate (such animals usually also have normal sight). It is based on the reflection of high-pitched sound emitted by the animal. Specialized receptor organs, such as the large external ears of bats, have evolved in order to amplify the returning echo.

eco- A word element meaning "ecology," originally derived from the Greek for "house." (see *ecotype*)

ecocline (ecological gradient) A change in the relative abundance of a species or group of species along a line or over an area. (see *cline*)

E. coli (see *Escherichia coli*)

ecological community (community) A group of populations of different species living in the same local area and interacting with one another. A community is the living portion of an *ecosystem*.

ecological gradient (see *ecocline*)

ecological island An area that is biologically isolated so that a species occurring within the area cannot easily mix with another population of the same species. They may range from city parks to large, geologically isolated regions such as islands.

ecological niche The functional position of a species in its habitat and environment. With respect to other species, a niche is often defined with regard to resource competition. If two species occupy the same niche, one will usually be eliminated. A similar niche may be occupied by different species in different areas. For instance, the woodpecker finch of the Galapagos Islands occupies the same niche as true woodpeckers in other parts of the world. (see *competitive exclusion*; compare *habitat*)

ecological pyramid (see *pyramid of numbers, pyramid of biomass, pyramid of energy*)

divided into six stages:

1. *nudation* (the exposure or clearing of a new surface)
2. *migration* (the influx of seeds and *spores*)
3. *ecesis* (*germination* and early growth)
4. *competition*
5. *reaction* (the effect of plants on the habitat)
6. *stabilization* (establishment of the climax community). (see *sere, clisere, primary succession, secondary succession*)

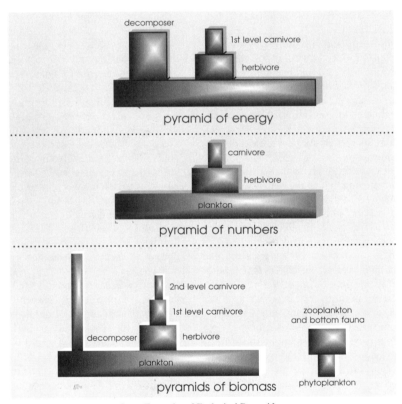

Some Examples of Ecological Pyramids

ecological succession (succession) A progressive series of changes in the vegetation of a site, from initial colonization to the establishment of a *climax community*. The term may also be applied to *sessile* (attached) animals, for instance in aquatic *ecosystems*. Succession can be

ecology The study of the relationships of organisms to each other and to their surroundings.

ecosystem The sum total of all living and nonliving components of a particular area that interact and exchange materials with each other. It is sometimes defined as the *ecological community* of organisms plus the environment with which they interact. Energy flow and nutrient cycling are regulated within a particular ecosystem and are studied as indicators of its overall health. In order to be defined as an ecosystem, the rate of flow of biologically important components must be greater within the system than between it and another system. (see *food web, pyramid of biomass*)

ecotone The narrow transition zone between two diverse *ecological communities* such as forest and grassland or land and water. They are present naturally but may also reflect human intervention, such as the clearing of forested areas for agriculture.

ecotype A local variant of a species that has adapted to specific environmental factors. Although the differences between ecotypes are genetic, they may still successfully interbreed. Differences between ecotypes may be physiological and/or morphological. (see *adaptive radiation, speciation*; compare *biotype, ecad*)

ectocyst In ectoprocs, the nonliving outer layer of the body wall that has been secreted by the animal. (compare *endocyst*)

ectoderm The outermost of the three *embryonic germ layers* in *triploblastic* animals. It develops into the *epidermis* (for example, skin) and its derivatives, such as hair and feathers. The lining of invaginated body cavities such as the mouth and the *nervous system* including the sense organs, brain, and spinal cord are also derived from ectodermal tissue. (compare *endoderm, mesoderm*)

ectomycorrhizae (see *ectotrophic mycorrhizae*)

ectoparasite An organism that lives on the outer surface of another organism and benefits from the relationship at the expense of its *host*. An example is the flea. (see *symbiosis, parasitism*; compare *endoparasite*)

ectopic expression The expression of a *gene* in a tissue in which it is not normally expressed. (see *oncogene*)

ectopic integration The insertion of an artificially introduced *gene* at a site other than its usual *locus*. (see *transgenic*)

ectoplasm In plant cells and some protists, the outer gel-like layer of cell *cytoplasm* that lies immediately beneath the *plasma membrane* and contains packed layers of *microtubules*. (compare *endoplasm*)

ectoproct (see *Bryozoa*)

Ectoprocta (see *Bryozoa*)

ectothermic (poikilothermic) Animals that lack an internal system for body temperature regulation, thus tend toward the temperature of their environment. They have evolved a wide array of behavioral mechanisms that enable them to control their temperature by using environmental cooling and heating. This situation is found in most animals other than birds and mammals. They have been called "cold-blooded" because their body temperature is often, though not always, cool relative to *endotherms*.

ectotrophic mycorrhiza, ectotrophic mycorrhizae (pl.) *Mycorrhizae* (fungal-plant *symbioses*) that are external to the plant root cells. They are basidiomycete based and are found mainly on woody trees and shrubs.

edaphic factors Soil conditions that influence plant distribution. These include physical, chemical, and biological characteristics of the soil, such as water content, pH, organic matter, and soil texture.

EEG A graphic recording of the electrical activity in the brain.

effector
1. An organ or cell by which an animal responds to internal or external stimuli, often via the *nervous system*. Effectors include *glands*, muscles, *cilia,* and *chromatophores* (color change cells). The *Cnidocytes* (stinging cells) of cnidarians are one of the few organismal effectors that respond without the intermediate stimulation of a nervous system.
2. Within a cell, any small molecule that induces *allosteric* changes in the shapes of proteins.

efferent The conveyance of impulses or substances from central regions of the body toward the periphery. It may apply to *motor nerve impulses*, blood, or other fluids. (compare *afferent*)

egestion The removal of undigested material, along with associated *endogenous* gut *flora*, through the anus. In contrast to *metabolized* material that is *excreted*, egesta is transmitted directly through the *alimentary canal* without ever entering body cells.

egg A female *gamete*. It is nonmotile, contains abundant *cytoplasm*, and is often larger than a male gamete (see *ovum, embryo sac*)

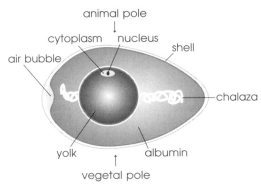

Bird or Reptile Egg

egg activation A series of events initiated by sperm penetration into an egg at *fertilization*. In mammals, it includes changes in the *vitelline membrane* preventing the penetration of additional *spermatozoa* and initiates completion of the second *meiotic* division. The point of entry also effects a rearrangement of the egg *cytoplasm*, establishing the symmetry of the developing egg. (see *oogenesis*)

egg membrane (see *vitelline membrane*)

egg yolk (see *yolk*)

elaio- A word element derived from Greek meaning "oil." (see *elaioplast*)

elaioplast A *plastid* in which lipids are stored. They are common in some non-vascular plants (for example, liverworts) and monocotyledons. (compare *aleuroplast, amyloplast*)

elastin A fibrous *structural protein* found in animal *connective tissue.*

electric organ A specialized organ found in fish that are able to produce an electric current. It is composed of modified muscle cells (*electrocytes*) that generate an ionic current rather than contracting. They may produce a high voltage charge, such as in the electric eel, or a series of low voltage pulses used for locating prey, obstacles, and mates. (see *action potential, muscle contraction*)

electrocytes The modified muscle cells that make up the *electric organ* of certain fish.

electroencephalogram (see *EEG*)

electrolyte Any chemical or compound that, when dissolved in water, produces ions that are able to conduct an electric current. Electrolytes such as calcium (Ca^+), sodium (Na^+), and potassium (K^+) ions are essential to normal body function. (see *electrolyte balance*)

electrolyte balance The delicate equilibrium between various *electrolytes* in the body that must be maintained for normal function. (see *kidney, adrenal cortex, pancreas, pituitary*)

electron micrograph A photograph made through an *electron microscope.*

electron microscope A microscope based on visualization using a beam of electrons rather than a beam of visible light. Since electrons have a much shorter wavelength, the resolving power can be as much as 400 times that of a light microscope. In *transmission electron microscopy*, specimens are prepared as thin sections and usually impregnated with a heavy metal to improve electron-scattering ability. In *scanning electron microscopy*, an electron beam scans the surface of an object to produce a relief image. (see *embedding*; compare *light microscope*)

electron transport chain In *aerobic respiration*, the final series of steps during which the energy from high energy electrons produced in the Krebs cycle is harvested. The *terminal electron acceptor* is often molecular oxygen (O_2), which is reduced to form water (H_2O). Alternative molecules such as elemental sulfur (S_2), nitrates (NO_3^-) or phosphates (PO_4^{3-}) may act as terminal electron acceptors in the diverse systems found in bacteria. They produce, respectively, hydrogen sulfide (H_2S), ammonia (NH_3), and phospine (PH_3). In the *light reactions* of *photosynthesis*, an electron transport chain links *photosystem II* to *photosystem I*. The electron acceptor is *NADP$^+$* which is reduced to *NADPH* and funnels high energy electrons (as part of the hydrogen atom) into the *dark reactions*. In either case, high energy electrons are passed through a series of membrane bound carrier molecules (*cytochromes*) to *proton pumps* embedded within the *mitochondrial* or *chloroplast* inner membranes. As the electrons arrive at the proton-pumping membrane channels, their energy is harvested to transport protons (H^+) out across the membrane, creating a proton gradient. The protons diffuse back into the mitochondrial or chloroplast matrix through special channels associated with *ATP synthase* Their release of energy as they travel down the concentration gradient powers the *chemiosmotic* synthesis of *ATP*. (see *NADH, FADH$_2$, heme, redox reaction*)

electrophoresis A laboratory technique used to separate *macromolecules* on the basis of charge and size. The solution containing the material is generally applied to a porous medium (gels of *starch, agarose,* or *acrylamide*) and placed in an electric field. Charged molecules migrate to one electric pole or the other with their speed, hence final location, roughly based on size. The characteristics of the system, including choice of matrix material, concentration of matrix, running buffer, and voltage, may be varied to achieve optimal separation. Electrophoresis is typically used to separate and characterize proteins and *nucleic acids*. (see *DNA analysis, serology, Southern blot, northern blot, western blot*; compare *chromatography*)

elephant (see *Proboscidea*)

elephant shrews (see *Macroscelidea*)

Eleutherozoa A subphylum of echinoderms containing the brittle stars, sea stars, sea urchins, sand dollars, and sea cucumbers. They are distinguished by their attachment to a substrate on the same side as their mouth.

eluate A liquid fraction that had initially been retained on a *chromatography* column then released through a change in conditions and collected.

elytron, elytra (pl.) The hardened forewings found in beetles and in some other insects. They protect the delicate hindwings when these are not in use and often give the animal its characteristic appearance, such as in the lady beetle.

Embden-Meyerhoff pathway (see *glycolysis*)

embedding The sealing of biological tissue in a solid block of paraffin wax prior to sectioning by a *microtome* for microscopy. Tissue prepared for *electron microscopy* may be embedded in Arldite and cut with a diamond knife.

Embioptera The order of insects comprising the web spinners.

embryo The general stage of development that follows *fertilization* and *zygote* formation. In animals, the term embryo applies to any stage after *cleavage* and before hatching or birth; in the later stages it is also called a *fetus*. In plants, the embryo refers to a young *sporophyte*, a stage between zygote formation and *germination*. (see *embryo sac*)

embryo culture The development and growth of plant *ovules in vitro*. The technique is used for research into variables affecting the process of *embryo* maturation and also in the development of new crops by genetic manipulation.

embryology The study of the development, particularly of animals, between *fertilization* and hatching or birth.

embryonic membranes (see *extraembryonic membranes*)

embryo sac The female *megagametophyte* (*haploid sexual* cell) of flowering plants (*angiosperms*), essentially analogous to an animal *egg*. It is enclosed within the *nucellus* and contains a number of *nuclei* (usually eight) derived from division of the *megaspore nucleus*. It is the site of *fertilization* and subsequent *embryo* development. The embryo sac commonly consists of a nucleus, two *synergid nuclei* at the *micropylar* end, three *antipodal* cells at the opposite *chalazal* end, and two central *polar nuclei* that fuse to form the *primary endosperm nucleus*. During a unique process called *double fertilization,* one male nucleus (from a *pollen* grain) fuses with the egg nucleus to form the *zygote* while the second male nucleus fuses with the *primary endosperm nucleus* to form a *triploid* cell that later gives rise to the *endosperm* (a food storage structure).

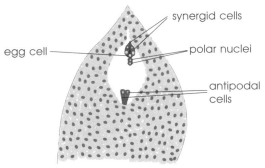

A Generalized Plant Embryo Sac

EMIT (enzyme-multiplied immunoassay technique) (see *immunoassay*)

enamel The hard white outer coating of vertebrate teeth. It is constructed from hexagonal crystals of calcium phosphate salts, carbonate, and fluoride, all bound by *keratin* fibers.

encyst To become *dormant* and form a protective coating or cyst. *Parasitic* mi-croorganisms often encyst in the body of an intermediate *host,* awaiting transfer to a *primary host.* (compare *excyst*)

endangered species A species that imminently faces potential *extinction* in a very short time. The United States Endangered Species Act of 1972 defines an endangered species as "one in danger of extinction through all or a significant portion of its range." (compare *rare species, threatened species*)

endemic

1. In general, all factors confined to a defined geographical region. It is often used to describe a population or species that is native to a particular area and not native elsewhere. (compare *indigenous, exotic*)

2. Any disease that occurs frequently in a particular geographical region. (see *endogenous*)

endergonic A chemical reaction to which energy from an outside source must be added before the reaction proceeds. (compare *exergonic*)

endo- A word element derived from Greek meaning "within." (for example, *endothelium*)

endobiotic (see *endosymbiotic*)

endocarp (see *pericarp*)

endocrine gland (ductless gland) A *gland* that has no duct or specific opening to the exterior. It generally produces *hormones* that diffuse directly into the bloodstream and are distributed via the *circulatory system*. It is one of several bodily communication systems. (compare *exocrine gland, nervous system*)

endocrine system The ductless (*endocrine*) *glands* and their *hormone* products that regulate and integrate body activities and maintain *homeostasis*. (see *hypothalamus, pituitary gland, thyroid gland, adrenal gland, pancreas, pineal gland, thymus, parathyroid glands, ovary, testis*)

endocrinology The study of *endocrine glands* and their *hormonal* secretions.

endocyst In ectoprocts, the inner living tissues of the body wall. (compare *ectocyst*)

endocytosis The mechanism by which a

cell ingests material in bulk. Termed *pinocytosis* (liquids) or *phagocytosis* (solids), it is accomplished by extension of the *cell membrane* to engulf the substance and then invagination to form a *vesicle* or *vacuole*, which is pinched off to the interior. Endocytosis is used by both *multicellular* and *unicellular* organisms for transport, feeding, and defense. (see *lysosome, pinocytosis, phagocytosis*; compare *exocytosis*)

endoderm The innermost of the three *embryonic germ layers* in *triploblastic* animals. It develops into the lining of the digestive cavity, its associated *glands* such as the liver and pancreas, and much of the respiratory tract. (compare *ectoderm, mesoderm*)

endodermis The innermost layer of the *cortex* of plant tissue, consisting of a single layer of cells that control the passage of water and solutes between the cortex and the innermost *vascular cylinder*. (see *Casperian strip*)

endogenote In bacterial genetics, the complete *genome* residing in the cell on the receiving end of a *conjugation* event. (compare *exogenote, merozygote*)

endogenous Any substance that is produced or originates within an organism. (compare *endemic, exogenous*)

endolithic Any organism living inside rock, often limestone, and that is capable of active penetration, such as microorganisms capable of dissolving minerals. (compare *chasmolithic*)

endolymph A fluid filling the *inner ear* of vertebrates. (see *labyrinth, sacculus, utriculus, semicircular canals*) (compare *perilymph*)

endometrium The glandular *mucous membrane* lining the internal surface of the *uterus* in mammals. It passes through cyclical periods of development and regression, synchronously with *ovulation*, so that it is ready to receive and nourish a fertilized *ovum*. If *fertilization* does not occur, it either returns to a quiescent state, or in the case of humans and some primates, breaks down and is discharged by the process of *menstruation*. (see *estrus cycle*)

endomitosis The duplication of *chromosomes* without *nuclear division*. *Spindle* formation and other normal *mitotic* apparatus formation is bypassed as is dissolution of the *nuclear membrane*. It occurs primarily in the *macronucleus* of *ciliates* where the duplicated *chromatids* separate within the nucleus causing *polyploidy*, and in certain insect *larvae* where the duplicated chromatids fail to separate leading to *polytene* chromosomes. The polytene chromosomes of *Drosophila melanogaster* (fruit fly) *larvae* have played an important part in genetic research. (compare *mitosis, amitosis*)

endomycorrhizae (see *endotrophic mycorrhizae*)

endonuclease Any enzyme that *catalyzes* the *hydrolysis* (breakage) of the internal *phosphodiester bonds* (as opposed to the free ends) of *nucleic acid*, usually *DNA*. *Restriction endonucleases* are important examples.

endoparasite An organism that lives inside the body of another organism and benefits from the relationship at the expense of the *host*. An example is the malarial parasite *Plasmodium*. (see *symbiosis, parasitism*; compare *ectoparasite*)

endoplasm In plant cells and some protists, the inner layer of *cytoplasm* within which are embedded the principal cell *organelles*. (compare *ectoplasm*)

endoplasmic reticulum (ER) An extensive system of flattened membranous sacs (*cisternae*) traversing the *cytoplasm* of all *eukaryotic* cells and continuous with the *nuclear envelope*. *Rough ER* is covered with *ribosomes* and provides a transportation system for the delivery of synthesized proteins to other parts of the cell or for secretion to the exterior via the *Golgi apparatus*. *Smooth ER* (lacking ribosomes) is involved with lipid synthesis, including *steroids*. In muscle cells, a specialized form of ER called the *sarcoplasmic reticulum* is present.

endopodite (see *biramous appendage*)

endopolyploidy An increase in the number of *chromosome* sets caused by *replication* without *cell division*.

endopterygote The group of winged insects that undergo complete *metamorphosis*. The wings develop within the *larva* and a true *pupal* stage is seen. Most insects, including members of the largest and most successful orders, display complete metamorphosis. Examples include beetles; fleas; true flies; butterflies and moths; and bees, ants, and wasps. (compare *exopterygote*)

endorphins (enkephalin) Sometimes referred to as *"endogenous* opiates," a group of *neuropeptides* produced in the brain that block the perception of pain and are released after injury. Their chemical structure is mimicked in narcotic analgesics such as morphine and heroin, which are derived from the opium poppy. Pain relief from acupuncture may be due to the stimulated production of endorphins.

endoskeleton An internal skeleton such as the bony or *cartilaginous* skeleton of vertebrates. It gives shape and support to the body, protects vital organs, and provides a system of rigid levers to which muscles can attach in order to effect movement. An endoskeleton also allows for the continuous growth of an animal and increases in size concomitantly. Simple endoskeletons are also present in echinoderms and other invertebrates. (see *ossification*; compare *exoskeleton*)

endosperm The nutritive tissue that surrounds the *embryo* in flowering plants (*angiosperms*). It develops from the *primary endosperm nucleus* and is therefore *triploid* (three sets of *chromosomes*). It frequently serves as a store of food materials that are used as an energy source during *germination*. Many endospermic seeds such as corn, beans, and wheat are cultivated for their nutritional reserves. (see *embryo sac, double fertilization*)

Endospora The phylum containing the *Gram-positive endospore*-producing bacteria. It includes members of the distinctive rod-shaped genus *Bacillus* (formerly aeroendospora) responsible for Anthrax as well as *Clostridium* (formerly fermenting bacteria), which causes gangrene and botulism. The reclassification has occurred based on DNA *homologies* in the small (16S) *ribosomal RNA (rRNA)* subunit.

endospore In bacteria, a thick-walled *spore* formed around their *chromosome* and a small portion of *cytoplasm*. Endospores are formed under unfavorable conditions and are resistant to heat, desiccation, and X-rays; they may remain viable for several centuries. An example is the species responsible for botulism (*Clostridium botulism*).

endostyle A shallow groove along the *ventral* wall of the *pharynx* (throat) of *Amphioxus* and other primitive *chordates*. It consists of several tracts of *mucus*-producing and *ciliated* cells. Food particles are trapped in the mucus and transported by the *cilia*.

endosymbiosis An organism that lives internally in its *host*. Endosymbiosis usually refers to a *mutualistic* situation in which both organisms benefit. A number of cell *organelles*, including *mitochondria, chloroplasts,* and probably *kinetosomes* and *centrioles*, seem to derive from ancient *prokaryotes* that were captured by pre-*eukaryotic* cells, and became incorporated into their *metabolic* machinery. Cyanobacteria are thought to be the precursors of chloroplasts, nonsulfur purple bacteria of mitochondria, and spirochetes of kinetosomes and centrioles. Internal *parasites* are not usually thought of as endosymbionts.

endothelium, endothelia (pl.) The single layer of thin, flat cells lining vertebrate blood and lymph vessels and the heart. In *capillaries,* it is the only layer separating the blood and *intercellular* fluid. (compare *epithelium*)

endothermic (homeothermic) Animals that are able to regulate their body temperature by internal mechanisms, regardless of environmental conditions. This enables them to inhabit areas with more extreme environments. However, this involves the generation of a large amount of *metabolic* heat, utilizing considerable amounts of *ATP*. Endotherms use about 80% of the calories consumed to maintain body temperature. Behaviors that take advantage of environmental conditions are also employed, such as a dog panting to expend heat and a cat lying in the sun to store heat. Birds and mammals are homeothermic, "warm-blooded" in the vernacular, because their body temperature is usually warmer than their surroundings. (compare *ectothermic*)

endotoxins Toxic substances formed inside the *cell walls* of certain types of bacteria (*Gram-negative*) and released upon *lysis* of the cell. They are heat stable *glycolipids* complexed with protein. An example is *Salmonella typhi*, which causes typhoid fever. (compare *exotoxin*)

endotrophic mycorrhiza, mycorrhizae (pl.) *Mycorrhizae* (fungal-plant symbioses) that actually invade the *cortex* cells of the root. They are zygomycete based and are found mostly in association with broadleaf trees and herbs.

end plate A flattened nerve ending occurring at the junction (*synapse*) of a *motor neuron* and a muscle cell. Through *neurotransmitters*, it transmits the *nerve impulses* that activate or inhibit the muscle cell. (see *neuromuscular junction*)

end plate potential An *action potential* propagated to an *end plate* for transmission to a muscle cell.

end product The substance produced at the end of a *biochemical pathway*. (see *feedback inhibition*)

energy flow The movement of energy through an *ecosystem*: from the environment, through a series of organisms, and back to the external environment. A large amount of energy is lost at each *trophic level* in a *food chain*.

enhancer A *regulatory sequence* in mammalian cells that serves to maximize levels of *mRNA transcription* from a particular *promoter*. Enhancers can act on

promoters over a distance of several thousand *base pairs*, although they are usually much closer. They may be located either upstream or downstream of the promoter. Like promoters, they serve as recognition sites for *trans-acting* factors. They are widely utilized in determining tissue specific *gene expression*. (see *UAS, CCAAT box, GC box*)

enkephalin (see *endorphin*)

enol base One of the rare isomeric forms of a *nucleic acid base*. (see *tautomer*; compare *keto, imino*)

enteric bacteria (see *Proteobacteria*)

enterokinase (see *enteropeptidase*)

enteropeptidase (enterokinase) A *peptidase* enzyme (breaks *peptide bonds* in proteins) secreted by the small intestine that converts *trypsinogen* to *trypsin*, a digestive enzyme that itself digests proteins.

enthalpy (H) A thermodynamic term describing the energy locked in the chemical bonds of a molecule. Breakage of these bonds, for instance by enzymes, releases *free energy (G)* that is then available to do work, such as driving other biochemical reactions. During this process, enthalpy may be lost in the form of heat (*exothermic*) or gained and stored as new chemical bonds (*endothermic*). (see *entropy*)

entomology The study of insects.

entomophilic Flowering plants (*angiosperms*) that are *pollinated* by insects.

Entoprocta A phylum of marine invertebrates characterized by having both their mouth and anus inside a tentacle cresent. They make up *colonial, sessile* mats on various organic and inorganic *substrata* (seaweed, rocks, shells). Individuals are linked by horizontal *stolons*. They reproduce both by *fertilization* and *budding*. Some scientists believe that entoprocts and ectoprocts should be united into a single phylum.

entropy (S) A thermodynamic term describing the disorder or randomness in the universe. The Second Law of Thermodynamics states that "disorder in the universe constantly increases; Energy spontaneously converts to less organized forms." In biochemical reactions for instance, every time a chemical bond is broken to release *free energy (G)* for work, a little leaks away as the kinetic energy of motion and is lost to the system. The transformation is manifest as heat energy, temporarily increasing the temperature, and is no longer available to do work. (see *enthalpy*)

enucleate A cell lacking a *nucleus.*

enveloped viruses A group of viruses that carry lipid-rich envelopes. The lipids present are determined by the genetic machinery of their particular *host* and contribute to the specification of their infective properties. (see *arboviruses*)

environment The totality of the surroundings in which an organism lives. The term is often used to denote the surroundings in which all organsims live, in other words the air, water, and soil of the earth.

enzyme Any molecule that *catalyzes* (promotes) biochemical reactions without itself being changed or consumed. Enzymes work by lowering the *activation energy* for a chemical reaction and are extremely specific for particular reactions. Their specificities are defined by the molecular shape and charge at the *active site*, which fits the *substrate* like a lock and key. Enzymes are almost exclusively protein based. Recent discoveries (such as an *RNA* molecule that catalyzes its own *splicing*) have challenged this dogma. (see *coenzyme, cofactor, competitive inhibitor, feedback inhibition, constitutive, inducible, ribozyme*)

enzyme cascade (see *cascade*)

enzyme-cofactor complex The complex formed between an enzyme and a *cofactor*, often a metal ion. The metal ion effects an *allosteric* (shape-changing) change in the enzyme.

enzyme inhibition The temporary or permanent arrest of normal enzyme activity. *Allosteric inhibition* of an enzyme early in a biochemical pathway is achieved by binding of the final product to a modula-

tor site (*feedback inhibition*) that closes the *active site* by changing its shape. When levels of the final product are reduced, it disengages, reenabling the synthesis pathway. In *competitive inhibition*, an inhibitor competes with the normal substrate for an enzyme's active site. The level of inhibition is dependent on the relative concentrations of the substrate and the inhibitor. In *uncompetitive inhibition*, an inhibitor molecule combines with the *enzyme-substrate complex*, preventing it from completing the reaction. In *noncompetitive* inhibition, the inhibitor prevents dissolution of the enzyme-substrate complex by binding at a modifier site and effecting a deformation of the active site.

enzyme-multiplied immunoassay technique (EMIT) (see *immunoassay*)

enzyme-substrate complex The complex formed during the interaction of an enzyme with its specific *substrate.* The substrate fits into the *active site* of the enzyme like a lock and key.

Eocene The second oldest geological epoch of the tertiary period, about 55 million to 38 million years ago. It is characterized by the predominance of early hoofed mammals, the ungulates (perissodactyls and artiodactyls), and the presence of other mammals including *carnivores*, bats, and whales. Birds were also present, as were an abundance of marine protoctists.

eon The largest unit of geological time. Two eons have been named. In chronological order, they are the *Precambrian* and *Phanerozoic.*

eosinophil A circulating white blood cell (*leukocyte*) involved in bodily defense, particularly against larger *parasites* such as worms. Eosinophils work by fusing with the *plasma membrane* of cells and releasing a toxic protein. They may also have a role in dampening the *histamine* response associated with asthma and hay fever by releasing *antihistamines.* Eosinophils are characterized by a lobed *nucleus* and *cytoplasmic* granules that stain with an acidic dye from which they derive their name. (see *immune system, granulocyte*; compare *basophil neutrophil*)

ephemeral A plant that completes its *life cycle* very rapidly. In favorable environments, ephemerals may *germinate*, bloom, and set seed several times in a single year. Ephemerals are particularly characteristic of desert environments. Many weed species are also ephemerals.

Ephemeroptera An order of *exopterygote* insects containing the mayflies. They have long-lived aquatic *nymphs* that may molt over 20 times. The adults live from a few minutes to at most a day since they have only rudimentary mouthparts and neither eat nor drink.

ephyra, ephyrae (pl.) The free-swimming organism produced by *asexual budding* from the sedentary *polyps* of some cnidarians and ctenophores. It is common in *plankton* and develops into the adult *medusa*. (compare *planula*)

epi- A word element derived from Greek meaning "near," "on," or "over." (for example, *epidermis*)

epibiosis An association of organisms in which one lives on the surface of another. (see *ectoparasite*)

epiboly The stage during *embryonic* development of amphibians and bony fish where cells from the *animal pole* of the *blastula* spread to engulf the *vegetal* part of the egg, including its yolky cells.

epicalyx In plants, a whorl of *sepal*-like structures or *bracts* that resembles the *calyx* but is located outside and below it.

epicarp (see *pericarp*)

epicotyl In *germinating* plant seeds, that portion of the *embryonic* shoot that extends above the *cotyledons* (first leaves). (compare *hypocotyl*)

epidemiology The study of the distribution of disease or disease factors within the population.

epidermis The outermost layer of cells of any *multicellular* organism. In invertebrate animals, it consists of a single layer of *ectodermal epithelium* and sometimes secretes a protective, noncellular *cuticle*.

In higher animals, the epidermis may be composed of several layers of cells, the outermost ones often dead (*cornified*). Protective products produced by the vertebrate epidermis include hair, claws, nails, hooves, horns, feathers, beaks, scales, and shells. In plants, the epidermis is one cell thick and in aerial parts may secrete a noncellular cuticle. (see *stratum corneum*)

epididymis A long, narrow, coiled tube attached to the *testis* in reptiles, birds, and mammals. It acts as a temporary storage organ for *spermatozoa* until their release during mating.

epigamic A characteristic or feature of an animal that an individual of the opposite sex finds attractive during courtship. Examples include the bright male coloration in birds and fish and the song of birds.

epigeal Any growth or occurrence above ground level.
1. In plants, it refers to seed *germination* in which the *cotyledons* (first leaves) form the first *photosynthetic* organs above the ground. (compare *hypogeal*)
2. In animals, it refers to the inhabitation of exposed land surfaces as opposed to underground.

epigenesis The process by which a developing organism increases in complexity. It is governed primarily by the genetic program inherent in the *nuclear genome*. The genetic program is modified by interactions with the organized *cytoplasm* of the developing egg, and, potentially, also with the environment. The organization of cytoplasm within the un*fertilized* egg is genetically predetermined by the mother's genes.

epiglottis A flap of tissue in mammals that closes the *glottis* (the passage between the *larynx* and *pharynx*) during swallowing in order to prevent food from entering the windpipe and the lungs.

epigynous A flower in which the *calyx*, *corolla*, and *stamens* are inserted near the tip of the *ovary*, producing an *inferior* ovary. (compare *perigynous, hypogenous*)

epilimnion In freshwater lakes and ponds, the warmer water that accumulates as the upper layer during the summer. The abrupt temperature difference between it and the lower *hypolimnion* is called a *thermocline*. (see *fall overturn*)

epinastic The curving of a plant structure away from the axis, caused by enhanced growth on the upper (*adaxial*) surface. A common example is the downward bending of certain leaves. This response is facilitated by plant growth substances such as *auxin* and *ethylene* and may also be induced by *herbicides*. (see *nastic movements*; compare *hyponastic*)

epinephrine (see *adrenaline*)

epiphyseal plate (see *epiphysis*)

epiphysis The ends of the long bones in mammals that articulate to form joints. They are formed separately from the central shaft (*diaphysis*) and are separated from it by a *cartilaginous* plate (*epiphyseal plate*) that is responsible for growth in bone length. When maximum bone growth is reached, it *ossifies*, and the diaphysis and epiphysis fuse.

epiphyte Any plant growing upon or attached to another plant or object merely for physical support. Water and nourishment are obtained from rain and aerial debris. Examples of tropical epiphytes are ferns and orchids. Examples of temperate epiphytes are lichens, mosses, liverworts, and algae. (see *Raunkiaer's plant classification*; compare *parasite*)

episome A *DNA plasmid* (circular *extrachromosomal* genetic element) found in some bacteria that can reversibly *integrate* (insert) itself into the *chromosome* of its *host*. When integrated, it behaves as part of the chromosome, *replicating* with it. Integration and *excision* of the element is achieved by *homologous recombination*. (see *bacteriophage, F factor, transposon, lysogenic, prophage, induction*)

epistatic The interaction between different *gene products* in which one modifies the *phenotypic* expression of the other. This situation often manifests in multistep biochemical pathways, where a nonfunc-

tional *mutation* in an enzyme early in the pathway prevents any subsequent steps regardless of the state of the later *genes* or gene products.

epithelium, eplithelia (pl.) A sheet of closely packed coherent cells lining interior body cavities or covering exposed surfaces of the body. Individual epithelial cells may be characterized as *cubical, columnar, ciliated,* or *squamous* and may be *secretory* (form *glands*). Epithelial cells may be further organized into a single sheet (*simple*) or into multiple layers (*stratified*). (see *endoderm, ectoderm*)

epitope (antigenic determinant) The region of an antigen molecule that is unique, therefore responsible for its specificity in an *antigen-antibody reaction. Tertiary structure* and charge distribution seem to be most important. The epitope is recognized by the *antigen-binding site* of a specific *antibody* molecule that binds to it, forming an *antigen-antibody complex*. (see *antibody specificity, immunoglobulin, antibody-binding site*)

epoch (see *geological time scale*)

EPP (see *end plate potential*)

EPSP (see *excitatory postsynaptic potential*)

equational division (equatorial division) A *cell division* in which the number of chromosomes remains the same from the parent to the daughter cells. It is typical of *mitosis* and the second division of *meiosis*. (compare *reduction division*)

equatorial division (see *equational division*)

equatorial plate The imaginary plane at the equator of the *nuclear spindle* upon which the *centromeres* of duplicated *chromosomes* are aligned during *nuclear division*. (see *meiosis, mitosis, metaphase*)

equilibrium potential In a *nerve cell*, the point at which the electrical gradient for a particular ion exactly balances the concentration gradient, resulting in no net flow across the *cell membrane*. The equilibrium potential for potassium ions (K^+)

is usually slightly more negative internally than the *resting potential* of the cell. For sodium ions (Na^+), it is more negative externally. Resting cells are more permeable to potassium than to sodium. The resting potential is thus much nearer to and determined by the equilibrium potential for potassium. (see *sodium-potassium pump, voltage-gated ion channel, action potential, depolarization, hyperpolarization*)

Equisetophyta (see *Sphenophyta*)

ER (see *endoplasmic reticulum*)

ergosterol A sterol present in plants that, when ingested by animals, is converted to vitamin D_1 by ultraviolet radiation.

erythroblast The *bone marrow* cell from which *erythrocytes* (red blood cells) develop. During their development (*erythropoiesis*), they accumulate the red-colored *hemoglobin* that confers their ability to bind and transport oxygen. In mammals, they lose their *nuclei* during development. Erythroblasts provide a continual source of new erythrocytes.

erythrocyte (red blood cell) The blood cells that contain *hemoglobin* and are responsible for the transport of oxygen in the blood. Mammalian erythrocytes are circular, biconcave discs packed with hemoglobin and lacking *nuclei*. Since they have lost their nuclei in the interest of maximizing hemoglobin content, red cells are not able to produce or replenish proteins, hence they have a very short life (about 4 months). Spent cells are removed and destroyed by the liver. They are continually replaced by *erythropoiesis* of *erythroblasts* in the *bone marrow*. Erythrocytes also contain the enzyme *carbonic anhydrase,* which is important in transporting carbon dioxide and maintaining blood pH (acid/base balance). *Polysaccharides* attached to the outside of the cell membrane are used in determining blood type. In other vertebrates (for example, birds and reptiles), erythrocytes are oval and retain their nuclei. (see *bilirubin, blood group, serology*)

erythropoiesis The process by which *erythroblasts* in the *bone marrow* develop into mature *erythrocytes* (red blood cells). During this process, they accumulate the red-colored *hemoglobin* that confers their ability to bind and transport oxygen. In mammals, they lose their *nuclei*. When oxygen levels in the blood fall below normal levels, increased amounts of the hormone *erythropoietin* are secreted into the blood. This stimulates erythropoiesis, increasing the oxygen-carrying capacity of the blood. This process is important at high altitudes, for instance, where lower oxygen pressure requires compensation by an increased number of erythrocytes.

erythropoietin A *hormone* secreted by the kidney and other organs that stimulates the development of mature erythrocytes (*red blood cells*) from *erythroblasts* in the *bone marrow* (*erythropoiesis*).

***Escherichia coli* (*E. coli*)** A *Gram-negative*, rod-shaped bacterium present in the intestinal tract of animals, soil, and water. Though *E. coli* is normally present in a *symbiotic* relationship in the colon of mammals, some strains can cause disease. Because of its simple genetic system and rapid growth characteristics, it has been used extensively in bacterial genetics and molecular biology research. It was the first organism used for *cloning* and propagating the genes of other species. (see *colon, vitamin K, insulin*)

-esis A word ending derived from Greek used to add motion or process to a noun. (for example, *electrophoresis*)

esophagus The muscular tube leading from the *pharynx* (throat) to the stomach in all animals with a *digestive cavity*.

essential amino acid (see *amino acid*)

essential element A chemical element that is required for the normal growth, development, and maintenance of a living organism. Some are required in relatively large quantities, and may be involved in several different *metabolic* reactions. Examples include carbon, hydrogen, oxygen, nitrogen, sulphur, phosphorus, potassium, magnesium, and calcium. Others are required only in trace amounts, such as iron, manganese, molybdenum, boron, zinc, copper, cobalt, iodine, and selenium. These are important principally as *cofactors* for different enzymes.

essential fatty acids (see *fatty acids*)

essential minerals Various elements required for the health and growth of an organism. Some, such as calcium and phosphorus, are required in somewhat larger amounts, and others are needed in trace amounts. (see *trace elements*)

ester A compound formed by reaction of a *carboxylic acid* with an alcohol:

$$RCOOH + HOR_1 = RCOOR_1 + H2O$$

Glycerides are esters of long-chain *fatty acids* and *glycerol*.

estivation

1. The arrangement of parts in a flower bud before opening.
2. The *dormancy* of some animals during a hot, dry season. For example, lungfish bury themselves in a mud bottom to await emergence at the start of the rainy season.

estradiol The strongest, naturally occuring *estrogen*.

estriol A naturally occuring, rather weak *estrogen*.

estrogen A collective term for the female *hormones*, the strongest of which is *estradiol*. They are produced in the *ovaries* (and in small amounts in the *testes* and *adrenal glands*). In women, estrogens function in the development of *secondary sex characteristics* and in regulating the *menstrual cycle*. In particular, it prepares and maintains the uterine lining for pregnancy. (compare *progesterone*)

estrone An *estrogen*.

estrus The period when most female mammals are sexually receptive or "in heat." Only human females and some apes exhibit no *estrus cycle* and are continually sexually receptive. (see *menstrual cycle*)

estrus cycle The periods of *estrus* (sexual receptivity) corresponding to *ovulation*

events during a periodic cycle in mammals. In small mammals, they tend to be more frequent than in larger mammals. (see *menstrual cycle*)

ethene (see *ethylene*)

ethidium bromide A compound that fluoresces under ultraviolet light and is used in the visualization of *double-stranded DNA*. It works by *intercalating* into the helical structure of the DNA molecule, which also makes it extremely *mutagenic*. Ethidium bromide also binds to single-stranded DNA and *RNA*, but to a much lesser extent. (see *double helix*)

ethology The study of natural animal behaviors.

ethylene (ethene) A gas (C_2H_4), naturally produced by plants, that functions as a *growth substance* in the control of processes such as *germination*, cell growth, fruit ripening, *senescence,* and *abscission*. It is also involved in the response of a plant to gravity and stress. Ethylene produced by ripening fruits stimulates other fruits to ripen, hence the trick of putting ripe bananas in a paper bag with green tomatoes. Ethylene is also used commercially in fruit ripening and horticultural plant flowering. Ecologically, ethylene production increases rapidly following exposure of a plant to toxic chemicals such as *ozone*, temperature extremes, drought, or attack by diseased organisms or animals. This can accelerate the dropping of leaves or fruits that have been damaged by these stresses. It may also be a signal for plants to activate their defense mechanisms, including the production of molecules that are toxic to the animals or pests attacking them.

etiolation The type of growth exhibited by green plants when grown in darkness. Such plants are pale due to the absence of *chlorophyll*, their stems are abnormally long, and their leaves reduced in size. Such growth maximizes the chances of the shoot reaching light, on which it depends for *photosynthesis* and energy production. Modified *chloroplasts* called *etioplasts* are formed under these conditions, containing an abnormal membrane structure and lacking in *polysomes* and *starch*. Adequate lighting restores normal growth.

etioplast An undeveloped *plastid* that lacks *chlorophyll*. It is typical of plants grown in the dark. (see *etiolation*)

Eubacteria A subkingdom of bacteria. All of its members are *prokaryotes*. As such, they lack a membrane bound *nucleus,* structured *chromosomes*, and any complex *cell organelles* or internal compartmentalization. Reproduction is predominantly *asexual*, although *conjugation* is seen. Any motility is achieved by a *flagella* containing a single fiber of a *flagellin* protein. Eubacteria have been separated from Archaea based on their *DNA* sequences (usually the 16S *rRNA* subunit) as well as other traits.

euchromatin (see *chromatin*)

eugenics A field that deals with efforts to change the genetic characteristics of the human population by artificial selection. Aside from the obvious moral implications, such efforts are essentially impossible to implement in any plausible human time frame.

euglenid (see *Discomitochondria*)

Euglenophyta (see *Discomitochondria*)

Eukarya (Eukaryotae) One of the two superkingdoms of life, the other being Prokarya. It comprises all the *eukaryotes*, organisms that arose from evolutionary *symbioses* (*symbiogenesis*) and contain membrane bound *organelles*, including a *nucleus*.

eukaryote A cell characterized by membrane bound *organelles*, including the *nucleus*. The *DNA* is structured into *chromosomes* that include both proteins and *RNA*. Animals, plants, fungi, and protoctists are all eukaryotes. (compare *prokaryote*)

Eumetazoa The animal subkingdom comprising the more sophisticated animals that possess definite symmetry and differentiated *tissues*. Except for the cnidarians and ctenophores, they have three body layers.

Euphausiacea An order of shrimplike, marine crustaceans known as krill. They occur in dense swarms and feed on *diatoms*. They themselves comprise the main food source of food for *filter feeding* (*baleen*) whales and are an important constituent of the oceanic *food web*.

euploid A cell containing an exact multiple of the normal *haploid* (one set) number of *chromosomes*. For example, if the haploid number is 3, a euploid number would be 6, 9, 12, 15, and so on, with each chromosome represented in equal number. For many organisms, in particular some plants, this is a normal state. (compare *aneuploid*)

Euryarchaeota The phylum of archaebacteria containing the methanogens and halophilic bacteria. The methanogens are *obligate anaerobes* and are poisoned by oxygen. They produce marsh gas (primarily methane [CH_4]) and are a major source of the methane found in natural gas reserves. Methanogens are also responsible for mobilizing *photosynthetically*-produced carbon in the ground. The methane (CH_4) they produce reacts with atmospheric oxygen (O_2) to form carbon dioxide (CO_2). Some methanogens have been incorporated as *symbionts* into *methanogenic* ciliates that live in *anoxic* environments such as sea water, sediment, and sewage. The halophils, or salt-requiring bacteria, are tiny organisms that thrive in the mud flats in salt marshes and are poisoned by freshwater. They boast special *cell membrane* components, such as derivatives of glycerol, that allow them to withstand the extreme salt concentrations. Halophilic bacteria are obligate *aerobes*. They are colored pink or orange from the *carotenoid bacteriorhodopsin*, chemically similar to the *rhodopsin* found in the vertebrate *retina*.

eustachian tube A tube connecting the *middle ear* with the back of the throat. It maintains atmospheric pressure on both sides of the eardrum (*tympanum*). Any external change in pressure is equalized by swallowing or yawning, which opens the tube permitting air exchange with the middle ear. (see *gill slit*)

Eustigmatophyta A somewhat obscure phylum of *unicellular photosynthetic protoctists*, the eustigmatophytes or eustigs. They are yellow-green and contain a single long *xanthoplast*, like that of xanthophytes. They are characterized by an *eyespot* containing a form of *rhodopsin*, a *photoreceptor* molecule also present in the vertebrate eye. This eyespot functions to direct the cell toward optimally lighted environments for maximum photosynthetic efficiency. Eustigs reproduce only by *asexual reproduction.*

Eutheria (placental mammals) The subclass of mammals that nourish their developing *embryos* within the body of the mother for the majority of development by means of a *placenta.* They are the dominant group of mammals in the modern world.

eutrophic Ponds and lakes that are rich in nutrients and organic debris. Consequently they are able to support microorganisms such as *plankton* and algae, which establish a base for the *food web. Eutrophication* is a process whereby fresh water becomes enriched in nutrients, thus beginning the cycle of *ecological succession.* When this happens as a result of sewage or fertilizer runoff, the concentrated overstimulation of algal growth results in a *bloom.* When the excess dead algae are decomposed by *aerobic* bacteria at an abnormally high rate, oxygen is depleted from the water, causing aquatic animals such as fish to die of suffocation. (compare *oligotrophic*)

even-aged stands A forest area where all live trees began growth from seed or roots planted at about the same time. It is typical of *second growth forests.* (compare *uneven-aged stands*)

evergreen Plants that retain their leaves year round. (compare *deciduous*)

evolution (see *biological evolution, microevolution, macroevolution*)

evolutionary taxonomy The most common approach to the classification of biological organisms, based on physical characteristics and *phylogenetics.* As more techniques, such as *DNA analysis,* become available to investigate phylogenetic relationships, classification based on evolutionary taxonomy continues to be refined.

exartate In insects that undergo *metamorphosis, pupae* in which the wings and legs are free and, therefore, that are capable of limited movement. (compare *coarctate, obtect*)

excision The event in which an *integrated plasmid,* virus, or *transposon* exits the *host genome* and either assumes a free existence or integrates elsewhere. (see *lysogeny, temperate virus, episome*)

excitatory postsynaptic potential (EPSP) An event occuring at an excitatory *nerve synapse* that may cause the second *neuron* to fire. At excitatory synapses, the binding of *neurotransmitter* molecules to the postsynaptic *receptors* increases the membrane permeability to sodium ions through *chemically gated ion channels.* This produces a *depolarization,* thus encouraging an *action potential.* It may also occur at the *end plate* at a *neuromuscular junction,* triggering *muscle contraction.* (see *acetylcholine, sodium-potassium pump, facilitation, summation;* compare *inhibitory postsynaptic potential*)

excitatory synapse A nerve *synapse* in which the *postsynaptic* receptor protein is a *chemically gated ion channel* specific for sodium ions (Na^+) and is closed at rest. Binding of a *neurotransmitter* released from the *presynaptic* membrane opens the channel, initiating an *excitatory postsynaptic potential* that leads to *depolarization,* enabling a *nerve impulse.* In the vertebrate *nervous system,* the combination of different specific neurotransmitters, each with specific receptors, permits distinct responses to a *nerve impulse* at different junctions. (see *action potential, synaptic cleft;* compare *inhibitory synapse*)

exclusion chromatography (see *gel*

filtration)

excretion The process by which waste is eliminated from the body. Waste is defined as any substance that is a product of the *metabolism* of cells and would be toxic if allowed to accumulate. In *unicellular* and simple *multicellular* organisms, excretion is accomplished chiefly by diffusion through *cell membranes* and body surfaces, and by *exocytosis*. The simplest method of excretion is *passive diffusion*, the method by which carbon dioxide gas and water vapor are discharged at the surface of lungs or gills. *Urea* or *uric acid*, as well as various salts, are excreted through the kidneys in vertebrates, *Malpighian tubules* in arthropods, and the *labyrinth* in crustaceans. The quantity of water in which *nitrogenous* wastes are dissolved depends on the *physiology* and *environment* of each individual organism. For instance, mammals and especially birds have *nephridia* in the kidney that are extremely efficient at reabsorbing and conserving water. Conversely, freshwater fish constantly absorb water from their environment and continuously excrete all they ingest. The gut also serves as a route for excretory products. However, it is not considered an excretory organ since most solid waste is simply *egested* without being metabolized. (see *nephridium, nephron, glomerulus, Bowman's capsule, loop of Henle, renal tubule, ADH, aldosterone, hypothalamus, adrenal gland, ureotelic, uricotelic*)

excyst To emerge from a *dormant* protective *cyst*. (compare *encyst*)

exergonic A chemical reaction that releases energy. They tend to proceed spontaneously, although energy is required to initiate them in biological systems. (compare *endergonic*)

exine The decay resistant outer coat of a *pollen* grain or *spore*. The exine is characteristic of plant families, genera, and sometimes species. With the advent of the *scanning electron microscope*, exine has become an important identification characteristic. Exine characteristics are used in the identification and quantitative analysis of the vegetation composition of *peats* and other organic sedimentary deposits.

exo- A word element derived from Greek meaning "outside" or "outer." (for example, *exogenous*)

exocarp (see *pericarp*)

exocrine gland Any *gland* that employs a duct for secretion. Ducts may pass to the body surface (for example, *sweat* and *mammary glands*) or may open to an internal *epithelial* layer such as the mouth, stomach, or intestine. (compare *endocrine gland*)

exocytosis The bulk transport of materials out of the cell across the *cell membrane*. Undigested food, *excretory* materials, or *secretory* materials are encased within membranes, forming a *vacuole* that is transported to the cell surface. The vacuole fuses with the *plasma membrane*, discharging its contents to the outside.

exodermis (see *hypodermis*)

exogenote In bacterial genetics, the usually incomplete *genome* derived from an *Hfr cell* during a *conjugation* event. (compare *endogenote, merozygote*)

exogenous Any substance that is produced or originates outside an organism. (compare *endogenous*)

exon A segment of a *eukaryotic gene* that is both *transcribed* into *RNA* and *translated* into protein. An exon is always a *coding region* of DNA. Exons are characteristic of *eukaryotes*, where they are interspersed with noncoding regions (*introns*). (see *splicing, hnRNA, primary transcript*)

exon shuffling The hypothesis that the arrangement of *eukaryotic genes* into discrete *coding regions* (*exons*) allows for more rapid evolution of proteins by providing an easy way of rearrangement.

exonuclease An enzyme that catalyzes the *hydrolysis* (degradation) of *nucleic acids* starting from the ends of the molecules, thus sequentially removing terminal nucleotides. (see *DNase*; compare *endonuclease*)

exopodite (see *biramous appendage*)

exopterygote The small group of winged insects that undergo incomplete *metamorphosis*. The wings develop externally to the body. No rapid metamorphosis or true *pupal* stage is seen. Successive *larval* stages, called *nymphs*, become progressively more adultlike. Grasshoppers, aphids, and termites are examples of exopterygotes. (compare *endopterygote*)

exoskeleton A protective, supportive skeleton covering the outside of the body and providing attachment for muscles. In arthropods (insects and crustaceans), it is found as a *cuticle* secreted by the *epidermis*. Growth of the body may occur only in stages, by a series of *molts*. In vertebrates (for example, tortoises and armadillos), the exoskeleton consists of bony plates beneath the epidermis. The scales and *denticles* of modern fish are evolutionary remnants of bony skin plates. The term is sometimes also applied to other hard, external structures such as the shells of mollusks. (see *metamorphosis*; compare *endoskeleton*)

exosporium, exosporia (pl.) (see *exine*)

exotic species A species that is newly introduced into an area with an existing *ecology*, either through natural processes or by direct or indirect human intervention. (compare *endemic*)

exotoxin Any *toxin* released by a microorganism into the surrounding growth medium or tissue. They are generally heat labile proteins produced mainly by *Gram-positive* bacteria and may be neutralized by *antibodies*. Diseases caused by exotoxins include tetanus, diphtheria, and botulism. (compare *endotoxin*).

explantation The *culture* of isolated animal tissues in an artificial environment (*in vitro*). (compare *implantation, transplantation*)

exploratory behavior A behavior in which an animal examines territory or objects with which it is unfamiliar, for instance to find food or nesting material. It is often seen in young animals as they learn about their *environment*. (see *appe-titive behavior*)

exponential growth A type of population growth in which the rate of increase is dependent only upon the number of individuals and the potential net reproductive rate. During exponential growth, resources are not a limiting factor, nor is there any inhibitory effect between individuals. Such situation is characteristic of the initial growth phase of microorganisms in culture. It is also seen when organisms are introduced into regions where neither food or space is limiting and where natural controls such as predators and *parasites* are absent. When environmental factors such as food availability, waste production, or competition between individuals become limiting, the *growth curve* levels out. Thus, a typical exponential growth curve is sigmoid (S-shaped).

expressivity In genetics, the variable expression of particular *genes* in an organism, resulting in a range of *phenotypes* (physical manifestation). (compare *penetrance*)

extensor One of the pair of *antagonistic* (*agonistic*) muscle sets that mediate voluntary movements. The extensor relaxes when the joint is flexed and contracts to extend the joint. (compare *flexor*)

external fertilization The union of *gametes* outside the mother's body in an aquatic environment. It was the first type of *sexual reproduction* to evolve. It usually involves the release of eggs and sperm in the same general area at approximately the same time. One obvious problem is the dilution of gametes, which is counteracted by the production of large numbers, and release of both eggs and sperm at the same time. The release is often cued by the lunar cycle. (compare *internal fertilization*)

extinction The disappearance of a life form, usually a species, from existence. (see *global extinction, local extinction*)

extracellular Any structure or process situated outside a cell.

extrachromosomal DNA Any *DNA* found

outside the *nucleus* of a cell and *replicating* independently of the *chromosomal* DNA. In *prokaryotic* cells and some *eukaryotes* such as yeast, extrachromosomal DNA is often found in the form of *plasmids* In eukaryotic cells, it is generally found in self-replicating *organelles* such as *mitochondria, chloroplasts, and centrioles.* Cells and organsims containing extrachromosomal DNA exhibit *cytoplasmic inheritance.* (see *uniparental inheritance, maternal inheritance, episome*)

extrachromosomal element (see *extra-chromosomal DNA*)

extraembryonic membranes The membranes developed by many vertebrate *embryos* for the purpose of nutrition and protection. They are not directly involved in the development of embryonic structures. (see *allantois, amnion, chorion, yolk sac*)

extranuclear Any material residing outside the *nucleus* of a *eukaryotic* cell. (see *extranuclear genes*)

extranuclear genes In *eukaryotes*, any *genes* residing in the *cytoplasmic organelles* such as *mitochondria, chloro-*

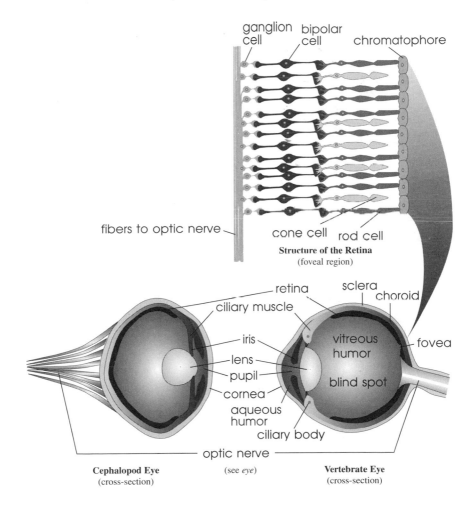

Structure of the Retina
(foveal region)

Cephalopod Eye
(cross-section)

(see *eye*)

Vertebrate Eye
(cross-section)

plasts, and *centrioles.* (see *mitochondrial DNA, endomitosis, uniparental inheritance, maternal inheritance, cytoplasmic inheritance*)

extrorse A flower in which the *anthers* are turned outward, away from the female *gynoecium.* This situation favors *cross-pollination.* (compare *introrse*)

eye An organ of sight or light perception. Most invertebrate eyes are simple *photoreceptors,* sensitive only to the direction and intensity of light. Higher mollusks and arthropods have *compound eyes* that form actual images. Cephalopod mollusks have eyes that bear a striking resemblance to vertebrate eyes; they probably evolved by analogous evolution. The vertebrate eye is a spherical structure, connected to the brain by the *optic nerve.* It has an outer white *sclerotic* coat with a transparent front, the *cornea.* This is lined by the *vascular* pigmented *choroid,* continuous with the *ciliary body* and the *iris* in front. In the center of the iris is a hole, the *pupil,* through which light enters to be focused by the *lens* onto the retina. This is the innermost layer and contains the light and color sensitive *rod* and *cone* cells. (see *rhodopsin*)

eye socket (see *orbit*)

eyespot (stigma)

1. A light sensitive organ that occurs in some *photosynthetic* protists (for example, *Euglena*), enabling them to sense and move toward light conditions optimal for *photosynthesis.* It contains *rhodopsin,* the same pigment used in vertebrates vision. (see *phototaxis*)

2. A light sensitive pigmented spot found on the heads of some primitive animals (for example, flatworms) and connected with the *nervous system.* It enables the detection of light and movement away from it.

eye tooth (see *canine tooth*)

F_1 Shorthand for "first filial generation." It is the first generation of offspring resulting from a particular genetic cross or set of parents. (see *monohybrid, dihybrid, dominant, recessive, sex-linked*; compare *P*)

F_2 Shorthand for "second filial generation." It is the generation of offspring resulting from a particular genetic cross or set of parents. Genetic characteristics of interest are often revealed in the F_2 generation. (see *monohybrid, dihybrid, dominant, recessive, sex-linked*; compare *P*)

facial nerves (cranial nerve VII) The pair of nerves that supply the facial muscles. They also carry *sensory nerve fibers* from the *taste buds* and *autonomic* nerve fibers to the *salivary glands* and *lacrimal* (tear) glands. (see *cranial nerves*)

facilitated diffusion The carrier-assisted transport of molecules across a *cell membrane*, down a concentration gradient. The membrane would be otherwise impermeable to the specific molecule. It is differentiated from *passive diffusion* by the use of *transmembrane carrier proteins* which serve only to accelerate the rate of equilibrium attainment; no energy is expended in this process. Examples include the transport of *glucose* across the *cell membranes* of *skeletal muscle* fibers and *adipose* cells. (compare *osmosis, active transport*)

facilitated transport (see *facilitated diffusion*)

facilitation
1. The increase in the responsiveness of a *neuron* or *effector* cell to successive stimuli, each one rendering the *postsynaptic* membrane more responsive to the next impulse. Eventually, the membrane will become *depolarized* to a degree large enough to trigger an *action potential*. (compare *summation*)
2. The intensification of a behavior that is caused by the presence of another animal of the same species.
3. In *ecological succession*, the prepara-

tion of the environment or other ways in which the entry of subsequent species is facilitated.

factor I (see *fibrinogen*)

factor II (see *prothrombin*)

factor III (see *thrombokinase*)

factor IV (see *blood clotting*)

factor VIII (see *blood clotting*)

facultative An organism able to adopt an alternate mode of living. A facultative *anaerobe* may survive *aerobically* (in the presence of oxygen) or *anaerobically* (without oxygen) when conditions necessitate. (compare *obligate*)

facultative mutualism (see *protcooperation*)

FAD (flavin adenine dinucleotide) A *coenzyme* important in *redox reactions*, particularly in the *Krebs cycle* and the *electron transport chain* in *oxidative phosphorylation*. It is derived from the vitamin *riboflavin*. (see *flavoprotein*)

fairy ring A ring of mushrooms that is formed by radial growth from an underground *mycelium*. It is seen mainly in basidiomycctc fungi. (see *fruiting body*)

fairy shrimps (see *Anostraca*)

Falconiformes An order of birds comprising the birds of prey including vultures, ospreys, hawks, eagles, and falcons.

falcons (see *Falconiformes*)

fall overturn The annual mixing of the warmer upper layer (*epilimnion*) with the cooler lower layer (*hypolimnion*) of freshwater lakes and ponds. In autumn when the epilimnion cools to the same temperature as the hypolimnion, a thermal inversion occurs, mixing the oxygen dissolved in the surface layer with the nutrients formed in the lower layers. (see *upwelling*; compare *spring overturn*)

fallopian tube One of a pair of *oviducts* in female mammals that convey *ova* (eggs) from the *ovary* to the *uterus*. The egg is released from the ovary into a *ciliated* funnel at the beginning of each fallopian tube, which transports it with the aid of muscular and *ciliary* action.

false scorpions (see *Pseudoscorpiones*)

family In the taxanomic classification of organisms, a collection of similar *genera*. Similar families are grouped into an *order*. Familial suffixes normally end in -*"ceae"* in botany and -*"idae"* in zoology.

fascia, faciae (pl.) A sheet of *connective tissue*. The layer of *adipose tissue* under the human *dermis* and the sheets of tough connective tissue around muscles are types of faciae.

fat The major and most efficient form of lipid storage in higher animals and some plants. *Adipose tissue* is composed of cells filled with little else but fat globules. Fat consists mainly of *triglycerides* of long-chain *fatty acids* (*carboxylic acids*).

fat body A mass of *adipose tissue* forming a definite structure within the body cavity of some animals. In amphibians and reptiles, a pair of solid fat bodies are attached to the kidneys or near the rectum and act as a food source during *hibernation* and breeding. In insects, the bodies are more diffuse and include protein and *glycogen* stores as well as fat. They may be used as energy reserves during *metamorphosis* or in adults that do not feed during this stage.

fate map A diagram showing the developmental fate of each region of an *embryo*, usually at the *blastula* stage. These divisions are genetically determined by information contained in the maternal *chromosomes*. (see *differentiation, determined*)

fatty acid Organic acids that form the building blocks for lipids, oils, and waxes. Essential fatty acids are those required for growth and health that cannot be synthesized by the body. Therefore, they must be included in the diet. *Linoleic acid* and *gamma-linolenic* acids are the only essential fatty acids in man. They are required for *cell membrane* synthesis and fat *metabolism*. Essential fatty acids occur mainly in vegetable seed oils.

fauna Animal life.

FB elements A family of *mobile genetic elements* found in *Drosophila*. They range in size from a few hundred *bp* (base pairs) to several *kb* (kilobase pairs) and are somewhat heterologous in *nucleotide* sequence. They are able to mediate their own *transposition* to a distant location in the *genome*. (see *transposon*; compare *TY element, copialike elements, P element, LINEs*)

F$^+$ cell In *E. coli*, a cell carrying a free *F factor* (fertility factor). It is considered a male cell.

F$^-$ cell In *E. coli*, a cell lacking an *F factor* (fertility factor). It is considered a female cell.

Fd (ferrodoxins) A group of proteins found in green plants and cyanobacteria that contain non*heme* iron in association with sulphur. They function as electron carriers in *photosynthesis* and possibly also in *nitrogen fixation*. A ferrodoxin is the last element in the *electron transport chain* of *photosystem I* before the formation of *NADPH*.

feathers The body covering of birds. They are *epidermal* outgrowths of *keratin* that provide thermal insulation, streamline the body, and assist in flight. They are often brightly colored and are important in *display*. Each feather has a base *quill*, which grows out of the feather *follicle* in the skin, and continues into the *vane* of the feather as the *rachis*. A series of *barbs* and *barbules* hook the fibers together to solidify and stiffen the feather structure. Distinct types of feathers include *down, intermediate, filoplumes,* and *bristles*. Worn feathers are renewed annually by *molting*.

feather stars (see *Crinoidea*)

feces Solid or semisolid material, consisting of undigested food, bacteria, dead cells, *mucus, bile,* and other secretions, that is expelled from the *alimentary canal* through the anus.

fecundity (see *fertility*)

feedback inhibition The inhibition of an initial or early enzyme in a *biochemical pathway* by its *end product* in order to regulate the cellular concentration. When an excess amount of end product is accumulated, it binds to an *allosteric site* on the enzyme, changing its shape so that it can no longer bind with its *substrate*. When the end product concentration drops again, it no longer binds to the enzyme to any significant degree and normal production is reestablished. (see *feedback regulation*)

feedback regulation A general organismal control mechanism involving the regulation of any factor by its own magnitude. It is sometimes called a feedback loop. Examples include the *homeostatic* regulation of enzyme and *hormone* levels, ion concentration, and temperature. (see *feedback inhibition, negative feedback, positive feedback*)

femur The long bone forming the upper bone or thigh bone of the hindleg in ani-

mals, extending from the hip to the knee.

fenestra ovalis (oval window) A membrane that separates the *middle* from the *inner ear*. It conveys vibrations from the *stapes*, an ear *ossicle* of the *middle ear*, to the liquid of the *inner ear*.

fenestra rotunda (round window) A membrane between the *middle* and *inner ear*. It moves back and forth to compensate for pressure changes in the *perilymph* that are propagated by vibrations of the *fenestra ovalis* (oval window).

fermentation The *anaerobic metabolism* of organic substances by microorganisms. It is a form of *anaerobic respiration*, and is typically seen in certain bacteria and yeasts. Chemically, it is defined as respiration in which the *terminal electron acceptor* is an organic molecule rather than molecular oxygen (O_2). The intermediate products of fermentation,

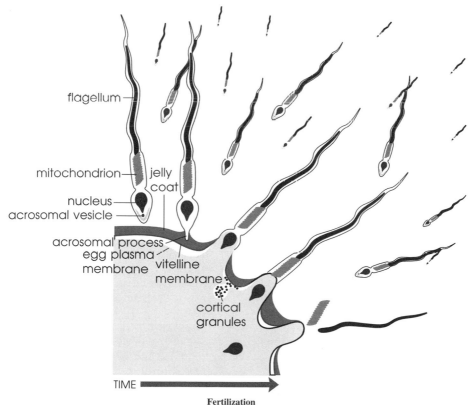

Fertilization
(Sperm Penetration of a Sea Urchin Egg)

ethanol and carbon dioxide, are important in the brewing and baking industries. (see *glycolysis*; compare *aerobic respiration*)

Fermenting bacteria (see *Endospora, Saprospirae*)

ferns (see *Filicinophyta*)

ferrodoxins (see *Fd*)

fertility The capacity of an organism to produce offspring.

fertility factor (see *F factor*)

fertilization The fusion of two *haploid gametic nuclei* to form a *diploid zygote nucleus*. This process, leading to the exchange of genetic information, is the heart of *sexual reproduction*. External fertilization is typical of lower plants and aquatic animals, where gametes are simply expelled from the parental bodies in great numbers and drift over each other in either wind or water currents. Internal fertilization takes place within the body of the female and is primarily an adaption to a terrestrial environment. It also provides increased protection for the *embryo* in mammals and seed plants by the parent's body and in birds, reptiles, and amphibians by the egg shell. (see *capacitation*)

fertilization membrane (see *vitelline membrane*)

Fe-S center (iron-sulfur center) An important motif repeated numerous times in the *electron transport chain* of *photosystems I* and *II* as well as in *aerobic respiration.*

fetal Any structure or function relating to a *fetus.*

fetal membrane (see *extraembryonic membranes*)

fetus An unborn or unhatched *embryo* that has passed through the earliest developmental stages and attained its basic structural plan. In humans, a developing individual is considered a fetus from about the second month of *gestation* until birth.

F factor (fertility factor) A *plasmid* found in some bacteria (for example, *E. coli*) that is capable of transferring a copy of it-self from its *host* cell (F^+) to one lacking an F factor (F^-). When an F factor is *integrated* into the bacterial *chromosome* (*Hfr cell*), it is capable of mobilizing the transfer of a *replicated* copy of the entire chromosome to an F^- *cell* (in practice, only a portion of the chromosome is usually transferred). The *F plasmid* contains *genes* that direct the formation of a hollow tube called a *pilus*, forming a *conjugation* bridge between the two bacterial cells and allowing transfer of the donor chromosome. Genes for *antibiotic* resistance are often carried on F factors. (see *F', extrachromosomal DNA element, episome, rolling circle replication*)

F' factor In *E. coli*, an F factor into which a few genes from the bacterial *chromosome* have become incorporated.

fiber Any long narrow structure.
1. A nonliving biological substance that is secreted by cells and performs a structural function. Examples include the *collagen* fibers in skin, cartilage, and tendon, *fibrin* at the site of a wound, and spider web silk.
2. A cell that has an elongated shape as is appropriate to its function. Examples included *nerve* and *muscle fibers.*
3. A specialized plant cell found in woody tissue (*sclerenchyma*) that has thickened walls and serves a support function. When mature, the cells die, leaving only the support structure in place. Such fibers may be economically important, such as flax.

fibril Any thread-shaped solid structure or filament.

fibrin An insoluble, fibrous *structural protein* that is essential in the formation of *blood clots*. It forms at the site of a wound from the soluble precursor *fibrinogen* by the action of the enzyme *thrombin*. A meshwork of fibrin fibers rapidly *polymerizes*, closing the wound.

fibrinogen A protein found in blood that is the soluble precursor of *fibrin*, the structural element in *blood clots*. Fibrinogen is converted into fibrin by the action of the enzyme *thrombin*.

fibroblast A type of *connective tissue* cell. Fibroblasts secrete *structural proteins*, most commonly *collagen*, forming a matrix in which they become embedded.

fibrous protein (see *structural protein*)

fibula One of two long bones in the lower hindlimb of animals between the knee and the ankle. It forms the prominence on the outer side of the ankle joint.

field capacity The amount of water held in a given soil after any excess has been drained by gravity.

fight-or-flight Alternative responses of an animal to a crisis situation, such as attack by a predator. The physiological manifestations that ready the animal for either response include the release of large amounts of *adrenaline* by the *sympathetic* portions of the *autonomic nervous system* leading to an increase in heartbeat and respiration and redirection of the blood supply from the internal organs to the *skeletal muscles*. Response to stress in the modern environment is essentially an inappropriate manifestation of the fight-or-flight response.

filament Any long, thin structure.
   1. Examples include the *actomyosin* filaments in muscle, the shaft of a down feather, and the *vegetative* body of filamentous algae.
   2. In a flower, the stalk of the *stamen* that bears the *anther.*

filamentous bacteria (see *Actinobacteria*)

Filicinophyta (Pterophyta, Pterodatina) The plant phylum comprising the ferns (except the whisk ferns). They are seedless, *vascular plants* in which the *sporophyte* generation bears characteristic divided feathery leaves called *fronds*; they differ botanically from *angiosperm* leaves. Ferns reproduce by way of *spores*, often found on the underside of the fronds or on a separate stalk. Motile *spermatozoa* fertilize eggs that developed into characteristic heart-shaped *gametophytes*. Many of the species are tropical.

filoplumes Hairlike *feathers* consisting of a bare shaft (*rachis*) tipped with a few *barbs* that are not held together by *barbules.*

filopodium, filopodia (pl.) A specialized *psuedopodium* used by certain *amebas* in food capture and locomotion.

filter feeding (ciliary feeding) A feeding method common in aquatic invertebrates and some whales in which suspended food particles are strained from the surrounding water. Some animals simply take advantage of natural water currents, while others produce a directed flow with *cilia. Baleen* whales have developed large plates to strain huge quantities of *krill*, their sole food source.

filtration A technique used to separate solids from liquid. The solution is applied to a filter made of paper or various synthetic materials with pores of the appropriate size. Separation may be accomplished by gravity or by application of a vaccum.

fingerprint In protein analysis, a characteristic pattern of bands or spots produced by the *electrophoretic* separation of *polypeptide* fragments. A reproducible set of fragments is obtained by the *degradation* of a protein with *proteolytic* enzymes. (see *stratum basal, DNA analysis, DNA profile*)

fins Flattened appendages found in aquatic organisms that are used for swimming, steering, and balance. The skinfold forming the fin membrane is supported by *cartilaginous*, horny, or bony fin rays that can range from soft to inflexible (spines). Most fish have paired *pectoral* and *pelvic* fins, *homologous* to the forelimbs and hindlimbs of terrestrial vertebrates, that are used for controlling the angle of ascent or descent. In some species, the pelvic fins are modified for *copulation*. Some fish also have a dorsal fin(s) and/or ventral (anal) fin(s) used for preventing rolling and sideways movement and a *caudal* (tail) fin used to propel the animal in a continuous forward motion.

FISH (fluorescent *in situ* hybridization) A method of locating a particular *gene* or region on a *chromosome* by *hybridizing* a fluorescently *labeled probe* to intact *chromosomes.* Probes to different *genes* or *genetic markers* can be labeled with differently colored fluorescent tags. A complete *karyotype* can now be performed, with each chromosome highlighted in a different color (*chromosome painting*). (see *in situ hybridization*)

fish (see *Osteichthyes, Chondrichthyes*)

fish lice (see *Branchiura*)

fission (see *binary fission*)

fitness In an evolutionary context, the ability of an organism to produce a large number of offspring that survive to reproduce themselves. (see *natural selection, Darwinism*)

five prime (5′) The end of a strand of *nucleic acid* (*DNA, RNA*) at which a phosphate group is attached to the 5′ carbon of the sugar residue. (see *ribose, deoxyribose*; compare *three prime* [*3′*])

fixation

1. In population genetics, the situation in which only one particular genetic variant (*allele*) is found at a particular *chromosomal* location (*locus*). All individuals would be *homozygous* (contain the same variant on both chromosomes) for that allele. Fixation is more likely to occur in small populations where inbreeding is common. (see *allele frequency, Hardy-Weinberg equilibrium, population substruture, random assortment*)

2. (see *nitrogen fixation, carbon fixation*)

fixed allele A genetic trait for which all members of a population are *homozygous* (same variant), so that no other *alleles* for this *locus* exist in the population.

fixing The initial step in preparing biological specimens for microscopic analysis. Fixatives such as formaldehyde (light microscopy) or osmium tetroxide (electron microscopy) help to accurately preserve cellular structure.

flagellate Any cell or organism bearing a *flagellum* (*undulipodium*) or flagella. (see *mastigote*)

flagellin A *polymeric* protein that is the main constituent of bacterial *flagella*.

flagellum, flagella (pl.) A long, thin cellular *organelle* that protrudes from the surface of a cell and is used in locomotion. It is common in *protists* and motile *gametes.*

1. In prokaryotes, it is a single fiber of *flagellin* protein, usually capable of rotary motion.

2. The *eukaryotic* flagellum has a characteristic structure, the *axoneme,* consisting of nine pairs of *microtubules* surrounding an inner pair. It is anchored to the cell by a *basal body* (*kinetosome*). Margulis and coworkers have proposed that the eukaryotic flagellum (and *cilium*) be renamed an *undulipodium,* a nomenclature that eliminates confusion.

flame cell (solenocyte) A specialized cell that functions in the *excretory system* of some invertebrates such as flatworms and rotifers. It is a cup-shaped cell containing *cilia* that collect and draw fluid waste products toward exit pores. (see *nephridium*)

flamingos (see *Phoenicopteriformes*)

flatworms (see *Platyhelminthes*)

flavin Either of the *nucleotide coenzymes* (*FAD, FMN*) derived from *riboflavin* (*vitamin B2*). They are important electron carriers in *oxidative phosphorylation.* (see *flavoprotein*)

flavin adenine dinucleotide (see *FAD*)

flavin mononucleotide (see *FMN*)

flavonoid One of a group of plant compounds responsible for non*photosynthetic* pigments, including yellow *chalones* and *aurones,* pale yellow and ivory *flavones* and *flavonols,* and the red, blue, and purple *anthocyanins.* They also include the colorless *isoflavones, catechins* and *leucoanthocyanidins.* They are water soluble and usually located in a cell *vacuole.*

flavoprotein A conjugated protein in which a *flavin* (*FAD, FMN*) is joined to a protein component. Some flavoproteins additionally contain either *heme* or metal ions. Flavoproteins are ubiquitous in cells and serve as electron carriers in the *electron transport chain* in *oxidative phos-*

phorylation. (see *aerobic respiration*)

fleas (see *Siphonaptera*)

flexsor One of the pair of *antagonistic* (*agonistic*) muscle sets that mediate voluntary movements. The flexor contracts to flex the joint and relaxes when the joint is extended. (compare *extensor*)

flies (see *Diptera*)

flocculation The adherence of small particles into aggregates. It is used to describe the aggregation of soil particles in clay soil, which coarsens soil texture and facilitates cultivation.

flora Plant life.

floret One of the small multiple flowers making up an *inflorescence.*

flower The structure in flowering plants (*angiosperms*) concerned with *sexual* reproduction. It usually consists of a *receptacle* on which four different sorts of modified leaves are inserted. The outermost *sepals* (collectively the *calyx*) enclose and protect the other flower parts in the bud stage. Within the sepals are the *petals* (collectively the *corolla*), often conspicuous and brightly colored. Within the petals are *stamens*, consisting of a *filament* (*stalk*) bearing an *anther*, in which *pollen grains* (male *gametes*) are produced. In the flower center is the *gynoecium* (female reproductive structure), with one or more *carpels*, each composed of an *ovary*, *style,* and *stigma* (receptacle for *pollen grains*). The ovary contains a varied number of *ovules* (female *gametes*) that develop into seeds after fertilization. Stamens and carpels are known as essential flower parts since they alone are needed for the process of reproduction. The remaining flower structures are

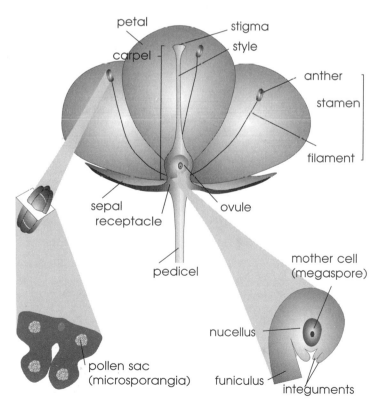

Generalized Flower Structures

nonessential and are extremely variable and show numerous adaptions to promote *pollination* (*fertilization*) and seed dispersal. (see *inflorescence*)

fluid-mosaic model A generalized model for the structure of *plasma membranes*. It consists of a *phospholipid bilayer* in constant lateral motion in which proteins are embedded, either confined to one layer or permeating both (*transmembrane protein*).

microscope configured to detect fluorescent light. It is used in situations where the specimen emits fluorescent light, either naturally or as a result of specific staining.

fluorescent *in situ* hybridization (see *FISH*)

fluorine (see *trace element*)

flycatchers (see *Passiformes*)

flying lemurs (see *Dermoptera*)

fMet (see *N-formylmethionine*)

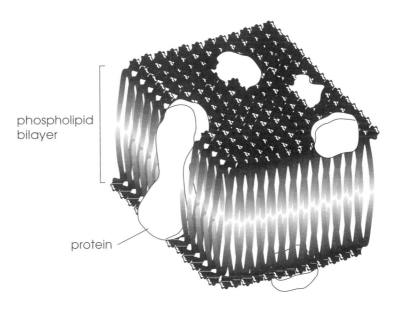

phospholipid bilayer

protein

Fluid-Mosaic Model of the Cell Membrane

fluke
1. The tail of *cetaceans* (whales).
2. (see *Trematoda, Monogenea*)

fluorescence The immediate reemission of visible light as a result of the excitation of a substrate by low-energy radiation. The energy level of the reemitted light is always lower than the exciting radiation and the wavelength concomitantly longer. (compare *phosphorescence*)

fluorescence microscope A type of light

FMN (flavin mononucleotide) A *coenzyme* important in *redox reactions*, particularly in the *electron transport chain* in *oxidative phosphorylation*. It is derived from the vitamin *riboflavin*. (see *aerobic respiration, flavoprotein*)

folic acid A member of the *vitamin B complex* that functions as a *coenzyme*. It participates in protein *metabolism* and *hemoglobin* synthesis. It is also essential in the synthesis of certain *nitro-*

genous bases (*purines* and *thymine*).

follicle
1. Any small cavity, sac, or *gland* within an organ or tissue. For instance, *Graafian follicles* within the *ovary* contain developing *ova*, *hair follicles* produce individual hairs, and feather follicles produce feathers.
2. A dry fruit derived from a single *carpel* that splits along one side only to release its seed(s).

follicle-stimulating hormone (FSH) A *hormone* that stimulates the growth and maturation of *gametes* in vertebrates. In particular, it regulates the *Graafian follicle* and its enclosed *ova* in the *ovary*, and *spermatozoa* in the *testes*. FSH is a *gonadotrophic* hormone produced by the *pituitary gland*. (see *oogenesis, spermatogenesis, menstrual cycle, estrus cycle, leuteinizing hormone, endocrine gland*)

fontanelle A gap in the newborn skull where bone has not yet formed, the space being covered by tissue and skin. The *frontal* and *parietal* lobes of the *cranium* grow together to close the gap after about 18 months.

food chain A succession of organisms that gain energy by feeding on each other. For example in a grazing food chain, *herbivores* eat green plants and are eaten by *carnivores*. *Producers*, mostly *photosynthetic* organisms, are found at the bottom of the food chain, and most *consumers*, herbivores and *carnivores*, are at the top. A special class of consumers, mostly microorganisms, recycles dead organic matter (*decomposers*). Each food chain commonly consists of three or four steps (*trophic levels*), the loss of energy being so great at each step that very little usable energy is retained in the system after it has been incorporated successively into the bodies of several organisms. At each successive trophic level, *biomass* tends to decrease, and the size of each individual animal tends to increase. The shorter the food chain, the less energy is lost. Energy transfer in the system is thus more effi-

cient if middle levels are bypassed. Practically, this means that higher carnivores, such as people, have more total food energy available to them as a population through a plant-based diet. This concept will be increasingly important as world population increases and food supplies decrease. Another way in which the succession of the food chain is manifest is in the accumulation of pesticides and other chemicals. Any animal at the top of the food chain (including people) incorporates many of the chemicals (in particular those stored in fat) that have been eaten at each trophic level before it. (see *food web, pyramid of energy, pyramid of numbers, pyramid of biomass, carbon cycle, nitrogen cycle, food chain concentration*)

food chain concentration (biomagnification) The fate of certain fat soluble chemicals, in particular *pesticides* such as DDT, that accumulate in the *adipose tissue* of organisms rather than being excreted. They are concentrated at each successive step in a *food chain*.

food web The interrelationship of organisms in any natural *ecological community* through which energy is transferred. Most organisms feed on and are eaten by more than one kind of organism, thus cre-

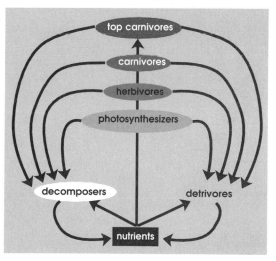

Generalized Food Web

ating a web of interaction. The food web is made up of all the *food chains* that interact in a given community. (see *ecosystem*)

foot In mollusks, the muscular, spadelike structure that is used in both burrowing and surface locomotion.

footprinting A set of methodologies for identifying specific *DNA* sequences that are bound by proteins either *in vitro* or *in vivo*. Detection methods include protection of the bound site from a *nuclease*, a shift of mobility upon *electrophoresis*, or the binding under specific conditions to a *nitrocellulose* filter.

foramen, foramina (pl.) A natural opening in any animal organ or structure. For example, foramina in bones permit nerves and blood vessels to enter and leave.

foramen magnum The opening in the skull through which the *spinal cord* passes.

Foraminifera (see *Granuloreticulosa*)

forebrain (prosencephalon) The most anterior of the three basic anatomical regions of the vertebrate brain. It consists principally of the *cerebrum* with its pair of greatly enlarged lateral outgrowths (*cerebral hemispheres*), the *thalamus,* and the *hypothalamus.* (compare *hindbrain, midbrain*)

formalin A compound used in the preservation of biological specimens. It consists of formaldehyde, methyl alcohol, and water.

formic acid The simplest carboxylic acid (HCOOH), it contributes to the sting of both ant bites and nettles.

forward mutation A *mutation* in *DNA* that converts a *wild-type* (normal) allele to a mutant allele.

fossil The preserved remains of, or impressions left by, organisms that lived prior to the end of the last glacial period (about 10,000 years ago). They include skeletons, tracks, impressions, borings, and casts. Fossils are usually found in hard rock but may also be recovered from areas of permafrost. Fossils provide much of the scant information known about prehistoric life forms.

fossil fuels The incompletely decomposed, carbon-containing remains of organisms that are burned as fuel. They originate from dead, organic matter, mostly plants and marine organisms, that were buried and escaped oxidation. Instead, they were converted by a series of complex chemical reactions to concentrated *hydrocarbons.* Examples are coal, oil, and natural gas.

founder effect The situation in which a new, isolated population is started by a few individuals. The genetic traits (*alleles*) they now carry form a significant fraction in the new population, and formerly rare alleles and allele combinations may be enhanced. This situation has produced new and interesting species on islands such as the Galapagos and Hawaii. (see *gene pool, genetic drift, gene flow, Hardy-Weinberg equilibrium*)

fovea The point of most acute vision on the vertebrate *retina*. The image of an object falls on the fovea, which lies directly opposite the center of the pupil and lens. It consists entirely of densely packed *cones* for sharp, color, daylight vision. No blood vessels or *nerve fibers* are retained in the fovea region, between the cones and incoming light. Some birds have two foveas for sharp forward and lateral vision.

foxes (see *Carnivora*)

F plasmid A bacterial plasmid containing *genes* that direct the formation of a hollow tube called a *pilus*. It forms a *conjugation* bridge between two cells, allowing transfer of all or part of the bacterial *chromosome*. This provides one mechanism for genetic exchange in *prokaryotes.*

frameshift mutation The insertion or deletion of a *nucleotide* pair or pairs within a *gene*, causing a disruption of the normal *translational reading frame* of the *genetic code* for a protein. It may cause the *mRNA* message to be reduced to gibberish or a protein to be aborted partway through synthesis. (see *protein synthesis*)

fraternal twins (see *dizygotic twins*)

free energy (G) The energy in a system that is available to do work. In a biochemical system, this is the energy available to form chemical bonds. (see *enthalpy, entropy*)

free-living

1. An organism that lives unattached to a *substrate*.
2. An organism that does not depend on another organism. (compare *parasite*)

free-living flatworms (see *Turbellaria*)

freemartin The sterile female member of unlike-sexed twins in cattle or, occasionally other *ungulates*. The sterility is usually the result of *hormonal* influences from the male twin that leak through their shared *placenta in utero*.

free radical (superoxide radical) Any molecule possessing unpaired electrons. Free radicals are produced as a result of bombardment by high-energy ionizing radiation such as X-rays and γ-rays. Free radicals are chemically reactive. In the body, they may be highly damaging, in particular by producing *double strand breaks* in *DNA*. Unrepaired or misrepaired DNA damage may result in *mutations*. (see *DNA repair, superoxide dismutase*)

freeze etching A laboratory technique used in *electron microscopy* for examining the outer surfaces of membranes.

freeze fracture A laboratory technique used in *electron microscopy* for examining the inner surfaces of *membranes*. Material is frozen rapidly (for example, in liquid nitrogen) and fractured. Fracture lines tend to follow lines of weakness such as that between *phospholipid bilayers* of *plasma membranes*.

freshwater The *biome* consisting of freshwater lakes, ponds, rivers, and streams. They make up only a small portion of the Earth's surface but are indispensable as a drinking water supply for all animals, including humans. They support a diverse collection of life. Freshwater biomes are commonly exploited as recreational areas, and are easily *polluted* beyond repair.

frigatebirds (see *Pelecaniformes*)

frogs (see *Anura*)

frond A large, well-divided leaf, in particular those of palms and ferns. It is also used to describe a divided *thallus* that resembles a leaf, as in seaweed.

frontal lobe The front portion of the mammalian *cerebrum*, anatomically separated from the three other lobes by deep grooves. Functionally, it contains the centers for associative activity and voluntary *motor* control. (see *cerebral cortex, telencephalon*; compare *parietal lobe, temporal lobe, occipital lobe*)

fructose A simple sugar of the structure ($C_6H_{12}O_6$) found in fruit juices, honey, and cane sugar.

fruit The ripened *ovary* of a plant. More loosely, the term is extended to the ripened ovary and seeds together with any structure in which they are contained. For example, the core of an apple, containing the seeds, is the true fruit and is surrounded by succulent flesh. Fruits are classified according to the fleshy or hard development of the ovary wall (*pericarp*) and how the seeds are released. Fruits play an important role in the dispersal method of the seeds they contain. (see *dehiscent, indehiscent*)

fruiting body In basidiomycete and many ascomycete fungi (except yeasts), the structure in which *meiosis* occurs and *sexual spores* are produced. It may manifest as a toadstool or mushroom.

FSH (see *follicle-stimulating hormone*)

fucoxanthin An accessory *photosynthetic pigment* found along with *chlorophyll* in diatoms, brown algae, and other non-green algae. It is a brown *carotenoid*, a *xanthophyll*.

fulmars (see *Procellariiformes*)

fumaric acid An important intermediate in several plant biochemical pathways including the *Krebs cycle*, *purine* pathways, and the *nitrogen cycle*.

Fungi One of the *phylogenetic* kingdoms into which all life forms are divided for academic discussion. It contains a group of *eukaryotic* organisms that develop from *spores*, have *chitin*-based *cell walls*, and are nonmotile since they lack *undulipodia* at all life cycle stages. Fungi are separated from plants, in particular, by their lack of *chlorophyll*. They obtain food by secreting enzymes into a *substrate* on which they live and absorbing the digested materials; thus, they are important in the food chain as *decomposers*. This *saprophytic* (*osmotrophic*) mode of nutrition is distinctive of the members of this kingdom. Fungi may be *unicellular* (yeasts) or composed of *multicellular* filaments (*hyphae*) that together comprise the fungal body or *mycelium*. Exchange of genetic material is accomplished by *fertilization* and *meiosis*, but no *embryo* forms. Although many fungi are harmful and cause decay or may even be extremely poisonous, others are commercially invaluable. The production of bread, cheese, alcoholic beverages, and some antibiotics all depend on various fungi. Certain *genetically engineered* yeasts, in particular, are used as biological factories to produce medically important *gene products* and are additionally being exploited in toxic cleanup efforts. (see *fermentation, mycorrhizae, lichen, Ascomycota, Basidiomycota, Zygomycota, fungi imperfecti, fruiting body*)

fungicide Any substance that is destructive or lethal to fungi.

fungi imperfecti (see *Deuteromycota*)

funicle The stalk attaching a plant *ovule* to the *placenta* in an *ovary*.

G (see *guanine*)

GABA (γ-aminobutyric acid) A *neurotransmitter* found at *inhibitory synapses.*

GAG (see *glycosaminoglycans*)

Gaia The entire living system of the Earth, including all living organisms existing in the *biosphere.*

Gaia hypothesis The hypothesis put forth by James Lovelock and Lynn Margulis, that the Earth is essentially a superorganism with interdependent parts, communication between the parts, and the ability to self-regulate. The concept was named after the Greek goddess Gaia (Mother Earth).

galactose A sugar that is a common constituent of plant *polysaccharides* (for example, *pectin*) and animal *glycoproteins* and *glycolipids.*

gallbladder In many vertebrates, a muscular *bladder* extending from the *bile duct* that serves as a temporary store for *bile.* Stored bile is released in response to the presence of food in the *duodenum* and controlled by the *hormone cholecystokinin.*

Galliformes An order of birds comprising the gamebirds, including pheasants, quail, turkeys, curassows, and moundbuilders.

gam-, -gamy, -gamic, -gamous A word element derived from Greek meaning "unite" or "marry." It is usually used in words having to do with *gametes* or *sexual reproduction.* (see for example, *gametophyte*)

gamebirds (see *Galliformes*)

gametangium, gametangia (pl) A reproductive organ in which *gametes* are produced in seedless plants, fungi, and protoctists. These complex *multicellular* structures have been lost in seed-bearing plants (see *oogonium, archegonium, antheridium*)

gamete A *haploid* reproductive cell capable of fusing with another reproductive cell (*fertilization*) to produce a *diploid zygote* that develops into a new organism. In *sexual* reproduction, each gamete transmits half of each parental *genome* to the progeny. Gametes are generally *haploid* and fuse to form a *diploid* zygote, but exceptions exist, particularly in plants. Occasionally, just the nucleus of a cell functions as the gamete. In simple organisms, gametes typically have a similar size and structure (*isogamous*). In higher organisms, they are often dissimilar (*anisogamous*). The male gamete is often smaller, motile, and produced in large numbers. The female gamete, in contrast, is generally large due to the presence of food reserves, immotile, and produced in relatively small numbers. (see *ovum, egg cell spermatozoon, pollen, gametophyte, meiosis, sexual reproduction*; compare *spore, propagule*)

gametocyte A *sexual* stage in the *life cycle* of apicomplexans, in particular *Plasmodium*, which is responsible for malaria. In the human bloodstream, *asexual merozoites* develop into gametocytes, which are capable of producing *gametes*, but only after extraction by a mosquito through a blood meal. (compare *sporozoite*)

gametogenesis The formation and development of *gametes* (reproductive cells). (see *ova, spermatozoa, oogenesis, spermatogenesis*)

gametophyte In plants, the *life cycle* generation that is *haploid* and during which *gametes* develop *mitotically.* It arises from a haploid *spore* produced by *meiosis* from a *diploid sporophyte.* The gametes may fuse during *fertilization* to produce a new diploid *zygote* that will develop into a new sporophyte body. In lower plants (mosses and liverworts), the gametophyte is the main, conspicuous generation, and the sporophyte is *parasitic* on it. Conversely, in *vascular plants*, gametophytes are always much smaller than sporophytes, and in most groups, are nutritionally dependent on them. Gametophytes are the sporophytes of vascular plants enclosed within the sporophyte tissues. Their only remaining function is to produce gametes (egg cells and *pollen grains*). (see *alternation of generations, meiosis; gametangia*)

gammaaminobutyric acid (γ-aminobutyric acid) (see *GABA*)

gamma globulin (see *immunoglobulin*)

gamma-interferon (γ-interferon) One of the early chemical signals in the *immune response*. It is a *monokine* secreted by *macrophages* that have come in contact with a virus. It functions to stimulate maturation of more macrophages from precursor *monocytes*.

gamont An adult *life cycle* stage that produces *gametes*. (compare *agamont*)

Gamophyta (Conjugophyta) A phylum of freshwater green algae, the gamophytes. They are immotile and contribute to the formation of "pond scum." Gamophytes are distinguished from other green algae by their method of *sexual reproduction, conjugation*. This consists of the mating and nuclear fusion of two more or less equal *vegetative* bodies; specialized *gametes* are not produced.

ganglion, ganglia (pl.) A bundle of *nerve cell bodies* where the integration of information takes place. In invertebrates, ganglia occur along the major *nerve cords* and are the centers for coordination of *nerve impulses*. In vertebrates, the ganglia are bound by *connective tissue* and located mostly outside the *central nervous system*, in the *peripheral* and *autonomic nervous systems*. The major centers have developed into the spinal cord and brain.

ganglion cell A specialized *nerve cell* in the vertebrate and cephalopod eye. Its *dendritic* end synapses with one or more *bipolar cells*, and its *axon* joins the *optic nerve*. (see *retina*)

gannets (see *Pelecaniformes*)

ganoid scales A fish scale characteristic of some primitive fish and found today in the skin of sturgeon and gars. They consist of a superficial layer of *enamel*-like *ganoin*, a middle layer of *dentine,* and a bony inner layer.

gap junction A type of *intercellular junction* involved in communication. Proteins called *connexins* form channels connecting adjacent cells through which ions and small molecules such as sugars and amino acids may pass. Gap junctions also provide for electrical communication between cells.

gas chromatography A *chromatographic* technique for separating molecules in which the stationary phase is liquid and the carrier, in which the questioned substance(s) is conveyed, is gaseous.

gastric Any structure, function, or substance associated with the stomach.

gastric juice The collective term for various enzymes, acids, and *hormones* that are secreted by the stomach to aid in the breakdown of food. It includes the enzymes *pepsin* and *rennin, mucus*, and the hormone *gastrin* which circulates in the bloodstream and stimulates the gastric *glands* to secrete hydrochloric acid and *pepsin*.

gastric pits Cavities deep in the underlying *mucosa* of the *epithelial* lining of the stomach. Cells forming *exocrine glands* are found at the base of each pit. (see *chief cells, parietal cells, gastric juice*)

gastrin A *peptide hormone* secreted by the stomach lining that stimulates secretions of hydrochloric acid and *pepsin* in the stomach and some enzymes in the pancreas. It is released in response to the presence of food in the stomach as well as by taste, smell, and the thought of food.

gastrodermal Any structure or function relating to the thin layer of tissue that lines the digestive tract of marine animals.

Gastropoda A large class of mollusks containing the terrestrial slugs and snails as well as marine and freshwater members such as limpets and pond snails. They are characterized by a well-developed head with tentacles and eyes, a single shell, and a large flat foot. The shell is often spirally coiled.

Gastrotricha A phylum of wormlike animals, the gastrotrichs, that are characterized by numerous *cilia* on their ventral surfaces, hence the name. They are *hermaphroditic* and exhibit *protandry* in that they produce their sperm earlier than their eggs, encouraging *cross-fertilization*. Freshwater gastrotrichs lack males altogether and develop via *parthenogenesis*. They lay two types of egg, optimizing their reproductive strategies. One must undergo desiccation or extreme temperature to initiate development, the other begins to divide immediately upon being laid.

gastrovascular cavity (see *coelenteron*)

gastrula The stage of *embryonic* development in higher animals in which the embryo consists of three layers, the *ectoderm, mesoderm,* and *endoderm*. They surround the beginnings of a digestive cavity (*archenteron*) with one opening (*blastopore*). (see *gastrulation*)

gastrulation In the *embryos* of higher animals, the transformation of a single-layered *blastula* into a three-layered *gastrula* consisting of *ectoderm, mesoderm,* and *endoderm* that surrounds a cavity (*archenteron*) with one opening (*blastopore*). During gastrulation in higher animals, developmental paths are solidified by the interactions of tissues with each other, (*committed*) and with chemical gradients established in the *ova*. In lower animals such as nematodes, gastrulation is simply a mechanical folding in of the *gut* and other internal structures.

gas vacuole A specialized, gas-filled vesicle found in most cyanobacteria. They are supported by protein ribs or spirals and function principally in providing buoyancy.

gating A general mechanism of controlling the opening and closing of *transmembrane ion channels*, particularly in *neurons* and *muscle cells*. Three specific types include *voltage-gated, stimulus-gated,* and *chemically gated ion channels*. (see *sodium-potassium pump, resting potential, equilibrium potential,* *action potential, nerve impulse, excitatory synapse, inhibitory synapse, sensory neuron*)

Gaviiformes The order of birds comprising the divers, including the loons.

GC box A **GC**-rich *DNA* motif often found in *eukaryotic promoters*. It is located 110 bp (base pairs) *upstream* from the start of *RNA transcription*. (see *CCAAT box, TATA box*)

GC content The percentage of an organism's *DNA* that is made up of the *base pair guanine:cytosine* (**GC**). The **GC** content is characteristic of a species and is used as an indicator of such. It also determines the *melting temperature* of the DNA. **GC** base pairs are held together by the stronger triple *hydrogen bond*, while the alternate **AT** (adenine:thymine) base pairs are held together by a weaker double *hydrogen bond*. Thus, DNA with a higher **GC** content will melt at a higher temperature than DNA with a lower **GC** content.

geese (see *Anseriformes*)

gel A flexible slab or sheet of material frequently used in the separation of *macromolecules* in the laboratory. It is commonly formed from *polymeric* substances that provide a structural framework with pores of varying sizes. Molecules migrate through these pores, typically in response to an electric field. (see *agarose, polyacrylamide*)

gel electrophoresis (see *electrophoresis*)

gel filtration (exclusion chromatography) A laboratory technique for separating *macromolecules* on the basis of size. A narrow column is packed with porous gel particles of chosen characteristics. A solution containing the molecules to be separated poured over the column and allowed to flow through by gravity. Smaller molecules are hindered in their passage down the column because they tend to penetrate the pores of the gel particles. Molecules too large to penetrate the pores are excluded and flow rapidly down and out of the column. By analyzing consecutive drips out of the bottom of the column, information can be gleaned about the molecules in the mixture. By retaining various fractions, separation is achieved. A common gel packing material is the *polymer* Sephadex™. (see *chromatography*)

gender The sum of the behavioral and physiological traits that identify potential *mating types* (for example, *sexual* part-

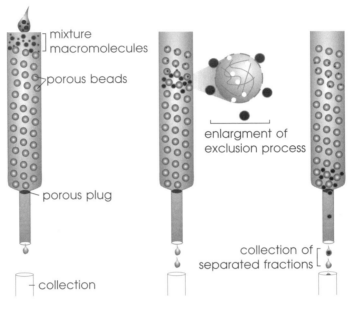

mixture
macromolecules
porous beads

enlargment of exclusion process

porous plug

collection of
separated fractions

collection

Gel Filtration
(Exclusion Chromatography)

gemma, gemmae (pl.) **(gemmule)** A *propagule* structure produced in some seedless plants that functions in *vegetative reproduction*. They often form in groups in receptacles called gemma cups and eventually become detached from the parent to form new plants. The term may also be applied to a thick-walled, *asexually* derived *spore* formed from the *hyphae* of some fungi.

gemmation A type of *asexual reproduction* typically seen in some seedless plants. It involves the production of a group of cells (a *gemma*) that develops into a new individual before or after separation from the parent. (see *budding*)

gemmule (see *gemma*)

ners or *conjugants*) prior to a sexual union.

gene Most simply, a unit of heredity. Traditionally, a *gene* is defined as a segment of *DNA* from which a single *mRNA* molecule (which is *translated* into a single *polypeptide*) or a functional RNA molecule (*rRNA, tRNA*) is *transcribed*. In this way, genes code for all of the structures and functions of an organism, both those that make it similar to others of its kind and those that make it unique. The definition is sometimes expanded (or returned) to include that of simply a physical location (*locus*) on a *chromosome*. This definition encompasses marker regions in DNA that do not produce a protein product but are analyzed directly in disease locus tracking and genetic identification. Genomic regions that do not code for proteins are better termed *genetic markers*. Genes may also be found on *extrachromosomal elements* such as *plasmids* and in *cytoplasmic* cell *organelles* such as *mitochondria* and *chloroplasts*. (see *allele, polymorphism, dominant, recessive, recombination, mutation*)

gene amplification The process by which the number of copies of a specific *chromosomal DNA* segment is increased in a *somatic* (body) cell. Gene amplification is utilized when a large amount of *gene product* is needed by the cell in a short time span. Examples include the *rRNA genes* in the *lampbrush chromosomes* of developing amphibian *oocytes* and the

chorion (egg shell) genes in *Drosophila*. (see also *amplification*)

gene bank A collection of *DNA* fragments from the *genome* of a particular organism that have been inserted into *cloning vectors*. They may then be screened for particular characters and individual clones selected for further manipulation. (see *shotgun cloning, restriction endonuclease*)

gene cloning A laboratory technique whereby a *gene* is isolated from its native *DNA* environment and inserted into a specialized DNA fragment (cloning *vector*), often a *plasmid* or *bacteriophage*. This allows the gene to grow in a foreign *host*, usually bacteria or yeast. Cloning allows the isolation of the gene from its natural environment for purposes of research. Cloning also allows production of

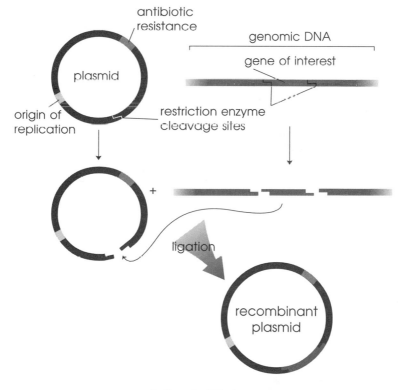

An Example of Cloning

the gene in large amounts, both for research and in the production of commercially and clinically important *gene products*. An example is the production of human *insulin* in the bacteria *E. coli*, used in the treatment of *diabetes*. Traditionally, the DNA fragment containing the gene has been isolated using *restriction endonuclease* enzymes. These cut the DNA at specific sites, allowing a particular fragment to be reproducibly isolated. This fragment is then inserted into a cloning vector by *ligating* (attaching) the ends together. The composite DNA molecule is inserted (*transformed*) into a host in which it can multiply as an *extrachromosomal element*. Increasingly more common is the isolation of DNA fragments by the *PCR (polymerase chain reaction)* process in which a defined segment of DNA is located and *amplified in vitro* by a special *DNA polymerase*. This

segment may then also be inserted into a cloning vector. Successful transfer of the composite plasmid into a living host must be detected through some obvious *phenotypic* property of the gene itself or through the phenotype of another "tag" gene. This "tag" gene is either artificially associated with the gene of interest, or is present in the cloning vector and disrupted by insertion of the foreign gene. (see *clone, genetic engineering*)

gene conversion A genetic phenomenon whereby a donor *DNA* sequence replaces a similar sequence on a *homologous* stretch of DNA. This is a normal error repair process that occurs during *meiosis* in *eukaryotes* when *sister chromatids* are aligned in the *synaptonemal complex*. It is mediated by enzymatic *DNA repair* mechanisms in the cell that recognize *base pair mismatches*. Gene conversion can occur with or without *reciprocal re-*

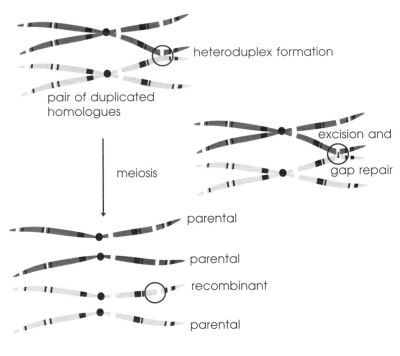

Gene Conversion

combination (exchange) of the sequences on either side of the converted region. It can also occur in *prokaryotes* when homologous duplicated DNA sequences are present. (see *homologous recombination*)

gene disruption (see *knockout*)

gene dosage The number of copies of a particular *gene* present in the *genome* of an organism.

gene duplication (see *tandem duplication*)

gene expression The *transcription* of *mRNA* from a *gene* and in *structural genes*, its subsequent *translation* into a functional protein. Variable gene expression in different *somatic* tissues is the primary way that development, growth, and *differentiation* of an organism takes place. It is also a mechanism by which a cell or organism may regulate its *metabolism*. (see *totipotent, determined, operon, regulatory genes, inducible enzyme*)

gene family A related set of *genes* in one *genome* all descended from the same ancestral gene.

gene flow The movement of *alleles* that results from mating and genetic exchange with immigrant and emigrant individuals with regard to interbreeding populations. The exchange of genetic markers may be unidirectional or bidirectional. (see *Hardy-Weinberg equilibrium, genetic drift, gene, gene pool, allele frequency*)

gene frequency (see *allele frequency*)

gene inactivation (see *knockout*)

gene interaction The collaboration of several different *genes* to produce a *phenotype* or set of *phenotypes*.

gene knockout (see *knockout*)

gene library (see *gene bank*)

gene mobilization The physical aggregation and propagation of a group of previously dispersed *genes* because together they confer some selective advantage on the organism. Initially, the movement of these genes often occurs through *transposable genetic elements*. Bacterial *plasmids* containing multiple *antibiotic* resistance genes (*resistance transfer factor*) are compiled in this fashion. A number of highly resistant disease strains were generated when the administration

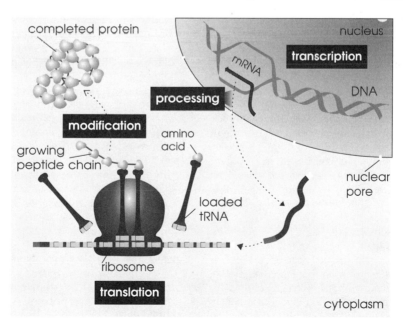

Gene Expression

of multiple antibiotic drugs was common practice.

gene mutation A change involving one or a few *nucleotides* that occurs within the fine structure of a *gene*.

gene pair The two copies of a particular *gene* present in a *diploid* cell, one on each *homologous* chromosome. They may be exactly the same (*homozygous*) or embody slight differences (*heterozygous*).

gene pool The totality of genes existing within a breeding population or species at a given point in time. (see *gene flow, genetic drift, sexual reproduction*)

gene product The protein or *polypeptide* whose synthesis is specified by a particular *gene*. It may also undergo various posttranslational chemical modifications such as the addition of carbohydrate (*glycoprotein*) or lipid (*lipoprotein*) groups. (see *transcription, translation, polysomes, amino acid, genetic code, central dogma*)

generation time The time that is required for a cell to complete one full growth cycle. It may also be considered as the average time between *cell divisions* in a population or the doubling time of a cell population. (see *mitosis*)

generative A cell, *propagule*, or organism that is capable of further growth and/or reproduction.

generative nuclei In seed plants, the two *nuclei* found in the *pollen tube* that function as male *gametes*. One fuses with the egg cell to form a *zygote*, while the other either degenerates or, in certain *angiosperms*, fuses with the *polar nuclei* to generate the *triploid primary endosperm nucleus*. (see *embryo sac, double fertilization*)

gene sequencing (see *DNA sequencing*)

gene splicing (see *splicing*)

gene synthesis Generation of the *nucleotide* sequence for a known *gene* by artificial synthesis in the laboratory.

gene therapy The introduction of healthy *genes* into particular cells in order to cure a disease caused by a defect in their *DNA*.

Gene therapy for diseases such as cystic fibrosis and diabetes are currently in the clinical testing stages and show great promise. When healthy genes can be permanently inserted into the *genome* of a person, a true cure will have been achieved. (see *gene cloning, genetic engineering, human genome project*)

genetic Any mechanism, structure or other element related to *genes* or *DNA*.

genetic code The formula by which the *nucleotide* sequence in *DNA* determines the *amino acid* sequence in proteins. The four DNA *bases, adenine (A), thymine (T), guanine (G), and cytosine (C),* are organized into triplets (*codons*) of which there are 64 different combinations. Most triplets specify an *amino acid*. However, three code for stop signals. The system is redundant, because 20 common amino acids are specified by the 64 different codons. The order of codons in a gene determines the sequence of amino acids in the resulting protein. (see *central dogma, cistron, transcription, translation, polysome*)

genetic disease Any disease caused by abnormalities in *genes* or *chromsomes* that are passed from one generation to the next. They may or may not be manifest in any particular generation, depending on their inheritance pattern. Examples include cystic fibrosis, Duchenne muscular dystrophy, and sickle-cell anemia. (see *dominant, recessive, sex-linked, inborn errors of metabolism*; compare *somatic genetic disease*)

genetic dissection The process of uncovering the hereditary components of any specific biological process. Genetic variants are used to assign a particular *gene* to a function or structure. (see *mutant hunt, mutation analysis, saturation mutagenesis*)

genetic diversity The total number of genetic characteristics, generally applied to a species, subspecies, or group of species. (see *biological diversity*)

genetic drift The random fluctuations in the frequencies of genetic *alleles* (variants) in a given population over time. It is due to an imperfect representation of parental gene frequencies in offspring. Genetic drift is more prevalent in a small population where the number of alleles is limited and relatively little genetic exchange takes place due to the small number of matings. It may also occur due to catastrophic effects obliterating a large portion of the mating population, hence the *gene pool*. (see *Hardy-Weinberg equilibrium, allele frequency, meiosis, gene flow*; compare *mutation, migration, natural selection*)

genetic engineering (recombinant DNA technology) A technology that encompasses a wide variety of techniques for introducing foreign *DNA* into a *host* organism. Usually, *gene cloning* is employed and the composite DNA molecule inserted into a new host. The first host organisms to be used were bacteria (usually *E. coli*) because of their simple genetic structure and high growth rate. For instance, human *insulin* was cloned into *E. coli* and produced commercially as a therapy for *diabetes*. This was accomplished in much greater quantity and at much lower cost than the previous technique of purifying it from physiological tissues. As knowledge increased, yeasts (usually *Saccharomyces cerevisiae*) were employed more often because their *eukaryotic* makeup was more compatible with cloning and *expressing* genes from other eukaryotic organisms. In particular, yeast has been the organism of choice for cloning human genes for study in the *Human Genome Project.* More recently, techniques for creating *transgenic* plants and animals, including tomatoes and mice, have been perfected. Such artificial hybrids have begun to have a commercial impact. Tomatoes that have had their "softening" *gene* removed and potatoes that are pest-resistant have been approved by the FDA. Human *gene therapy* is currently in the clinical testing stages. Early histrionics about possible hazards of recombinant DNA technology have so far proven unfounded, although the technology is nevertheless subject to strict regulation.

genetic fingerprinting (see *genetic typing*)

genetic linkage The phenomenon in which *genes* or *genetic markers* are inherited together with some defined frequency. The degree of linkage is determined by the physical distance between the *loci* (locations). Loci that are close together are inherited together more frequently, those farther apart less frequently. Genes far apart on the same piece of *DNA* or *chromosome* behave as if they were not linked at all, in other words, on physically separate chromosomes. The frequency with which genetically unlinked loci are inherited either together or separately is equal and due only to random chance.

genetic map (chromosome map, linkage map) The linear arrangement of *genes* on a *chromosome* or sites within a gene. Because chromosomal exchanges (*reciprocal recombination*) during *meiosis* are more frequent between genes that are physically farther apart, the percentage of *crossovers* between *homologous* chromosomes in *eukaryotes* can be used as an indication of genetic distance. One *genetic map unit* (*centimorgan, cM*) is defined as the distance between gene pairs for which one out of one hundred products of meiosis is *recombinant* (that is, equals a *recombination frequency* of 1%). The same basic technique can be used to investigate genes on homologous *DNA* fragments introduced either naturally or artificially in *prokaryotes.* However, the calculation is slightly different in order to account for the lack of four meiotic products. More recently, physical chromosome maps have been constructed using *restriction enzymes* to cut DNA into reproducible sets of different size pieces and using tagged, known genes as *probes* to establish the physical order of genes on chromosomes. The newest technique applied to gene mapping is *PCR*. It amplifies the DNA adjacent to a known site so that it can be studied more readily. Much knowledge has been gained by comparing the genetically and physically derived maps. (see *recombination, F factor, Human Genome Project, electrophoresis, RFLP, genetic typing, chromosome walking, chromosome jumping*)

genetic map unit (m.u.) (map unit, centimorgan, cM) The unit measurement of distance in a *genetic map* It is defined as the distance between *gene* pairs for which one out of one hundred products of *meiosis* is *recombinant* (that is, equals a *recombination frequency* of 1%). (see *genetic linkage*)

genetic marker Any defined location (*locus*) on a *chromosome* that can be identified and followed using either genetic or physical methods. (see *DNA analysis, RFLP, PCR, genetic map, DNA typing, Human Genome Project*)

genetic polymorphism (see *polymorphism*)

genetic recombination (see *recombination*)

genetic risk The relatively high possibility of propagating detrimental genetic changes in small populations. Factors contributing to this include reduced genetic variation, *genetic drift,* and *mutation*. It is used particularly in discussing *endangered species.*

genetics The study of *genes* and heredity.

genetic self-incompatibility (see *self-incompatibility*)

genetic typing The analysis of genetic *loci* for the purposes of identification and comparison. The term "*DNA* fingerprinting" was originally applied to this type of analysis, but is, in fact, a misnomer. (see *DNA analysis genetic marker, RFLP, PCR, Southern blot, blood group*)

genome The total genetic makeup of an organism. This includes all the native *DNA* in each cell. The *nuclear genome* would be all the DNA normally contained in the *nucleus* of a cell. The *human genome* is all the DNA normally contained in a human cell.

genomic DNA *DNA* that is contained

within the native *genome* of an organism. It is distinguished from artificial constructs such as *cloned* DNA or *cDNA* that has been artificially synthesized from *mRNA*. Sometimes, the *nuclear genome* is distinguished from DNA found in cell organelles such as *mitochondria* and *chloroplasts*. (see *mtDNA, cpDNA*)

genotype The fundamental genetic makeup of an organism. It is distinguished from physical appearance or *phenotype*, which is determined by the relationships and interactions between *genes*. The term may also refer more specifically to the set of *alleles* at a single *locus*.

genus A *taxonomic* category used in the classification of organisms. It consists of a collection of similar species. Similar genera are grouped into families. It is the first part of the two part name used to identify a species in *binomial nomenclature* and is always italicized.

geological time scale A system of measuring the history of the earth by dating the *fossils* contained in the rocks of the earth's crust. Each layer is calibrated by measuring the spontaneous decay of *radioactive isotopes* locked within the fossils when the rock was formed. Based on these measurements, a time scale has been devised based on two eons (*Precambrian, Phanerozoic*), each of which is subdivided into eras. The three Phanerozoic eras (*Cenozoic, Mesozoic, Paleozoic*) are subdivided into *periods* (Cenozoic: *Quaternary, Tertiary*; Mesozoic: *Cretaceous, Jurassic, Triassic;* Paleozoic: *Permian, Carboniferous, Devonian, Silurian, Ordovician, Cambrian*). Each period is further subdivided into early middle and late *epochs*.

geophyte (see *cryptophyte*)

geotaxis Directional movement in which gravity is the stimulating force.

geotropism (see *tropism, gravitropism*)

germ cell Any cell from which *gametes* (including *ova* and *spermatozoa*) are derived. (see *germline, meiosis, oogenesis, spermatogenesis*)

germinal epithelium The *epithelial* layer of cells in an *ovary* or *testis* from which the reproductive cells (eggs and *spermatozoa*) first begin to differentiate.

germinal vesicle The enlarged *nucleus* of a mammalian *oocyte* that persists until *fertilization* triggers the completion of *meiosis*. (see *oogenesis*)

germination The initiation of growth of a seed, *spore,* or other structure. It usually follows a period of *dormancy* and generally occurs in response to the return of favorable external conditions, most notably warmth, moisture, and oxygen. The internal biochemical conditions must also be conducive to growth. (see *epigeal, hypogeal*)

germ layer The three major body layers—*ectoderm, mesoderm, and endoderm*—that develop in the *embryos* of most animals during *gastrulation*. Simpler, *diploblastic* animals have only two of the layers, ectoderm and endoderm. More complex *triploblastic* animals (the majority) contain all three. These layers determine the development of all of the major body systems and organs.

germline The cell line that becomes differentiated (*determined*) from the remaining *somatic cell line* early in the development of many animals and alone has the potential to undergo *meiosis* and form *gametes* (including *ova* and *spermatozoa*). (see *germ cell, oogenesis, spermatogenesis, ovary, testes*)

gestation The period of time between *fertilization* and birth in a *viviparous* (live birth) animal. The length of the gestation period tends to vary with the type of *placenta* and the size of the species. Those with a highly developed placenta and a smaller size have a shorter gestation. In humans, it is about nine months. (see *embryo*)

giant chromosome (see *polytene*)

giant fiber (giant axon) A *nerve fiber* that has an unusually large diameter, enabling the rapid conduction of *nerve impulses*. Giant fibers occur in the relatively simple *nervous systems* of many invertebrates.

They usually supply muscles used in reflex protective responses, such as the end-to-end contraction in earthworms. The giant fibers of lobsters have played an important role in neurological research because of the relative ease in making electrical measurements on them. (see *patch clamp*; compare *saltatory conduction*)

gibberellic acid (see *gibberellin*)

gibberellin A group of plant *growth substances* that stimulate the growth of leaves and shoots. Unlike *auxins*, they tend to affect the whole plant rather than inducing localized bending movements. Their best known effect is on the elongation of plant stems, specifically by enhancing cell elongation. Additionally, gibberellins help regulate *germination* in some seeds by initiating synthesis of *hydrolytic* enzymes, such as α-*amylase*, that mobilize *endosperm* food reserves in the *aleurone layer*. It is not yet clear whether they act at the DNA level or as enzyme inducers. About thirty different gibberellins have been isolated, the best known is *gibberellic acid*. (see *dormancy*)

Giemsa stain A reagent that differentially stains *chromosomes*, producing regions of dark and light banding. The band pattern is consistent within and characteristic of a species. The basic difference is thought to reside in the *base pair ratio*. The lighter bands appear to be **GC** (*guanine:cytosine*) rich and the darker bands **AT** (*adenine:thymine*) rich.

gill

1. An organ in aquatic animals in which the exchange of respiratory gases (oxygen and carbon dioxide) takes place between the blood or body fluid and the surrounding water. Gills are usually greatly convoluted in order to increase the surface area for gas diffusion and are highly *vascularized*. Organisms with external gills, such as those in many amphibian *larvae*, rely on body movement and water circulation to renew the local *oxygenated*

water supply. The internal gills of most fish are ventilated by forcing water from the *pharynx* (throat) past the gills and out through the *gill slits* (*ram ventilation*). The success of the gill's operation depends on water passing in the opposite direction of blood circulation in the *lamellae* so that respiratory gases can be efficiently exchanged. (see *gill slit, gill arch, gill bar, gill filament, gill raker, countercurrent exchange;* compare *lung*)

2. (lamella) In fungi that produce *basidiocarps* (toadstools and mushrooms), one of the thin radiating structures on the underside of the cap (*pileus*). *Spores* are produced within the gills.

gill arches Structural elements in jawed fish, *larval* amphibians, and some lower chordates, (*Amphioxus*) that are located between successive *gill slits* and contain the *gill bars*.

gill bars Skeletal structures in the wall of the *pharynx* (throat) in jawed fish, *larval* amphibians, and some lower chordates (*Amphioxus*) that support the tissue separating successive *gill slits*. They bear numerous fine *gill filaments* that bear the *lamellae* into which oxygen diffuses from the surrounding water. (see *gill arch*)

gill books (see *book gills*)

gill cleft (**visceral cleft**) An opening through which water leaves the body of a *cartilaginous* fish, such as a shark, after passing through the gills.

gill filament The two rows of thin membranous plates that make up each of the two *gills* in bony fish. They project out into the flow of water and carry rows of thin, disklike *lamellae* containing a dense *capillary* network.

gill-footed shrimps (see *Branchiopoda*)

gill rakers Skeletal projections of the inner margins of *gill bars* of fish. They are particularly elongated when modified for use as a *filter feeding* apparatus.

gill slits Openings in the *pharynx* (throat) of aquatic vertebrates leading to the gills. In some organisms (sea squirts, some

fish, squid, and octopus), they are used for filter feeding. Traces of gill slits appear in the *embryos* of all vertebrates but persist only as the *Eustachian tube* in adult terrestrial vertebrates. (see *gill rakers*)

gingiva The gum tissue that covers the jaw bones and is continuous with the lining of the mouth. It contains nerves, *lymph vessels,* and many *capillaries* which give it a pink color.

Ginkgo (see *Ginkgophyta*)

Ginkgophyta A phylum of *gymnosperms* containing only one ancient species, the ginkgo (*Ginkgo bilboa*). It is a *deciduous* tree with small, fan-shaped leaves and is *dioecious*. Like cycads, the *spermatozoa* are *undulipodiated*. In Ginkos, they are carried to the vicinity of the egg by a *pollen tube*. Ginkgo extract has become popular as an herbal remedy.

giraffes (see *Ruminantia*)

gizzard Part of the *alimentary canal* in animals that are unable to chew their food. It is located between the *crop* (a food storage organ) and the *intestine*. The gizzards of birds contain small stones or grit, and the muscular contractions of the walls grind the food between the stones prior to the main digestive process. In arthropods and earthworms, projections made of *chitin* perform the same function. (see *proventriculus*)

gland A cluster of specialized *secretory* cells that synthesize a specific chemical substance that is exported either through a duct (*exocrine*) or directly into the bloodstream (*endocrine*). Glands are comprised of *epithelial* tissue. *Sweat glands* are an example of an ectocrine gland and the *thyroid* is an example of an endocrine gland.

-glia- A word element derived from Greek meaning "glue." (for example, *neuroglia*)

glial cells Cells surrounding vertebrate *neurons* that perform supportive and protective functions. *Astrocytes* attach *neurons* to blood vessels, facilitating nutritional sustenance. *Oligodendrocytes* form the insulating *myelin sheaths* that surround each *axon* in the *central nervous system,* and *Schwann cells* perform the same function in the *peripheral nervous system.*

global extinction The disappearance of a species from the face of the earth. (compare *local extinction*)

global warming The rise in mean global temperature due to the changing composition of the earth's atmosphere. The process by which it takes place, the *greenhouse effect,* results from industrial *pollutants* such as carbon dioxide, methane, nitrous oxide, ozone, and chlorofluorocarbons. These absorb heat given off from the earth instead of radiating it out into space. For instance, carbon dioxide is released in large amounts when *fossil fuels* are burned and also when large tracts of forest are destroyed. The termites that then invade the forest remnants also excrete large amounts of methane, which propagates the cycle. Even apparently small changes of one or two degrees may have huge climactic effects including a rising sea level, major environmental shifts, and greatly altered weather patterns.

globin The *globular protein* that complexes with *heme prosthetic groups* in the oxygen carrying proteins *hemoglobin* and *myoglobin.*

globular proteins A class of soluble proteins that includes all enzymes, *antibodies,* and various transport and storage molecules. Structurally, at least one *polypeptide chain* of the protein is folded three-dimensionally. (see *tertiary structure, alpha-helix, amino acid*; compare *structural protein*)

globulins A group of *globular proteins* that occur particularly in plant seeds and vertebrate *blood plasma. Immunoglobulins* (*antibodies*) play an important role in the *immune system.*

glomerulus In the vertebrate kidney, a small knot of *capillaries* where water and dissolved waste products are filtered out for excretion. It is particularly well-developed in freshwater fish, which must excrete large amounts of water to maintain proper *osmotic balance.* The glomerulus is surrounded by the cup-shaped *Bowman's capsule.* Together they form a *Malpighian corpuscle* that is part of the *nephron.*

glossopharyngeal nerves (cranial nerves IX) A pair of mixed *sensory* and *motor nerves* that arise from the vertebrate hindbrain to supply the mouth cavity, including the tongue. They are essential to taste, palate sensation, *parotid gland* secretion (saliva), and swallowing. (see *cranial nerves*)

glottis The opening through which air passes from the *pharynx* (throat) to the *trachea* (windpipe). It sits in front of the opening to the *esophagus.* (see *epiglottis*)

glucagon A *peptide hormone* produced by cells of the *islets of Langerhans* in the pancreas. It causes an increase in blood sugar by promoting the breakdown of *glycogen* stored in the liver. (see *endocrine,* compare *insulin*)

glucan (see *glycan*)

glucocorticoid A type of *steroid hormone* produced by the *adrenal cortex.* Glucocorticoids (for example, *corticosterone* and *hydrocortisone*) help the body respond to stress and fatigue by increasing *metabolism* and inhibiting the *inflammatory* response. (see *corticosteroid, gluconeogenesis*)

glucosamine A *nitrogenous polymer* of *glucose* found in some *polysaccharides,* particularly, *chitin* and *hyaluronic acid.* (see *exoskeleton, vitamin C*)

glucose (dextrose) A biologically ubiquitous *monosaccharide* sugar of the general structure $C_6H_{12}O_6$. It is the primary energy source for most, if not all, cells and is *polymerized* for storage as *glycogen.* It is also found as a subunit of *starch* and *cellulose.* (see *carbohydrate, glycolysis, amylase, dextran, diabetes*)

glume A specialized *bract* found in grasses and reeds.

glutamic acid (glu) An *amino acid.*

glutamine (gln) An *amino acid.*

glutathione A *tripeptide* of *glutamic acid,*

cysteine, and *glycine* universally found in animal tissue. It functions as a *coenzyme* (for example, for *glyoxylase*) and as an *antioxidant.*

gluten A mixture of proteins found in wheat flour. The amount of gluten in a flour influences the texture of bread dough.

glycan A *polysaccharide* consisting of a single type of sugar unit. Examples are *cellulose, chitin* (structural proteins), *glycogen,* and *starch* (energy storage). The most common glycan, containing only *glucose* units, is called a *glucan.*

glyceride A molecule containing *glycerol* and one or more *fatty acids.* They may be mono-, di-, or *triglycerides* depending on the exact structure of the molecule. Mammalian fat stores consist mainly of *triglycerides.* These are *metabolized* to provide energy when carbohydrate levels are low. (see *lipase, adipose tissue, fat*)

glycerol A component of many lipids, notably *glycerides.* Purified glycerol is also used in laboratory solutions, in particular to maintain the integrity of frozen cells in long-term storage.

glycine (gly) An *amino acid.*

glycogen A *polysaccharide* that is the main carbohydrate store for animals. It is composed of *glucose* units linked so as to be readily *hydrolyzed* to *monosaccharide* units that can then enter into the metabolic pathways of the cell (*glycolysis*). Glucose obtained from food is converted to glycogen and stored in the liver and muscle (*glycogenesis*). The concentration of glucose in the blood is regulated by the conversion of glycogen back to glucose (*glycogenolysis*). (see *glucan, glycan, glucagon, insulin, diabetes*)

glycogenesis The process of converting blood *glucose* into its *polymeric* form of *glycogen,* which is stored mainly in the liver and muscles. It is mediated by the *hormone insulin.* (compare *glycogenolysis*)

glycogenolysis The process of converting *glycogen* stores into *monomeric glucose* units to be used as an energy source. It is

mediated by the hormone *glucagon.* (compare *glycogenesis*)

glycolipid A lipid covalently attached to a carbohydrate moiety (*mono-* or *oligosaccharide*). Glycolipids are found particularly in the outer half of *phospholipid bilayers* of *cell membranes.* Many are combined with proteins in *cell surface receptors.* (see *glycosylation, glycoprotein, blood group*)

glycolysis (Embden-Myerhoff pathway) The *anaerobic* breakdown of *glucose* into two molecules of *pyruvate* with the net generation of two molecules of *ATP* and two of *NADH.* It is one of two main energy-producing pathways in the cell. Glycolysis is utilized independently in *anaerobic* (no oxygen) situations and in conjunction with the *Krebs cycle* under *aerobic* (with oxygen) conditions. In some cells, such as certain *prokaryotes* lacking *mitochondria,* glycolysis is their primary source of energy. In *facultative anaerobes* (for example, some bacteria) and in vertebrate *striated muscle* fibers, glycolysis is utilized exclusively when there is a shortage of oxygen, such as during heavy exercise. In *anaerobic respiration,* no additional energy is liberated from the *pyruvate* generated by glycolysis; it is processed into waste products such as *ethanol* (for example, in yeasts) and *lactic acid* (in muscles). If *aerobic respiration* is possible, glycolysis is followed by the *Krebs cycle.* (see *fermentation*)

glycoprotein A protein molecule with carbohydrate side chains. Many *antigens,* enzymes, and *hormones* are glycoproteins.

glycosaminoglycans (GAGs) Long, unbranched *polysaccharides* that always contain an amino sugar (for example, N-acetylglucosamine). Along with protein constituents, they form various *extracellular* matrices of tissues, for example *hyaluronic acid.*

glycosidic bond The particular chemical bond (-O-) joining *monosaccharides* into an *oligosaccharide* or *polysaccharide*

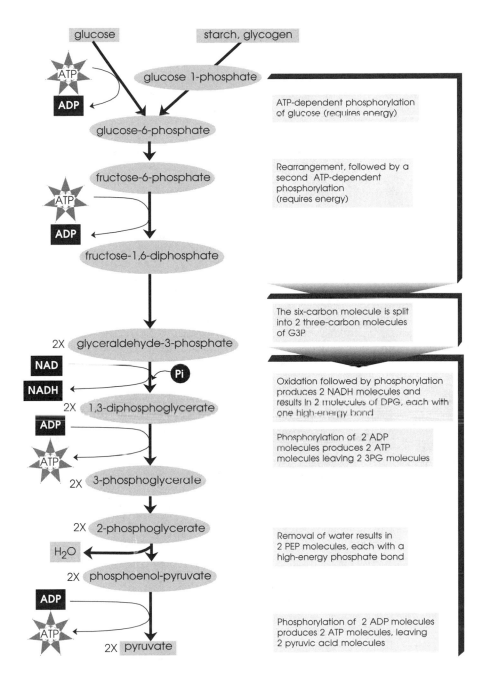

Glycolysis

chain. They are found in *glycogen* and *starch,* for instance.

glycosylation The chemical bonding of a sugar moiety to another organic compound. (see *glycoprotein, glycolipid*)

glyoxylate cycle A modification of the *Krebs cycle* occurring in many microorganisms, algae, higher plants, and, in particular, in *germinating* fat-rich seeds. The glyoxylate cycle exploits *fatty acids* as a sole carbon source, eventually forming carbohydrates. (see *glyoxysomes*)

glyoxysomes *Organelles* in plants that contain enzymes that convert fat to carbohydrate (*glyoxylate cycle*) and enzymes that destroy harmful *peroxidases* (*catalases*). (see *peroxisomes, microbody*)

Gnathostomulida A phylum of animals, the gnathostomulids or jaw worms, that inhabit the *anoxic* environment created by sulfur-producing bacteria in some black ocean sands. These microscopic *acoelomates* are *facultative anaerobes*, but the details of their *metabolism* are obscure. They are *hermaphroditic* but do *cross-fertilize.* In fact, they exhibit a primitive form of internal *fertilization.*

Gnetophyta A phylum of *vascular plants* including the Mormon tea plants and the species of *Ephedra* from which ephedrine is isolated. They are *cone-bearing,* nonflowering seed plants (*gymnosperms*), and their *spermatozoa* are immotile.

goats (see *Artiodactyla*)

goblet cell A cell specialized for the production of *mucus.* They are found in the *bronchi* and lining of digestive organs and in the skin of animals such as earthworms.

Golgi apparatus (see *Golgi complex*)

Golgi body An *organelle* present in *eukaryotic* cells that functions as a collecting and packaging center for substances that the cell manufactures for export. As such, Golgi bodies are particularly abundant in secretory cells. They are associated with the *endoplasmic reticulum.* They consist of stacks of membranous

sacs (*cisternae*) that are pinched off as *Golgi vesicles* for delivery to the exterior. (see *Golgi complex*)

Golgi complex (Golgi apparatus) A collective term for the numerous *Golgi bodies* in a cell.

Golgi vesicle A small portion of a *Golgi body* containing secretory material. It is pinched off for delivery to the exterior.

gon-, -gonia A word element derived from Greek meaning "sexual" or "reproductive." (for example, *oogonia*)

gonad The reproductive organ of animals. It produces reproductive cells (*gametes*) and sometimes *hormones.* The female gonad, the *ovary*, produces *ova*. The male gonad, the *testis*, produces *spermatozoa.* Some invertebrates (*hermaphrodites*) possess both male and female gonads.

gonadotrophic hormone (see *gonadotrophin*)

gonadotrophin Any *hormone* that acts on the *gonads* (*ovary* and *testis*) to regulate reproductive activity. Examples include *follicle-stimulating hormone (FSH)*, *luteinizing hormone (LH)*, and *prolactin.* (see *chorionic gonadotrophin*)

gophers (see *Rodentia*)

Gordian worms (see *Nematomorpha*)

G-proteins A family of signal-coupling proteins that serve as intermediaries between activated cellular *receptors* and effectors. They interact with a complementary family of *transmembrane protein* receptors to interconvert between an inactive *GDP* form and an active *GTP* form. G-proteins participate in a wide variety of cellular functions including translocation of *macromolecules* in *protein synthesis*, regulation of cell proliferation, transduction of *hormonal* signals and sensory stimuli, *leukocyte chemotaxis*, and even simple forms of learning in invertebrates.

Graafian follicle A fluid-filled vesicle in the *ovary* of a mammal that contains an *oocyte* attached to its wall and produces *estrogen.* One of the numerous follicles present in the ovary at birth matures periodically during a female's active reproductive years and bursts, releasing the oocyte (now an *ovum*) into the *fallopian tube*. The Graafian follicle then solidifies and begins to secrete *progesterone*, which maintains the uterine lining until *menstruation*, or if the ova is *fertilized*, during pregnancy. (see *chorionic gonadotrophin, luteinizing hormone, menstrual cycle, estrus cycle*)

graft The removal and reattachment of a piece of viable tissue, either to another location within the same individual or to another individual. In plants, grafting is an important horticultural technique and is often used to improve rootstock, especially in roses and fruit trees. In animals, grafting is often used to replace skin areas on burn victims. (compare *transplant*)

grain (see *caryopsis*)

Gram-negative (see *Gram's stain*)

Gram-positive (see *Gram's stain*)

Gram's stain A staining procedure used for differentiating morphologically similar bacteria and that divides them into *Gram-negative* and *Gram-positive* groups based on their color. Gram-positive bacteria take up a crystal violet stain and turn purple; Gram-negative bacteria exclude the crystal violet and counterstain, instead, with stains such as safranin, eosin red, or brilliant green. The uptake of particular stains by bacteria is ultimately determined by the physiological properties of their *cell wall*. As such, Gram-positive bacteria are more susceptible to *penicillin*, Gram-negative bacteria to body defenses such as *antibodies* and *complement*.

granellare The body of a xenophyophoran protoctist together with its surrounding branched tubes.

granulocyte (polymorphonuclear leukocyte) Any white blood cell (*leukocyte*) that is distinguished by granules in the *cytoplasm*. As a group, they are part of the body's defense system and contribute to the *inflammatory* response, in particular. They are *ameboid* and push out through the walls of *capillaries* at the site of an injury. Granulocytes are divided into three groups based on their staining properties with an acid/base stain: *basophils, eosinophils* and *neutrophils*. (see *immune system*; compare *lymphocyte, monocyte, macrophage*)

Granuloreticulosa (foraminiferans, forams, reticulopods) A phylum of *unicellular heterotrophic* marine protists made up almost exclusively of the foraminiferans, or forams, and a few groups of reticulomyxids. Each foram cell forms a pore-studded shell called a *test* from organic material combined with a variety of inorganic material, such that they vary widely in appearance. They are often multichambered, brightly colored, and tend to have a spiral shape. The largest may be mistaken for snails. Most live in sand or attached to other organisms, but a few are free-swimming *planktonic* organisms. Thin *podia* emerge through openings in the test and are used for swimming, gathering test material, and feeding. Although most are predacious *heterotrophs*, many forams harbor *symbiotic* algae or even foreign *chloroplasts* and may, thus, secondarily participate in *photosynthesis*. Their *life cycle* is complex, involving *alternation of generations*, and not yet well understood. The reticulomyxids are essentially naked forams. Granuloreticulosans are important geological markers and have contributed to limestone deposits all over the world, most famously in the English Cliffs of Dover.

granum, grana (pl.) A stack of the membranous disks (*thylakoids*) that contain the *photosynthetic pigments* in a *chloroplast*. The stacks of grana resemble piles of coins.

grasshoppers (see *Orthoptera*)

gravitropism (geotropism) A directional movement of a plant in response to the stimulus of gravity. For instance, primary roots (*tap roots*) grow toward gravity (*positive gravitropism*), and primary shoots grow away from gravity (*negative gravitropism*). Other structures may grow laterally or at other angles to the force of gravity. Gravitropic responses are mediated by *growth substances*, in particular *auxins*. A probable method of gravity detection involves *statocyte* cells that may be sensitive to the gravity-determined position of large starch grains (*statoliths*) situated within them. (see *tropism*)

gray matter Nerve tissue present in the *cerebral cortex* and *spinal cord* that consists mainly of *nerve cell bodies* and un-*myelinated neuron* connections. The lack of myelin imparts a gray cast. (compare *white matter*)

grebes (see *Podicipediformes*).

green algae (see *Chlorophyta*)

green gland (see *labyrinth*)

greenhouse effect The rise in atmospheric temperature resembling heat reflection by greenhouse glass. Ultraviolet radiation originating from the sun is reflected by the earth in longer infrared wavelengths that are absorbed by some atmospheric components as heat. Products of *fossil fuels*, such as carbon dioxide (CO_2), and other industrial *pollutants*, such as carbon dioxide, methane, nitrous oxide, ozone, and chlorofluorocarbons, tend to absorb and hold heat more efficiently. They may contribute to *global warming* via the greenhouse effect.

green revolution The name attached to post-World War II programs that have led to the development of new strains of agricultural crops with higher yields, better resistance to disease, or better ability to grow under poor conditions. The ultimate success and use of this direction of research has yet to be established. Some of the high-production crops require large amounts of fertilizer and water, and have other environmental drawbacks.

grex The *multicellular*, migratory, slug phase of some cellular slime molds.

grooming The mammalian equivalent of preening in birds. It involves licking and cleaning of the fur. Mutual grooming is an important behavior in group social structure.

gross primary productivity (see *primary productivity*)

ground meristem The *primary meristematic* tissue that differentiates into the *ground tissue* constituting the main body of *vascular plants*. (compare *procambium, protoderm*)

ground tissue The tissue comprising the main body of a *vascular plant* in which the *vascular tissue* is embedded. It is covered by the *epidermis*. (see *ground meristem*)

growth A general term denoting an increase in either size or quantity. It is one of the qualities that all living organisms possess.

growth curve A graphical depiction of the growth kinetics of a population of organisms. Its shape is altered by both the intrinsic biology of the organism (*biotic potential*) and its interaction with other

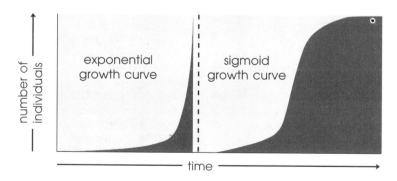

Types of Population Growth

individuals and the environment (*carrying capacity*). Unrestrained growth produces a characteristically shaped exponential growth curve. The addition of limiting factors tends to produce a sigmoid growth curve.

growth factor Any of the *peptide hormones* that regulate the growth of various cells or tissues.

growth hormone (somatotrophin) A *peptide hormone* secreted by the *pituitary gland* in mammals that regulates body growth by controlling the deposition of bone and cartilage. It is regulated by *somatostatin,* which is secreted by the *hypothalamus.* Growth hormone effects the release of *somatomedin* from the liver which, in turn, mediates its effects at the cell level. Insufficient growth hormone results in *dwarfism* and overproduction in *acromegaly.* Human growth hormone has recently been *cloned* and has the potential for effective therapy as well as possible misuse.

growth ring (see *annual ring*)

growth substance (plant hormone) Any compound, natural or artificial, other than a nutrient that promotes, inhibits, or otherwise modifies plant growth. The term "growth substance" is preferred because of the diverse nature of the compounds and their mode of production. (see *auxin, abscisin, cytokinin, ethylene, gibberellin*; compare *hormone*)

Gruiformes The order of birds containing the rails, coots, cranes, limpkins, and bustards.

guanine A *nitrogenous* base found in *nucleic acids* and *nucleosides*. It is based on a *purine* ring structure.

guano The *nitrogenous* excretion, in particular, of birds and terrestrial reptiles. It is composed of *uric acid* concentrated into a semisolid paste. Droppings from bat and seal colonies are also called guano. The nitrogen-rich guano may contribute to the natural eutrophication of ponds and lakes. Guano has also been harvested as plant fertilizer, disrupting its participation in the normal *nitrogen*

cycle. (see *ureotelic, uricotelic*)

guanophore One of several types of pigment-containing cells used in animals for camouflage. (see *chromatophore*)

guanosine A *nucleoside* of *guanine* linked to D-ribose with a β-glycosidic bond. *Guanosine triphosphate (GTP)* is derived from guanosine.

guard cell A specialized type of cell in plant leaves that uses *osmotic pressure* to control water loss and gas exchange. Two guard cells surround each *stoma* (pore) and control the opening and closing of its aperture by the level of *turgidity*. When a plant's internal moisture level is high, the guard cells become enlarged by *osmosis* from the surrounding cells and assume a bowed shape, creating an opening between them. When moisture levels are low, the cells are relaxed and rest against one another, closing the stoma to water loss. The process is also influenced by carbon dioxide levels and temperature.

guinea pigs (see *Hystricognathi*)

gullet An oral cavity situated just past the mouth.

gulls (see *Charadriiformes*)

gut The *alimentary canal* of an animal.

guttation The extrusion of water and sometimes salts from plants due to water pressure from the roots (*root pressure*). It may take the place of *transpiration*, particularly at night and in tropical plants where high humidity inhibits transpiration. It occurs through *hydathodes*, specialized cells at the edges of leaves. Guttation is only effective in relatively short plants because it depends on hydrostatic pressure to force the liquid to the surface.

Gymnophionia (see *Apoda*)

gymnosperms An informal designation for the four phyla of seed plants in which the *ovules* are carried naked on the cone scales (*strobili*). They contrast with flowering plants (*angiosperms*) in which ovules are enclosed by an *ovary*. Gymnosperms dominated the *flora* of the *carboniferous* period but have since been progressively displaced by the angiosperms. Conifers are the most familiar group. (compare *Anthophyta*; see *Cycadophyta, Ginkgophyta, Gnetophyta, Coniferophyta*)

gyn-, gyno-, -gynous A word element derived from Greek meaning "female." (for example, *gynoecium*)

gynodioecious A plant species that bears female and *hermaphroditic* flowers on separate plants. (compare *androdioecious, gynomonoecious*)

gynoecium, gynoecia (pl.) The collective term for the female reproductive structures of a flower, the *carpels* (*pistil*). (compare *androecium*)

gynomonoecious A plant species that bears female and *hermaphrodite* flowers on the same plant. (compare *andromonoecious, gynodioecious*)

gyttja A watery organic sediment that consists mainly of excretory material from animals that live on the sea bed (*benthic*).

habitat The local *environment* in which an organism is usually found. Defining factors may include variables such as climate, vegetation, and geology. (compare *ecological niche*)

habitat diversity The number of different kinds of *habitats* in a unit area. (see *biological diversity*)

habituation A form of learning in which there is a decrease in response to a repeated stimulus that has no consequences. Habituation is important in learning to ignore unimportant stimuli in a complex environment. (compare *sensitization*)

hagfish (see *Agnatha*)

hair An outgrowth from the skin of mammals consisting of dead cells embedded in a *keratin* matrix, often containing pigment granules.

hair cell (sensory hair) Any *ciliated sensory receptor* cell. Hair cells are found particularly in *statocysts* in the vertebrate *inner ear* where they are involved in the perception of vertical position relative to gravity, in the *macula* where they function in the detection of motion, and in the *organ of Corti*, which detects sound.

hair follicle An *epidermal* sheath enclosing the root of a hair and housing the cells that produce it. Muscles attached to the outside of the follicles can erect the hairs to increase thermal insulation. Nerve endings and *sebaceous* (oil) *glands* may also be present.

hallux The innermost *digit* of the hind foot in animals. It forms the big toe in man and in most birds, faces backwards for perching.

Halophilic bacteria (see *Euryarchaeota*)

halophyte Any terrestrial plant that is adapted to grow in high concentrations of salt, such as in salt marshes.

halosere A characteristic sequence of plant *communities* associated with the developmental stages of *ecological succession* in salt marshes or salt desert.

haltere One of a pair of club-shaped structures protruding from the sides of the *thorax* of flies (*Diptera*). They are highly modified hind wings and act like gyroscopes to maintain balance during flight.

haplodiploidy The unusual system of *sex determination* found in social insects, including bees, wasps, and ants. The males are all *haploid* and exist only to help rear their *diploid* sisters, only a few of whom will, in turn, reproduce. This system maximizes the *genetic fitness* of the entire colony at the expense of the individual. (see *amphoteric, altruism*)

haploid A cell or organism containing only one representative of each *chromosome*, the haploid chromosome number being represented as (n). Haploid cells are commonly produced by *meiosis* of *diploid* (2n) cells in the process of *gamete* formation. They may divide *mitotically* but may not undergo meiosis. Haploid gametes typically fuse to restore a *diploid* (2n) organism (*zygote*), which then develops into an adult. Some protoctists and fungi may also spend large portions of their *life cycle* in a haploid state. *Prokaryotes* are normally haploid throughout their *life cycle*. (see *fertilization, haplont*; compare *diploid, polyploid, aneuploid, euploid, diplont*)

haplont An organism representing the *haploid* stage of a *life cycle*. (compare *diplont*)

Haplospora A phylum of *unicellular*, *heterotrophic* protoctists, the haplosporidians, that are *parasitic* on marine or freshwater animals, primarily mollusks. They may be benign or *pathogenic*, a state believed to vary as a function of water salinity and also probably temperature. Some are extremely destructive and cause serious epidemics in oysters and other shellfish. Haplosporidians live between the cells in animal tissue as *uni-* or *multinucleate* unwalled cells called *plasmodia* and propagate primarily by *asexual spores*. They contain *haplosporosomes*, membrane bound *organelles* of unknown function.

haplosporosome A spherical, membrane bound *organelle* found in haplosporidians and possibly myxozoans and paramyxians whose function has not yet been elucidated.

haponasty (see *nastic movements*)

hapten A small molecule that has *antigenic* properties when attached to a *macromolecular carrier* and elicits *antibody* formation in the body.

hapteron The bottom part of some *multicellular* algae, attaching the plant to its substrate.

Haptomonada A phylum of *photosynthetic, unicellular* protoctists containing the prymnesiophytes and coccolithophorids. They are essentially single-celled algae found among the marine *plankton*. Like the freshwater chrysophytes, they contain chrysoplasts, *chloroplasts* filled with a yellow-brown pigment (*xanthophyll*). Some also form a distinctive resting stage called the *coccolithophorid*, a single cell covered with calcium carbonate plates called coccoliths. Along with forams, prymnesiophytes contributed to the formation of the White Cliffs of Dover in England. Coccoliths are distinguished by their characteristic geometrical shapes.

haptotropism (see *thigmotropism*)

hard palate (see *palate*)

hardwood A commercial designation of the wood from *dicotyledons*. It has little to do with the actual hardness of the wood.

Hardy-Weinberg equilibrium The situation for two particular genetic *loci* in a defined *population* where *allele frequencies* at the loci are constant in the population over time and no statistical correlation exists between particular alleles of each locus possessed by individuals in the population. Such a condition is approached in large randomly mating populations in the absence of *selection*, *migration*, and *mutation*. The Hardy-Weinberg *equation* ($p^2 + 2pq + q^2 = 1$) is derived from the frequencies of *alleles* (p and q) in the population.

hares (see *Lagomorpha*)

Hatch-Slack pathway (see C_4 *carbon fixation*)

haustorium, haustoria (pl.) A specialized outgrowth found at the ends of some fungal *hyphae* and *saprophytic* plants. It penetrates into and withdraws nutrients from the cells of the *host* plant.

Haversion canals A series of fine, interconnecting canals that permeate the compact *bone* (such as limb bones) of vertebrates. They contain the blood and nerve supply. Each canal serves a series of concentric rings of bone units (*lamellae*) that surround it. Each series of lamellae with its canal is termed a Haversian system. (see *Volkmann's canals*)

hawks (see *Falconiformes*)

HCG (see *chorionic gonadotrophin*)

HDL (high density lipoprotein) A blood *plasma* protein that serves to transport *cholesterol* and other lipids from the bloodstream to various tissues. It has a relatively low lipid content compared to other *lipoproteins*, thus a higher protein content. Because protein is more dense than fat, it has been given the designation "high-density." According to some studies, a high blood level of HDL is associated with a lowered risk of cardiovascular disease.

heart A muscular organ present in higher animals that pumps blood to all parts of the body. In mammals, it consists of four chambers (two upper *atria* and two lower *ventricles*) with the right and left sides of each completely separate. *Deoxygenated* blood is carried to the right atrium via the *venae cavae* and *oxygenated* blood from the lungs is carried to the left atrium by the *pulmonary vein*. Contraction of both atria forces blood into the respective ventricles that, in turn, contract, forcing blood into *arteries*. The *pulmonary artery* carries *deoxygenated* blood from the right ventricle to the lungs and the *aorta* transports oxygenated blood from the left ventricle to the body. The rhythmic contractions of the heart (*cardiac*) muscle are regulated by an internal *pace-*

maker (*sinoatrial node*). Mollusks and fish have two-chambered hearts, amphibians two atria and one ventricle, and reptiles one ventricle. Crocodiles have a septum that completely separates the ventricle into two. Crustaceans have a single heart with openings possessing valves, and annelids have a number of contractile vessels known as hearts.

heartwood (duramen) The dead, woody center of a tree trunk made up of *xylem* vessels that are no longer involved in water transport. Such vessels are often blocked by substances that give the wood a darker color. (compare *sapwood*)

heavy chain The *polypeptide* subunit that occurs in duplicate in each *antibody* molecule and determines into which one of the five classes it falls. Heavy chains also carry *variable* and *hypervariable regions* that together with *light chains*, contribute to *antibody specificity*. (see *antibody diversity, immunoglobulin*)

hedgehogs (see *Insectivora*)

heliotropism (see *tropism, phototropism*)

Heliozoa (see *Actinopoda*)

helix (see *double helix, alpha helix*)

helix-loop-helix A structural motif found in some *DNA binding proteins*, particularly those involved in *differentiation*, and in some *proto-oncogenes*. Helix-loop-helix proteins form dimers of identical protein units (homodimers) that facilitate their binding to the DNA *double helix*. (see *transcription factor*; compare *zinc finger, leucine zipper, helix-turn-helix*)

helix-turn-helix A structural motif found in many *DNA binding proteins*, in particular *regulatory proteins*. The two α-helical domains of the protein are separated by a turn of 34 Å, the pitch of a *DNA helix*. It has been postulated that the α-helices fit into the *major groove* of DNA. (compare *zinc finger, leucine zipper, helix-loop-helix*)

helocrene A seepage area where springwater percolates thorough a substratum of mosses, leaf litter, and detritus.

helophyte Any *perennial* plant in which

the *perennating* (overwintering) bud lies in soil or mud below the water level. Helophytes are typical of marshy or lake edge environments.

helper T-cell (see *T-cell*)

hemagglutinin A surface protein found on many viruses that allows it to gain access to the interior of the *host* cell. This protein becomes altered in flu viruses through periodic *recombination* of its *genes*. The virus is then able to present a novel *antigen* that is not detected by *antibodies* to previous flu infections, resulting in new flu outbreaks. (see *antigen-shifting*)

hematocrit The fraction of the total volume of blood that is occupied by red blood cells (*erythrocytes*). In humans, this is typically 45%.

hematopoiesis (see *hemopoiesis*)

heme An organic ring molecule (*porphyrin*) containing an iron atom at the center. It is the *prosthetic group* in *hemoglobin*, *myoglobin*, and *cytochromes*. (compare *chlorophyll*)

Generalized Structure of a Heme Group (Hemoglobin, Myoglobin, Cytochrome C)

hemelytron, hemelytra (pl.) The leathery forewings found in true bugs (Hemiptera). They protect the delicate hind wings when these are not in use.

hemicelluloses A heterogeneous group of *polysaccharides* that, together with *pectins*, form the matrix within which *cellulose* fibers are embedded in plant *cell walls*. Their heterogeneous sugar subunit composition is characteristic.

Hemichordata An ancient, small phylum of marine invertebrates including the acorn worms. They exhibit a trisegmented body and inhabit U-shaped burrows on the sea bottom. They are *coelomate deuterostomes*. Hemichordates are linked to chordates in their *embryonic* pattern of development in which the *blastopore* develops into the adult anus. (see *tornaria*)

hemichordates (see *Hemichordata*)

hemicryptophyte A nonwoody *perennial* plant whose *perennating* (overwintering) buds are at ground level. The aerial shoots die at the onset of unfavorable conditions. (see *Raunkiaer's plant classification*)

Hemiptera The order of insects including the true bugs such as aphids, bed bugs, cicadas, leafhoppers, and scale insects. They have two pairs of wings and their mouthparts are specialized for piercing and sucking. Many are *vectors* of *pathogens* and are themselves destructive. They undergo incomplete *metamorphosis* (*exopterygote*).

hemizygous Genetic material in a *diploid* (paired *chromosomes*) organism that exists in a *haploid* (unpaired) situation. It may refer to a whole chromosome, chromosomal segments, or single *genes*. A normal example is found in the *sex chromosome* pair of many animals where little if any genetic material is duplicated between them. The situation may also occur as the result of chromosome loss or *deletion*. Higher organisms lacking large portions of genetic material are generally nonviable. (see *heterogametic*)

hemocoel (see *coelom*)

hemocyanin A copper-containing protein that functions as a *respiratory pigment* in mollusks and arthropods. It is blue when *oxygenated*, colorless when *deoxygenated*. Unlike *hemoglobin,* it does not contain a *porphyrin* structure. (compare *hemoerythrin*)

hemocytometer A microscopic grid de-

signed for estimating the number of particles in suspension, in particular microorganisms and blood cells. Generally, they are designed so that a fixed, known volume of fluid is contained beneath the cover slip.

hemoerythrin An iron-containing protein that functions as a *respiratory pigment* in a few minor groups of invertebrates. Unlike *hemoglobin*, it does not contain a *porphyrin* structure and only functions *intracellularly*. (compare *hemocyanin*)

hemoglobin The *respiratory pigment* in vertebrate red blood cells (*erythrocytes*). It carries oxygen from the lungs to other tissues and less specifically carries carbon dioxide back to the lungs. The *globular protein globin* is linked with four *heme prosthetic groups*, each containing an iron atom that is able to reversibly bind one molecule of oxygen. In the *oxygenated* state, *oxyhemoglobin* lends blood its characteristic bright red color. Carbon dioxide combines with hemoglobin at its amino groups. Hemoglobin binds oxygen more tightly when the partial pressure is high, and releases it to tissues when less oxygen is available. An increase in dissolved carbon dioxide, leading to higher blood acidity, also increases the affinity of hemoglobin for oxygen. Carbon monoxide (CO) binds to hemoglobin irreversibly, causing suffocation at the cellular level. Variations in globin structure occur between species, and in fact during different stages in human development. (compare *hemoerythrin, hemocyanin*)

hemolymph The circulatory fluid or blood of animals with open circulatory systems.

hemolysis The rupture of red blood cells (*erythrocytes*) with the subsequent release of *hemoglobin*. It may be caused by toxins, an incompatible blood transfusion, or other conditions.

hemophilia An inherited condition resulting in a deficiency in *blood clotting*, causing even small injuries to bleed profusely. The *gene* for *clotting factor VIII* is recessive and located on the *X chromo-*

some. Therefore, males exhibit the disease if they inherit a copy of the defective gene from their mother, whereas females exhibit the disease only if they inherit a copy of the defective **X** chromosome from each parent (*homozygous*). (see *sex chromosomes, sex-linked*)

hemopoiesis (hematopoiesis) The formation and development of blood cells. In the vertebrate fetus, hemopoiesis occurs in the spleen and liver. In the vertebrate adult, it occurs in *bone marrow* and *lymphoid tissue*. (see *stem cells, erythrocytes, leukocytes*)

heparin A substance that inhibits *blood clotting* by preventing the conversion of the *clotting factor prothrombin* to the active enzyme *thrombin*. It is normally present in low levels in the tissue of mammals and is secreted by some blood-sucking animals.

Hepatophyta A phylum of non*vascular plants*, the liverworts. In some species, the dominant *gametophyte* generation may be relatively undifferentiated and grow flat along the ground. Other species may have simple leaves and stems that resemble mosses in their complexity. The *sporophyte* generation is *parasitic* on and is usually held within gametophyte tissue until its *spores* are shed.

herb A herbaceous plant

herbaceous Non woody plants in which *secondary growth* is limited or absent. They send up new stems above ground every year, producing them from underground *perennating structures* (*perennial*), or *germinate* and flower just once (*annual, biennial*). (compare *woody*)

herbaceous perennial (see *perennial*)

herbicide Any chemical, either natural or artificial, that kills vegetation.

herbivore An animal that eats plant material as a main or sole food source. Various modifications may be associated with this diet, for instance to the teeth and *digestive system*. (see *food chain, food web, crop, gizzard*; compare *carnivore, omnivore*)

heredity The transmission of characteristics from parent to offspring through the genetic material encoded in *DNA*. In *sexual reproduction*, this process takes place through *gametes*. Heredity is the one defining characteristic of life.

hermaphrodite (bisexual) An organism bearing both male and female reproductive organs (in plants, on the same flower). In animals, the situation is generally confined to lower life forms (for example, earthworms, many *parasites*, and a few fish), and mating, nonetheless, often occurs between two different individuals. In plants, the situation is rather common, as *stamens* and *carpels* are often both present in a single flower. (compare *unisexual, monoecious, dioecious*)

herons (see *Ciconiiformes*)

herpes viruses A group of *double-stranded DNA* viruses. They are responsible for a number of diseases including cold sores (herpes simplex 1), genital herpes (herpes simplex 2), chickenpox, shingles, mononucleosis (Epstein-Barr), and cytomegaloviruses, which cause systemic infections of newborns.

herpetology The study of reptiles and amphibians.

hesperidium, hesperidia (pl.) The berry of a citrus fruit. Its flesh is divided into segments and surrounded by a separable skin.

-het-, hetero- A word element derived from Greek meaning "different" or "other." (for example, *heterokaryon*)

heterocercal tail The type of tail found in *cartilaginous* fish (for example, sharks) in which the *vertebral column* extends into the tail. (compare *homocercal, diphycercal*)

heterochromatin (see *chromatin*)

heterochrony In organizational development, the dissociation during development of factors influencing shape and size, from sexual maturity. This leads to reproductive ability in an otherwise juvenile animal (*paedogenesis*) such as the axolotl. (see *allometry, neoteny, paedomorphosis*)

heterocyst In filaments of adhering cyanobacteria, an enlarged, specialized cell in which *nitrogen fixation* occurs. Some of the nitrogen fixation *genes* are apparently rearranged during heterocyst formation in order to activate them. These organisms exhibit one of the closest approaches to true *multicellularity* among the bacteria. (compare *somatic recombination*)

heteroduplex A *double-stranded structure* formed by *annealing* single *DNA* strands that are mostly *homologous* except for a few differing *base pairs*. It is a natural part of various *recombinational* mechanisms and may also be generated *in vitro*. (see *duplex*)

heteroecious A term applied to a *parasitic* organism (for example, rust fungus) in which part of the *life cycle* occurs obligatorily in one *host* and the remaining part obligatorily in another. (compare *autoecious*)

heterogametic In species with dissimilar *sex chromosomes*, the *gender* that contains two different chromosomes. For instance in humans, females contain two **X** chromosomes (**XX**) and males contain one **X** and one **Y** chromosome (**XY**) making males the heterogametic sex. In birds, reptiles, and butterflies where females are the heterogametic sex, their chromosomes are named **Z** and **W**, and the *homogametic* males have a genotype of **ZZ**. (see *sex determination, sex-linked*; *heteromorphic*, compare *homogametic sex*)

heterogamy (see *anisogamous*)

heterogeneous nuclear RNA (see *hnRNA*)

heterokaryon A cell that contains *nuclei* of two or more different *genotypes* (genetic types). Heterokaryotic nuclei are often found in *fungal hyphae* and *mycelia*. (see *dikaryon*; compare *homokaryon*)

heterokont Any motile cell that bears two *undulipodia* of unequal length or unlike form, often directed in opposite directions.

heterologous gene expression The *RNA*

transcription and *translation* of a *gene product* in a different organism from which it occurs naturally.

heteromorphic Literally, differing in form. For example, the different forms of an organism in *alternating generations* of a *life cycle* are described as heteromorphic. The term is also used to describe a *chromosome* pair that differ in size and shape, such as the **XY** *sex chromosomes* in humans. (see *heterogametic*)

heterophilic A plant having more than one form of foliage leaf on an individual. It may occur as a result of the normal developmental process or in response to environmental changes.

heteroplasmon (see *cytohet*)

heterosporous The production of small and large *spores* on the same plant (for example, ferns). *Microspores* develop into male *gametophytes* (*gamete*-producing generation) whereas *megaspores* produce female gametophytes. Heterospory occurs only in *vascular plants* and is seen as a significant stage in the development of the seed habit. (compare *homosporous*)

heterostylic A *dimorphic* variation in flowering plants (*angiosperms*) in which the *styles* (female reproductive structure) and *stamens* (male reproductive structure) in flowers of the same species are found in two different lengths. An example is the primrose where the "pin-eyed" variant has long styles and short stamens while the "thrum-eyed" variant has short styles and long stamens. This situation encourages *cross-fertilization* and discourages *self-fertilization*, promoting genetic variation within the species.

heterothallic In certain fungi and protoctists, the situation in which *sexual reproduction* requires individuals of strains exhibiting different *mating types*. This situation occurs in mutants of *homothallic* strains normally found in nature where individuals can switch mating types, enabling self-*diploidization*.

heterotrophism A method of obtaining nutrients by feeding on other organisms.

Heterotrophic organisms are *chemotrophic*, obtaining both their energy and carbon atoms by degrading ingested organic compounds (*chemoheterotrophism*). At least 95% of the organisms on earth (all animals, all fungi, and most bacteria and protists) live by feeding on the chemical energy fixed into carbon compounds by *photosynthesis*. *Autotrophs* (green plants, algae, and photosynthetic bacteria) are the *primary producers* of organic carbon for all *heterotrophic* organisms. (see *trophism*; compare *autotrophism*)

heterozygosity A measure of the genetic variation in a population. With respect to one *locus*, it is stated as the frequency of *heterozygotes* for that locus.

heterozygous The situation in a *diploid* organism when different *alleles* (genetic variants) are present on *homologous chromosomes* at a particular genetic *locus* (location). The individual is called a heterozygote.

hexacanth The *embryo* of a tapeworm. Many hexacanths are contained in each *proglottid* (segment). They are surrounded by shells and emerge from the proglottid through a pore or the ruptured body wall. They leave the *host* with the feces and are deposited in places from which they may be ingested by another animal.

hexose Any simple sugar molecule (*monosaccharide*) containing six carbon atoms. Examples include *glucose, fructose*, and *galactose*.

hexose monophosphate shunt (see *pentose phosphate pathway*)

Hfr cell A bacterial cell in which an *F factor* is *integrated* in the *host chromosome*, mobilizing it for transfer to a cell lacking an *episome* (F^-). The acronym stands for high frequency *recombination*. (see *pilus, conjugation*)

hibernation A physiological adaption of some mammals, enabling survival during periods of low temperature and food scarcity. Body temperature and *metabolism* are decreased, and fat stores supply the minimal energy needed for survival. (see *adipose tissue*; compare *dormancy*)

high density lipoprotein (see *HDL*)

high-energy phosphate bond The chemical bond between two phosphate groups or sometimes between a phosphate group and another molecule that releases large amounts of energy when it is split. An additional criteria is the ability of the bond-splitting event to be coupled to another reaction in a *biosynthetic* pathway.

hilum, hila (pl.) The point on a *fungal spore* marking its prior attachment to its *sporophore*, or on a plant seed marking its attachment to the *ovary* wall.

hindbrain (rhombencephalon) The posterior region of the brain consisting of the *medulla oblongata*, the *pons*, and the *cerebellum*. (compare *forebrain, midbrain*)

hindgut The last part of the *alimentary canal* of arthropods and vertebrates. In vertebrates, it consists of the last part of the *colon*. (compare *foregut, midgut*)

hip girdle (see *pelvic girdle*)

hippocampus The center deep in the vertebrate brain that is responsible for emotional responses and also is involved in memory recall. It is part of the *limbic system*.

Hippomorpha A suborder of odd-toed ungulates that includes horses.

hippopotamuses (see *Artiodactyla, Suina*)

Hirudinea The class of annelids containing the leeches. They occur mostly in freshwater. Their *coelom* is continuous throughout the body rather than *segmented*. Leeches have suckers at one or both ends of the body that may be used for movement, and they lack *setae*. Most leeches are predators or scavengers. A number live as external *parasites*, sucking blood from vertebrates. A few are parasitic on crustaceans. They are *hermaphroditic* but generally *cross-fertilize*. The medicinal leech, *Hirudo medicinalis*, is used as a source of *anticoagulant* and as a result of continued harvesting, is becoming scarce in certain areas.

histamine A chemical generated by *mast*

cells in response to injury or *allergic response*. Histamine stimulates *smooth muscle* contraction and gastric secretions of hydrochloric acid and *pepsin*. It also increases local blood vessel permeability, leading to swelling, and dilates blood vessels, decreasing *blood pressure*. It is a factor in the itching, sneezing, and *inflammation* associated with allergy, and in the most severe cases (anaphylactic shock), may lead to asphyxiation and death. The effects are reduced by *antihistamine* drugs.

histidine (his) An *amino acid.*

histocompatibility antigens Any *cell surface antigen* that influences the acceptance of grafts or transplants. (see *MHC proteins, H-Y/H-W antigen*)

histology The study of tissues and cells at the microscopic level.

histones A group of eight, small, basic proteins found in combination with *nucleic acids* in *eukaryotic chromosomes*. They are essential in forming *nucleosomes*. (see *tertiary structure*; compare *polyamine*)

histophage An organism that feeds on the tissue of dead animals, as opposed to a *parasite* that feeds on a live *host*. (see *saprotrophism, saprophage*)

HIV (human immunodeficiency virus) The virus responsible for *AIDS*. It is a *single-stranded RNA virus* that specifically infects T_4 *lymphocytes* (white blood cells). This is significant because these cells play a key role in bodily defense. The *double-stranded DNA* copy made by the viral enzyme *reverse transcriptase* inserts into the *host* cell *genome* where it may remain for a long period (typically eight years or longer) before active *transcription* and viral *replication* are initiated.

HLA complex (human leukocyte-associated antigen complex) The *MHC* (major histocompatibility) complex of humans. It is located on *chromosome* VI and consists of a group of highly *polymorphic* (variable) linked genetic *loci* that determine the set of *glycoproteins*

found on the surface of all *nucleated* cells in the body. They are unique to each individual and aid the body in identifying "self" from "nonself." In *autoimmune* diseases, this mechanism fails, and the body attacks itself. The HLA complex is also important in determining *transplantation* compatibility. (see *β-microglobin, MHC proteins, histocompatibility*)

HLA genes (see *HLA complex*)

hnRNA A diverse assortment of *RNA* types found in the *nucleus* of a *eukaryotic* cell as they are *transcribed* directly from *nuclear DNA*. It includes *mRNA* molecules that undergo further processing, often including *splicing*, before being *translated* into protein.

holdfast (see *hapteron*)

holoblastic cleavage A term describing the type of *cleavage* (cell division) seen in eggs that have very little *yolk*. It produces cells in the *blastula* that are of a uniform size. (compare *meroblastic*)

Holocene The geological epoch covering the last 10,000 years.

holoenzyme An active enzyme complex consisting of a *core enzyme* or *apoenzyme* and a *cofactor, coenzyme,* or additional protein subunits. An apoenzyme is catalytically inactive without its cofactor or coenzyme. A core enzyme has an inherent activity that can be modified or narrowed by additional subunits.

Holothuroidea The class of echinoderms containing the sea cucumbers. They are soft, sluglike organisms with a much-reduced calcareous skeleton that inhabit the bottom of the ocean. Tentacles around the mouth may secrete a mucus net used to capture *plankton*. Some ingest bottom sediment and leave castings similar to earthworms. They have a respiratory tree that arises from the anal cavity and distributes gases to the body by contractions of the *cloaca*. The sexes are most often separate, but a few are *hermaphroditic.*

holothuroids (see *Holothuroidea*)

homeobox A family of similar *nucleotide* sequences that encodes sequence-specific *DNA-binding proteins*. They are essential for establishing the developmental fate of various cell groups and also for maintaining the *differentiated* state. Homeobox sequences were first detected in *Drosophila* as a result of studying *homeotic mutations*. They have *homologues* in the *genomes* of many organisms, including humans and some plants. (see *homeodomain, conserved sequence, helix-turn-helix*)

homeodomain A protein motif encoded by the *homeobox DNA* sequences. It forms a *helix-turn-helix* structure that binds to DNA in a sequence-specific manner and either activates or represses the *transcription* of specific target *genes*. Homeodomains are particularly found in *genes* with important developmental roles in many different organisms. (see *transcription factor, homeotic mutation*)

homeologous chromosomes *Chromosomes* that share partial *homology*, usually indicating an ancestral relationship. They are typical of *allopolyploids*.

homeosis The replacement of one body part by another. It can be caused by *mutation* or by environmental factors leading directly to developmental anomalies. (see *homeotic mutation*)

homeostasis The maintenance of a constant *physiological* environment by an organism. It is often mediated by *feedback regulation* involving the *nervous system* and various *hormones*.

homeothermic (see *endothermic*)

homeotic mutation Any of a family of *mutations* in *Drosophila* that can change the gross developmental fate of particular body segments. The study of homeotic gene mutants led to the discovery of the *homeobox* in DNA. The homeobox is a *conserved sequence* that encodes the *homeodomain* protein motif, an important developmental regulator in both animals and plants.

hominid Any primate in the human family *Hominidae* of which *Homo sapiens* (modern man) is the only living representative.

hominoid A group encompassing the apes together with *hominids* (humans and their direct ancestors).

Homo A genus of the family *Hominidae* of which modern man, *Homo sapiens*, is the only surviving species. *H. sapiens neanderthalensis* (Neanderthal man) is regarded as a subspecies of *H. sapiens*.

homocercal tail The type of tail found in adult bony fish in which the *vertebral column* stops short of the end of the tail and the *caudal* fin has two lobes of equal size. (compare *diphycercal tail, heterocercal tail*)

Homo erectus The second species of the genus *Homo* to evolve, *Homo erectus* used fire and made characteristic stone tools. They appeared in Africa about 1.7 million years ago and migrated throughout Eurasia. They disappeared from Africa about 500,000 years ago and from Asia about 250,000 years ago.

homogametic sex In organisms with dissimilar *sex chromosomes*, the *gender* that contains the two *homologous chromosomes*. For instance in humans, males contain one *X* and one *Y chromosome* (**XY**) and females contain two *X chromosomes* (**XX**). Females are considered the homogametic sex. In birds, reptiles, and butterflies, the males are the homogametic sex and contain two *Z chromosomes* (**ZZ**). (compare *heterogametic sex*)

homogamous The condition in plants in which the *anthers* and *stigmas* mature at the same time, encouraging *self-pollination*. (compare *dichogamous*)

homogamy (see *isogamous*)

Homo habilis The first known members of the genus *Homo*. They appeared about 2 million years ago in Africa and persisted for about half-a-million years. *Homo habilis* was almost certainly derived from *Australopithecus*. They represented the first genus of the *Homo* line who used tools.

homokaryon A cell that contains multiple *nuclei* of similar *genotypes* (genetic types). Homokaryotic nuclei are often found in *fungal hyphae* and *mycelia*. (see *monokaryon*; compare *heterokaryon*)

homologous Structures or organs that are similar or have a similar evolutionary origin. For instance, the forelimbs and hind limbs of all land vertebrates are said to be homologous, having evolved on the same *pentadactyl* (five digit) pattern. The term may also be applied at the protein and *DNA* levels. (see *homologous chromosomes;* compare *analogous*)

homologous chromosomes In a *diploid* organism, the duplicates of the same *chromosome* that make up a pair. In general, excepting *mutation*, they are the same size and shape, and carry mostly the same genetic material. Homologues pair during *meiosis* and separate from each other at the first meiotic division. When *gametes* fuse to form a *zygote*, each carries one homologue of each chromosome contributed by the parent. *Sex chromosomes* are an exception. They have very little *homology* but still pair during *meiosis*. (see *homozygous, heterozygous*)

homologous recombination The rearrangement of genetic material by the breakage and reunion of *DNA* molecules at sites of similar *nucleotide* sequence. Classically, this applies to exchange between *homologous chromosomes*, in particular during *meiosis I*. Similar mechanisms are used in *plasmid* and *bacteriophage integration* and *excision* as well as various types of *site-specific recombination* including the generation of *antibody* diversity in *immune receptor genes*. Homologous recombination is also one mechanism used to introduce foreign genes into the genome of an organism. (see *gene conversion, crossing-over, somatic recombination, transposable genetic elements*)

homopolymeric tailing The *in vitro* addi-

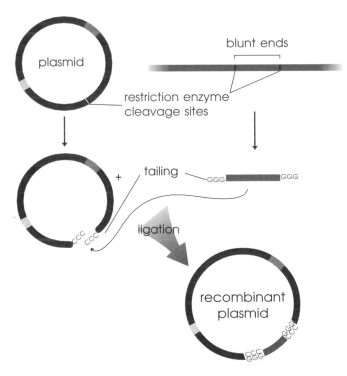

Homopolymeric Tailing

tion of *single-stranded nucleotide* tails onto *blunt-ended DNA* fragments. Each tail is of a single type of *base* and is *complementary* to the tail added to the other fragment. In this way, *sticky ends* can be generated and the molecules easily joined by *ligation*. It was an early *cloning* technique that is mostly defunct.

Homoptera An order of insects including the leafhoppers and cicadas. They are true bugs and have two pairs of wings. Their mouthparts are specialized for piercing and sucking. Many are *vectors* of *pathogens* and are themselves destructive. They undergo incomplete *metamorphosis* (*exopterygote*).

Homo sapiens The species of modern humans that appeared about 500,000 years ago, probably in Africa. It is likely that *Homo sapiens* is descended from *Homo erectus* with *Neanderthal* man comprising a parallel line that became extinct.

homosporous The production of only one type of *spore* of uniform *morphology* from which both male and female *gametophytes* (gamete-producing generation) develop. This type of development is common in mosses and liverworts. (compare *heterosporous*)

homostylic The usual condition found in plants in which the *styles* (female reproductive structure) and *stamens* (male reproductive structure) of all flowers of the same species are approximately the same length. This situation encourages *self-fertilization*. (compare *heterostylic*)

homothallic In certain fungi and protoctists, the situation in which individuals can switch *mating types*. This enables the formation of a colony containing opposite mating types, leading to fusion and *zygote* formation. Homothallism permits the ability on a single cell to self-*diploidize*. In certain yeasts, the mechanism is particularly well known. It involves a switch at the genetic level by the replacement of *DNA* segments at the mating type locus. Mutations in this mechanism produce *heterothallic* strains that have a fixed mat-

ing type. (see *double-strand break*)

homozygous The situation in a *diploid* organism when the same *allele* (genetic variant) is present on both *homologous chromosomes* at a particular genetic *locus* (location). The individual is called a homozygote.

homunculus The mythical, preformed, miniature person once believed by supporters of the *preformation theory* to be present in the sperm head or egg.

horizontal cell One of the *neuron* types in the vertebrate *retina*. They conduct signals laterally between *rods* and *cones* without firing *action potentials*. (see *amacrine cell*)

hormone A chemical messenger produced by an *endocrine gland* that is transported in the bloodstream and performs its function at a distant location. Hormones are recognized by specialized *receptors* in the cells of the target organ and may have diverse chemical structures including *peptide*, *steroid*, and amine. They regulate many important body processes including growth, *metabolism*, and *sexual* reproduction. In plants, hormones are more properly referred to as *growth substances*. (see *second messenger,* individual hormones)

hornbills (see *Coraciiformes*)
hornworts (see *Anthocerophyta*)
horsehair worms (see *Nematomorpha*)
horses (see *Hippomorpha*)
horseshoe crabs (see *Merostomata*)
horsetails (see *Sphenophyta*)
host

1. The organism supporting another *parasitic* organism in or on its body and to its own detriment. In a primary (*definitive*) host, a parasite reproduces *sexually* or becomes sexually mature. A secondary (*intermediate*) host generally houses one or more *larval* stages of the parasite. (see *symbiosis*)

2. A cell into which a fragment of foreign *DNA* has been inserted. The introduced DNA may be self-replicating or *integrated* into the host *genome*. (see *gene cloning, genetic engineering*)

host range The spectrum of species that a *parasite* can infect.

housekeeping genes *Genes* that are expressed in most or all tissue types of an organism. They run the basic functions of the cell.

human chorionic gonadotrophin (see *chorionic gonadotrophin*)

Human Genome Project The worldwide project to decipher and catalogue the entire *nucleotide* sequence of the human *genome*.

human immunodeficiency virus (see *HIV*)

human leukocyte-associated antigen complex (see *HLA complex*)

humerus The long bone of the upper forelimb in animals, the upper arm bone in man.

hummingbirds (see *Apodiformes*)

humoral immune response The second, longer-term, bodily defense system mediated by *B-cells* and their *antibody* products. Initially, an *antigen*-presenting *macrophage*, a *helper T-cell* and a *B-cell*

bind together to form a complex that initiates the response. The binding is mediated by *MHC-II proteins* present on these three cell types exclusively. Like the *T-cell-mediated immune response*, B-cell proliferation is activated via *lymphokines* released by *helper T-cells*. Only B-cells presenting *B-receptors* specific to the foreign *antigen* proliferate and mature into a large *clone* of antibody-secreting *plasma cells*. The initial production of antibodies (*primary immune response*) incorporates class M *heavy chains* followed by a second, more extended wave (*secondary immune response*) bearing class G heavy chains. B-cells release antibodies to mark invading *pathogens* for destruction by the otherwise nonspecific mechanisms including *complement*, *macrophages*, and *K-cells*. *Suppressor T-cells* terminate the antibody response after several weeks. A few B-cells are retained as *memory cells*, providing an accelerated secondary immune response to a later encounter with the same antigen. (see *immunoglobulin,*

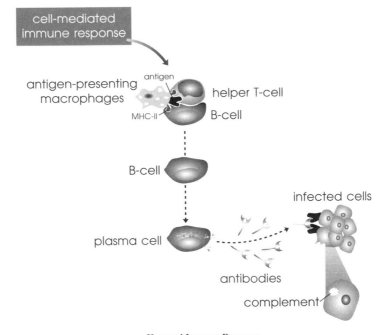

Humoral Immune Response

immune receptor genes, somatic recombination, hemopoiesis, bone marrow, stem cell)

humus Decomposed organic matter present in the top layer of some soils.

hyaluronic acid A viscous *glycan* found in *extracellular* matrices, particularly in *connective tissue*. It is a derivative of *vitamin C* and acts as an *intercellular* cement.

hybrid An organism resulting from a cross of genetically dissimilar parents. In animals, hybrids from different species are usually sterile due to the failure of *chromosomes* to pair and *segregate* properly during the *meiotic* formation of *gametes*. This characteristic contributes to a definition of species boundaries. Plants may often overcome this deficiency by a doubling of the overall chromosome number, thus providing each chromosome with a partner. In fact, many important agricultural crops are *polyploid*. (see *postzygotic isolating mechanisms*)

hybrid arrest of translation A technique for identifying *recombinant DNA* molecules containing part or all of a *nucleotide* sequence coding for a specific *polypeptide* or sequences necessary for *translation*. When a portion of an *mRNA* molecule is hybridized to its *DNA complement*, *translation* of the encoded *polypeptide* is blocked. Failure to synthesize a protein in a *cell-free translation* system provides positive identification of a *cloned* DNA fragment.

hybrid dysgenesis A phenomenon seen when females from laboratory stocks of *Drosophila* are crossed with males from wild populations. The high rates of *mutation* and sterility are caused by a mobilization of *P elements*, a family of *transposons* that has been lost from most laboratory strains.

hybridization
1. The production of offspring from genetically dissimilar organisms. (see *hybrid, postzygotic isolating mechanisms, prezygotic isolating mechanisms*)

2. A technique used in molecular biology whereby two *complementary, single-stranded DNA* and/or *RNA* molecules are allowed to reanneal, forming a *double-stranded* molecule. This takes place only under specific conditions (*stringency*) that can be varied to specify the degree of *base pair mismatching* that will be tolerated. Usually, one of the molecules is *labeled* and used as a tracer in order to identify its complement in a mixture of fragments that are bound to a solid support. (see *in situ hybridization, Southern blot, northern blot, electrophoresis, C_oT curve*)

3. (see *somatic cell hybridization*)

hybridoma An artificial cell type produced in the laboratory by fusing a *clonal* population of *antibody*-producing *B-cells* with *cancer* cells. The resulting cell line is immortal, a property gained from the cancer cells. It ceaselessly produces a single *antigen*-specific antibody, in other words *monoclonal antibodies*. Monoclonal antibodies are used in research, for purifying medically important antigens such as *interferon*, and offer promise as (for instance) vehicles to target and selectively kill, cancer cells in the body.

hybrid selection The *in vitro* purification of a particular *mRNA transcript* by *hybridization* to its *complementary DNA* sequence. The fragment containing the known *gene* is generally immobilized on a solid support so that the hybrid species can be easily recovered.

hybrid vigor The apparent advantages conferred on offspring resulting from a *cross* between genetically disparate parents. It results from increased *heterozygosity* at *loci* that may have been *homozygous* for deleterious *recessive* genes in more *inbred* parental lines.

hydathode A specialized leaf structure involved in the removal of excess water from plants (*guttation*), often a modified *stoma* that remains open. Hydathodes are found at the leaf tips or along the leaf margins.

hydatid cyst The *bladderworm* stage of *parasitic* tapeworms. It is fluid filled and very large, often containing several inverted heads, together with protective *connective tissue* secreted by the host.

hydranth A *polyp* of a *colonial* hydrozoan, specialized for feeding. Food taken in by hydranths is shared with the reproductive *polyps* of the colony. (see *Hydrozoa*)

hydrarch (see *hydrosere*)

hydras (see *Hydrozoa*)

hydrate To add water.

hydrocarbons A group of compounds composed primarily of hydrogen and carbon. Some examples include natural gas, simple fuels, petroleum products, and components in secondhand cigarette smoke. Many of these types of hydrocarbons are *toxic* or may be converted to toxic, *carcinogenic*, and/or *mutagenic* compounds both in the atmosphere and within the body. (compare *chlorofluorocarbons*)

hydrocortisone (cortisol) A *steroid hormone* produced by the *adrenal cortex*. It has *glucocorticoid* activity and is particularly used as an anti-inflammatory agent.

hydrogen An element essential to life processes. With oxygen it forms water, in which all biochemical reactions take place. Hydrogen is synthesized, along with oxygen and carbon, into carbohydrates and fats as a form of energy storage.

hydrogen bond A weak, extremely polar type of chemical bond responsible for the properties of water and essential to the integrity of large biological *polymers*, such as proteins and *nucleic acids.*

hydrolase Any enzyme that catalyses a *hydrolysis* reaction (splitting of a compound involving water). They are found particularly in digestion where food material must be broken down into units that are soluble and can be transported in the bloodstream.

hydrolysis A chemical reaction in which a molecule, for instance a *polypeptide*, is cleaved with the addition of a water molecule. (compare *condensation reaction*)

hydrophilic Any substance having an affinity for water. They are generally polar molecules that readily form *hydrogen bonds* with water, leading to high water solubility.

hydrophilous Plants that use water movement as a means of *pollination.*

hydrophobic Any substance that tends to exclude water molecules. They are generally nonpolar molecules such as fats and oils. Hydrophobic interactions are responsible for much of the *tertiary structure* of proteins as well as the basic structure of the *phospholipid bilayer* that forms *plasma membranes.*

hydrophyte A plant adapted to grow in water or very wet environments. (see *Raunkiaer's plant classification*)

hydroponics The growth of plants in liquid culture rather than soil.

hydrosere The characteristic sequence of *communities* reflecting the developmental stages of an *ecological succession* that begins in open water. (see *sere*; compare *xerosere*)

hydroskeleton The mechanism of producing muscle force in soft-bodied invertebrates that have neither a rigid *endoskeleton* nor an *exoskeleton.* In worms, for instance, the relative incompressibility of water within their bodies is used as a stationary anchor against which muscles push to produce movement. In jellyfish, expelling water produces an opposite force, causing the animal to move.

hydrostatic The pressure exerted by a volume of fluid at rest.

hydrothermal vents A recently discovered *biome* occuring in the deep ocean where plate tectonics create vents of hot water with a high concentration of sulfur compounds. They provide an energy basis for *chemosynthetic* bacteria, which grow in great abundance, and support giant clams, worms, and other unusual life-forms. The temperatures range from the boiling point in the waters of the vents, to freezing in the surrounding waters.

hydrotropism A *tropism* (plant bending) in which the stimulus is water. For instance, roots that grow toward water are positively hydrotropic. (see *tropism*)

hydroxytryptamine (see *serotonin*)

Hydrozoa A class of cnidarians, the hydras, in which *colonial polyps* are dominant but *alternate generations* with free-swimming *medusae*. The polyps reproduce *asexually,* forming either new polyps or *sexually* reproducing medusae. Although most species are marine, the best-known example is the freshwater *Hydra*, which exists as a solitary polyp with no medusa phase.

hygroscopic Any substance that tends to absorb water.

H-Y/H-W antigen A sex-specific *histocompatibility antigen* located on the *Y chromosome* of most vertebrates and the *W chromosome* of birds. (see *MHC proteins*)

hymenium, hymenia (pl.) A layer of regularly arranged, *spore*-producing structures found in the *fruiting bodies* of many fungi (for example, *ascomycetes, basidiomycetes*).

Hymenoptera A large order of insects containing the bees, wasps, and ants. Most are *carnivorous*. The female's *ovipositor* (egg-laying organ) may be modified as a saw, drill, or sting. The *larvae* are either caterpillarlike or legless and helpless, being cared for by adults, and undergo *complete metamorphosis* (*endopterygote*). The order includes many social insects (for example, honeybees, ants, and termites) that live in highly organized colonies.

hyomandibular A pair of bones in fish that attach the ends of the upper and lower jaws to the rest of the skull. In tetrapods, it is modified to form an *ear ossicle*. (see *hyostylic jaw suspension, stapes*)

hyostylic The type of jaw suspension found in most fish whereby the upper jaw is suspended from the skull by means of the *hyomandibular* bone. (compare *amphistylic, hyostylic*)

hypermorph A *mutant* with more than the normal amount of some *gene product*.

hyperplasia Enlargement of a tissue due to an increase in the number of cells. For example, if part of the liver is removed, the remaining part may undergo hyperplasia in order to regenerate. (compare *hypertrophy*)

hyperpolarization An occasional increase in polarization across the membrane of a *neuron* or *muscle cell* over the *resting potential* so that the inside becomes even more negatively charged in relation to the outside. The greater the degree of polarization, which is effectively the result of the potassium gradient, the stronger the stimulus needed to evoke a response. (see *sodium-potassium pump, action potential, nerve impulse, inhibitory postsynaptic potential*; compare *depolarization*)

hypertonic A solution with greater *osmotic pressure* (more dissolved solutes) relative to another solution (*hypotonic*). When separated by a *semipermeable membrane* (*cell membrane* or laboratory *dialysis* membrane), water tends to move from the hypotonic solution into the hypertonic solution. (compare *isotonic*)

hypertrophy Enlargement of a tissue or organ due to an increase in the size of its cells or fibers. An example is the enlargement of muscles as a result of exercise. (compare *hyperplasia*)

hypervariable region
1. A small section or sections of the *variable region* in each of the many *immune receptor genes* that shows particularly extreme variation. The combination of hypervariable regions in the final multisubunit protein determines the specificity of an *antibody* or *T-receptor*.
2. Any region in the *genome* of an organism that shows extreme variation in the *nucleotide* sequence. Such regions are generally nontranscribed, as most proteins can tolerate little variation without losing function. *DNA* not under selective pressure tends to accumulate *mutations*. Collections of hypervari-

able regions are used in *DNA analysis*, in particular for personal identification. (see *sequence variation, length variation, DNA profile*)

hypha, hyphae (pl.) In fungi, a filament that spreads to form a loose network collectively termed a *mycelium*. The hyphae are divided only incompletely into cells, encouraging rapid and efficient nutrient distribution. The mycelium may cover vast areas. It is often hidden underground, evidenced only by the relatively few *fruiting bodies* it produces above ground during the *sexual* phase of its *life cycle*. It digests and penetrates the substrate it is living on, resulting in an unusual *symbiotic* relationship termed *saprophytic*. The hyphal walls are generally composed of *chitin*. Hyphae are also found in protoctists such as oomycotes and some kelps.

Hyphochytriomycota A phylum of *heterotrophic* protoctists, the hyphochytrids, that have historically been classified as fungi because of their fuzzlike appearance. Unlike fungi, they produce motile *zoospores* and lack the *life cycle* details of the well-characterized fungi. The zoospores have the capacity to *differentiate* and propagate *asexually*.

hypocotyl In *germinating* plant seeds, the part of the *embryonic* shoot located below the *cotyledons* (first leaves) and above the root. (compare *epicotyl*)

hypodermis (exodermis) In the leaves of some plants, the layer(s) of cells found immediately below the *epidermis* (outer layer). The hypodermis may store water, as in succulents (for example, cacti) and the aerial roots of *epiphytes*, or may be mechanically strengthened, such as in pine needles.

hypogeal Any growth or occurrence below ground level.

    1. In plants, seed *germination* in which the *cotyledons* (first leaves) remain underground. (compare *epigeal*)

    2. In animals, the inhabitation of spaces below ground.

hypoglossal nerves (cranial nerves XII) A pair of nerves that arise from the *medulla oblongata* in the brain of higher vertebrates and supply the muscles of the tongue. (see *cranial nerves*)

hypogynous A flower in which the *calyx*, *corolla*, and *stamens* are located below and separated from the *ovary*, giving a *superior ovary*. (compare *epigynous, perigynous*)

hypolimnion In freshwater lakes and ponds, the cooler water that accumulates as the lower layer during the summer. The abrupt temperature difference between it and the upper *epilimnion* is called a *thermocline*. (see *fall overturn*)

hypomorph A *mutant* with less than the normal amount of some *gene product*.

hyponastic The curving of a plant structure toward the axis, caused by enhanced growth on the lower (*abaxial*) surface. A common example is the upward bending of certain leaves. (see *nastic movements*; compare *epinastic*)

hyponeuston Organisms living just below the air/water interface of a water body.

hypopharynx A short, tonguelike structure found in chewing insects. The *salivary glands* often open at or near the hypopharynx.

hypothalamus The portion of the vertebrate *forebrain* that is concerned primarily with regulating the physiological state of the body and maintaining *homeostasis*. It controls the *endocrine system*, the *autonomic nervous system*, and regulates blood temperature and chemical composition. The hypothalamus responds to signals such as thirst and hunger. (see *feedback regulation*)

hypothesis A working explanation or idea based on accumulated data that can be tested experimentally and either proved or disproved. It is the nature of scientific advancement that tested, proven hypotheses are continually disproved when new information becomes available. (compare *theory*)

hypotonic A solution with lesser *osmotic pressure* (fewer dissolved solutes) relative to another solution (*hypertonic*). When separated by a *semipermeable membrane* (*cell membrane* or laboratory *dialysis* membrane), water tends to move from the hypotonic solution into the hypertonic solution. (compare *isotonic*)

hypoxia A dearth of dissolved oxygen.

Hyracoidea An order of placental mam

mals containing a single family of primitive ungulates, exemplified by the hyraxes.

hyrax (see *Hyracoidea*)

Hystricognathi An order of placental mammals, the hystricognath rodents, including porcupines, cane rats, mole rats, and guinea pigs.

◆ I ◆

IAA (indole acetic acid) A naturally occuring *auxin* (plant growth substance).

IAN (indole-3-acetonitrile) A naturally occuring *auxin* (plant growth substance).

ibises (see *Ciconiiformes*)

ichthyology The study of fish.

ICSH (interstitial cell-stimulating hormone) (see *luteinizing hormone*)

identical twins (see *monozygotic twins*)

ileum The longest part of the *small intestine* in mammals, between the *duodenum* and *colon*. Digestion of food and absorbtion of the soluble products takes place in the ileum.

imaginal disk An *embryological* segment in flies that becomes committed to a specific fate early in development. *Homeotic mutations* can change the fate of an imaginal disk, causing gross anomalies in the development of body parts. (see *homeosis*)

imago The adult, *sexually* mature stage of an insect. (see *metamorphosis*)

imino base One of the rare isomeric forms of a *nucleic acid* base. (see *tautomer*; compare *keto, enol*)

immediate immune response (see *nonspecific immune response*)

immune receptor genes (immunoglobulin genes) The library of *genes* in vertebrates from which millions of different *immune receptors* and *antibodies* can be generated, each specific to a different *antigen*. Only a few hundred different variations of these genes exist in the *nuclear genome* of most body cells. Within the *bone marrow*, immune receptor genes in the *stem cells* destined to become *B-cells* and *T-cells* undergo directed somatic recombination. Different segments of the *variable regions* mix and match to increase the pool of *antibody diversity* exponentially. Each *light chain* or *heavy chain* gene is assembled from different combinations of the variable (V), diversity (D), joining (J), and constant (C) segments. The *transcribed* proteins are then assembled into a completed receptor. (see *antibody diversity, clonal selection, immune response, hypervariable regions*)

immune response In vertebrates, a specific defensive reaction of the body to invasion by a foreign substance or organism. It is mediated by the cells and cell products of the *immune system*. The invader is recognized as foreign or "non-self," and a multilevel defense is mounted. This defense includes direct attack by *T-cells* and recognition by *B-cell*-generated *antibodies* that complex with the invader, marking it for elimination by *complement* and *macrophages*. The initial production of antibodies against a specific antigen is called the *primary immune response*. A second exposure to the same antigen results in an amplified production of antibodies called the *secondary immune response*. (see *cell-mediated immune response, humoral immune response, immunity, lymphocytes, MHC complex, HLA complex, MHC proteins, phagocyte*; compare *allergic response*).

immune serum *Blood serum* containing *antibodies* against a specific *antigen(s)*. It is obtained from the blood of a human or other animal that has been exposed to the specific antigen(s) either naturally through disease or artificially through *immunization*. (see *antiserum*)

immune system A multilevel defense system that protects vertebrates against foreign invaders, mostly viruses, bacteria, fungi, and protists. Technically, the immune system includes only the cells (*leukocytes*) and organs of the immune system. However, a number of nonspecific defenses are activated prior to mounting a specific *immune response*. The skin forms the first level, acting as a physical barrier. Nonspecific responses in the bloodstream that act to recognize and destroy any organism or substance recognized as "non-self" are mediated by *macrophages* and *NK-cells* (natural killer cells). A protein complex called *comple-*

190

ment is also triggered, binding to and disrupting invading cells. The *inflammatory response* is a generalized response to infection. It is characterized by the release of *histamines* and *prostaglandins* that produce redness and swelling as a result of increased circulation and fluid leakage. The increase in body temperature, mediated by *progens,* is also a general response. It inhibits microbial growth and stimulates various aspects of the bodily defense systems. The specific immune response consists of a further multilevel defense, including direct attack by *T-cells* and recognition by *B-cell*-generated *antibodies.* Antibodies complex with the invading entity and mark it for elimination by *complement, macrophages,* or *K-cells* (killer cells). Antibodies also serve to protect the body from future infection (*active immunity*). (see *cell-mediated immune response, humoral immune response, lymphocytes, B-cells, thymus, spleen, bone marrow, lymphatic system, immunity, MHC complex, HLA complex, MHC proteins*)

immunity The resistance of an organism to a *pathogenic* microorganism or its products. Immunity may be *passive* or *active,* and may be *indigenous,* acquired, or artificially induced. (see *immune system, immune response, immune serum, antiserum, antibodies*)

immunization The injection of a mild or noninfective strain of a *pathogenic* microorganism into an animal. The intent is to stimulate the production of *antibodies* so that they will already be in place in the event of exposure to the *virulent* pathogen. (see *immunity, memory cells, vaccination, antiserum*)

immunoassay A laboratory assay used to measure minute amounts of specific molecules, such as *hormones,* enzymes, *nucleic acids,* or *steroids,* in a biological system precisely. It is based on an immunological *antigen-antibody reaction.* The molecule to be measured serves as an *antigen* or *hapten.* The assay measures the ability of native, unlabeled antigen to

competitively inhibit binding of a known concentration of enzymatically or radioactively labeled antigen by the antibody. A laboratory sample of the labeled antigenic molecule and specific antibodies are introduced into the system containing the antigen to be measured. After incubation under the proper conditions, the antibody-bound labeled antigen is separated, assayed, and compared to a set of standards. The amount of *competitive inhibition* directly reflects the original concentration of the molecule in question in the biological system. The test may also be conducted with labeled antibodies. When a radioactive tag is utilized, the assay is called a radioimmunoassay (RIA). Another variation is EMIT or enzyme-multiplied immunoassay technique.

immunoelectrophoresis A laboratory technique whereby *antigens* are separated by normal *electrophoretic* processes and subsequently identified by their reaction with known *antibodies.* The antibodies are placed into wells in the same *agar* slab used to run the electrophoresis, and the antigens and antibodies diffuse toward each other. Where an antibody recognizes an antigen, a visible white precipitin band forms.

immunofluorescence The use of fluorescently-labeled *antibodies* to detect the location of specific *antigens.* The technique may be used on tissue samples, in or on cells, or on isolated biochemical fractions. (see *Western blot*)

immunoglobulin genes (see *immune receptor genes*)

immunoglobulins The five major classes of *antibodies,* IgA, IgD, IgE, IgG, and IgM (α, δ, ε, γ, and μ). Each molecule consists of a pair of identical *heavy chains* that determine class and carry *variable regions,* and a pair of identical *light chains* that also carry *variable regions.* The *antigen-binding sites* are formed by a complex of both the heavy and light chains. The combination of the *hypervariable segments* within the variable regions of both the heavy and light chains determines *antigen*-binding specificity. IgM antibodies with "M" heavy chains are secreted by *B-cells* in the initial *humoral immune response*. As the response continues, B-cells switch to secreting antibodies with G heavy chains. E heavy chains insert antibody molecules into *mast cells*, mediating the *inflammatory response*. A heavy chain antibodies are seen in both mother's milk where they may mediate immune protection in nursing children and in mucus membranes where they work chiefly against invading viruses. The function of D chain antibodies is largely unknown. (see *immune system, immune response, immune receptor genes*)

immunohistochemistry The use of *antibodies* or *antisera* as tools for identifying patterns of protein distribution within a tissue or organism. The probe molecule is labeled with a tag, historically radioactive or colored and now more commonly fluorescent.

immunological tolerance (see *acquired immunological tolerance, natural immunological tolerance*)

implantation The attachment of the developing mammalian egg (*blastocyst*) to the wall of the uterus. In humans, the blastocyst penetrates the surface cell layer by breaking *intercellular junctions* using *proteolytic* enzymes. Implantation initiates the induction of *hormones*, in particular *HCG*, that maintain the uterine lining and prevent *menstruation*. (see *estrus cycle, corpus luteum*)

imprinting A form of learning that occurs soon after birth in which social bonds are formed. The most common example is the recognition of a parent by a young animal as exemplified by the experiments of Konrad Lorenz. By presenting himself in place of a mother goose soon after birth, he was able to imprint himself on the goslings as their parent. Imprinting is particularly important in the rescue and care of wild animals. A well-intentioned rescue effort may be for naught if the an-imal imprints on humans, thus effectively destroying any possibility of rereleasing the animal into the wild. Lorenz shared the Nobel prize for physiology or medicine in 1973.

inborn errors of metabolism Genetic defects in which particular enzymes in *metabolic* pathways are missing or *mutant*. Examples include Tay-Sachs, phenylketonuria, and hemophilia. (see *genetic disease*)

inbreeding Breeding between closely related individuals or in its most extreme form, between *gametes* from the same individual or with the same *genotype*. This type of *sexual reproduction* tends to result in *homozygosity* (two copies of the same *gene*) at genetic *loci*, which is disadvantageous if, for instance, a normally *recessive* and deleterious gene is exposed. Although *self-fertilization* is common in plants (self-*pollination*), mating between close relatives in animals is more unusual. In human populations, cultural restraints normally exist to discourage this possibility. If it occurred, inbreeding would result in a decrease in the *genetic diversity,* and, therefore, the genetic health of a population. (compare *outbreeding, cross-fertilization*)

inbreeding coefficient The particular increased probability of *homozygosity* at a specific genetic *locus* in an individual produced from a mating of individuals who are related to some degree.

incisor One of a pair of teeth in the front of the jaw of a mammal. They usually have sharp cutting edges and are used for biting off a portion of food or in rodents, for gnawing.

incomplete dominance (see *codominance*)

incomplete metamorphosis (see *exopterygote*)

incus (anvil) The anvil-like bone forming the middle *ear ossicle* in higher vertebrates.

indefinite inflorescence (see *racemose inflorescence*)

indehiscent A term that describes a fruit or

fruiting body that employs mechanisms other than bursting to release its seeds. Common mechanisms include decay of the fruit, releasing seeds at the location of the fallen fruit, dispersal by wind or water, or digestion and dispersal by birds, mammals, or insects. (compare *dehiscent*)

independent assortment (random assortment) A fundamental genetic principle describing the behavior of non*homologous chromosomes* in the generation of *gametes* by *meiosis*. Any one of a homologous pair associates with any one of another pair randomly. Thus, genes on different chromosomes (or far apart on the same chromosome) are distributed to individual gametes independently of each other. The number of possible chromosomal combinations in a *haploid* (one copy of each chromosome) gamete is then 2^n, (n) being the number of chromosome pairs characteristic of the organism. (see *Mendel's laws*)

independent segregation A fundamental genetic principle describing the obligate separation of *homologous chromosomes* from each other at the second division of *meiosis*. This ensures that each *gamete* receives an exact *haploid* (one copy of each chromosome) complement. *Fertilization* involving two haploid gametes then restores a normal *diploid* (two copies of each chromosome) number of chromosomes in the new organism. (see *Mendel's laws*)

independent variable In scientific experimentation, the factor controlled or changed by the investigator. It may or may not influence the *dependent variable*, the factor being measured in the experiment. (see *scientific method*)

indeterminate In plants, growth that continues indefinitely. A good comparative example is found in tomato plants. The tall, leggy varieties that continue to grow are indeterminate, the short bush type are *determinate*.

indicator (pH indicator) A substance that changes color with the acidity or alkalinity of a solution. Examples are litmus and phenolphthalein. A *universal indicator* (a natural example is red cabbage juice) changes to different colors over a wide range from acid to base and can be used to estimate the pH.

indicator species A species that requires a defined, relatively narrow set of environmental conditions and, as such, is indicative of them. For instance, the presence of certain *diatoms* enables inferences about the environmental history of a lake. The absence of a particular species (*lichens* are often indicative) may suggest an increase in the levels of various *pollutants*. In plant community classification, the definition of *indicator species* may be broadened simply to mean "typical of" a certain environment. (see *ecological community*)

indigenous A species or population native to a particular geographic area as opposed to an introduced species. Such species may also be native to other areas. (compare *endemic, endogenous*)

indirect deforestation The death of trees from *pollution* originating outside the forest, for instance by *acid rain*.

indole acetic acid (see *IAA*)

induced fit The change in the three-dimensional configuration of the *active site* of some enzymes to accommodate the *substrate* when binding occurs.

induced mutations Changes in *DNA* that are produced when an organism or cell is exposed to various *mutagens*. Such events typically occur at much higher frequencies than do *spontaneous mutations*. (see *base analogue, alkylating agent, ethidium bromide, UV radiation, ionizing radiation*)

inducer The local agent, usually an enzyme *substrate*, that triggers *RNA transcription* at an *operon* coding for an *inducible enzyme*. For the *lac operon* in *E. coli*, the inducer is any β-galactoside. (see *induction*)

inducer T-cell (see *T-cell*)

inducible enzyme An enzyme that is produced only in the presence of its *substrate* or a suitable analogue. The regulation takes place at the *DNA* level. An example is β-galactosidase, coded for by the *lac operon* in *E. coli*. It is efficiently *transcribed* only in the presence of a β-galactoside such as *lactose*. (see *operon, inducer;* compare *constitutive enzyme*)

induction

1. In *embryological* development, the interaction between two groups of cells in which a chemical signal from one group causes the other to change its developmental path. Inductions between one of the primary tissues, *endoderm, mesoderm* or *ectoderm*, are referred to as primary inductions. An example is the *differentiation* of the *central nervous system* during *neurulation* by the interaction of dorsal ectoderm and dorsal mesoderm to form the *neural tube*. Differentiation of the lens of the vertebrate eye from ectoderm by interaction with tissue from the central nervous system is an example of secondary induction. (see *determined, committed*)

2. The resumption of a *lytic cycle* by a *bacteriophage* or virus that has *integrated* into the *host genome*. (see *lysogeny, temperate phage, prophage*)

3. The initiation of *RNA transcription* for a *gene* or set of genes that are normally *repressed*. (see *operon, inducible enzyme, inducer, repressor, promoter*)

indusium, indusia (pl.) The covering that encloses the mass of developing *spores* (*sporangia*) in the *sorus* of a fern.

industrial melanism The natural selection for darker forms of some insects (in particular moths) in regions where high industrial pollution (soot and sulphur dioxide) tends to darken surfaces. The matching of the insect's coloring to its environment increases its camouflage, hence decreases detection by potential predators. The best-known example is the peppered moth. In the mid-19th century, it rapidly changed from a light-speckled color similar to the *lichen*-covered trees of its natural environment to dark gray, the color of the soot that came to cover the trees and kill the lichens. The evolution was particularly rapid in this instance because the dark *gene*, although initially rare in the population, was dominant. Even *heterozygous* moths (containing only one dark gene) were dark. With the advent of pollution controls, these trends are reversing. (see *adaption*)

infauna Animals that burrow and live in sediment or another organism.

infection Generally, the invasion of one species by another, usually to the detriment of the first. A specific example is the entrance and subsequent establishment and multiplication of *pathogenic* microorganisms into an individual. It is usually recognized by a diseased state.

inferior The position of a structure or organ as "below" relative to another one. In botany, it applies to the position of the *ovary* in *epigynous* flowers where the *sepals, petals,* and *stamens* rise above it. (compare *superior*)

inflammatory response A defensive reaction of animal tissues to injury, infection, or irritation, characterized by redness, swelling, heat, and pain. *Histamines* and *prostoglandins* released by the damaged cells dilate *capillaries* and cause them to leak *plasma* as well as *phagocytes* that attack any infecting microorganisms. The inflammation associated with, for instance arthritis, is an example of a misdirected *immune response*. (see *immune system*)

inflorescence A multiple-flowered structure. The many types of inflorescences are characterized mainly by the branching structure. (see *racemose inflorescence, cymose inflorescence*)

inhibitory postsynaptic potential (IPSP) The action at an inhibitory *nerve synapse* that hinders the second nerve from firing. At inhibitory synapses, the binding of *neurotransmitter* molecules to the *postsynaptic receptors* increases the membrane permeability to potassium ions (K^+) through *chemically gated ion channels,*

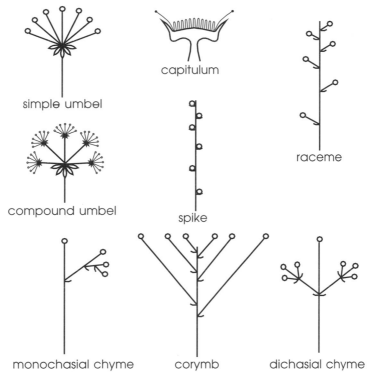

simple umbel

capitulum

compound umbel

spike

raceme

monochasial chyme   corymb   dichasial chyme

Types of Inflorescence

leading to the ions exiting the cell. This produces a *hyperpolarization*, thus inhibiting an *action potential*. (see *GABA, sodium-potassium pump, facilitation, summation*; compare *excitatory postsynaptic potential*)

inhibitory synapse A nerve *synapse* in which the *postsynaptic* receptor protein is a *chemically gated ion channel* specific for potassium ions (K^+), or in some cases chloride ions (Cl^-), that is closed at rest. Binding of a *neurotransmitter* released from the *presynaptic* membrane opens the channel, initiating an *inhibitory postsynaptic potential*. This leads to *hyperpolarization*, effectively inhibiting the initiation of any *nerve impulse*. In the vertebrate *nervous system*, the combination of different specific neurotransmitters each with specific receptors permits distinct responses to a *nerve impulse* at different junctions. (see *action poten-*

tial, synaptic cleft; compare *excitatory synapse*)

innate behavior Any behavior that is inherited genetically and does not need to be learned.

inner ear The innermost region of the vertebrate ear, where *sensory* information is received. It is filled with a fluid (*perilymph*) in which the membranous *labyrinth* is suspended. The organs responsible for hearing and balance reside in the labyrinth. (see *cochlea, macula, sacculus, semicircular canals, utriculus*)

inoculation

  1. The addition of a few starter cells to a microorganismal culture in order to grow a large batch.

  2. (see *immunization*)

inorganic Originally, any substance other than carbon-containing compounds produced by living organisms. The term must now also exclude the complex carbon-containing molecules produced in the laboratory.

inosine A rare *base* most commonly found in the *anticodon* of a *tRNA*.

inositol A member of the *vitamin B complex* and required for growth in certain animals and microorganisms. It is an important constituent of animal membranes, muscle, and brain tissue.

Insecta The largest class of arthropods and the largest group of organisms on earth. Most are terrestrial, have three pairs of legs, and one or two pair of wings enabling flight. The body is characteristically divided into a head, *thorax,* and abdomen. The head bears a pair of *antennae*, *compound eyes,* and simple eyes (*ocelli*). Mouthparts are elaborate and *mandibulate*, usually consisting of *mandibles* (jaws), *maxillae* (secondary mouthparts), a *labium* (lower lip), a *labrum* (upper lip), and a *hypopharynx* (tonguelike organ). The type and extent of *metamorphosis* varies. Many insects are beneficial, *pollinating* flowers and acting as natural predators of pests. Others are harmful, destroying crops, clothes, furniture, and buildings and carrying disease.

Insectivora An order of small, primitive *insectivorous* or *omnivorous* and generally *nocturnal* mammals including the shrews, hedgehogs, and moles. Their small, tapering snout containing sensitive bristles and numerous small teeth are adapted for finding and crushing insects. The tree shrews and elephant shrews that were once grouped with the other insectivores are now classified as separate orders, *Macroscelidea* and *Scandentia,* respectively.

insectivore Any animal (and a few plants) that feed exclusively or mainly on insects.

insertion A *chromosomal mutation* in which extra genetic material is added at a particular site. Insertions are usually deleterious if they interrupt an *expressed* gene. They may occur by various *recombinational* mechanisms including

unequal crossing-over of *homologous chromatids* during *meiosis* (resulting in a tandem duplication) and the *integration* of *transposons.*

insertional inactivation The inactivation of a functional *gene* by interruption with a *transposable genetic element.* It is thought that many spontaneous *mutations* occur in this manner.

insertion sequence elements (see *IS element*)

in situ Literally, "in place." It usually refers to *in situ hybridization.*

in situ **hybridization** A method of locating a particular *gene* or region on a *chromosome* by *hybridizing* a *labeled probe* to intact *chromosomes.* (see *FISH,* compare *DNA analysis*)

instar (nymph) Any of the stages between successive *molts* (shedding of the *exoskeleton*) of insects that develop by simple *metamorphosis* (*exopterygote* for example, locusts and aphids).

instinct An involuntary response to an external stimulus. No prior experience is necessary for an animal to exhibit such a response. Instinctual behaviors tend to be the same in all individuals of a species and seem to be inherited. An example would be the annual upstream *migration* of salmon to spawn. (compare *conditioning, habituation, imprinting*)

insulin A *hormone* produced by the *islets of Langerhans* in the pancreas that controls the *metabolism* of *glucose.* High levels of blood glucose, such as after a meal, stimulate its secretion which, in turn, facilitates its uptake by various tissues and its conversion into *glycogen* and *fat* for storage. A deficiency in insulin production or lack of its *receptors* leads to the disease *diabetes mellitus.* (see *endocrine*, compare *glucagon*)

integration

1. The insertion of extraneous DNA into a *host genome.* In the common case of circular *plasmid DNA,* the event proceeds through *homologous recombination* involving *nucleotide* sequences that are shared by both the plasmid and

the *genomic chromosome*. Noncircular fragments of *DNA* can also sometimes be integrated into a *chromosome*; an event that requires a double *cross-over*. *Transposable genetic elements* often have specific and complex integration and excision mechanisms. (see *lysogeny, episome*; compare *excision*)

   2. The process by which various excitatory and inhibitory *nerve impulses* tend to cancel or reinforce each other at a neural *synapse*. (see *excitatory postsynaptic potential, inhibitory postsynaptic potential, summation, facilitation*)

integument
   1. In animals, any protective body covering. Examples are *cuticle* and skin.
   2. In plants, the covering(s) of the *ovule* that develops into the seed coat.

intercalary meristem In plants, a region of actively growing primary tissue clearly separated from the *apical meristem*. It occurs at *internodes* and leaf sheath bases (joints) of many *monocotyledons* including grasses. (see *meristem*; compare *apical meristem, lateral meristem*)

intercalating agent A chemical that can insert between the stacked *base pairs* at the center of the *DNA* molecule. Such molecules (for example, *ethidium bromide*) are commonly used in the laboratory for visualizing DNA, but because of their properties are also highly *mutagenic*.

intercellular The spaces between cells in an organism including the substances found or processes occuring there. (compare *intracellular*)

intercellular junction (junctional complex) Any of a variety of *intercellular* adhesion mechanisms, particularly abundant in animal *epithelial* cells. The three

most common are *desmosomes* that are principally adhesive, *gap junctions* involved in *intercellular* communication, and *tight junctions* that occlude the intercellular space, thereby restricting the movement of solutes.

interchromosomal recombination The exchange or rearrangement of genetic material that takes place between different *chromosomes*. Some investigators also use the term to refer to *crossing-over* or other interactions between the *chromatids* of *homologues* during *meiosis*. It may also indicate the *independent assortment* of all chromosomes in a *genome* that normally occurs at meiosis. (compare *intrachromosomal recombination*)

intercostal muscles The set of muscles connecting adjacent ribs.

interference In *ecological succession*, the prevention of the entry of *late succession species* by an early species. For instance, some grasses are so dense that they prevent tree seeds from *germinating*. While the grasses persist, later stages cannot proceed.

interferon In vertebrates, a class of proteins released by viral-infected cells that stimulate the proliferation of *killer T cells* and *macrophages*. They also diffuse out to other cells and inhibit the ability of viruses to infect them. γ-interferon, in particular, self-stimulates the production of more macrophages. Various interferons have been *cloned*. Attempts are being made to adapt the genetically engineered product for clinical use, particularly against diseases such as cancer and AIDS. (see *immunity, immune system*)

intergrade The production of individuals who are intermediate in appearance between two different *subspecies* or *vari-*

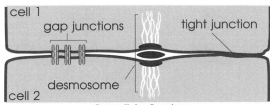

Intercellular Junctions

eties of a species that inhabit the same area. In contrast, species do not intergrade but may occasionally *hybridize* with each other.

interleukins Proteins secreted by cells of the *immune system* that mediate the *immune response*. Interleukin-1, a *monokine*, is secreted by *antigen*-sensitized *macrophages* and activates any *helper T-cells* that they encounter. The helper T-cells, in turn, secrete the *lymphokine* interleukin-2 that stimulates the proliferation of all *T-cells* that have encountered the foreign *antigen*. (see *cell-mediated immune response*)

intermediate feather In birds, a *feather* type showing a combination of characteristics from both *down* (juvenile) and *contour* (outer adult) feathers.

intermediate filament A *polymeric* protein filament that is part of the *cytoskeleton* of *eukaryotic* cells. Intermediate filaments are constructed from *vimentin* monomers that are arranged in a triple helix. They provide structural reinforcement to the cell. Occasionally, other proteins may form intermediate filaments such as *keratin* in hair and nails and *desmin* in *muscle cells* and *desmosomes*.

intermediate host (see *host*)

internal fertilization The union of *gametes* inside the protected environment of a mother's body. Internal fertilization solves the problems of dilution of *gametes* in seawater and the exact timing of their release so that they are not washed away before *external fertilization* can occur. It is a necessary adaption to terrestrial life, counteracting desiccation of the gametes. Internal fertilization also permits fewer eggs to be produced while maintaining a reasonable probability of encountering *spermatozoa*. (compare *external fertilization*)

interneuron A *neuron* that connects *sensory neurons* and *motor neurons*. They are generally located in the *central nervous system*.

internode
1. The *myelinated* region of a *nerve fiber* between two *nodes of Ranvier* (which are unmyelinated). (see *nervous system*)
2. In plants, the stem region between the points of leaf attachment (*nodes*).

interphase In the cell cycle, the period between *mitotic cell divisions*. It includes the G1 and G2 phases of cell growth and the S phase when *DNA* is synthesized (*replicated*). A cell spends most of its time in interphase.

interspecific The comparison of any structure, function, or behavior between species. (compare *intraspecific*)

interstitial In general, the space between defined bodies or regions.
1. The spaces between cells in any *multicellular* organism.
2. The microhabitat formed in the spaces between submerged grains of sand or gravel at the bottom of an ocean, lake, pond, or stream.

interstitial cells Cells in the *testes* of male mammals that secrete *androgens* (male *hormones*) in response to *luteinizing hormone*.

interstitial cell-stimulating hormone (see *leuteinizing hormone*)

intertidal (see *littoral*)

intervening sequence (see *intron*)

intervertebral disc A cartilaginous disc found between each of the *vertebrae* of the *spinal column* except the first two (*atlas* and *axis*).

intestine The part of the *alimentary canal* between the stomach and the anus. It is a muscular tube in which digestion and absorbtion of food is completed, water is absorbed, and feces are produced. In vertebrates, the intestine is divided into the small and large intestine.

intracellular The interior space of a cell and the substances found or processes occuring there. (compare *intercellular*)

intrachromosomal recombination The exchange or rearrangement of genetic material that takes place within one *chromosome* or sometimes between *homologous* chromosomes. Some investigators interpret the *crossing-over* or other interaction between the *chromatids* of homologous chromosomes during *meiosis* as intrachromosomal. Recombination between different parts of the same chromosome is a much rarer, usually *mitotic*, event. (compare *interchromosomal recombination*)

intragenic Any structure or function, such as *recombination* or *complementation*, occurring within the *chromosomal* segment defined by a single *gene*.

intraspecific The comparison of any structure, function, or behavior within a species. (compare *interspecific*)

intromittent organ The male copulatory organ in animals with internal *fertilization*. It is used to introduce *spermatozoa* into the reproductive tract of the female. Examples are *claspers* and the *penis*.

intron (intervening sequence) A segment of *DNA* that is *transcribed* but excised before the final *mRNA* molecule leaves the *nucleus* for *translation* into protein. Several introns are often interspersed with the *exons* (protein-coding portions) of a *gene*. Introns are only found in *eukaryotic* DNA and, in fact, comprise the bulk of most eukaryotic genes. (see *hnRNA*)

introrse A flower in which the *anthers* are turned inward toward the female *gynoecium*. This situation favors *self-pollination*. (compare *extrorse*)

introvert A contractile organ found in some marine invertebrates that can turn inside out and be extruded from the body while feeding. (see *Sipuncula, Echiura*)

intussusception The incorporation of *cellulose* into the existing wall of a plant cell, resulting in an increase in the wall area. (compare *apposition*)

in utero Literally, "in the uterus." It refers to anything that takes place while a mammalian baby develops in the mother's body.

invagination The local infolding of a layer of tissue so as to form a depression or pocket opening to the outside. In particular, invagination occurs during the *gastrulation* of animal *embryos*.

inversion A type of *mutation* in which a section of *chromosome* is excised, turned 180°, and reinserted into the chromosome. This results in the *genes* on that particular segment of chromosome being positioned in reverse order to the rest of the chromosome. Inversion tends to produce at least partial sterility, as *homologous* chromosomes may fail to pair correctly at *meiosis*. If a *zygote* does result from *fertilization* with a *gamete* containing a chromosomal *inversion*, it may or may not be viable depending on the specific situation. (see *meiosis*)

invertebrate Any animal that does not posses a *vertebral column* (spine). The term is not a technical form of classification.

inverted repeat *Nucleotide* sequences that are the same but are oriented in opposite directions close together on the *chromosome*. They often comprise or flank *transposable genetic elements*. (see *IS element*; *transposon*, compare *tandem repeat*)

in vitro Literally, "in glass," a term describing biological reactions that occur in a laboratory apparatus or test tube rather than in a living organism. The isolation and study of reactions *in vitro* is an important research tool. (compare *in vivo*)

in vitro **mutagenesis** The production of either random or specific *mutations* in a piece of *cloned DNA*. Typically, the DNA will then be repackaged and introduced into a cell or an organism to assess the results of the mutagenesis.

in vivo Literally, "in life." A term describing processes that occur within a living cell or organism. (compare *in vitro*)

involuntary muscle (see *smooth muscle*)

involution

1. The decrease in size of an organ, either as a result of normal aging processes or following enlargement, as in the shrinking of the uterus after pregnancy. (see *atrophy*, compare *hypertrophy, hyperplasia*)

2. In the *gastrulation* of some vertebrate *embryos*, the migration of cells from

the *dorsal lip* of the *blastopore* to the interior.

iodine A trace element essential in animal diets mainly as a constituent of *thyroid hormones*. It is not essential to plant growth although it is accumulated in large amounts by certain plants, notably brown algae (sea kelp).

ion Any atom or molecule that carries a positive or negative electric charge.

ion channel A protein tube through which ions may pass in order to cross the *hydrophobic plasma membrane*. They do not require an input of energy (*ATP*), thus engage in *passive transport*. When open, they form a continuous pathway through the membrane. Ion channels may be described by noting the kinds of ions that can pass through them and by describing factors that control the opening of a channel, termed *gating*. Sodium (Na$^+$) and potassium (K$^+$) channels are particularly important in the transmission of *nerve impulses*. (see *phospholipid bilayer, transmembrane protein, membrane channel, voltage-gated ion channel, stimulusgated ion channel, chemically gated ion channel, patch clamp*; compare *membrane pump*)

ion exchange chromatography A technique for separating *macromolecules* that is based on their electric charge at a particular pH. A column is prepared from one of various inert materials that are either negatively or positively charged at neutral pH (7.0). The mixture to be separated is passed over the column and either negatively or positively charged molecules will be retained, whichever is the opposite of the column material. Fractions are collected by washing the column with buffers of different salt concentrations or pH. This has the effect of gradually decreasing the binding strength of the different components. The *eluates* are then analyzed.

ionizing radiation High-energy radiation, such as X-rays and γ-rays, that causes electrons to be ejected from atoms, leaving behind *free radicals*. Ionizing radiation is highly *mutagenic*.

IPSP (see *inhibitory post synaptic potential*)

iris The pigmented circular area in the front of vertebrate and cephalopod eyes. The *involuntary muscles* around the iris act as a diaphragm to vary the size of the central *pupil*. This controls the amount of light entering the eye, and the iris acts as an adjunct focus to the *lens*.

iron An essential nutrient for animal and plant growth. Its primary function is in *hemoglobin* where it binds oxygen for transport in the bloodstream of vertebrates. It is also found in the *cytochromes* of the *electron transport chain* in *aerobic respiration* and *photosynthesis*. It is required as a *cofactor* for some enzymes. Certain iron-containing proteins (for example, *leghemoglobin*) are essential for *nitrogen fixation* by bacteria.

iron-sulfur center (see *Fe-S center*)

irritability (see *sensitivity*)

IS element (insertion sequence element) A *transposable genetic element* in bacteria that consists only of the sequences necessary for *transposition*. It carries no expressed *gene* but can disrupt the function of a gene by inserting into it. It consists of a pair of *inverted repeat sequences*.

islets of Langerhans Cells of the pancreas that produce the *hormones insulin* and *glucagon*, both of which control sugar *metabolism*. Insulin is produced by the β cells and reduces blood sugar after a meal by encouraging uptake and storage. *Glucagon* is produced by the α cells and raises blood sugar between meals by mobilizing liver *glycogen* stores. (see *diabetes*)

isoaccepting tRNAs A group of *tRNA* molecules that carry the same *amino acid* but differ in their own *nucleotide* sequence and are *transcribed* from different *genes*.

isoantigen An *antigen* that induces *antibody* formation in genetically diverse individuals or species.

isoelectric point The pH of a specific solution at which the net charge on an *amino acid* or protein molecule tends to be neutralized. This value is put to practical effect in the laboratory by using it to precipitate proteins out of solution. It is also used to establish a pH gradient in an *electrophoretic* field, allowing the protein to migrate to the position corresponding to its isoelectric point. (see *amphoteric*)

isoenzyme A variant of an enzyme that exists in different structural forms within a single species. Each has the same *substrate* specificity but often has different substrate affinities. The different *isomeric forms* (*isozymes*) all have the same or similar *molecular weights* but may differ in configuration and charge. Different isozymes can often be distinguished by the appropriate form of *electrophoresis* and detection. They are used to identify and characterize both individual organisms and species.

isogamous (homogamy) The *sexual* fusion of *gametes* of similar size and form. It occurs in some fungi, algae, and other protoctists. (compare *anisogamous*)

isoleucine (ile) An *amino acid.*

isometric The type of *muscle contraction* in which constant length is maintained. It is the type of force generated when, for instance, one tries to lift an immovable object. (compare *isometric*)

isomorphic An organism in which *alternating generations* of the *life cycle* are morphologically identical. It is seen in some algae. (compare *heteromorphism*)

Isoptera The order of insects containing the termites and white ants. They have characteristic *mandibulate* (chewing) mouthparts and undergo *incomplete metamorphosis* (*exopterygote*). They have an elaborate social system built on *castes*, each colony founded by a winged male and female. They feed on plant material, the *cellulose* of which is digested by *symbiotic* protists or bacteria in their gut. When they colonize wooden structures, they are often extremely destructive.

isotonic

1. The situation in which two solutions have equal *osmotic pressure* (dissolved, charged particles). It is often used to describe the relation of a cell to its environment—if they are isotonic, water does not move either into or out of the cell. (compare *hypertonic*, *hypotonic*)

2. The type of muscle contraction in which constant tension is maintained. It is the type of force generated when, for instance, an object is lifted by shortening the muscle. (compare *isotonic*)

isotope A variant of a chemical element, differing in the number of neutrons. Many isotopes are *radioactive* and are used for purposes such as *radioactive dating* (^{14}C) and labeling of biological molecules in the laboratory (for example, ^{14}C, ^{32}P, ^{35}S). (see *radioactive labeling, DNA analysis*)

isozyme (see *isoenzyme*)

◆ J ◆

Jacob-Monod (see *operon*)

jaegers (see *Charadriiformes*)

jawless fish (see *Cyclostomata*)

jaw worms (see *Gnathostomulida*)

jellyfish (see *Scyphozoa*)

joining regions In *immune receptor genes*, a small region that contributes to diversity, mainly in *T-receptors* (12 to 100 different copies) and, to a lesser extent, in *B-receptors* (1 to 6 copies). (compare *constant regions, variable regions, diversity regions*)

joint The points where bones meet each other. They are cushioned by pads of *cartilage* and held together by bands of *ligaments.*

jugular vein One of a pair of *veins* in mammals each with an internal and external portion that carries *deoxygenated* blood away from the head and neck, and ultimately back to the heart.

jumping gene (see *transposable genetic element*)

junctional complex (see *intercellular junctions*)

Jurassic The middle period of the *Mesozoic* geological era about 190 million to 135 million years ago. It is characterized by the dominance of dinosaurs as well as the first known examples of birds and mammals.

juvenile hormone A *hormone* found in insects that prevents *metamorphosis* into the adult form and maintains *larval* characteristics. It functions antagonistically to the hormone *ecdysone*, which promotes *molting* and *differentiation* into the adult. Juvenile hormone is produced in specialized secretory cells (*corpus allatum*) near the insect brain.

kangaroo (see *Diprodontia*)

karyogamy The fusion of two *nuclei* within the same cell. It is a normal part of *fertilization* that occurs after the fusion of two *gametes* and may be followed by *meiosis*. It also occurs within various *multinucleate plasmodia*, for example in plasmodial slime molds *(Myxomycota)*. (compare *plasmogamy*)

karyotype The gross physical characterization of the *chromosome* set of an organism, as viewed in a light microscope after special staining procedures. Chromosomes for a karyotype are obtained from cells arrested in *metaphase* of *meiosis I* at the time when they are maximally condensed, thus easily visible. Each chromosome has been duplicated but the *chromatids* remain attached at the *centromere*, presenting the appearance of four arms. The results are organized in a photograph or diagram showing the chromosomes arranged as pairs and arranged in order of size. Gross *mutations* such as large *deletions, insertions, inversions,* and *translocations* may often be discovered in this manner, making it a useful tool in prenantal analysis. (see *FISH*)

kb (kilobase) An abbreviation for 1000 *DNA base pairs*.

K-cell (killer cell) A *lymphocyte* involved in the *humoral immune response*. K-cells recognize foreign invaders that have been marked by *antibodies* and destroy them by piercing holes in them using a set of proteins similar to *complement*. (see *immune system*; compare *NK-cell*)

keel A projection from the *sternum* of modern flying birds (carinates) and bats that provides for the attachment of wing muscles.

keratin A sulphur-containing *structural protein* found particularly in mammal hair, horns, nails, feathers, and other *epidermal* outgrowths. Two types of keratin exist. The α-keratins have a unique coiled structure held together by *disulfide*

bonds, and the β-keratins have a *beta pleated sheet* structure. Keratin may sometimes *polymerize* into *intermediate filaments*, for instance in hair and nails.

keratinization (cornification) A process by which the protein *keratin* replaces the *cytoplasm* in vertebrate *epidermal* cells. For example, the cornified outer layer of the *epidermis* of the skin consists of dead keratinized cells. Hair and nails also consist of keratinized cells.

keto base The normal isomeric form of a *nucleic acid base*. (see *tautomer*; compare *imino, enol*)

keystone species A species that has a particularly large effect on an *ecosystem*. Its removal would change the basic nature of the *ecological community* and the interactions of its other inhabitants.

kidney The major excretory organs of vertebrates. They also function in *osmoregulation*. Kidneys occur in pairs. They consist of excretory units (*nephrons*) that are responsible for the filtration and selective reabsorbtion of water, mineral salts, and glucose and the subsequent production of waste. A collecting duct, the *ureter*, conveys excess water, salts, and nitrogenous wastes (*urea, uric acid*) as urine from each kidney to the *bladder* for excretion to the exterior. (see *excretory system*)

killer cell (see *K-cell*)

kilobase (see *kb*)

kin-, -kine, -kinesis

    1. A word element derived from Greek meaning "movement." (see *kinesis*)

    2. A word element derived from Greek meaning "related." (see *kingdom*)

kinase An enzyme that catalyses a reaction involving the transfer of a phosphate group from a high-energy compound, often *ATP*, to another molecule, often an enzyme, for the purpose of activating it for a specific function. (compare *phosphorylase, phosphatase*)

kinesin The integral protein of the *ATP* driven molecular engine of *eukaryotic* cells. It moves *vesicles* and *organelles* along *microtubule* tracks and participates in the movement of *chromosomes* along the *mitotic spindle.*

kinesis A change in the rate of movement of an organism or cell in response to specific environmental conditions. The direction of motion is random and is not related to the stimulus. Kineses tend to stimulate an animal to remain quiescent under favorable conditions and move to another location under unfavorable conditions. (compare *taxis, tropism, nastic movements*)

kinetid A structural unit found in *undulipodiated* cells comprising a *kinetosome* and its associated *microtubules* and fibers.

kinetochore (see *centromere*)

kinetoplast An *organelle* present in some zoomastigote protists such as trypanosomes. It is apparently a modified *mitochondrion* containing unusually large amounts of *DNA* and is commonly situated near the origin of an *undulipodium,* ostensibly to provide immediate energy.

kinetoplastids (see *Discomitochondria*)

kinetosomal fiber A proteinaceous fiber found associated with the *kinetosomes* underlying all *undulipodia,* including *cilia.* It is diagnostic for ciliate protist taxonomy.

kinetosome (basal body) A cellular *organelle* composed of *microtubules* and found at the base of all *undulipodia.* It acts as an organizing center for the 9 + 2 microtubule structure located there. Kinetosomes are identical in structure to the *centrioles* that organize microtubules into a *spindle* during *nuclear division.* Kinetosomes contain their own *DNA* and are self-reproducing. They are probably evolutionarily derived from *endosymbiotic spirochete* bacteria.

kingdom The penultimate unit of taxonomic classification. Five kingdoms are recognized by most biologists and grouped into two superkingdoms, eukaryotes and prokaryotes. Each kingdom is divided into *phyla.* For plants and fungi, this hierarchical grouping has traditionally been called a *division.* (see *Animalia, Plantae, Bacteria, Fungi, Protoctista*)

kingfishers (see *Coraciiformes*)

kinin (see *cytokinin*)

Kinorhyncha A obscure phylum of animals, the kinorhynchs, that are marine worms with spiny heads. They are *pseudocoelomates* and are covered with a segmented *cuticle,* which they molt during growth. *Fertilization* is external.

kinorynchs (see *Kinorhyncha*)

kin selection Any behavior that tends to favor the survival of genetically related individuals within a population or species. The strength of the behavior may be proportional to the relatedness of the organisms. (compare *altruism*)

kiwis (see *Dinornithiformes*).

kneecap (see *patella*)

knockout (*targeted gene knockout*). The targeted disruption of function in a *cloned* gene. It is usually reintroduced back into the organism in such a way as to replace a normal gene copy in order to analyze the effect of the *mutation.* By definition, knockouts are *null alleles* and they replace an *endogenous* copy by *homologous recombination.* The gene is often disrupted by inserting an unrelated but *expressed* gene in the middle in order to select and follow the *recombinant* construct.

koala bears (see *Marsupiala*)

Krebs cycle (citric acid cycle, tricarboxylic acid (TCA) cycle) In *aerobic respiration,* the set of oxidative reactions that take place after *glycolysis.* In *eukaryotes,* the process occurs in the *mitochondria.* Pyruvate, produced previously in *glycolysis,* is oxidized to *acetyl CoA* (with the liberation of a CO_2 molecule and the reduction of NAD^+ to *NADH*), providing entrance to the Krebs cycle pathway. *Coenzyme A* is then liberated, producing citrate, which is further oxidized through a series of steps. Each of these steps reduces another NAD^+ to

NADH (or in one instance, *FAD*⁺ to *FADH₂*) and one of which produces *ATP*. The end product of the Krebs cycle, oxaloacetate, is then ready to accept another acetyl group from acetyl CoA, making citrate, which repeats the cycle. The total energy yield from the Krebs cycle is 2 *ATP* molecules and 12 reduced electron carriers. The high-energy electrons stored temporarily in NADH and FADH$_2$ are subsequently introduced into the *electron transport chain* where the bulk of the *ATP* is formed. (see *aerobic respiration*)

krill (see *Euphausiacea*)

k strategist Organisms that tend to be limited by the *carrying capacity* of their environment rather than their own reproductive capacity. They are generally large, slow-growing organisms with long generation times. They reproduce late in their *life cycle* and have relatively few offspring that must receive a great amount of parental care. These factors render them highly vulnerable to *extinction*. Some of the more extreme examples include whales, whooping cranes, and California redwoods. (compare *r strategist*)

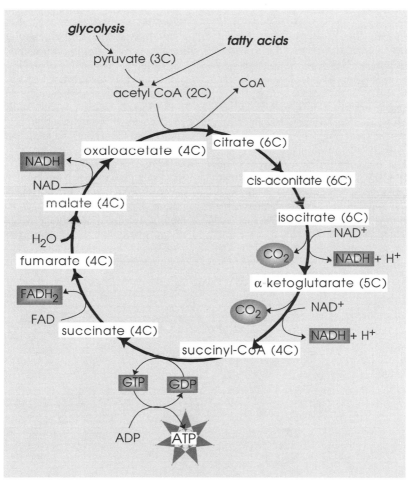

The Citric Acid Cycle
(Krebs Cycle, Tricarboxcylic Acid Cycle
(see *Krebs cycle*)

labeling The technique of tagging biological molecules with reporter molecules so that they can be detected indirectly. Most commonly, *radioactive isotopes* have been used, and the presence and location of the labeled molecules detected either using X-ray film or a radioactive counter. More recently, the sensitivity of color change reactions and, particularly, *chemiluminescent* tags has been improved, and they have begun to supplant radioactive techniques. (see *DNA analysis, radioactive labeling*)

labium, labia (pl.)
1. In insects and crustaceans, a plate forming the lower lip. (see *mouthparts*)
2. In female mammals, the folds of skin protecting the exterior opening of the vagina.

labrum, labra (pl.) In insects and crustaceans, a plate forming the upper lip. (see *mouthparts*)

labyrinth
1. The intricate system of cavities and channels that occurs in the vertebrate *inner ear*. It consists of two cavities (the *utriculus* and the *sacculus*), three *semicircular canals* that act as organs of balance, and a spirally coiled canal (the *cochlea*) containing the *organ of Corti*.

2. An organ involved in the excretion of *nitrogenous* wastes in crustaceans. The osmotic and ionic composition of the blood is regulated by a pair of *glands* located in the head. They consist of a bladder, a tubule, and a spongy mass called the labyrinth or *green gland*. Salts, amino acids, and some water are extracted as the waste products pass along the tubule and are excreted through the labyrinth.

Labyrinthulata A phylum of *colonial heterotrophic* protoctists, the slime nets. They produce a membrane bound *ectoplasmic* network from specialized *organelles* at the cell surface called *sagenogens* or *bothrosomes*. Individual cells migrate within this network, but the mechanism is unknown. They are found in shallow seabeds associated with algae, sea grasses, or organic-rich sediments. Slime nets disperse via motile *zoospores* that leave the net colony and settle elsewhere.

lacewings (see *Neuroptera*)

lachrymal gland A gland found in the eye in many vertebrates. It is situated beneath the upper eyelid and produces antiseptic tear fluid that continually washes the front of the eye, draining through the *lachrymal duct* into the nose.

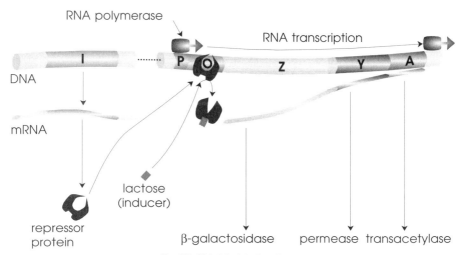

Simplified Model of the Lac Operon

lac operon A classic *operon* in *E. coli* that was originally used to dissect out the elements and functions of bacterial *gene expression*. It contains three *structural genes*, permease (lacY), β-glactosidase (lacZ), and transacetylase (lacA). Both lacY and lacZ code for enzymes involved in *lactose* metabolism. The lac operon is regulated by both *negative* and *positive control* circuits. *RNA transcription* is induced by the presence of *lactose,* which binds the *repressor* protein otherwise blocking *transcription* by sitting on the *operator* region. *RNA polymerase* is then able to gain access to the *promoter* and the structural genes are transcribed. A positive control circuit (*catabolite repression*) also operates when a large amount of *glucose* is present in the cell, preventing the lactose-*metabolizing* enzymes from being expressed until needed. (see *polycistronic, inducible enzyme, allosteric, CAP, genetic dissection*)

lactase An enzyme that breaks down the sugar lactose, found in milk. A deficiency of lactose, or the inability to produce lactase in the human small intestine, results in the inability to digest dairy products This condition is called lactose intolerance.

lactation The secretion of milk from the *mammary glands* of female mammals. During pregnancy, the *hormones estrogen* and *progesterone* act to increase the amount of milk-producing tissue. After birth, *prolactin* and *oxytocin* stimulate milk production and secretion.

lactic acid A common end product of *glycolysis*. In animals, it is produced during *anaerobic respiration* and is responsible for muscle soreness after overexertion. Certain bacteria always *ferment* carbohydrates to lactic acid (in the presence or absence of oxygen) and are used in the production of food products such as yogurt, cheese, and sauerkraut.

lactose The sugar found in milk. It is a *disaccharide* composed of *glucose* and *galactose* units.

lacuna, lacunae (pl.) Any small cavity such as those containing bone cells (*osteocytes*) or cartilage cells (*chondrocytes*).

Lagomorpha An order of *herbivorous* mammals that includes the hares and rabbits. They resemble *rodents*, differing principally in their dentition. Like *ruminants*, they employ intestinal, *cellulase*-secreting bacteria in order to help digest the *cellulose* in their food. Unlike ruminants, however, their *cecum* is positioned behind their stomach, precluding regurgitation, rechewing, and redigestion. Instead, they eat their feces, passing the cellulose through their *digestive system* a second time.

Lamarckism Lamarck's theory of *evolution* (proposed in 1809) postulating that acquired characteristics can be inherited and effect permanent changes in populations. The most infamous example is that of giraffes acquiring long necks by attempting to reach the leaves in tall trees. Theories such as this were subsequently dropped in favor of *Darwin's* theory of *natural selection* and subsequent refinements.

lamella, lamellae (pl.) In general, any flat, thin structure.
  1. The layers of calcified matrix in bone.
  2. The *spore*-bearing structures in some fungi.
  3. The *gill* membranes of fish.
  4. (see *stroma lamellae*)

lamellipodium, lamellipodia (pl.) **(ruffled membrane)** The sheetlike extension of the leading edge of many *ameboid* vertebrate cells during movement.

lamina, laminae (pl.) The flat, expanded portion of a leaf blade or petal. It may also refer to the expanded leaflike structures in brown algae (sea kelp).

lampbrush chromosome The large furry-looking *chromosomes* found in amphibian *oocytes* (developing egg cells) that are particularly obvious at the *diplotene* phase of *meiosis*. Their appearance is caused by laterally extruded loops of *DNA* along the length of the chromosome. They are *transcriptionally* active and may be an adaption to serve a relatively large cell from a single *nucleus*. (see *gene amplification*; compare *polytene chromosome*)

lamprey (see *Agnatha*)

lamp shells (see *Brachiopoda*)

lancelets (see *Cephalochordata*)

large intestine The last part of the *alimentary canal* in animals, consisting of the *colon* and the *rectum*. It has a larger diameter than the preceding *small intestine* and receives the undigested remains of food which it prepares for evacuation through the *anus*. (see *digestive system*)

larva, larvae (pl.) The immature form in which some animals hatch from the egg. Larvae are capable of independent existence but are usually incapable of *sexual* reproduction. They often differ appreciably from the adult form in *morphology*. The term is usually applied to insects such as the caterpillars of moths and butterflies, to the *planktonic* larvae of many marine species, and to the frog tadpoles.

larynx (voice box) The *cartilaginous* structure that contains the *vocal cords* and is responsible for sound production in vertebrates. It is situated between the *pharynx* (throat) and the *trachea*.

lasso cell (see *cnidocyte*)

latent period The time between the application of a stimulus to an *irritable* tissue (for example, nerve or muscle) and the first detectable response. (compare *reaction time*)

lateral bud A bud that gives rise to side branches off the main stem of a plant. It may be located in a leaf *axil*.

lateral line system The *sensory* system of fish and aquatic amphibians. It consists of *sensory* cells on or near the surface of the body that respond to pressure waves (sound) in the surrounding water. The *nerve impulses* are transmitted via the *lateral lines*, a pair of *sensory* canals running the length of each flank. (see *neuromast*)

lateral meristem A group of actively dividing cells found in the *cambium* layer of *vascular plants*. It gives rise to secondary tissues such as the *vascular cambium* and *cork cambium*. (see *meristem, secondary growth*)

late successional species Species that occur only in, primarily in, or are dominant in the late stages of *ecological succession*. They are typically slower growing and long-lived plants. (compare *early successional species*)

latex A thick, white liquid found in some flowering plants (*angiosperms*), such as dandelions, that is exuded from cut surfaces. It is harvested commercially (often from plants in the fig family) for use in the manufacture of rubber and other products. Opium is extracted from the latex of the opium poppy.

lawn A continuous layer of bacteria or other microorganism on the surface of a solid *agar* growth medium.

LD$_{50}$ The median lethal dose of a toxin at which 50% of the exposed organisms are killed. It is used as a standard measure of toxicity.

LDL (low density lipoprotein) A blood *plasma* protein that serves to transport *cholesterol* and other lipids from the bloodstream to various tissues. It tends to deposit cholesterol in the walls of arteries, leading to *atherosclerosis*. It has a relatively high lipid content compared with other *lipoproteins*, thus a lower protein content. Because fat is less dense than protein, it has been given the designation "low-density." According to some studies, a high blood level of LDL is associated with an increased risk of cardiovascular disease, in particular blockage of the *coronary* arteries.

leaching The removal of soil materials that become dissolved in water percolating downward through it. The upper layer of soil becomes increasingly acidic and deficient in plant nutrients. Water leaching agricultural land may also accumulate large amounts of dissolved nitrates and contribute to groundwater *pollution*.

leaf A thin, flattened appendage extending from the stem of many green plants. Each leaf arises at a *node* on the stem of a plant and is typically comprised of a *petiole* (stalk) and *lamina* (blade). The leaves are the main site of *photosynthesis* in green plants.

leaf base (see *phyllopodium*)

leaflet Part of a *compound leaf.*

leaf scar The marking where a leaf was formerly attached to the stem.

leaky mutant A *mutant* that results from a partial rather than a complete inactivation of the *wild-type* (normal) function.

learning The alteration of behavior by an individual as the result of experience. It is most pronounced in animals with a long lifespan and a long period of parental care. (compare *instinct*)

lecithin (phosphatidylcholine) A *phospholipid* that is widely distributed in higher plants and animals, particularly as a component of *cell membranes.*

lectins A group of proteins and *glycoproteins* extracted primarily from plants, in particular legumes, that shows an ability to interact with some human *cell surface markers*. For example, H-lectin exhibits a pseudo*antibody* activity against O *antigen* and is thus able to *agglutinate* red blood cells (*erythrocytes*), a characteristic exploited in ABO *blood group* typing. Some lectins stimulate *capping*, a phenomenon where particular *cell surface markers* are drawn to one pole of the cell. Others are *mitogenic*, stimulating the initiation of *cell division*. (see *concanavalin A, phytohemagglutanin*)

leeches (see *Hirudinia*)

leghemoglobin An oxygen-carrying protein found in the *root nodules* formed by

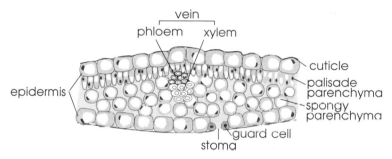

Cross-Section of a Leaf

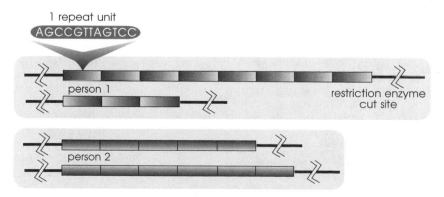

Length Polymorphisms in DNA

symbiotic nitrogen-fixing bacteria on legumes. It consists of a *heme prosthetic group*, which lends a pink color to the nodules attached to a remarkably *globin*-like protein. Leghemoglobin is an example of *convergent evolution.*

legume The fruit of the plant family now called *Fabaceae* (formerly *Leguminosae*), such as peas and beans They are characterized as a dry fruit formed from a single *carpel* that liberates its seed by splitting open into two parts. The seed coat of legumes is virtually impenetrable to water and oxygen, contributing to their long shelf life. Legumes are known particularly for their *symbiotic* association with *nitrogen-fixing bacteria* that form characteristic nodules on their roots. (see *root nodules*)

lemma, lemmae (pl.) The lower member of a pair of *bracts* (specialized leaves) surrounding not only the flower but also the other bract.

lemmings (see *Rodentia*)

lemur (see *Scandentia*)

length polymorphism A variation in the *DNA* of an organism that is detectable as a difference in the length of a defined fragment produced by cleavage with a *restriction enzyme*. The difference may be due to the addition or loss of a *restriction site*, or a change in the number of *tandem repeats* located between two restriction sites. Length polymorphisms produced by tandem repeats may be highly *polymorphic* and are used in *DNA analysis* for personal identification. (see *DNA profile, RFLP, electrophoresis, Southern blot*; compare *sequence polymorphism*)

lens A transparent, biconvex disc in vertebrate and cephalopod eyes. It serves to focus light on the *photoreceptor* cells of the *retina* to form an image. (see *ciliary body*)

lentic An environment created by standing water for instance lakes, ponds, and permanent or temporary pools.

lenticel A raised pore in the bark of a woody stem that allows gaseous exchange between internal tissues and the

atmosphere. Large numbers of lenticels form spongy areas on the surface of some plants.

Lepidoptera A large order of *endopterygote* insects containing the butterflies and moths, characterized by a covering of scales, often brightly colored, over their wings and bodies. They have a *proboscis* for sucking nectar or fruit juices. The *larvae* are mostly *herbivorous*, including some serious plant pests. Butterflies are *diurnal*, have slim bodies and clubbed *antennae*, and rest with their wings folded over their back. Moths are mostly *nocturnal*, never have clubbed antennae, and rest with the wings in various positions.

leptotene In the first *meiotic* division, the stage in early *prophase I* when *chromosomes* start to condense and become visible. Although *DNA replication* has occurred, *sister chromatids* are not usually visible as separate entities.

lethal gene A *gene* whose expression results in the death of the organism.

leucine (leu) An *amino acid.*

leucine zipper A structural motif in some *DNA binding proteins* that allows the formation of protein dimers through *leucine* interactions. Dimer formation is required for DNA binding by the *lysine-* and *argenine*-rich DNA binding domains flanking the leucine zipper. The possibility of forming heterodimers, zippered pairs of nonidentical proteins, benefits the organism by increasing the variety of protein combinations that can be used to regulate *gene expression*. A number of *proto-oncogenes* encode leucine zipper proteins. (see *transcription factor*; compare *zinc finger, helix-turn-helix, helix-loop-helix*)

leucoplast In plants, a *plastid* lacking any *photosynthetic pigment* such as *chlorophyll* and therefore colorless. It functions in the *metabolism* and storage of starches (*amyloplasts*), proteins (*aleuroplasts*), and oils (*elaioplasts*).

leukemia *Cancer* of the white blood cells (*leukocytes*).

leukocyte (white blood cell) Any white blood cell. Leukocytes include a diverse array of *nucleated*, non*hemoglobin*-containing blood cells mostly concerned with bodily defense. They are continually produced in the *bone marrow* from *stem cells* and are considered *connective tissue*. Leukocytes are divided into two groups, *granulocytes* (*basophils, eosinophils, neutrophils*) and *agranulocytes* (*lymphocytes and monocytes*). (see *macrophage, mast cell, B-cell, T-cell*)

LH (see *luteinizing hormone*)

lice (see *Anoplura*)

lichen (Mycophycophyta) A *symbiotic* association between a fungus (usually an *ascomycete*) and a *photosynthetic* organism (usually green algae or cyanobacteria). They may colonize areas too inhospitable for other plants, such as the surfaces of rocks or the aerial parts of trees. As such, they are good examples of *mutualism*. Lichens reproduce *asexually* by *soredia* or by bits blown by the wind to new locations. Many lichens are extremely sensitive to atmospheric *pollution* and are used as *indicators*. Some researchers consider lichens as a subphylum of ascomycete fungi.

life No single, comprehensive definition of life exists. All carbon-based life forms on earth generally exhibit the properties of growth, *metabolism*, cellular organization, reproduction, *heredity*, and death. Heredity, in particular the ability to transmit characteristics encoded in *DNA* to offspring, appears to be a unifying concept. Viruses are not generally considered to be alive because they cannot reproduce autonomously but depend on the cellular machinery of the *host* cell that they *parasitize*. Additional properties commonly exhibited by living organisms include movement and response to stimuli (*sensitivity*).

life cycle The sequence of phases in the growth and development of an organism that describe genetic and cytological changes from *zygote* to new *gamete formation*. In some lower organisms including most plants, the *life cycle* may involve *alternation of generations*, usually between *haploid* and *diploid* forms, and often of differing *morphology*. (see *haplont, diplont, dimorphic*)

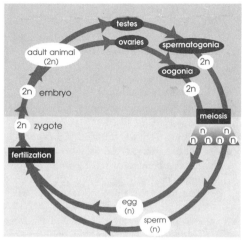

Generalized Animal Life Cycle

life history The sequence of events throughout the development of an organism that describes visible changes in morphology such as *spore* formation, transformation from *unicellularity* to *multicellularity*, and *alternation of generations*.

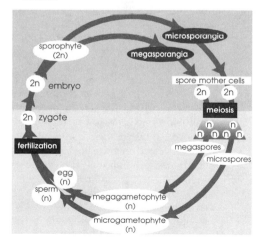

Generalized Higher Plant Life Cycle

ligament A tough band of *connective tissue* that joins two bones together at a joint.

ligand Any molecule that is physically bound by another. Examples include the *substrate* of an enzyme and the activators of *cell surface receptors* either in the *plasma membrane* (for example, *peptide hormones*) or in the *cytoplasm* or *nucleus* (for example, *steroid hormones*).

ligase (see *DNA ligase*)

ligation Chemical bond formation between two molecules, usually facilitated by a *ligase* enzyme specific to the reaction. (see *DNA ligase*)

light chains Two identical *polypeptide* subunits that, together with two *heavy chain* subunits, form an *antibody* molecule. The two light chains are found on the arms of the Y-shaped molecule. They contain at their extremities *hypervariable regions* that, in addition to those found on the heavy chains, define a specific *antigen-binding site*. (see *immunoglobulin, immune receptor genes, B-cell, immune system, antigen-antibody reaction*)

light microscope A microscope that depends on visible light as the source of illumination for objects magnified through a lens or lenses (*compound microscope*). The resolution of the magnified object is limited by the wavelengths (λ) of visible light. (compare *electron microscope*)

light reactions The light dependent reactions of *photosynthesis* that convert solar energy into chemical energy. The initial event in these reactions is the excitation of an electron in a *photoreceptor* molecule (most often the green pigment, *chlorophyll*) by a photon of light. High energy electrons are then shuttled along a series of electron transport molecules embedded in the photosynthetic membrane (*electron transport chain*) until they arrive at a *proton pump*. This pump drives the *chemiosmotic* synthesis of *ATP* or reduction of $NADP^+$ to $NADH$. The electrons are recycled to the *photoreceptor*. (compare *dark reactions*)

lignin One of the main structural materials of *vascular plants* and, with *cellulose*, one of the main constituents of wood. It is a complex, cross-linked *polysaccharide* that is a component of plant *cell walls*. It is relatively inert and often survives in *fossils* of woody plants.

ligule
1. A scale-shaped, membranous outgrowth that covers the surface of some leaves in many grasses and some ferns.
2. In some flowers, a flattened, straplike *corolla*.

limbic system A group of linked structures deep in the brain that are responsible for emotional responses and also in the formation and recall of memories. They are linked to the *hypothalamus*. (see *amygdala, hippocampus*)

limiting factor Any factor in the environment that alone restricts the growth or behavior of an organism. The limiting factor functions as an *independent variable*. For instance, plant growth is limited by low temperature and increases with rising temperature to an optimum, beyond which growth rate decreases.

limnetic In a freshwater *ecosystem*, the zone between the *littoral* (surface) and *profundal* (deep) zones (about 6 to 10 meters). It is synonymous with the *oceanic* zone in which *pelagic* organisms live. Its depth is limited by the depth at which *photosynthesis* is equaled by respiration rate, below which plants cannot live. It is inhabited by *plankton* and other open water organisms.

limnology The scientific study of freshwater and its *flora* and *fauna*.

limpkins (see *Gruiformes*)

LINEs (long interspersed elements) A collective term for the families of long *repeated sequences* that are dispersed throughout mammalian *genomes*. They are usually from 1 to 5 *kb* (kilobase pairs) and show significant *nucleotide* sequence *homology* to *retroviruses*. They often include *genes* that code for enzymes used in their own *transposition*. (see *transposable genetic element, transposon;*

compare *SINEs, TY element, P element, copialike element*)

linkage (see *genetic linkage*)

linkage equilibrium The situation when two *genetic loci* or *genes* show no propensity to be inherited together. In other words, their association in the population is no greater than random. It is an important concept in *population genetics.* It occurs when genes are located on physically distinct *chromosomes* or far apart on the same chromosome. (see *Hardy-Weinberg equilibrium*)

linkage group A group of *genes* known to be inherited together. They are usually located close together on the same *chromosome.* (see *genetic map*)

linkage map (see *genetic map*)

linoleic acid (see *essential fatty acids*)

linolenic acid (see *essential fatty acids*)

lipase An enzyme that catalyses the *hydrolysis* (breakdown) of fats to *fatty acids* and *glycerol.* It is present in the stomach and intestines of vertebrates.

lipids A heterogeneous group of small organic molecules characterized by their insolubility in water and their relative solubility in various organic solvents. Included in this classification are fats, oils, and waxes as well as less obvious molecules such as *carotenoids* and *steroids.* These compounds function in energy storage and as *hormones, vitamins, photosynthetic pigments,* and *cell membrane* components.

lipo- A word element derived from Greek meaning "fat." (for example, *lipoprotein*)

lipoic acid (see *vitamin B complex*)

lipolysis The splitting of a lipid into its component *fatty acids* by *lipases.* It is part of the *digestion* and *catabolism* of fat molecules.

lipophore (see *chromatophore*)

lipopolysaccharide A *polysaccharide* molecule that contains a conjugated lipid group. They are common in the *cell walls* of bacteria.

lipoprotein A protein molecule that contains a conjugated lipid group. Lipoproteins are common in *cell membranes.*

lithotrophic Any organism that uses inorganic molecules as a source of hydrogen atoms (electron donors) in *cellular respiration.* In the case of green plants and algae, this is always water (H_2O). Some bacteria have the capability of using other inorganic molecules such as hydrogen sulfide (H_2S), hydrogen gas (H_2), ammonia (NH_3), or even metal ions (for example, manganese Mn^{++}, iron Fe^{++}) as a source of electrons. (compare *organotrophic*)

litmus paper Paper strips containing an acid-base *indicator.* In acids, blue litmus paper turns red. In alkali, red litmus paper turns blue. Litmus solution is obtained from a *lichen.*

littoral
1. In freshwater aquatic *ecosystems,* the shallow zone. The depth is defined by rooted vegetation. Light penetrates to this depth, enabling the existence of *photoautotrophic* and, consequently, *chemoautotrophic* organisms. (compare *limnetic, profundal*)
2. In marine *ecosystems,* the intertidal zone where periodic exposure and submersion is normal. (see *neritic*; compare *sublittoral*)

liver In vertebrates, a large *gland* arising from the intestine, the main function of which is to regulate the chemical composition of blood. Almost all substances entering the body pass through the liver. Decomposition into excretable products, including detoxification if necessary, is accomplished at this site. The liver is central to fat and carbohydrate *metabolism* and is the main storage site of *glycogen* as well as iron and some vitamins. It also produces *bile,* which aids in digestion. The liver removes damaged red blood cells from the blood and manufactures some blood proteins, including the *clotting factors prothrombin and fibrinogen.*

liverworts (see *Hepaticophyta*)

living fossil A living species that retains features characteristic of an extinct species known only from the fossil record. Examples are the deep sea coelacanth and the Ginkgo tree.

lizards (see *Squamata*)

loam A class of soil texture composed of sand, silt, and clay, producing an intermediate texture that is easily cultivated.

lobopodium, lobopodia (pl.) A blunt *pseudopodium* used for feeding and locomotion.

lobster (see *Crustacea*)

local extinction The disappearance of a species from only part of its range and its continued existence elsewhere. (compare *global extinction*)

locus, loci (pl.) The physical location of a *gene* on a *chromosome*. *Alleles* (variants) of the same gene occupy equivalent loci on *homologous* chromosomes in a *diploid* organism. (see *genetic mapping*)

locusts (see *Orthoptera*)

long-day plant A plant in which flowering is favored by long days and correspondingly short nights. The critical factor is actually the length of the dark period (see *photoperiodism*; compare *short-day plant*)

loons (see *Gaviiformes*)

loop of Henle In the mammalian kidney, the middle portion of the *renal tubule* that is folded into a hairpin loop where water is reclaimed from the waste filtrate by an ionic *countercurrent exchange*. Desert animals, for instance, have a very long loop of Henle in order to conserve as much water as possible before urine is excreted.

lophophorates The three phyla of marine animals, ectoprocs (bryozoans), brachiopods, and phoronids, that are characterized by a *lophophore*. They are attached to a substrate or move slowly, using the *cilia* on their lophophore to capture the *plankton* on which they feed.

lophophore A circular or U-shaped ridge around the mouth of some marine invertebrates, bearing either one or two rows of *ciliated*, hollow tentacles. It functions as a food collection organ and as a surface for gas exchange. (see *lophophorates*)

lorica

1. The thickened body wall of rotifers.

2. (see *test*)

Loricifera A relatively new phylum, introduced in 1983, of minute *pseudocoelomate* marine animals that inhabit the *interstitial* microhabitat in the spaces between grains of sand in the ocean bed. Their feeding apparatus consists of a unique flexible tube that can be telescopically retracted into the body. *Larvae* are propelled by unusual locomatory spines attached to a kind of ball-and-socket joint. The adults lack appendages and may be sedentary. They are, as yet, poorly characterized. (see *meiofauna*)

lories (see *Psittaciformes*)

lotic Environments formed by running water, such as streams and rivers.

low density lipoprotein (see *LDL*)

LTH (see *prolactin*)

luciferase The enzyme that catalyzes the *oxidation* of *luciferin* and consequent emission of light in *bioluminescent* reactions.

luciferin The compound responsible for emitting light in *bioluminescence*. (see *luciferase*)

lumbar vertebrae The bones of the lower back region of the *vertebral column* between the *thoracic* region and the *sacral* region.

lumen Any cavity inside a cell or body structure.

luminescence The production of light by mechanisms other than incandescence (high temperature). (see *bioluminescence, chemiluminescence*)

lung The respiratory organ of air-breathing vertebrates, including aquatic forms (for example, turtles and whales). Lung tissue is highly convoluted, providing a large surface area for respiratory gas exchange. The main exchange occurs at the finest branches of the lungs, the *alveoli* (air sacs), where they converge with *capillaries* returning *deoxygenated* blood from the bloodstream. The *mantle* and mantle cavity of terrestrial mollusks (snails) contains a highly *vascularized* region that is open to the air and also termed a lung.

(see *bronchiole, bronchus, respiration, book lungs*; compare *gill*)

lung books (see *book lungs*)

lutein In plants, the most common of the *xanthophylls*. It is found as an accessory *photosynthetic pigment* in green plants and also in certain algae.

luteinizing hormone (LH, interstitial cell-stimulating hormone) A *glycoprotein hormone* secreted by the *pituitary gland* under regulation of the *hypothalamus*. In female mammals, it stimulates the secretion of *estrogen* and, subsequently, *ovulation*. It promotes the formation of the *corpus luteum* from the ruptured *Graafian follicle* in the *ovary*, which maintains the uterine lining during pregnancy. In male mammals, it stimulates *interstitial cells* in the *testes* to secrete *androgens*.

luteotrophic hormone (see *prolactin*)

lycopene A *carotene photosynthetic pigment*.

Lycophyta A plant phylum including the club mosses and quillworts, or ground pines. They are primitive, *vascular*, nonseed plants found in *temperate forests* and as *epiphytes* in *tropical forests*. The *haploid gametophyte* generation is often subterranean, living in *symbiosis* with *mycorrhizal* fungi. It is connected to the aboveground *sporophyte* by a *rhizome*. The *sporangia*, in which *haploid meiotic spores* are produced, develops from modified leaves called *sporophylls*. Some lycopods bear their sporangia on club-shaped *strobili*.

lycopods (see *Lycophyta*)

lymph Fluid that is drained from the *intercellular* spaces via the *lymphatic system*. It is similar to *blood plasma* but contains *lymphocytes* and bacteria. (see *lymphatic system, lymph node, lymphoid tissue*)

lymphatic system In animals, an open *vascular system* that reclaims liquid *(lymph)* that has entered *intercellular* spaces from the bloodstream. It consists of lymph *capillaries* that begin blindly in the tissues and lead to a network of progressively larger vessels that empty into

the *vena cava*, eventually leading to the heart. It also includes the *lymph nodes, spleen, thymus,* and *tonsils.* In mammals, this is accomplished by muscular and respiratory movements. In vertebrates other than mammals, *lymph hearts* facilitate this circulation. The lymphatic system is also the main route by which fats reach the bloodstream from the intestine. (see *lymphoid tissue*)

lymphatic tissue (see *lymphoid tissue*)

lymph gland (see *lymph node*)

lymph heart (see *lymphatic system*)

lymph nodes (lymph gland) A large number of *glands* distributed within the *lymphatic system* and clustered in certain regions such as the neck, armpits, and groin. Lymph nodes are a site of *lymphocyte* (in particular *B-cell*) development and, consequently, *antibody* generation. They also contain *macrophages* that engulf bacteria and other foreign material from the lymph, sometimes becoming inflamed and enlarged as a result. (see *lymphoid tissue, immune system*)

lymphocyte A type of *leukocyte* (white blood cell) that is responsible for the *immune response*. Of the two principle classes, *B-cells* differentiate into *antibody*-producing *plasma cells* upon *antigen* presentation (*humoral immune response*), and *T-cells* interact directly with foreign invaders and are responsible for the *cell-mediated immune response*. Both arise from *stem cells* in the *bone marrow*. (see *immune system, immune response, hemopoiesis, lymphatic system*)

lymphoid tissue (lymphatic tissue) Tissue found in the *lymph nodes, tonsils, spleen, and thymus*. It consists of a network of cells through which lymph flows continuously and is the site of *lymphocyte* production. Lymphoid tissue also contains numerous *macrophages* that ingest foreign particles, especially bacteria, acting as a filter to remove them from the lymph. (see *lymphatic system, immune system*; compare *myeloid tissue*)

lymphokines Soluble proteins produced by *lymphocytes* that act as messengers in the *cell-mediated immune response*. In particular, *interleukin-2*, secreted by activated *helper T-cells*, stimulates the proliferation of all *T-cell* types that have been exposed to the foreign *antigen*. (see *immune system*; compare *monokines*)

lymphoma Cancer of the *lymph* tissue.

lysine (lys) An *amino acid.*

lysis The disintegration of a cell by rupture of its *cell membrane.*

lyso-, -lysis, -lytic A word element derived from Greek meaning "decompose." (for example, *hydrolysis*)

lysogenic The situation in which a *temperate bacteriophage* is able to integrate its *nucleic acid* into the bacterial *genome* after penetrating the cell. In this state (*lysogeny*), most *phage* genes are repressed, the *host* cell remains intact (not *lysed*), and the phage *genome* (now termed a *prophage*) is *replicated* along with the host *DNA*. Certain factors can *induce* the phage to excise itself from the host *chromosome* and resume the infective lytic cycle. (see *transduction, integration, excision*; compare *lytic*)

lysosome A cell *organelle* that contains digestive (*hydrolytic*) enzymes that are necessarily separated from the rest of the *cytoplasm* by a single layer membrane. The enzymes are released when appropriate, to remove *intracellular* debris including spent *organelles* and bacteria destroyed by *phagocytes*. Lysosomal enzymes are also used in some *protists* for food digestion.

lysozyme An enzyme present in saliva, tears, mucus, and egg white that destroys bacteria by *hydrolysis* and subsequent *lysis* of the *cell walls*. Lysozyme was accidentally discovered by Alexander Fleming in 1922 when he sneezed into some of his bacterial cultures. He shared the Nobel Prize in 1945.

lytic The situation in which a bacterial cell is infected by a *bacteriophage* that *parasitizes* it, using the *host* cellular machinery to replicate the phage *genome* and synthesize phage proteins. Eventually, mature phage particles are assembled and released, rupturing the *cell wall* and leading to the death of the bacterium. The replicated bacteriophages continue the infective lytic cycle. (compare *lysogenic*)

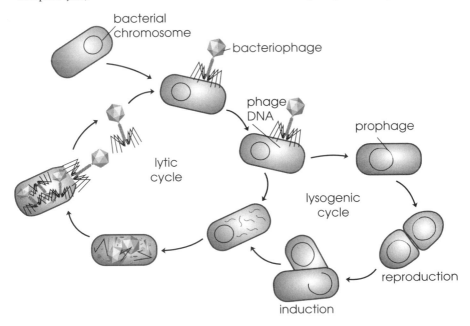

Lytic and Lysogenic Life Cycles of a Bacteriophage

macaws (see *Psittaciformes*)

macrocyst In some cellular slime molds, a giant cell resulting from the *sexual* fusion of a subset of individuals. The macrocyst is then able to undergo *meiosis*.

macroevolution Major evolutionary changes resulting in new species, genera, orders, and families. (compare *microevolution*)

macroinvertebrates Generally, invertebrates visible to the naked eye.

macromolecule A very large (high *molecular weight*) molecule, often a *polymer*. Macromolecules are characteristic of biological systems and include *polypeptides* (proteins), *nucleic acids*, *polysaccharides* (carbohydrates), lipids (fats), and complexes of them.

macronucleus, macronuclei (pl.) The larger of the two *nuclei* found in most *ciliate protozoans* (for example, *Paramecium*). In the macronucleus, the *DNA* is organized into smaller units than normal *chromosomes*, and certain *genes* are differentially amplified by replicating multiple copies. This mechanism may allow for control of a relatively large cytoplasmic volume by a single nucleus. The macronucleus functions during the nonreproductive (growth) phase of the *life cycle*, divides *amitotically* (unequally), and degenerates during *sexual reproduction* when the *micronucleus* becomes active. It is regenerated from the *micronuclear* chromosomes that produce the new *zygote*. (see *micronucleus*)

macronutrient A nutrient required in more than trace amounts by an organism. These include carbon, hydrogen, oxygen, nitrogen, phosphorus, and sulfur. (see *essential element*; compare *micronutrient, trace element*)

macrophage In vertebrates, a large *ameboid leukocyte* (white blood cell) that functions in bodily defense by engulfing, ingesting, and destroying bacteria and cellular debris. They are derived from *monocytes*, which convert into macrophages at a site of infection. Macrophages stimulate invading microbes to release *pyrogens* that act on the *hypothalamus* to elevate the body's temperature, causing a fever. They also activate *T-cells* by secreting *interleukin-1* and increase their own production by secreting γ-*interferon*. Additionally, they incorporate ingested foreign *antigens* into their own cell-surface membrane, presenting them in a form that is easily recognizable by other cells in the *immune system*. (see *phagocyte, lymphoid tissue, immune response*)

macrophagous Organisms that feed on particles that are large relative to their own size.

macrophyll (see *megaphyll*)

macrophyte A *macroscopic photosynthetic* organism growing in water. Most are *angiosperms*, but some are nonvascular plants or algae.

macropterous An organism bearing fully formed wings.

Macroscelidea An order of placental mammals comprising the elephant shrews.

macroscopic Any organism that can be seen with the unaided eye.

macrosporangium (see *megasporangium*)

macrospore (see *megaspore*)

macrosporophyll (see *megasporophyll*)

macula, maculae (pl.)

1. In the vertebrate eye, a region of acute vision surrounding the *fovea* (region lacking cells). It is rich in *cone cells*.

2. In the *inner ear* of most vertebrates, an organ that registers head movements with respect to gravity and acceleration. The *sacculus* and *utriculus* each contain a macula. *Hair cells* protrude into a gelatinous matrix containing crystals of calium carbonate called *statoliths* or *otoliths*. Movements of the crystals bend the *cilia* extending from the hair cells and are registered by the *afferent sensory nerves*.

madreporite (see *water vascular system*)

magainins A relatively new class of *antibiotics* isolated from the skin of the African clawed frog (*Xenopus laevis*). They were discovered serendipitously when the investigator noticed that the wounds of the animals he was operating on always healed perfectly, even though they were placed into old aquarium water filled with bacteria, fungi, and other *parasites*.

magnesium An essential element for plant and animal growth. It is integral to the *chlorophyll* molecule and is thus essential for *photosynthesis*. It is also a common structural component in bones and teeth and is found in smaller quantities in muscles and nerves. Magnesium is an essential *cofactor* for certain enzymes.

major groove The wider of the two parallel spiraling grooves that follow the structure of the *DNA double helix*. (compare *minor groove*)

major histocompatibility complex (see MHC)

Malacostraca The class containing the decapod crustaceans including prawns, shrimps, lobsters, crabs, and crayfish. All except the crabs have a long abdomen ending in a tail used for swimming backwards. The head and *thorax* are characteristically fused into a *cephalothorax* and covered with a *carapace*. Of the five pairs of walking legs, the first and second pairs often have pincers (*chelae*) used in feeding and defense.

malic acid An intermediate in the *Krebs cycle* in *aerobic respiration*. It is also found in fruits such as grapes and gooseberries.

malleus In the *inner ear*, the hammer-shaped bone (ear *ossicle*) attached to the *tympanic membrane* (eardrum). (see *incus, stapes*)

Malpighian body In the vertebrate kidney, a part of the excretory unit (*nephron*). Located within the *cortex*, it consists of the *glomerulus* (a knot of capillaries) and the surrounding *Bowman's capsule*. High pressure created within the glomerulus results in the filtration of water, salts, and nitrogenous wastes across the capillary walls into the capsule, leading to the *uriniferous tubule* for reabsorbtion or excretion.

Malpighian corpuscle (see *Malpighian body*).

Malpighian layer (see *stratum basal*)

Malpighian tubules The excretory organs in terrestrial arthropods that project from the digestive tract into the blood. They extract waste products from the surrounding blood, reabsorb the fluid, and pass them into the hindgut for discharge as a precipitate of uric acid or *guanine*.

maltose A sugar found in *germinating* cereal seeds. It is a *disaccharide* composed of two *glucose* units. Malt barley is important in the beer industry.

Mammalia The class of vertebrates, the mammals, that contain tetrapods (four-limbed animals) that are *homeothermic* (maintain body temperature), have at least some hair, and nurse their young. *Oxygenated* and *deoxygenated* blood are separated in a four-chambered heart (*double circulation*), and the brain is relatively large. Both terrestrial (for example, man and dogs) and aquatic (for example, whales and sea lions) organisms are included.

mammary gland The milk-producing *gland* in female mammals. There may be one or more pairs, depending on the species. Each gland consists of fatty tissue in which are embedded clusters of milk-producing *alveoli* that lead into tubules converging at the nipple or teat. Some primitive mammals (monotremes) such as the duck-billed platypus lack nipples and have mammary glands scattered over the abdomen.

manatee (see *Sirenia*)

mandible In general, a part of the mouth or feeding apparatus of an animal.
1. A pair of mouthparts found on most arthropods. (see *maxillae, mouthparts*)
2. The extended upper and lower jaws of birds that are extended to form a beak.
3. The lower jaw of vertebrates.

Mandibulata (Uniramia, Atelocerata) A

phylum of mandibulate arthropods (formerly a subphylum) including the centipedes, millipedes, and insects. Uniramians are characterized by their many body segments, each bearing one or more pairs of unbranched (*uniramous*) legs.

mandibulate Any arthropod possessing jaws or *mandibles*. (see *Mandibulata*)

manganese (see *trace element*)

mannitol A carbohydrate found widely in plants and forming a characteristic food reserve in brown algae.

mantle The outermost layer of the body wall of mollusks and brachiopods. It secretes the shell. In shell-less mollusks, it is tough and protective and encloses the mantle cavity that contains the respiratory organs. In squids, it has muscular walls that contract to force water out of the cavity as a means of propulsion.

Mantodea The class of insects containing the mantids and preying mantises. They are characterized by their large forelegs that are used for seizing prey.

map unit (m.u.) (see *genetic map unit*)

marginal meristems The primary plant tissue from which leaves are produced. Unlike the *apical meristems* of stems and roots, the marginal meristem ceases to function once a leaf is fully expanded. Thus, the growth of leaves, like flowers, is *determinate*.

marine The saltwater environment of the oceans and seas. It comprises a subset of the aquatic environments.

marl A soil containing a high proportion of calcium carbonate, such as in limestone areas.

Marsupiala A somewhat historical order of mammals, the *marsupials* (pouched mammals), containing the kangaroos, koala bears, and opossums. They are more primitive than placental mammals, and the brain is relatively small. The young are born in an extremely immature state. For a time after birth they continue to develop in an abdominal pouch (*marsupium*) where they are suckled. Marsupials have recently been divided

into numerous separate orders based on more discriminating (particularly genetic) data. (see *Appendix*)

marsupial carnivores (see *Dasyuromorphia*)

marsupial moles (see *Notoryctemorphia*)

marsupials (see *Marsupiala*)

marsupium, marsupia (pl.)

1. A pouch on the abdomen of marsupial mammals and monotremes. In marsupials, young born in an extremely immature state are transferred to the marsupium where they complete early development. In monotremes, the only egg-laying mammals, the eggs are hatched in the marsupium and the young are nourished and protected in the pouch. In both monotremes and marsupials, the abdominal surface within the marsupium contains *mammary glands*, which produce milk.

2. In some invertebrates, a space formed for incubation of *embryos* and *larval* stages.

masculinization The development of male traits in a female organism. This is often due to inappropriate exposure to male *hormones*. (see anabolic steroids)

mass A measure of weight that is independent of gravity.

mass flow The mechanism of movement of materials within the *phloem* of *vascular plants*. It depends on the difference in *osmotic potential* created by the relative amounts of synthesized sugars in the *sieve tubes* in various parts of the plant. In regions of *photosynthesis* such as the leaves, sugars are drawn into the sieve tubes, encouraging the concomitant uptake of water from the *xylem*. This is called a *source*. In regions where carbohydrate is unloaded from the sieve tubes into structures such as roots and fruits, water is drawn after the solutes. This is called a *sink*. Sugars and other dissolved solutes move passively with the water, creating a directional flow. Energy for uptake into and withdrawal from the sieve tubes is provided by *companion cells*. (see *solute potential*)

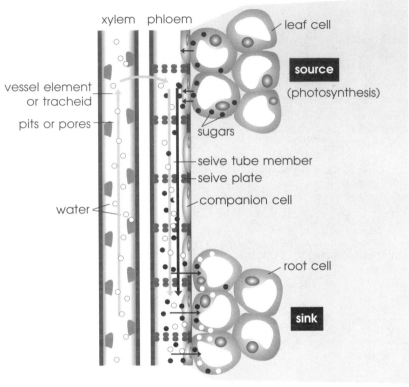

xylem phloem

leaf cell

source

(photosynthesis)

vessel element
or tracheid

pits or pores

sugars

seive tube member

seive plate

companion cell

water

root cell

sink

Mass Flow

mast cell A stationary cell involved in bodily defense and found embedded in connective tissue. Mast cells synthesize *histamine* and *seratonin,* which mediate the *inflammatory response* to injury or allergy. They also synthesize *heparin,* which inhibits *blood clotting.* (see *allergic response*)

mastigoneme A hairlike, lateral projection on *undulipodia.*

mastigote (flagellate) Any *eukaryotic* microorganism that achieves motility via *undulipodia.*

maternal inheritance The situation in which a particular trait or characteristic is inherited through *DNA* passed on only by way of the mother. Maternal inheritance is seen for cellular *organelles* that are present in the egg but not carried with the sperm head, which is usually too small to transmit anything more than *nuclear chromosomes. Mitochondria,* in particular, are lost when the head of the sperm enters the egg at *fertilization.* Therefore, *mitochondrial DNA* can be used to trace female ancestry. This general mode of *uniparental inheritance* holds true for any organism in which the male and female *gametes* are of extremely disparate sizes (*anisogamous*).

mating The contact or fusion exhibited by cells, *nuclei,* or individuals of complementary sexes.

mating types In some fungi and protoctists (ciliates), the two complementary *sexual* types that can mate with each other. The *sexual* cells are usually *haploid* and fuse to form a *diploid, asexual zygote.* The mating types are determined at the genetic level and manifest as *cell surface markers* of protein or *glycoprotein*. (see *homothallic, heterothallic*)

maxilla, maxillae (pl.) A pair or two of secondary mouthparts in many insects. They are located immediately behind the *mandibles*. (see *mouthparts*)

mayflies (see *Ephemeroptera*)

mechanism The view that all living processes can ultimately be explained by chemical reactions. Although first proposed in the 17th century, it began to gain general acceptance around the turn of the century. (compare *vitalism*)

mechanoreceptor A *sensory* receptor that responds to a mechanical stimulus such as touch, pressure, and sound waves. (see *baroreceptor*)

meconium The contents of the mammalian fetal intestine.

Mecoptera An order of *endopterygote* insects, the scorpionflies, that resemble scorpions although they do not sting. The *larvae* live in burrows and emerge at the surface as adults. They are mainly scavengers.

median eye An eyelike light receptor (*ocellus*) found in the middle of the head of some crustaceans, such as the microscopic pond animal *Cyclops*. Some insects, such as the locust, have a median ocellus as well as a pair of *compound eyes*. The New Zealand lizard has a median third eye that actually functions in vision (*pineal eye*) and is an evolutionary remnant.

mediastinum The space between the two *pleural* (lung) cavities containing the heart in its *pericardium* and the *trachea*, *esophagus*, and *thymus*.

medulla, medullae (pl.)
1. The central region of an organ or structure. It often differs in structure or function from the outer regions, such as in the kidney and *adrenal gland*. (compare *cortex*)
2. (see *medulla oblongata*)

medulla oblongata A region of the *hindbrain* that is actually an extension of the

spinal cord into the skull. It controls the most basic life functions, including respiration and heartbeat, as well as other *autonomic* activities.

medullary ray In plants, a plate of undifferentiated *parenchyma* tissue that extends from the *medulla* (*pith*) to the *cortex*, across the *vascular* region of young plant stems. It contains *starch* stores and sometimes also *tannins*.

medullated nerve fiber (see *myelinated nerve fiber*)

medusa, medusae (pl.) The jellyfish stage in the *life cycle* of many cnidarians. The free-swimming medusa is shaped like a bell or inverted saucer with a fringe of tentacles around the rim and a mouth beneath. It is the *sexual* stage of the *life cycle* and in *scyphozoans* (jellyfish), is the only form. *Spermatozoa* swim from the male to *fertilize* eggs in the female. (see *alternation of generations;* compare *polyp*)

megagametophyte In *heterosporous* plants, the female *gametophyte*. They are located within the *ovule* of seed plants and the *archegonia* of nonseed plants. (compare *microgametophyte*)

megakaryocytes The large cells within the *bone marrow* that regularly pinch off bits of their *cytoplasm* to become the circulating *platelets* necessary for *blood clotting*.

Megaloptera An order of *endopterygote* insects, the dobsonflies. The *larvae* are aquatic and *carnivorous*, and *pupation* takes place in a silken *cocoon*. Many adults are quite large, having a wingspan of about 15 centimeters.

meganucleus (see *macronucleus*)

megaphyll (macrophyll). A foliage leaf with a branched *vascular system* in the blade. It is typical of many higher plants. (compare *microphyll*)

megasporangium A *sporangium* in which female *meiotic* products, *megaspores*, are formed. In seed plants, it is the *ovule*.

megaspore The larger of the two *haploid spore* types in *heterosporous* plants. It develops into a female *gametophyte* (*gamete*-producing generation). (compare *microspore*)

megasporocyte A *diploid* cell in *heterosporous* plants that divides by *meiosis* to produce four *haploid megaspores* (the first cells of the female *gametophyte* generation). (see *spore mother cell*)

megasporophyll (macrosporophyll) A leaf or modified leaf bearing *megasporangia*. The *carpels* of *angiosperms* and the *ovuliferous scales* of *gymnosperms* are modified *megasporophylls*. (compare *microsporophyll*)

meiocyte A cell that is destined to undergo *meiosis*, usually to produce *gametes*.

meiofauna Animals adapted to living in the *interstitial* microhabitat of the spaces between grains of sand in the ocean.

meiosis The process of reductive *cell division* leading to progeny containing half the genetic complement of the parent cell. Meiosis is the key element in the production of *haploid gametes*; fusion of gametes (*fertilization*) restores the cor-

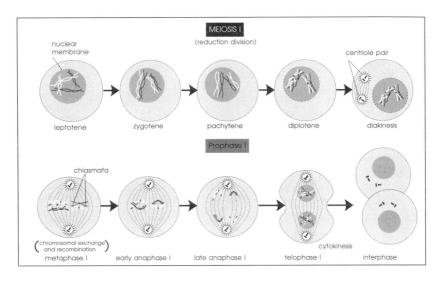

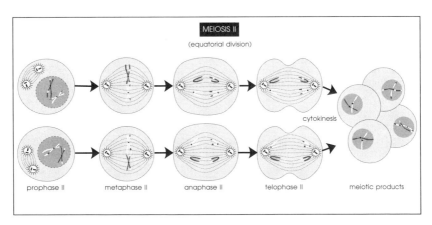

Meiosis in Animal Cells

rect *chromosomal* complement, usually *diploid*. Meiosis consists of two divisions (*meiosis I* and *meiosis II*) during which the *chromosomes* replicate only once. During the first meiotic *prophase*, *homologous* chromosomes become paired to form *bivalents*, at which time genetic material is exchanged between homologous *chromatids*. At the end of the first meiotic division, homologous chromosomes separate from each other, but the replicated chromatids remain attached. This results in a reduction of the chromosome complement from diploid to haploid. During the second meiotic division, each *nucleus* undergoes a second division. At that time, the chromatids separate, and each *haploid* set is packaged into one of four meiotic products that usually develop into *spores* or *gametes*. (see *double strand break*; compare *mitosis*)

meiosis I (reduction division) The first cellular division of *meiosis*. *Homologous chromosomes* consisting of replicated *chromatids* separate from each other after undergoing pairing (*bivalents*) and genetic exchange. Because the chromatids do not separate, the chromosomal complement is reduced from *diploid* to *haploid*. (compare *meiosis II*)

meiosis II The second cellular division of *meiosis*. The previously replicated *chromatids* contained in the two products of *meiosis I* separate from each other. They are subsequently packaged into a total of four *haploid* meiotic products. (compare *meiosis I*)

meiospore One of the four products of *meiosis* in flowering plants (*angiosperms*) that eventually gives rise to gametes. (compare *mitospore*)

meiotic cross-over A *cross-over* that results from pairing between two *homologous chromosomes* in a *diploid* cell during *meiosis*. (compare *mitotic cross-over*)

melanins A group of related pigments found in animals and plants. Melanins range from black through brown to orange, red, or yellow. In animals, they confer color to the hair, eyes, and skin, and protect against ultraviolet radiation. In plants, they give color to various seedlings and roots. Melanins are derived from the *amino acid tyrosine*. The absence of the enzyme *tyrosinase* leads to *albinism* in animals.

melanocyte-stimulating hormone (see *MSH*)

melanophores *Melanin*-containing cells in animals. (see *chromatophore*)

melatonin A *hormone* produced by the *pineal gland* that stimulates lightening of the skin in *embryonic* fish and *larval* amphibians. In humans, its function is not well understood but may be related to the establishment of *circadian rhythms* in response to daylight. Melatonin has been implicated in the winter depression syndrome SADS.

melting temperature In general, the temperature at which a solid turns to liquid. In *DNA* technology, it specifically refers to the temperature at which a *double-stranded* DNA molecule separates into single strands. The exact temperature is determined by the *GC content*. (see *complementary base pairing*)

membrane
  1. (see *plasma membrane*)
  2. Thin sheets of tissue surrounding or incorporated into an organismal structure.

membrane bone A bone formed by the *ossification* of *connective tissue* instead of cartilage. (see *osteoblast*)

membrane channel A *passive transport* system composed of multisubunit *transmembrane proteins*. They form continuous pores across the *phospholipid bilayer* that allow specific ion species to flow down their already established concentration gradients. The transition from closed to open may be mediated by voltage, chemicals, or stimuli. (see *ion channel, voltage-gated ion channel, stimulus-gated ion channel, chemically gated ion channel*; compare *membrane pump*)

membrane potential The voltage measured across the *cell membrane* in a nerve or muscle cell. It is established by local differences in ion distribution. These occur when there is a concentration gradient of a particular ion (for example, potassium [K$^+$]) and/or when there is a difference in the relative permeability of the membrane to two different ions (for example, K$^+$ and Sodium [Na$^+$]). It is a measure of electrical driving force and can cause ions to move, producing a current. (see *potential difference, resting potential, equilibrium potential, action potential, depolarization, hyperpolarization*)

membrane proteins Any protein that is embedded in the *plasma membrane* of a cell. They are not anchored in one place but tend to float within the *phospholipid bilayer*. Since the membrane itself is relatively impermeable to biological molecules, the different membrane proteins control the interactions of the cell with its environment. These interactions include *membrane channels* that admit ions, sugars and *amino acids*, and *membrane pumps* that perform a similar function but require energy to function. *Transmembrane receptors* transmit information from the exterior to the interior via *conformational* changes (for example, *hormone* receptors). They also help identify cells as self (belonging to the organism) and identify their function. (see *gating, ion channel, coupled channel, sodium-potassium pump, proton pump, calcium pump, second messenger, MHC proteins*)

membrane pump An *active transport* system that mediates the concentration of small molecules, in particular certain ions and fuel molecules, across the cell membrane. Like *membrane channels*, they are formed from multisubunit *transmembrane proteins*. They differ from channels in that membrane pumps require an input of energy, and they never form a continuous pore through the membrane. Rather, they undergo a cycle of *conformational* transitions that simultaneously change the orientation and

affinity of the binding site for the transported species. (see *proton pump, sodium-potassium pump, calcium pump*)

membranous labyrinth (see *labyrinth*)

memory cells A *clonal* population of *lymphocytes* that have previously encountered a particular foreign invader or *antigen* and persist in the body. They are thus able to mount a much swifter *secondary immune response* to a second encounter with the particular invader or antigen. Both *T-cells* and *B-cells* persist as memory cells. (see *immune response, immune system, cell-mediated immune response, humoral-mediated immune response*)

menaquinone (see *vitamin K*)

Mendelian inheritance Genetic elements, usually *nuclear chromosomes*, that are inherited in general accordance with *Mendel's laws*. (compare *non-Mendelian inheritance*)

Mendelian ratio (see *Mendelian inheritance*; compare *non-Mendelian ratio*)

Mendel's laws The two genetic laws formulated by Gregor Mendel, a 19th century Austrian monk, to explain the patterns of inheritance he observed in garden peas. The first, **The Law of Segregation,** can be paraphrased into the following three statements. (1) the alternative forms of a trait encoded by a *gene* are specified by alternative *alleles* (genetic variants) of that *gene* and are discrete entities. (2) In *heterozygous diploid* individuals, the two alternative *alleles* segregate from each other into different *gametes*. (3) each gamete has an equal probability of inheriting either member of an allele pair. The second genetic law, **The Law of Independent Assortment** can be paraphrased to state that *genes* located on different *chromosomes* sort independently of one another into individual gametes. In other words, the inheritance of alternative alleles of one gene is independent of the simultaneous inheritance of any particular allele of a second, unlinked gene. At the time, neither genes nor chromosomes had been identified. Mendel instead referred to "discrete characters" in his work. (see *independent assortment, independent segregation, genetic linkage*)

meninges The protective membranes that surround the brain and *spinal cord* in vertebrates. Man and other mammals have three: the outer *dura mater*, the middle *arachnoid membrane*, and the inner *pia mater*.

menstrual cycle A modified form of the *estrus cycle* found in humans, Old World monkeys, and anthropoid apes. The female does not come into "heat" as do lower mammals but is, instead, continually *sexually* receptive. Approximately once a month, an egg is released from the *ovary*. If it is not *fertilized*, the uterine lining is expelled in the menstrual flow. (see *oogenesis, ovum, oocyte, corpus luteum, follicle-stimulating hormone, Graafian follicle*)

menstruation (see *menstrual cycle*)

mer-, -mere A word element derived from Greek meaning "part." (for example, *chromomere*)

meristem In plants, a group of cells that are capable of dividing indefinitely and whose main function is the production of new growth. Meristematic tissue is found at the growing tip of a root or a stem (*apical meristem*), in the *cambium* (*lateral meristem*), and within the stem and leaves (*intercalary meristem*) of grasses. (see *procambium*)

meroblastic cleavage The type of incomplete *cleavage* that occurs after *fertilization* in eggs with a large yolk (for example, birds and sharks) in which the egg *cytoplasm* divides but the yolk does not. (compare *holoblastic cleavage*)

Merostomata A class of *chelicerate* arthropods, the horseshoe crabs. They live in deep ocean waters and migrate to shallow coastal waters every spring to mate in the sand on moonlit nights when the tide is high. *Fertilization* is external and respiration is by *book gills*. They swim by moving their abdominal plates and walk on their five pairs of legs.

merozoite A *life cycle* stage found particularly in apicomplexan protists such as the malarial *parasite Plasmodium sp.* It is the *mitotic* product of the *trophozoite* feeding stage found in the blood, and is the vehicle for reinoculation of a mosquito via a blood meal.

merozygote A partial *diploid* bacterial cell formed during a *conjugation* event. It consists of the complete *genome* (*endogenote*) of the receiving cell and a portion of the transferred genome (*exogenote*). (see *Hfr cell*)

mes-, meso- A word element derived from Greek meaning "middle." (for example, *mesoderm*)

mesencephalon (see *midbrain*)

mesenchyme
  1. A loose network of *mesoderm* cells in animal *embryos* that give rise to internal systems such as *connective tissue*, muscle, and blood.
  2. The gelatinous, protein-rich matrix situated between the inner and outer layers of the body wall in sponges.

mesentery
  1. A double membrane of *peritoneal* (abdominal) membrane that attaches the *alimentary canal* to the abdomen wall. The blood vessels, *lymph vessels*, and nerves that supply the alimentary canal lie between the two layers.
  2. One of the vertical partitions in the *coelenteron* (central cavity) of sea anemones.

mesocarp (see *pericarp*)

mesoderm The middle of the three *embryonic germ layers* in *triploblastic* animals. It gives rise to muscles, *connective tissues*, some organs, and the *vascular system*. (compare *endoderm, ectoderm*)

mesoglea The layer of jellylike material that separates and is secreted by the inner *endoderm* and outer *ectoderm* layers of the body wall in jellyfish.

mesokaryotic The type of *chromosome* structure, found in particular in the *nuclei* of dinomastigote protists, in which conventional *histones* are lacking and the chromosomes are permanently condensed. It appears to be an intermediate form between typical *prokaryotic* and *eukaryotic* chromosome structure.

mesophilic Microorganisms for which the optimum growth temperature is between approximately 25°C and 45°C. (compare *psychrophilic, thermophilic*)

mesophyll In plants, the specialized tissue in leaves in which *photosynthesis* takes place and where *starch* is stored.

mesophyte A plant type that is adapted to grow only under conditions of adequate water supply. In drought conditions, wilting is soon apparent because the plants have no water conservation mechanisms. Most *angiosperms* (flowering plants) are mesophytes. (see *Raunkiaer's plant classification*)

mesosome A complex invagination of the *plasma membrane* in *prokaryotic* cells that contains respiratory enzymes and is functionally similar to the *mitochondria* of *eukaryotes*. The bacterial *chromosome* is usually attached to the mesosome. The mesosome also appears to play a role in *cell division* as the initiation site of *cytoplasmic* separation. (see *binary fission*)

mesothelium, mesothelia (pl.) **(serous membranes)** In vertebrates, the type of membrane that lines interior body cavities (*coelomic* spaces) such as the *peritoneal and pleural* cavities. It is characteristically composed of a double-layered membrane that may contain either liquid or gas between the layers. It is derived from *mesoderm*.

Mesozoa A phylum of animals, the mesozoans, that contain only one organ, a *gonad*. They absorb nourishment from the urine of animals they inhabit, usually cephalopod mollusks, thus function without any other body systems. They *alternate generations*, undergoing their *sexual* phase on the inner surface of their *host's* kidney. The *larvae* escape in the urine to infect another host. They also reproduce *asexually*. (see *symbiotroph, commensalism*)

Mesozoans (see *Mesozoa*)

Mesozoic The middle era in the geological

time scale, about 230 million to 70 million years ago. It is divided into the *Triassic, Jurassic,* and *Cretaceous* periods. It is principally known for the dominance of dinosaurs, particularly during the Jurassic period.

messenger RNA (see *mRNA*)

met-, meta- A word element derived from Greek meaning "among" or "along with," and denoting change. (for example, *metamorphosis*)

metabolic water The water molecules produced as a byproduct of *oxidative respiration*. It is typical of heterotrophic *metabolism* where other organisms are ingested and their organic carbon molecules are oxidized to provide energy. The formation of metabolic water takes place as the final step of the *electron transport chain* where the electrons, having given up their energy in the formation of *ATP*, are donated to oxygen gas (O_2) to form water (H_2O). In some desert animals, metabolic water provides most of the water they require to live.

metabolism The totality of the physical and chemical processes occurring in a living organism. They may be partitioned into *anabolism*, the synthesis of complex molecules from simple ones that usually requires energy, and *catabolism*, the breakdown of complex molecules to simple ones that usually releases energy. It is often used in the context of a particular class of compounds, for instance "carbohydrate metabolism." (see *metabolite*)

metabolite Any substance that participates in a *metabolic* reaction either as starting material, intermediary, or final product.

metacarpal bones The bones in the lower forefoot of animals. They form the hand in *primates*. (see *pentadactyl*; compare *metatarsal*)

metacentric A *chromosome* having its *centromere* in the middle.

metacercaria In flukes, the *larval* form that *encysts* in an *intermediate host* awaiting ingestion by the *primary host*. (see *parasite*)

metachronal rhythm A pattern of movement shown by *cilia* in which each beats one after the other in regular succession giving the appearance of wave motion. (see *ciliated epithelium*)

metamere (segment) The similar repetitive body unit of animals exhibiting *segmentation*, such as in earthworms.

metameric segmentation (see *segmentation*)

metamerism (see *segmentation*)

metamorphosis A phase in the post*embryonic life history* of many animals encompassing a series of discontinuous *morphological* transformations that occur during the change of the *larva* into the adult form. It is widespread among invertebrates, particularly marine organisms, arthropods, and amphibians, and is generally under *hormonal* control. Most insects display complete metamorphosis (*endopterygote*), although a few undergo incomplete or simple *metamorphosis* (*exopterygote*). (see *nymph, pupa, chrysalis, imago, autolysis, ecdysone*)

metanephridium (see *nephridium*)

metaphase The stage in *nuclear division* in *mitosis* and *meiosis* when the *chromosomes* become aligned along the equator (*metaphase plate*) of the *nuclear spindle*. (compare *prophase, anaphase, telophase, interphase*)

metaphase plate During the *metaphase* portion of *nuclear division*, the imaginary plane bisecting the *chromosomes* as they are aligned in a ring following the inner circumference of the cell. The metaphase plate defines the plane in which the cell will divide.

metaplasia The transformation of one kind of adult cell in the body into another kind. It is seen when the normal mechanisms of maintaining *differentiation* are lost, such as in *cancer* and *tumor* formation.

metastasis, metastases (pl.) The spread of *tumor* cells to new sites in the body where they produce new tumors.

metatarsal bones The bones in the lower hindfoot of animals. They form the arch of the foot in *primates*. (see *pentadactyl*; compare *metacarpal*)

Metatharia A mammalian infraclass that for a long time has contained just one order, the *marsupials* (pouched mammals), and includes the kangaroos, koala bears, and opossums. They are more primitive than placental mammals (the great majority being Eutheria), and their brain is relatively small. The young are born in an extremely immature state and continue to develop in an abdominal pouch (*marsupium*) where they suckle. Metatherians possess a *cloaca* as well as a double uterus and vagina. As genetic classification has become more discriminating, "marsupial" has become an informal name and numerous distinct orders have been classified. (see Appendix)

Metazoa (see *Eumetazoa*)

methanogenic The generation of methane gas (CH_4). (see *methanogens*)

methanogens Any organism that produces methane (marsh gas) as a *metabolic* product. Primary methanogens are all bacteria, specifically archaebacteria, although some have been incorporated into ciliates as *endosymbionts*. (see *Euryarchaeota*)

methionine (met) An *amino acid*. In *eukaryotes*, it is always the initiating amino acid in a *polypeptide* chain, inserted in response to the first AUG *codon* encountered on the *mRNA*. In both *prokaryotes* and *eukaryotes*, it is also an important internal amino acid. (see *N-formylmethionine*)

methylation Modification of a molecule, often *DNA*, by the addition of a methyl group. (see *restriction modification*)

MHC (major histocompatibility complex) A group of mammalian *genes* coding for the *glycoprotein cell surface markers* that distinguish each cell as "self." In humans, the complex is called the *human leukocyte-associated antigen (HLA) complex* and resides on *chromosome* VI. The locus consists of two subregions each containing multiple clustered copies of the MHC-I and MHC-II genes. MHC-I proteins are present on every *nucleated* body cell, while

MHC-II proteins are present only on *macrophages* and some *lymphocytes*. The HLA genes are the most *polymorphic* expressed human genes known, with as many as 50 *alleles* (variants) each. Very few, if any, humans have the same combination of alleles. When the MHC alleles in a population become less polymorphic such as in *inbred* zoo populations or *endangered species*, they become much more vulnerable to attack by viral or bacterial mimics. (see *MHC proteins*)

MHC proteins (histocompatibility antigens, transplantation antigens) A set of *glycoprotein cell surface markers* that appear to be unique to an individual (or identical twins). They occur on most body cells but are particularly abundant on *T-cells* where they are important in the distinction of "self" from "nonself" in the *immune system*. For this reason, tissue and organ *transplants* are less likely to be rejected if the histocompatibility markers are at least similar, such as in close relatives. Two basic classes of MHC proteins exist and are designated MHC-I and MHC-II. MHC-I proteins are present on every *nucleated* body cell. MHC-II proteins are present only on *macrophages*, *B-cells*, and *helper T-cells* (T_4), which permits them to complex with each other during the initial stages of the *humoral immune response*. The basic structure of MHC proteins is similar in many respects to *immunoglobulins*. (see *HLA complex, β-microglobulin, H-Y/H-W antigen*)

mice (see *Rodentia*)

micelle A spherical particle, often formed from a *phospholipid* monolayer. In aqueous solution, it orients with the *hydrophobic* phospholipid heads toward the interior and the *hydrophilic* tails towards the water molecules. In cells, they are used to store and transport nonpolar materials. (see *colloid*; compare *phospholipid bilayer*)

micro- (μ) A word element denoting 1/1,000,000 (one-millionth) or 10^{-6}. For example, 1 microgram (μg) = 10^{-6} gram. The microgram is a common unit of measure for small amounts of *DNA* and protein.

microaerobic Environmental conditions in which oxygen is present in less than normal atmospheric concentrations.

microbe (see *microorganism*)

microbial mat A carpetlike community of microorganisms, usually cyanobacteria. It is the living precursor of *stromatolites*.

microbiology The study of microscopic organisms such as bacteria and viruses. (compare *molecular biology*)

microbiota The sum total of microbial organisms found in a given *habitat*.

Microbiotheria An order of marsupial mammals containing only one species, *Dromicops sp.*

microbody (peroxisome, glyoxisome) An enzyme-bearing, membrane bound vesicle found in *eukaryotic* cells. Microbodies may contain several sets of enzymes, including those for converting fat to carbohydrate and ones for processing harmful peroxides into harmless constituents (*catalase*). Microbodies serve to sequester such enzymes from the rest of the cell, organizing the cellular *metabolism*. In plants, microbodies are called *glyoxysomes* and are particularly important in the *glyoxylate cycle*. In animals, they are called *peroxisomes*.

microclimate The climate in a small local area. It may be as small as under a tree or near the surface of city streets. The San Francisco Bay Area is infamous for its microclimates.

microcosm A miniature world. Relating to biology, communities of organisms that can be visualized only by microscope.

microevolution Genetic change that occurs within populations of a species as a result of progressive adaption to a changing environment. Natural selection is the process by which microevolutionary change occurs. (compare *macroevolution*)

microfauna Animals that can only be seen with a magnifying lens or microscope.

microfilaments Filaments found in *eukaryotic* cells that are *polymers* of the protein *actin*, also found in muscle as *thin filaments*. Microfilaments, along with *microtubules* and *intermediate filaments,* form the *cytoskeleton*, influencing cell shape and facilitating movement of the cell and of components within it.

microgametophyte In *heterosporous* plants, the male *gametophyte*. They comprise the *pollen grains* of seed plants and are generated in the *antheridia* of non-seed plants. (compare *megagametophyte*)

microglobulin (see *beta-2-microglobin*).

micrograph A photograph taken through a microscope. *Photomicrographs* and *electron micrographs* are produced using optical microscopes and electron microscopes, respectively.

microinjection A laboratory technique involving the injection of *macromolecules*, usually *nucleic acid (DNA, RNA)*, through a glass needle into animal cells, often *oocytes*, in order to assess their effect.

microinvertebrate Any small *invertebrate* that must be viewed using a microscope.

micrometer (μm, micron) A unit of length equal to 10^{-6} (one-millionth) of a meter. It is the correct order of magnitude to measure distances at the level of a single cell.

micron (see *micrometer*)

micronucleus The smaller of the two *nuclei* found in most *ciliates* (for example, *Paramecium*) and the one that functions in *sexual reproduction*. During *conjugation,* the cells exchange a pair of *haploid* micronuclei that fuse in both cells to form a new *diploid* micronucleus while the *macronucleus* in each disintegrates. After fusion, the new micronuclei divide *mitotically*. One of the products of this division gives rise to more micronuclei. The other undergoes multiple rounds of *DNA replication* and becomes the new macronucleus that functions during the growth phase of the *life cycle*.

micronutrient A nutrient required by an organism in only trace amounts such as *trace elements* and *vitamins.*

microorganism Any organism, usually single or few celled, that must be viewed using a microscope. Microorganisms include members and life stages of bacteria, fungi, and protists as well as the smallest invertebrate animals. (see *microbe*)

microphagous Organisms that feed on particles that are small relative to their own size. Such feeding tends to occur rather continuously often by sieving methods, such as in *baleen* whales and *filter-feeding* bivalve mollusks.

microphyll A foliage leaf that has only a single, unbranched vein running from base to apex. It is typical of some mosses, primitive ferns, and horsetails.

micropyle In plant seeds, a pore in the *ovule* leading to the *nucellus* through which the *pollen tube* enters during *fertilization*. It also serves as a water pore during *germination*.

microsatellite DNA *Tandem repeats* of a DNA sequence that are very short, from two to about six nucleotides long. They are found in the *genomes* of *eukaryotes*. (see *satellite DNA*, compare *minisatellite DNA*)

microscope An instrument designed to magnify objects for viewing. It is important that resolution (the ability to distinguish between adjacent points) be increased along with magnification. Resolution is inversely correlated with the wavelength (λ) of the source irradiating the sample. Thus, an *electron microscope* gives a significant increase in both magnification and resolution over a *light microscope.*

microscopic Any organism that can be seen only using the aid of a magnifying lens or microscope.

microsomes Fragments of *endoplasmic reticulum* that form individual vesicles after the mechanical disruption of cells, such as homogenization. Microsomes from rough endoplasmic reticulum retain their *ribosomes* and can carry out *protein synthesis in vitro.*

Microspora A phylum of *heterotrophic protoctists*, the microsporans or microsporidians, once thought to belong to a larger grouping, the sporozoans. They are small *parasites* that lack *mitochondria* and live inside the cells of their *hosts*, which include arthropods and chordates. They are found in all classes of vertebrates. The microsporidian *life cycle* is typified by a *multicellular plasmodium* stage after the initial infection of a cell. The plasmodium then undergoes multiple *fission* events to produce offspring. The resulting mass resembles a single cell tumor and is called a *xenoma*.

microsporangium A *sporangium* in which male *meiotic* products, or *microspores*, are formed. In seed plants, it is the *pollen sac*.

microspore The smaller of the two *haploid spore* types in *heterosporous* plants. It will develop into a male *gametophyte* (*gamete*-producing generation). (compare *megaspore*)

microsporocyte A *diploid* cell in *heterosporous* plants that divides by *meiosis* to produce four *haploid microspores* (the first cells of the male *gametophyte* generation). (see *spore mother cell*)

microsporophyll A leaf or modified leaf bearing *microsporangia*. The *stamens* of *angiosperms* and the male *cones* of *gymnosperms* are modified *microsporophylls*. (compare *megasporophyll*)

microtome An instrument for cutting thin sections of biological material, sometimes embedded in a matrix, for microscopic examination.

microtubule organizing center (see *MTOC*)

microtubules Thin, hollow protein cylinders found in *eukaryotes* that are assembled from *globular monomers* of the protein *tubulin*. Microtubules always consist of 13 tubulin *protofilaments* arranged in a circular fashion around a central core. They assemble spontaneously *in vitro* but in a cell form only from a *basal body* (*kinetosome*) or *centriole*. They radiate from *MTOC*s (microtubule organizing centers), found on the periphery of *centrioles* (*centrosomes*) or

at the poles of a *mitotic spindle*. Microtubules, along with *microfilaments* and *intermediate filaments,* form the *cytoskeleton*, influencing cell shape and facilitating movement of the cell and components within it. Microtubules also transport the *chromosomes* during *nuclear division* and are further organized to provide the functional internal structure of *undulipodia*. (see *protofilament*)

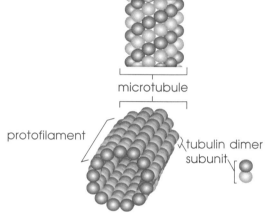

Microtubule Structure

microvillus, microvilli (pl.) External projections of the *plasma membrane* found especially in secretory and absorptive cells, such as in the small intestine and kidney. Microvilli serve to increase the surface area for exchange of molecules at the cell surface. (see *brush border*)

microwhipscorpions (see *Palpigrada*)

midbrain (mesencephalon) The middle of the three anatomical divisions of the vertebrate brain, connecting the *forebrain* and the *hindbrain*. It is the dominant center of the brain in fish and amphibians and is also prominent in birds. The midbrain is less well-developed in man.

middle ear (tympanic cavity) An air-filled cavity between the *outer* and *inner ear* in higher animals. It is connected to the back of the throat by the *Eustachian tube*. In mammals, it contains the three *ear ossicles* (*malleus, incus, and stapes*) that transmit vibrations from the *tympanic membrane* (eardrum) to the *inner ear* through the *oval window*. In other tetrapods, the middle ear contains only one ossicle, the *columella auris*.

middle lamella In plants, a thin membrane separating adjacent cells and also cementing them together. It is laid down at the *cell plate* during *cell division* and consists of *pectins* and other *polysaccharides*.

midgut The central part of the *alimentary canal* of arthropods and vertebrates responsible for digestion and absorbtion of nutrients. (compare *foregut, hindgut*)

migration
1. An instinctive regular movement of an animal population along well-defined routes, typically between birthing and feeding grounds. It is triggered by seasonal factors such as day length and temperature, and functions to assure an adequate food supply and breeding conditions. Many animals including birds, hoofed mammals, bats, whales, fish, and insects migrate, often covering immense distances. For example, the Arctic/Antarctic tern breeds on the northernmost coasts of Eurasia and America and winters in the Antarctic. Modern civilization has disrupted many normal migration routes, threatening the survival of numerous species.
2. The movement of an individual, population, or species from one geographical area to another.
3. In genetics, the movement of an animal, along with its *genes*, from one *population* to another. In addition to the movement of adult organisms, this includes, for instance, the dispersal of *gametes* and the random drifting of immature marine animals and plants. (see *gene flow, gene pool*)

milk sugar (see *lactose*)

milk teeth (see *deciduous teeth*)

milli- (m) A word element denoting one-thousandth or 10^{-3}. For example, 1 milligram (mg) = 10^{-3} gram. The milligram

is a common unit of measure for small amounts of protein and biological reagents.

millipedes (see *Diplopoda*)

mimicry The physical resemblance of one organism to another that benefits the mimic, usually by helping it escape predation. In *Batesian mimicry*, innocuous insects have evolved to resemble poisonous insects, thus avoiding predation. For example, the Viceroy butterfly resembles the poisonous Monarch butterfly, so predators avoid both. In Mullerian mimicry, two or more unrelated but similarly protected species resemble one another, thus achieving a kind of group defense. This is illustrated in the group of stinging insects that all resemble yellow jacket wasps. (see *aposematic*, compare *cryptic coloration*)

mineralocorticoid A type of *steroid hormone* (for example, *aldosterone*) produced by the *adrenal cortex* that control salt and water balance by their action on the kidneys.

minimal viable population The smallest number of individuals that have a reasonable chance of persisting as a population.

minisatellite DNA *Tandem repeat* groups of a DNA sequence that are of medium length, from about nine to thirty *nucleotides* long. They are found in the *genomes* of *eukaryotes* and serve no apparent function. Because the number of repeats in a given cluster is highly variable (*hypervariable*), they are exploited in *genetic typing* for personal identification. (see *satellite DNA*; compare *microsatellite DNA*)

minor groove The narrower of the two parallel spiraling grooves that follow the structure of the *DNA double helix*. (compare *major groove*)

Miocene An epoch of the Tertiary geological period, approximately 25 million to 27 million years ago, in which modern mammals began to evolve.

miracidium, miracidia (pl.) The *ciliated*, first stage *larva* of a *parasitic* fluke. They are contained within eggs and passed out of the primary *host* in the feces. If ingested by a snail, they transform into a *sporocyst* containing *embryonic germ cells*. (see *redia, cercaria*)

missense mutation A *mutation* in *DNA* that alters a *codon* so that it encodes a different *amino acid*. The complete protein is synthesized but may be nonfunctional. (see *translation, protein synthesis*)

mites (see *Acari*)

mitochondrial cytopathies Diseases caused by *mutations* in the *mitochondrial genome*. They are associated with a deficit of mitochondrial *ATP* production.

mitochondrial DNA (see *mtDNA*)

mitochondrial genome (see *mtDNA*)

mitochondrion, mitochondria (pl.) A *cytoplasmic organelle* found in all *eukaryotic* cells in which *aerobic respiration* and energy production take place. Almost all the *ATP* in nonphotosynthetic cells is produced in mitochondria. Each mitochondrion is surrounded by two *phospholipid bilayer* membranes, the inner of which forms projections into the interior matrix. The reactions of the *Krebs cycle* take place in the matrix. Those of the *electron transport chain* occur on the inner membrane, where the enzymes are highly organized. Mitochondria probably arose from nonsulfur purple bacteria that were captured by a pre-eukaryotic cell and then continued as *endosymbionts*. (see *respiration, mitochondrial DNA, serial endosymbiosis theory*)

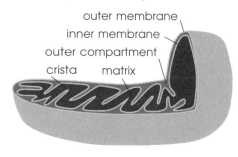

outer membrane
inner membrane
outer compartment
crista   matrix

Mitochondrion

mitogen Any compound that stimulates

the initiation of *mitotic cell division*. Certain compounds, called *lectins* extracted primarily from plant seeds, show this activity. (see *PHA, Con A*)

mitosis The ordered process by which a cell *nucleus* and *cytoplasm* divide into two identical progeny. It is separated into four phases: *prophase, metaphase, anaphase,* and *telophase. Interphase* is the period between *cell divisions* where the cell spends most of its time and during which *DNA replication* takes place. In prophase, the replicated *chromosomes* condense and become visible, the *nuclear membrane* disintegrates, and the *nucleolus* disappears. In metaphase, chromosomes line up on the equator of the *nuclear spindle*, a temporary structure made of *microtubules*. During anaphase, the *chromatids* (replicated chromosomes) split at their *centromere* and move to opposite poles of the spindle. This movement is accomplished by addition and subtraction of microtubular subunits such that one end grows while the other shortens. During telophase, the nuclear membrane re-forms around each of the newly replicated *genomes*. Most often, the *plasma membrane* pinches off, dividing the cytoplasm into two as well (*cytokinesis*). In plants, a *cell plate* forms, dividing the two daughter cells. (see *amitosis, endomitosis*; compare *binary fission, meiosis*)

mitospore A *spore* produced after *mitosis*. (compare *meiospore*)

mitotic cross-over A *cross-over* that results from pairing between two *homologous chromosomes* in a *diploid* cell during *mitosis*. It is far less frequent than *meiotic crossing-over* because, in contrast to meiosis, chromosomes do not systematically pair during mitosis.

mitotic spindle (see *spindle*)

mitral valve (bicuspid valve) In the heart, a valve situated between the left *atrium* and left *ventricle* in mammals and birds. When the ventricle contracts, blood is prevented from returning to the atrium by closure of the valve, thus preventing dilution of *oxygenated* blood with *deoxygenated* blood.

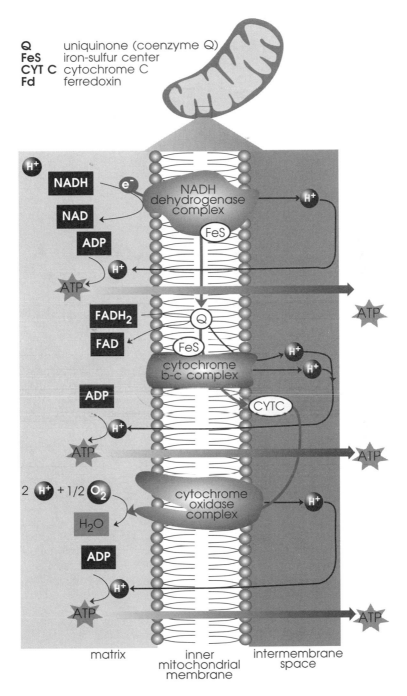

Electron Transport and Oxidative Phosphorylation in Mitochondria
(see *mitochondrion*)

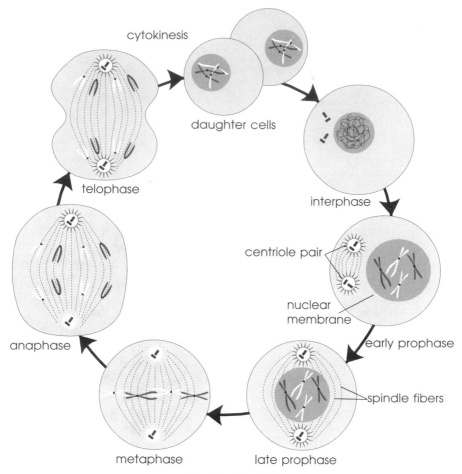

cytokinesis

daughter cells

telophase

interphase

centriole pair

nuclear membrane

early prophase

anaphase

spindle fibers

metaphase   late prophase

Mitosis in Animal Cells

Mn complex A motif essential to the initial splitting of water in *photosystem II*. It is a cluster of four manganese ions found at the center of the water-splitting enzyme, *protein Z*. (see *Z scheme*)

moas (see *Dinornithiformes*)

mobile genetic element (see *transposable genetic element*)

modifier gene Any *gene* that affects the *phenotypic* expression of another gene.

molar A large tooth, two or more of which are found at the back of jaws of mammals. They are used for crushing, chewing, and grinding food. In humans, the third molars found on each side of the

upper and lower jaws do not appear until later in life, and are therefore sometimes referred to as wisdom teeth.

molecular biology The marriage of biology and biochemistry. It is better defined as a set of tools and techniques, rather than an area of study, that allow investigation into the molecular processes of life.

molecular clock A tool for calculating the time and order of evolutionary divergence of different lines. It is based on differences in *amino acid* sequence that seem to accumulate at a relative rate in proteins. Information on evolutionary re-

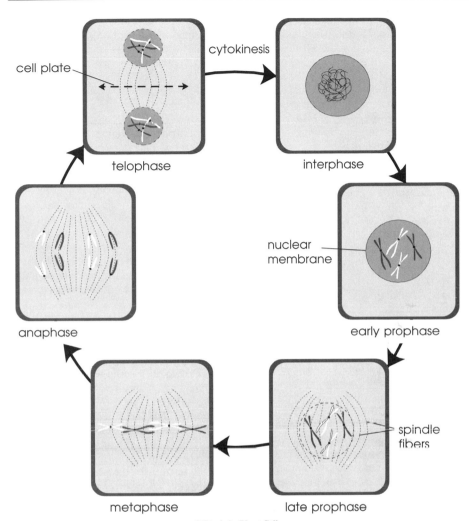

cell plate

cytokinesis

telophase

interphase

nuclear membrane

anaphase

early prophase

spindle fibers

metaphase

late prophase

Mitosis in Plant Cells

latedness obtained from *DNA analysis* continues to refine the evolutionary tree. (see *cladistics, phylogenetics, biochemical taxonomy, biosystematics, classification*)

molecular genetics The study of the molecular processes underlying *gene* structure and function.

molecular weight More stringently called molecular mass, it is the sum of the *atomic weights* of all the constituent atoms in a molecule.

molecular weight size marker Molecules of known *molecular mass* from which the size of an unknown molecule, usually *DNA* or protein, can be determined. They are often used as "marker ladders" in *gel electrophoresis*.

mole rat (see *Hystricognathi*)

moles (see *Insectivora*)

Mollusca A phylum of bilaterally symmetrical, unsegmented invertebrates. It includes the aquatic bivalves such as mussels and clams, terrestrial slugs and snails, and octopi and squids. Most but not all mollusks form a shell, either external or internal, secreted by the *mantle* in which most of the body organs are also housed. The head and a muscular foot are prominent and can often be retracted into the shell. Except for marine bivalves, a rasping, tonguelike *radula* is often present for feeding.

mollusks (see *Mollusca*)

molting (see *ecdysis*)

molting hormone (see *ecdysone*)

molybdenum (see trace *element*)

monad A single unit or single-celled organism.

monadelphous Flowers in which the *stamens* are united to form a tube around the *style*. (compare *diadelphous, polyadelphous*)

Monera An alternative name for the kingdom Bacteria.

mongolism (see *Down's syndrome*)

monkeys (see *Primates, Anthropoidea*)

mono- A word element derived from Greek meaning "one," "alone," or "single." (for example, *monomer*)

monocarpic Plants that flower once then die such as *annuals* and *biennials*. Monocarpic plants also include those such as bamboo and century plants that live for long periods of time at the end of which they produce massive amounts of flowers and seeds just before they die. (compare *polycarpic*)

monochasial cyme A *cymose inflorescence* (multiple-flowered structure) that consists of a single branch bearing flowers and ending in a single terminal flower. (compare *dichasial cyme*)

monocistronic mRNA An *mRNA* molecule that codes for one protein. This is the most common situation and the rule particularly in *eukaryotes*. (compare *polycistronic*)

monoclonal antibody The product of a laboratory construct in which an *antibody*-producing cell (*B-cell*) is fused with an immortal *cancer* cell, resulting in a hybrid (*hybridoma*) exhibiting particular

properties of both. The hybrid *clonal* population can then be propagated indefinitely *in vitro* for scientific research. The initial clone is selected to produce large amounts of only one specific antibody type, which may be used to identify and isolate a specific *antigen* within a mixture.

monocolpate pollen *Pollen grains* in which one furrow or pore is characteristic. They are found particularly in *monocotyledons* and primitive *dicotyledons*.

monocot (see *monocotyledons*)

Monocotyledons A class of *angiosperms* (flowering plants) characterized by a single *cotyledon* (first leaf) in the seed. They are usually *herbaceous* plants and bear flower parts in multiples of threes. Leaf veins are parallel, *vascular* bundles are scattered, and true secondary growth is rare. (compare *Dicotyledons*)

monoculture The exclusive cultivation of a single crop over wide areas. It is an efficient way to use certain soils, but ultimately reduces soil fertility It carries the risk of an entire crop being destroyed by a single pest or disease. (compare *crop rotation*)

monocyte The largest type of white blood cell (*leukocyte*). They convert into *macrophages* at a site of *inflammation* (often arising from injury), and ingest and destroy bacteria and other foreign particles. Monocytes are *granular* and arise from *stem cells* in the *bone marrow*. (see *hemopoiesis*)

monoecious Plants in which male and female reproductive organs are borne on separate flowers but on the same individual. (compare *dioecious, hermaphrodite*)

Monogenea A minor class of flukes, the monogeneans, sometimes classed with the trematode flukes. They are *ectoparasitic* on fish.

monohybrid A genetic term for an organism *heterozygous* (different *gene* variants) at one *locus* (*chromosomal* location) and obtained by a cross between parents each *homozygous* (same *gene* variants) for the different two *alleles* (gene variants) at this locus. This can be accomplished only in a *diploid* organism and is *phenotypically* obvious only when the genetic alleles exhibit clear *dominant* and *recessive* characteristics. For example in Mendel's cross between tall (TT) and dwarf (tt) peas, all offspring can have only a (Tt) genotype and show a *phenotype* of tall. A characteristic ratio of phenotypes (3:1) is obtained and the recessive phenotype revealed (see below) when a monohybrid is *self-crossed*, assuming a simple dominant/recessive relationship between the alleles. (compare *dihybrid*)

MONOHYBRID CROSS

	T	t
T	TT	Tt
t	Tt	tt

3	tall	TT, Tt (2)
1	dwarf	tt

monokaryon A cell containing a single *haploid nucleus*. In particular, it refers to *ascomycete* and *basidiomycete* fungal *mycelia* at an intermediate stage between the haploid and *diploid* phases of the *life cycle*. (see *homokaryon*; compare *dikaryon*)

monokines Soluble proteins produced by *macrophages* that signal the onset of the *immune response*. In particular, *γ-interferon* activates other *monocytes* to mature into macrophages, and *interleukin-1* activates *helper T-cells* to initiate the *cell-mediated immune response*. (see *immune system*; compare *lymphokines*)

monomer An individual unit of an entity that can associate to form multiunit structures (*polymers*). For instance, monomers of some *globular* proteins, such as *actin* and *tubulin*, polymerize to form long filamentous structures. (see *monosaccharide*; compare *polymer*)

monomorph In particular reference to *DNA*, a genetic *allele* that is constant in all or most individuals in a species. A particular monomorph is sometimes used as a diagnostic standard to check for variation caused by analytic factors when comparing alleles in *genetic typing*. (see *DNA analysis, RFLP, Southern blot*)

monopodial (indefinite growth, racemose branching). In plants, a type of branching in which lateral branches arise from a definite main central stem that continues to grow indefinitely. It is typical of the formation of a *racemose inflorescence*. (compare *sympodial*)

monosaccharide The simplest sugar unit. *Glucose* and *fructose* are examples. (compare *disaccharide, oligosaccharide, polysaccharide*)

monosomic A *diploid* organism in which one *chromosome* of a *homologous* pair lacks a partner. If the organism is viable, it is likely to manifest severe defects partly because any deleterious *mutations* on the remaining chromosome are obligately uncovered.

Monotremata An order of mammals, the monotremes, containing the only mammals that lay eggs. After hatching, the young are transferred to a pouch (*marsupium*) on the abdomen and nourished by milk secreted by primitive *mammary glands* whose ducts do not form nipples. Other primitive features include poor temperature control and possession of a *cloaca*. Monotremes comprise only the duck-billed platypus and the spiny anteaters.

monotremes (see *Monotremata*)

monotrichous Bacteria that posses only one *flagellum*, for example *Vibrio*.

monoxenous (see *autoecious*)

monozygotic twins (fraternal twins) Twins that have arisen from the splitting of one *zygote* into two separate entities early in development. They are genetically identical and always of the same sex. (compare *dizygotic twins*)

Mormon tea plant (see *Gnetophyta*)

morpho-, -morph, -morphic, -morphous A word element derived from Greek meaning "form." (for example, *monomorph*)

morphogenesis The development of the

form and structure of an organism. (see *morphology*)

morphological defenses One of the ways in which plants limit the activities of *herbivores* and insects. They include thorns, spines, prickles, and sticky plant hairs.

morphology The study of the form and structure of organisms and their parts. (compare *anatomy, physiology*)

morula A solid ball of cells (*blastomeres*) resulting from *cleavage* of the egg after *fertilization* prior to the *blastula* stage. Human *embryos* implant in the uterus at this stage.

mosaic An organism whose cells differ genetically although they have arisen from a single *zygote*. This may be caused by natural or artificial mutation during the early stages of development. At this time, one or a few cells that have acquired a *DNA mutation* will divide and grow to produce a patch or *clone* of cells different from the rest of the organism.

mosquitoes (see *Diptera*)

moss (see *Bryophyta*)

moths (see *Lepidoptera*)

motor In general, the term refers to *neurons, muscles* and *cells* involved in effecting a physiological response, often movement. Signals are transmitted via *efferent* neurons from the *central nervous system* or central *ganglion* of an organism. (see *motor cell, motor neuron, motor unit;* compare *sensory neuron, receptor*)

motor cells A pair of specialized cells in plants that mediate leaf movement. They flank a jointed structure at the base of the leaf, the *pulvinus*, and apply differential pressure depending on the *turgor pressure* (water intake) in each cell.

motor endplate (see *neuromuscular junction*)

motor neuron A *nerve cell* that transmits impulses from the brain or *spinal cord* to a muscle or intermediate *effector*. Motor neurons always carry impulses away from the *central nervous system* (*CNS*), thus are *efferent*. (see *voluntary, autonomic*; compare *sensory neuron*)

motor unit The set of *muscle fibers* inner-

vated by all branches of the *axon* of a single *motor neuron*. (see *neuromuscular junction, recruitment*)

moundbuilders (see *Galliformes*)

mousebirds (see *Coliiformes*)

mouthparts Various paired appendages on the heads of arthropods, modified in various ways to harvest food and gather *sensory* information. (*see chelicera, labium, labrum, mandible, maxilla, pedipalps, proboscis*)

mRNA (messenger RNA) *RNA transcribed* from *genes* that are to be *translated* into protein. In *eukaryotes*, mRNA is the final processed product, has had all *introns* removed, and is ready to be *translated* by *ribosomes* (see *hnRNA, exon*)

MSH (melanocyte-stimulating hormone) In reptiles and amphibians, a *pituitary hormone* that stimulates color changes in the *epidermis*. At present, it has no known function in mammals.

mtDNA (mitochondrial DNA) The *extranuclear DNA* found in the *mitochondria* of all *eukaryotes*. It is a circular *chromosome* and utilizes a *genetic code* that contains *prokaryotic codon* usage, supporting the probable origins of mitochondria as *endosymbionts*. As such, mitochondrial DNA is self-replicating. Since mitochondria are lost from the *spermatozoa* head that *fertilizes* an egg, mitochondrial DNA is inherited in a maternal lineage in animals. This trait has been exploited by researchers studying DNA for the purpose of tracing the evolution of man. Mitochondrial DNA is also present in hundreds to thousands of copies, making it ideal for the analysis of ancient remains such as Egyptian mummies. It also has a relatively high *mutation rate*, making it highly *polymorphic* (variable), thus useful in individual identification. (see *DNA analysis, maternal inheritance, uniparental inheritance*; compare *nuclear DNA, cpDNA*)

MTOC (microtubule organizing center) Any of the sites on the periphery of *centrioles* (*centrosomes*) or at the poles of *mitotic spindles* that function in catalyzing the growth of *microtubules*. Microtubules will *polymerize* spontaneously but at a much slower rate.

m.u. (see map unit)

mucilage A viscous fluid secreted by some plants and animals containing complex carbohydrates that adsorb and retain water.

mucilaginous Any structure or function relating to the production of *mucus*.

mucin (mucoproteins) *Glycoproteins* that are the main constituent of *mucus*. (see *glycosaminoglycan, proteoglycan*)

mucopeptides A group of complex *polypeptides* that are used to cross-link the main carbohydrate matrix of bacterial *cell walls*. They are not found in the cell walls of *eukaryotes*. (see *glycosaminoglycans*)

mucopolysaccharide (see *glycosaminoglycans*)

mucoproteins (see *mucin*)

mucormycosis An infection of humans by the zygomycete fungus *Mucor*. It is common in individuals with compromised *immune systems,* such as with leukemia or *AIDS*.

mucosa (see *mucous membrane*)

mucous membrane The *epithelial* tissue that lines many tracts in animals (for example, intestinal, respiratory) that open to the exterior. It contains the oftencilated *goblet cells*, specialized for secreting *mucus* that helps keep the tissue moist.

mucus (mucous) A viscous substance consisting mainly of *glycoproteins* (*mucins*) that lubricates and protects the surface on which it is secreted. It is produced by *goblet* cells in the *mucous membranes* of animals, as well as in many other situations.

mud puppies (see *Caudata*)

Mullarian duct The *oviduct* of vertebrates (excluding *cartilaginous* fish). It collects the egg (*ovum*) from the *ovary* and passes it to a point where it can be *fertilized* or in internal fertilization, provides a channel for *spermatozoa*. In *placental* and *marsupial* mammals including humans it is dif-

ferentiated into the *Fallopian tubes,* uterus, and vagina.

multiallelic The existence of more than two *alleles* (genetic variants) of a *gene* in a population. Under normal conditions, only one (*homozygous*) or two (*heterozygous*) variants are present in a single *genome*. An example is the ABO *blood group*, where three different alleles are present in the population. (compare *biallelic*)

multicellular Any organism consisting of more than one cell. *Multicellularity* is an evolutionary advancement that enables specialization and division of labor among the cells. (compare *unicellular*)

multifactorial inheritance (see *polygenic*)

multilocus probe A *DNA probe* that simultaneously detects genetic variation at multiple sites in the *genome* of an organism. An *autoradiograph* of a multilocus probe yields a complex, stripelike pattern originally miscalled a "DNA fingerprint." Probes that detect one locus at a time (*single locus probe*) are now more commonly used in *genetic typing*. (see *DNA analysis, Southern blot, RFLP, DNA profile*)

multinucleate A cell containing more than one *nucleus*. In organisms such as slime molds, this is a normal condition. It is also found, for instance, in vertebrate *striated muscle*. (see *syncytium, coenocyte, acellular*)

multiple hit hypothesis The proposal that a single cell must receive a series of *mutational* events in order to become *cancerous*.

murres (see *Charadriiformes*)

muscle Tissue consisting of bundles of elongated cells (*muscle fibers*) containing contractile fibrils. Muscles function to produce movement, both within and of the body. Body movement is produced by muscular force on various rigid body elements, the skeletal system in vertebrates, and the hydroskeleton in soft-bodied invertebrates. (see *cardiac muscle, skeletal muscle, smooth muscle, actin, myosin, actomyosin, troponin, tropomyosin*)

muscle cell A specialized cell containing numerous filaments of the intercalated *polymerized* proteins *actin* and *myosin* (*myofibrils*). The arrangement and the various accessory proteins present varies with the three main types. In *striated (skeletal) muscle*, the fibers are formed by the fusion of several cells end-to-end to form a long fiber with many *nuclei*. *Smooth muscle* fibers are long and spindle shaped, with a single nucleus. *Cardiac muscle* fibers, those found in the heart, also contain a single nucleus but are organized into a lattice of long branching chains that interconnect. (see *multinucleate, syncytium, troponin, tropomyosin, α-actinin, calmodulin, muscle contraction*)

muscle contraction.

1. (striated). Contraction is initiated by a *nerve impulse* arriving at a *neuromuscular junction* where the *neurotransmitter, acetylcholine*, passes across the muscle membrane and causes calcium *ion channels* to open. Calcium ions (Ca^{++}) that have been sequestered in resting muscle in the *sarcoplasmic reticulum* by an *ATP*-driven *calcium pump* are released into the *sarcoplasm*. This initiates contraction of the *myofibrils* by binding to *troponin* molecules. These, in turn, displace the *tropomyosin* filaments that cover the *myosin* binding sites on *thin filaments* in a muscle at rest. The myosin heads are now free to form crossbridges with opposing actin filaments that are anchored at *Z-lines*. With *ATP* expenditure, the myosin heads walk along the actin molecule, causing further overlap of the two different filaments and contraction of each *sarcomere*, hence the muscle as a whole.

2. (smooth). Contraction is often initiated by a *neurotransmitter* released from an *autonomic nerve*. Smooth muscle lacks sarcoplasmic reticulum, so calcium ions enter from the *extracellular* space to combine with the protein *calmodulin*. The Ca^{++}/calmodulin complex initiates a series of reactions, finally exposing the actin binding site on myosin heads. Thus, smooth muscle contraction differs fundamentally from that in striated muscle, where initiation involves the *thin filaments*.

3. (cardiac). The molecular mechanism of force generation is the same as in skeletal muscle because cardiac muscle is essentially striated. However, cardiac muscle cells are all electrically linked in a network, enabling them to contract as a unit. (see *pacemaker, AV node, bundle of His*)

muscle fatigue The decrease in the ability of a muscle to generate force as use con-

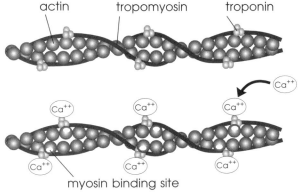

actin   tropomyosin   troponin

myosin binding site

**Initiation of Muscle Contraction
(Striated Muscle)**

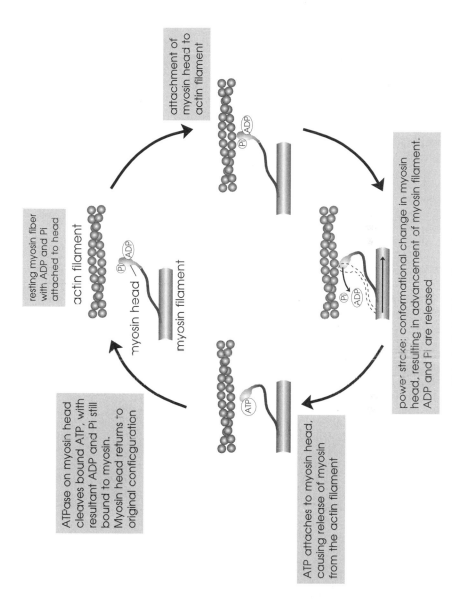

resting myosin fiber with ADP and Pi attached to head

actin filament

myosin head

myosin filament

attachment of myosin head to actin filament

power stroke: conformational change in myosin head, resulting in advancement of myosin filament. ADP and Pi are released

ATP attaches to myosin head, causing release of myosin from the actin filament

ATPase on myosin head cleaves bound ATP, with resultant ADP and Pi still bound to myosin. Myosin head returns to original configuration

Molecular Basis of Muscle Contraction

tinues. Initially, it may occur due to *lactic acid* buildup when the muscle continues to work under *anaerobic* conditions. The acidity inhibits the enzymes used in energy production and also interferes with the mechanism of *muscle contraction* itself. A more extreme situation occurs when *glycogen* stores in the muscle and liver are depleted and the body must burn fat as a sole energy source, with much reduced efficiency. Runners refer to this as "hitting the wall."

muscle fiber (see *muscle cell*)

muscle spindle A specialized muscle cell modified as a stretch *receptor* found in the muscles of all vertebrates except for bony fish. The end of a *sensory* (*afferent*) nerve is wrapped spirally around it and reports the state of muscle tension to the brain. They are involved in both involuntary movement (reflex, posture) and control of voluntary movement. (see *proprioreceptor*)

Muscophagiformes The order of birds containing the turacos and plantaineaters.

muscular system The system of *striated (skeletal) muscle, smooth muscle,* and *cardiac muscle* that produces movement within and of the body. The muscular system is directed by the *nervous system.*

muskox (see *Artiodactyla*)

mussels (see *Bivalvia*)

mutagen Any physical or chemical agent that can effect changes in *DNA*. These include chemical alteration, formation of *pyrimidine dimers, deletion, insertion, translocation, base substitution,* and others. Chemical agents include tobacco smoke and asbestos, many common laboratory substances such as *ethidium bromide,* and various industrial *pollutants.* Physical mutagens include *ultraviolet radiation* and *ionizing radiation* (X-rays and gamma-rays). All *carcinogens* (cancer-causing agents) are mutagenic, but not all mutagens are *carcinogenic.* (see *mutation*; compare *teratogen*)

mutant An organism or cell carrying a change from what is considered normal, or *wild type,* in its *genome.*

mutant allele A genetic variant at a particular *locus* in the *genome* that is different from the standard or *wild type.*

mutant hunt An accumulation of *mutations* in a particular *gene* or *chromosome* region for further study. By systematically perturbing each part of the system, the normal functions may be elucidated. (see *genetic dissection, supermutagen, saturation mutagenesis*)

mutation Any change in the *DNA* of an organism. Mutations in the *somatic cells* (non-*gamete*-producing cells) of an adult can give rise to *cancers* and other abnormalities. Only changes that affect the *germline* cells (sperm and eggs) are inherited in the next generation. Most mutations are deleterious but are often *recessive* and retained in the population in *heterozygous* form. Beneficial mutations may increase in the population due to *natural selection.* The spontaneous *mutation rate* is quite low (on the order of one in a million (10^{-6}) to one in a trillion (10^{-9}) but may be increased by various *mutagens.* Three major sources of mutational damage to DNA are high-energy radiation such as X-rays that cause physical breakage, low-energy radiation such as UV light that creates DNA cross-links, and chemical modification of DNA bases. The latter two ultimately lead to replication errors. (see *chromosome mutation, gene mutation, pyrimidine dimers, deletion, insertion, translocation, base substitution, double strand break, DNA repair*)

mutation analysis The study of *mutations* in various regions of a particular *gene* or group of genes in order to understand their normal functions. (see *genetic dissection, supermutagen, saturation mutagenesis*)

mutation frequency The proportion of *DNA mutations* occurring in individuals in a population. The frequency of a particular mutant may be of interest. (compare *mutation rate*)

mutation rate The number of *mutational* events per *gene* per unit time (for example, cell generation). Spontaneous mutation rates are on the order of one in a million (10^{-6}) to one in a trillion (10^{-9}) but may be increased by various *mutagens*. (compare *mutation frequency*)

mutualism The *symbiotic* relationship between two or more species in which mutual benefit is obtained. In obligatory mutualism, neither can survive without the other. This is exemplified in the algal/fungal partnership found in *lichens*. The term is sometimes extended to cover facultative mutualism (*protocooperation*) in which both species benefit but the relationship is not obligatory for survival. (compare *parasitism, commensalism*)

myc-, myco- A word element derived from Greek meaning "fungus." (for example, *mycology*)

mycelium, mycelia (pl.) The mass of filamentous *hyphae* forming the main vegetative body of fungi and some protoctists. In fungi, it is often hidden underground or permeating the substrate on which it lives. The mycelium produces the reproductive organs, which may sometimes be seen as *fruiting bodies* such as toadstools or mushrooms aboveground.

mycology The study of fungi.

Mycophycophyta (see *lichen*)

mycorrhiza, mycorrhizae (pl.) The *symbiotic* association between certain fungi and the roots of most plants from which both benefit. The fungus obtains organic nutrients from the plant and, in turn, increases the availability of inorganic nutrients, such as phosphorous, to the plant. *Ectotrophic mycorrhizae* are external to the plant root cells, enveloping the smaller roots. An example of this relationship is found in pine trees. *Endotrophic mycorrhizae* actually invade the *cortex* cells of the root. Some orchids, in particular, are completely dependent on endotrophic mycorrhizal relationship for survival. (see *mutualism*; compare *actinorrhizae*)

myelinated nerve fiber (medullated nerve fiber) A nerve *axon* surrounded by an insulating *myelin sheath*. In between each *Schwann cell* comprising the myelin

sheath is a short region of bare axon (*node of Ranvier*). In myelinated nerve fibers, the *action potential* literally jumps from one node of Ranvier to the next. This greatly increases conduction velocity over nonmyelinated fibers, where each small, adjacent area must be excited in order to propagate a *nerve impulse*. This mode of conduction is called *saltatory* (jumping). Myelinated nerve fibers are found in most, but not all vertebrate *neurons*. In nonmyelinated nerve cells, an equivalent increase in velocity can only be accomplished by an increase in axon diameter, as seen in the *giant fibers* of some invertebrates. (see *peripheral nervous system*)

myelin sheath An insulating covering that surrounds the *axons* of most but not all vertebrate *neurons*, shielding the electrical *nerve impulse* from leakage into the surrounding area. It is composed of the *cell membranes* of *Schwann cells* that are made of a fatty material, *myelin*, wound spirally around the axon. (see *myelinated nerve fiber, saltatory conduction, node of Ranvier*)

myeloid tissue A major site of white blood cell (*leukocyte*) development (*hemopoiesis*) in vertebrates. It is distinct from *lymphoid tissue* and in adults is found only in the *bone marrow*. In the *embryo*, myeloid tissue is also found in the liver and spleen. Leukocytes have a very short life span and are continually replaced from myeloid tissue. (compare *lymphoid tissue*)

myoblast A precursor cell of vertebrate *striated muscle fibers*. Myoblasts eventually fuse to form characteristic *multinucleate syncytia*.

myofibril The basic contractile unit in *striated muscle*. It is made of *myofilaments* organized into *sarcomeres*.

myofilament (actomyosin) A contractile filament composed of the proteins *actin* and *myosin*. It is found in various types of *muscle fibers*. (see *thin filament, thick filament, sarcomere, striated muscle, myofibril*)

myoglobin A hemoglobinlike protein complex found in the muscle tissue of vertebrates and some invertebrates. In contrast to hemoglobin, it has only one *polypeptide* chain associated with the *heme* group but performs essentially the same function of reversibly binding oxygen. It has a higher oxygen affinity (binds it more tightly) than hemoglobin. It releases oxygen only when the blood oxygen supply becomes limiting, thus serving as an extra oxygen store in the muscles.

myosin The predominant *structural protein* found in muscle and other contractile structures. *Globular* subunits of myosin *polymerize* to form *thick filaments* that form *myofilaments* (*actomyosin*) with the *thin filaments* of the protein *actin.* Myofilaments are the basic contractile unit of vertebrate *striated muscle.* Myosin is found in *eukaryotes* but not *prokaryotes* and is involved in organismal, cellular, and *intracellular* movement.

myotome The part of each *somite* of a vertebrate *embryo* that differentiates as a muscle block. It is of *mesodermal* origin.

Mystacocarida A minor class of crustaceans. They are *filter feeders* and live between the sand grains in the ocean. (see *meiofauna*)

mystacocarids (see *Mystacocarida*)

Mysticeti A suborder of whales comprising the baleen whales, including the blue and humpback whales. They feed on *krill*, a type of *plankton*, filtered from the sea by *baleen* plates in their mouth.

Myxobacteria (see *Saprospirae*)

Myxomycota The phylum of protoctists containing the plasmodial slime molds. They are *multinucleate*, lack *cell walls,* and exhibit conspicuous *cytoplasmic streaming.* They move as a *plasmodium*, engulfing and digesting bacteria, yeasts, and other small particles of organic matter as they move. The mass is able to flow through a fine mesh and rejoin on the other side. *Asexual reproduction* is accomplished by a *sporophore* that contains either a *sporangium* or *sporocarp.* *Diploid spores* produce either *haploid* mobile swarmer cells or haploid *ameboid* cells. These Myxamebas can fuse to produce a *zygote* that grows only by *nuclear division*, and bypasses *cytoplasmic division* to reform the multinucleate plasmodium. They are distinct from the cellular slime molds (*Acrasiomycota*).

Myxospora (Myxozoa, Myxosporidia) A phylum of *heterotrophic protoctists*, the fish *parasites*, once thought to belong to a larger grouping, the sporozoans. Unlike other former sporozoans that are single-celled protists (apicomplexans, microsporans, and paramyxeans), myxozoans are *multicellular.* Like the others, they are *parasitic* and attach themselves to their *host's* intestine with sticky filaments. In the gut, the *spore*like *propagules* burst, releasing the two *haploid* nuclei that eventually fuse in *sexual fertilization.* A *multinucleate plasmodium* then develops that contains both *haploid* nuclei and *diploid generative nuclei.*

myxovirus A group of *RNA*-containing viruses that cause diseases such as influenza, mumps, measles, and rabies.

NAD (nicotinamide adenine dinucleotide) An important *coenzyme* in electron transfer reactions, in particular in oxidation reactions of *aerobic respiration*. Its reduced form is *NADH*. NAD is derived from the vitamin *nicotinic acid* (niacin). (see *Krebs cycle, electron transport chain, mitochondrion*)

NADH The reduced form of the *coenzyme NAD*.

NADP (nicotinamide adenine dinucleotide phosphate) The phosphorylated form of the *coenzyme NAD*. It also functions in electron transport. In particular, it is reduced to *NADPH* in the *light reactions* of *photosynthesis*. It transfers its reducing power to the *Calvin cycle* (dark reactions) where it provides the energy for *carbon fixation.*

NADPH The reduced form of the *coenzyme NADP.*

nano- (n) A word element denoting one-billionth or 10^{-9}. For example, 1 nanogram (ng) = 10^{-9} gram. The nanogram is a common unit of measure for small amounts of *DNA.*

nares (see *nostrils*)

nasal cavity A paired cavity in the vertebrate head containing *olfactory* (smell) nerve endings. In mammals, the large area of *mucous membrane* warms and moistens air on its way to the lungs, and hairs near the nostrils filter dust and bacteria from the air.

-nastic, -nasty A word element derived from Greek meaning "pressed close" and indicating asymmetric cell growth in plants. (see *nastic movements*)

nastic movements The response of a plant part to a nondirectional stimulus such as light, temperature, or touch. Movement may be caused by discrete areas of cell growth, a relatively slow process, or very quickly by *osmotic* changes. For example in crocus and tulip flowers, opening and closing is controlled by disproportionate growth on either the upper (*epinastic*) or lower (*hyponastic*) surface of the plant

structure. Plants such as the Venus flytrap react quickly to touch (*haponstic*) by a loss of *turgor* in special cell groups (*pulvini*), causing the closure of the two halves of the leaf. (compare *kinesis, taxis, tropism*)

natural immunological tolerance The recognition of an organism's own tissues as "self" so they are not attacked by the *immune system.* (see *MHC complex, MHC proteins, autoimmunity*)

natural killer cell (see *NK-cell*)

natural selection The process by which organisms well adapted to their environment survive to produce offspring while those more poorly adapted may not. In this way, particular combinations of *genes* are propagated differentially. According to *Darwinism*, natural selection acting on a varied population results in *evolution.*

nature vs. nurture The colloquial expression for the two forces that affect any particular characteristic or set of characteristics of an organism. They are respectively synonymous for heredity and environment, and are the subject for much debate in areas such as intelligence. Only the heritable contribution will be passed on to future generations. (see *Lamarkism, instinct, learning*)

nauplius A free-swimming *larval* stage of many crustaceans. It hatches with only three body segments, approximating the eventual head region of the adult. Successive segments are added through several *metamorphoses* as it develops into the adult.

nautilus (see *Cephalopoda*)

Neanderthal man A subspecies of man, *Homo sapiens neanderthalensis*, that was replaced by modern man about 40,000 years ago. Neanderthals were dominant in Western Europe during the first stages of the last glaciation, dwelt in caves, and used fire and tools. (see *Homo*)

necrotroph (see *saprophage*)

nectar The fluid secreted by flowers, par-

ticularly those that are insect *pollinated.*
It contains sugars and other nutrients designed to attract insects. (see *nectary*)

nectary In plants, the *gland* in flowers that secretes *nectar.*

nectotrophic A form of nutrition in which a *parasitic symbiont* damages or kills the organism where it resides.

negative assortative mating (see *assortative mating*)

negative control Regulation of *RNA transcription* that is mediated by proteins that block or turn off expression of a particular *gene.* (see *lac operon*; compare *positive control*)

negative feedback One way of maintaining the status quo or *homeostasis.* It occurs at many biological levels including within organisms, within *ecosystems,* and globally. It is a type of *feedback regulation* in which a perturbation triggers processes that reduce or reverse the effects of a change. (compare *positive feedback*)

negative staining A staining method employed in microscopy in which the background is stained, separating it from the unstained specimen. It is particularly useful for studying three-dimensional and surface features.

nekton Free-swimming organisms inhabiting the *pelagic* zone (open waters) of a sea or lake, including fish, whales, and some mollusks. (compare *plankton*)

nematocyst A stinging structure found only in the *cnidocyte* cells of cnidarians. Within each cell is a coiled tube lined with barbs. Upon stimulation of its *cnidocil,* the tube is violently discharged, turning it inside out and exposing the prey or predator to the stinging barbs that may also inject *toxins.* The neurotoxins (affecting the *nervous system*) of cnidarians such as the Portuguese man-of-war can be fatal to humans. Some nudibranchs (sea slugs) can eat cnidarians, retaining and incorporating the nematocycsts for their own defense.

Nematoda A large phylum of marine, freshwater, and terrestrial invertebrates,

the roundworms. They have three body layers but are *pseudocoelomate.* Many are free living, but large numbers are *parasites* and cause diseases such as elephantiasis and trichinosis. Nematodes are unsegmented and lack a true *coelom* and any major body systems. They have unusual *embryonic* development and are not closely related to any other phylum. Reproduction is *sexual.*

Nematomorpha A phylum of *pseudocoelomate* wiry worms called either horsehair or Gordian worms after the complicated knots King Gordium of Phrygia was known to tie. They are only about a millimeter in diameter but can be as long as a meter. The *larvae* are *parasitic* on various species of terrestrial arthropods, and their eggs are released either in feces or from the dead *host.* The adult worms hatch and live in shallow oceans and freshwater lakes.

Nematomorphs (see *nematomorpha*)

Nemertina (see *Rhynchocoela*)

neo- A word element derived from Greek meaning "new" or "recent." (for example, *neomorph*)

neo-Darwinism Darwin's theory of evolution and *natural selection* as modified by the knowledge obtained through modern genetics. (see *Darwinism*)

Neognathae A superorder of modern birds containing all the flying and swimming birds. They are anatomically distinguished from the *Paleognathae* by palate structure.

Neolithic The recent Stone Age dating from about 10,000 years ago until the beginning of the Bronze age. It is characterized by a refinement of stone tools and the development of agriculture.

neomorph A *mutant* in which a novel substance or structure is found.

Neornithes A subclass of avians comprising all living birds.

neotenic The retention of ancestral *larval* or other juvenile features in an otherwise adult animal. This may include the attainment of sexual maturity in an apparently juvenile animal such as the axolotl. It is an important force in the evolution of new species. (see *heterochrony, paedomorphosis, paedogenesis*)

nephridium, nephridia (pl.) A primitive excretory organ present in many aquatic invertebrates including flatworms, rotifers, mollusks, segmented worms, and Amphioxus. Typically, one pair of tubular nephridia occur per body segment, and excretory products diffuse into them for conduction to the exterior. In *protonephridia, flame cells* collect waste from blind tubules and funnel it toward the exterior. In *metanephridia*, the tube opens directly to the *coelom*. (see *nephrostome*; compare *nephron*)

nephro-, nephri- A word element derived from Greek meaning "kidney." (see *nephron*)

nephron The excretory unit of the vertebrate kidney composed of a *Malpighian body* (*glomerulus* and *Bowman's capsule*) and a *uriniferous tubule* (*nephron tubule*). Water, salts, and nitrogenous wastes (*urea, uric acid*) are filtered across the walls of the glomerulus and collected by the Bowman's capsule. As the filtrate passes through the *uriniferous tubule* (in mammals, the *renal tubule*), water and *electrolytes* are selectively reabsorbed into surrounding *capillaries* that join a *renal vein*. In mammals, the remaining waste (*urine*) passes to the ureter for excretion to the outside. (compare *nephridium*)

nephrostome The collecting funnel of a *nephridium*, the *excretory system* of many invertebrates. From the nephrostome, a coiled tubule runs into an enlarged bladder that, in turn is connected to an excretory pore.

neritic The marine environment along the coasts, from low tide level to a depth of about 300 meters, corresponding approximately to the extent of the continental shelf. It makes up about 1% of the marine environment. Because light penetrates at least the shallow water, nutrients are relatively abundant. This part of the ocean is most densely populated by *benthic* organisms that live on or in the sea bed. Part of this zone, the intertidal or *littoral* zone, is exposed to the air whenever the tides recede. The part that remains submerged is the *sublittoral*. (compare *abyssal, oceanic, photic zone*)

nerve A bundle of *nerve fibers* (*axons*) with their accompanying supporting cells (*neuroglia, Schwann cells*) bound by *connective tissue*. Nerves are located in the *peripheral nervous system* and may contain only *sensory* or *motor* fibers, or both (mixed nerves). (compare *nerve tract*)

nerve cell (see *neuron*)

nerve cell body The portion of a *neuron* containing the cell *nucleus* from which the *axon* and *dendrites* extend. *Gray matter* in the *spinal cord* and brain is made of nerve cell bodies.

nerve cord An enclosed cylindrical tract of *nerve fibers* that forms a central route for conduction of *nerve impulses* within the body. Chordates have a single hollow nerve chord (the spinal cord). Invertebrates generally have two or more solid nerve cords with *ganglia* situated at intervals along its length. It is an integration center for both *sensory* and *motor* information.

nerve fiber The *axon* of a *neuron* along with its *myelin sheath*, if present. (see *giant fiber*)

nerve impulse The transient, self-propagating electrical signal transmitted along *neurons*. All nerve impulses are identical in form and strength. They are propagated by changes in the permeability of the *axon* membrane followed by the flow of ions across it. This produces a reversal in electrical potential that passes along the axon as the *action potential*. Since each impulse is an *all-or-none* event, the strength of the stimulus is signaled by the frequency and number of identical impulses. *Stimulus-gated ion channels* are responsible for propagating the nerve impulse from one nerve to the next. (see *node of Ranvier, myelin sheath, saltatory conduction*)

nerve net A netlike interconnection of *neurons* that is found in the body wall of certain invertebrates such as cnidarians and echinoderms. It is the most primitive type of *nervous system* and conducts impulses slowly away from the point of stimulation in all directions.

nerve tract A bundles of *nerve fibers* in the *central nervous system.* (compare *nerve*)

nervous system The organized communication system in animals that distributes both *sensory* and *motor nerve impulses.* It is found most fundamentally as a *nerve net* in the simplest animals. It has developed into nerve integration centers (*ganglia*) in lower invertebrates and into a single control center, the brain, that is highly enlarged in vertebrates (craniates). (see *autonomic nervous system, central nervous system, peripheral nervous system, spinal cord*)

net primary production (see *primary production*)

neur-, neuro- A word element derived from Greek meaning "nerve." (for example, *neurology*)

neural arch In vertebrates, an arch of bone or cartilage that arises from each vertebra, collectively forming a protective channel through which the spinal cord runs.

neural crest A strip of cells at the edge of the *neural plate* in vertebrate *embryos* that migrates to various locations and ultimately dictates their developmental fate (*induction*). In their migration, they actually pass through other tissues. Specific structures derived from the neural crest include the *gill arches* as well as the elements of the *nervous system* and its adjunct cells and organs. The neural crest has played a key role in vertebrate evolutionary development. (see *differentiation, neurula*)

neural groove In the development of chordate *embryos*, the first invagination of *ectodermal* cells over the *notochord* to begin formation of the *dorsal nerve cord.*

neural plate The area on the outer layer of vertebrate *embryos* that gives rise to the *neural tube* as determined by the *notochord* tissue, which lies beneath it. Some of the cells from the neural tube eventually differentiate into the *neural crest.* (see *differentiation, neurula*)

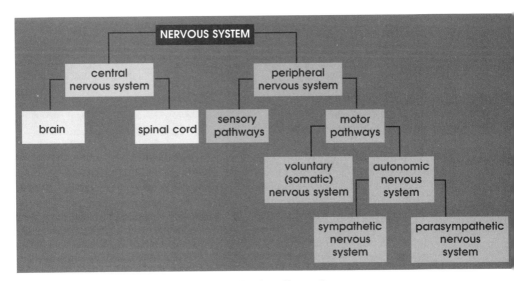

Organization of the Vertebrate Nervous System

neural tube In vertebrate *embryos*, the first formed element of the spinal cord and brain. It is usually formed by the *neural plate* closing to form a hollow tube that comes to lie beneath the surface. Some of the cells from the neural tube eventually differentiate into the *neural crest* that subsequently gives rise to the key elements of the vertebrate *nervous system*. (see *differentiation, neurula*)

neuraminidase A surface enzyme found on many viruses that allows replicated virus particles to break free of the *host* cell.

neuroblast An *embryonic* cell slated for development into nervous tissue, in particular *nerve cells*.

neuroendocrine system Any of the vertebrate body systems integrating both nervous and *endocrine* factors. For instance in the *pituitary gland, hormones* are secreted under direct nervous stimulation. (see *neurohormone, nervous system, endocrine system*)

neuroglia In the *central nervous system*, the various nonconducting cells that nourish and support the *neurons*. (see *glial cell*)

neurohormone A type of *peptide hormone* that is produced by specialized nervous tissue. Examples include *norepinephrine* produced at *postsynaptic* nerve termini, *serotonin* produced in the vertebrate brain, and *oxytosin* produced by the *pituitary gland*. (see *neuroendocrine system, neuromodulator*)

neuromast In fish and aquatic amphibians, the basic *receptor* units of the *acousticolateralis sensory* system. Their hairlike processes, located on cells on the head and along the body, detect and conduct vibrations transmitted through the water.

neuromodulators A class of *neuropeptides* characterized by the longer time scale in which they act as compared with *neurotransmitters*. They are also called *neurohormones*. They may act either *pre-* or *postsynaptically* and are usually medi-

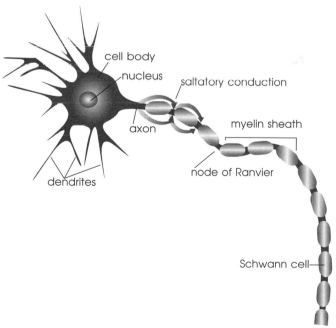

cell body

nucleus

saltatory conduction

axon

myelin sheath

dendrites

node of Ranvier

Schwann cell

A Typical Neuron

ated by *intracellular second messengers.* *Vasopressin, somatostatin,* and *oxytosin* are all now classed as neuromodulators as are *enkephalins* and *endorphins*, which function in decreasing the sensation of pain.

neuromuscular junction (motor end-plate) The specialized region (*end plate*) at which a *motor nerve* ending contacts a muscle. It is similar to a nerve *synapse* except that the *neurotransmitter* released (*acetylcholine*) *depolarizes* the muscle cell membrane (*sarcolemma*), permitting the entry of calcium ions that initiate a *muscle contraction.* (see *troponin, tropomyosin*)

neuron (nerve cell) A cell that is specialized for the transmission of *nerve impulses.* It consists of a *cell body* that contains the *nucleus*, has numerous branching extensions (*dendrites*), and has a single long extension called the *axon* or *nerve fiber.* In most vertebrate axons in the *peripheral nervous system*, the axon is covered with an insulating *myelin sheath.* Dendrites carry nerve impulses toward the cell body. The axon transmits them away, eventually contacting another nerve or an effector such as a muscle or gland. (see *motor neuron, sensory neuron, interneuron, nerve, nerve net, synapse, nervous system*)

neuropeptides Small *peptides* released by nervous tissue that act on the *neuroendocrine system.* (see *neuromodulators, neurohormone*)

Neuroptera An order of *endopterygote* insects containing the snakeflies, lacewings, and antlions. The larvae are *carnivorous* and sometimes *parasitic*, and *pupation* takes place in a silken *cocoon.* They are highly beneficial insects since both the *larvae* and adults are *predators* of "pest" insects including aphids, mites, and flies.

neurotransmitter The chemical signal by which electrical *nerve impulses* are relayed to an adjacent *neuron* or *effector* (muscle or *gland*). Neurotransmitters take the form of small *peptides* that are released from the *presynaptic* end of an *axon*, diffuse across the *synaptic cleft*, and bind to *receptors* on the far side (*postsynaptic*). Depending on the nature of the receptor, the result is either excitation (*depolarization*) or inhibition (*hyperpolarization*). Neurotransmitters are stored in *intracellular* vesicles near the *synapse* and are released upon arrival of a *nerve impulse.* The most widespread neurotransmitter is *acetylcholine.* (see *action potential; GABA;* compare *neuromodulator*)

neurovascular The situation in which a concentration of nerves is closely associated with *vascular* tissue (blood vessels). (see *carotid body*)

neurula The stage in chordate *embryos*, following *gastrulation*, when the *neural tube* is formed (*neurulation*). (see *neural crest, neural plate*)

neurulation In chordate *embryological* development, the formation of the *notochord* from *mesodermal* cells just below the dorsal surface. (see *neurula*)

neuston Organisms living at or very near the air/water interface of a water boundary.

neuter An animal lacking sex organs but otherwise normal. This can sometimes occur naturally due to mutation or may be accomplished artificially, such as the spaying of pets.

neutral mutation
1. A *mutation* that has no obvious effect.
2. A mutation that does not affect the Darwinian *fitness* of its carriers.

neutrophil A circulating white blood cell (*leukocyte*) that engulfs and digests bacteria and other foreign particles. They are *ameboid*, therefore able to pass through *capillaries* to a site of infection. They also release chemicals (identical to household bleach) that kill surrounding bacteria and themselves in the process; dead neutrophils form pus. They comprise about 70% of all leukocytes and derive their name from a lack of staining with either basic or acid dyes. (see *immune system, immune response, granulocyte;* compare *eosinophil, basophil*)

newts (see *Anura*)

nexin One of the proteins found in *eukaryotic undulipodia*. It forms links between adjacent *microtubule* doublets. (see *dynein*)

N-formylmethionine (fMet) In *prokaryotes*, a specialized *amino acid* that initiates a new *polypeptide* chain in *protein synthesis* but is inserted nowhere else in the protein. The N-formyl group blocks the amino terminus, preventing it from forming any bonds in that direction. It is inserted by the initiator tRNA^{fMet} but in response to the same *codon* as *methionine*, AUG.

niacin (see *nicotinic acid*)

niche (see *ecological niche*)

nicking A *nuclease* action that severs the *sugar-phosphate backbone* in one *DNA* strand of a *duplex*.

nick translation A specific procedure for incorporating radioactive *nucleotides* into *double-stranded DNA*, usually for use as a *probe*. Single strand *nicks* are produced at random sites in a DNA fragment by a *DNase*, and *DNA polymerase* is used to incorporate the ^{32}P-labeled nucleotides. The technique has been largely supplanted by more efficient and specific labeling techniques such as *primer extension* often involving *PCR* (polymerase chain reaction). (see *DNA analysis, Southern blot*)

nicotinamide adenine dinucleotide (see *NAD*)

nicotinamide adenine dinucleotide phosphate (see *NADP*)

nicotinic acid (niacin) A member of the *vitamin B complex*. It is a precursor to the *coenzymes NAD* and *NADP*, essential components of *respiration* and *photosynthesis*. A deficiency in man causes pellagra.

nictitating membrane The "third eyelid" in amphibians, reptiles, birds, some mammals (for example, rabbit), and sharks. When closed, it reduces the amount of light entering the eye. It can also be used to wash the *cornea*.

nidicolous Birds, in particular, that are born naked, blind, and helpless. Although they are initially completely dependent on their parents, they mature quickly. (see *altricial*; compare *nidifugous*)

nidifugous Birds, in particular, that are born alert, covered with down, with well-developed legs, and with open eyes. Although independent almost immediately from birth, nidifugous birds tend to reach full maturity slowly. (see *precocial*; compare *nidiculous*)

nightjars (see *Caprimulgiformes*)

ninhydrin A chemical reagent used to test for the presence of amino groups, indicating the presence of *amino acids* and potentially *polypeptides* and proteins.

nitrification The conversion of ammonia to nitrite and subsequently to nitrate. The different stages are carried out by different *chemoautotrophic* soil bacteria that make nitrogen available to plants in the form of nitrate. It is a critical step in the *nitrogen cycle*. (compare *denitrification, nitrogen fixation*)

nitrocellulose membrane A paperlike material that has been used extensively in the laboratory as a solid support for *nucleic acids (DNA, RNA)*. Nitrocellulose is positively charged, so it helps to bind nucleic acids, which are negatively charged because of their phosphate backbone. It has been generally replaced by more durable synthetic materials such as nylon. (see *DNA analysis, Southern blot*)

nitrogen An essential element found in all *amino acids*, therefore all proteins, and in various other critical organic compounds such as *nucleic acids*. (see *nitrification, denitrification, nitrogen fixation, nitrogen cycle*)

nitrogen base (see *base*)

nitrogen cycle The circulation of nitrogen between living organisms and the *environment*. Atmospheric gaseous nitrogen (N_2) enters the biological cycle through fixation to ammonia, nitrites, and nitrates. This is accomplished by specialized *chemoautotrophic* soil bacteria that release these *metabolic* products into the soil by excretion and decay. Atmospheric nitrogen may also be fixed by lightening. Plants generally take up the inorganic nitrogen only in the form of nitrates. These are converted into organic compounds (*amino acids* and proteins) and may then be assimilated into the bodies of animals. Certain plants have formed a *symbiotic* association with *Rhizobium* bacteria and other *root-nodule*-forming organisms (actinomycete bacteria) through which the plant gains access to locally fixed nitrogen. *Excretion* and microbial action on dead organisms break down nitrogenous organic compounds and return nitrogen atoms to the inorganic state, completing the cycle. The use of artificial nitrogen fertilizers and the industrial emission of nitrous oxides also influence the nitrogen cycle. (see *nitrogen fixation, nitrification, denitrification*)

nitrogen fixation The formation of *nitrogenous* compounds from atmospheric nitrogen (N_2). In nature, this is achieved by the normal *metabolism* of specialized *chemoautotrophic* soil bacteria and by lightening. In *leguminous* plants such as beans and peas (*Fabaceae*), the *symbiotic* bacteria *Rhizobium* form characteristic *root nodules* and supplies the plant with usable nitrate, while itself obtaining carbohydrates. One genus of actinomycete bacteria performs the same function is some temperate forest trees such as alders. In industry, the Haber process is used to make ammonia from nitrogen and hydrogen, chiefly for incorporation into fertilizers. (see *nitrogen cycle*; compare *denitrification*)

nitrogen-fixing aerobic bacteria (nitrogen-fixing bacteria) A group of bacteria that grow *symbiotically* on the roots of *leguminous* plants. The bacteria fix inorganic nitrogen, making it available to the plant. In turn, the plants provide carbohydrates that the bacteria can *metabolize*. They were at one time considered a sep-

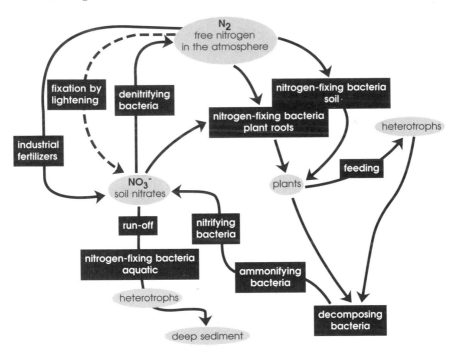

The Nitrogen Cycle

arate phylum but have now been combined into the large bacterial phylum *Proteobacteria*.

nitrogenous Any compound containing the element nitrogen.

NK-cell (natural killer cell) A *lymphocyte* involved in the initial *nonspecific immune response* to foreign invaders. NK-cells recognize body cells that have become infected by viruses. NK-cells destroy them by piercing holes in them using a set of proteins similar to *complement*. (see *immune system*; compare *K-cell*)

nociceptors Internal body *receptors* that are sensitive to pain.

nocturnal Organisms that are active primarily at night.

node
1. A swelling or thickening in any anatomical structure, for example a *lymph node*.
2. In plants, the part of the stem where one or more leaves arises.

node of Ranvier In *myelinated neurons*, the region of bare *axon* that occurs between successive *Schwann cells*. In the propagation of a *nerve impulse*, the *action potential* jumps from one node of Ranvier to the next, greatly increasing conduction speed. This mode of propagation is called *saltatory conduction*. (see *myelin sheath, nervous system, peripheral nervous system*)

nondisjunction In *nuclear division*, the failure of *homologous chromosomes* (in *meiosis I*) or *sister chromatids* (in *mitosis* or *meiosis II*) to separate and move to opposite poles at *anaphase*. If this occurs in meiosis, the resulting *gametes* will not receive the proper chromosomal complement (*aneuploidy*). *Zygotes* formed from aneuploid *gametes* lacking a chromosome are usually inviable, while those acquiring an extra chromosome or chromosomal segment are often abnormal. In humans, Down's syndrome is associated with *trisomy* (three copies) of chromosome 21. Mitotic nondisjunction is most often seen in laboratory *cell culture*.

non-Mendelian inheritance The generational transmission of genetic traits in a fashion not in accord with *Mendel's laws*. This normally occurs for *extranuclear* genetic elements such as *episomes, plasmids,* or the *DNA* in *subcellular organelles* such as *mitochondria* or *chloroplasts*. (see *mitochondrial DNA, maternal inheritance, uniparental inheritance*; compare *Mendelian inheritance*)

non-Mendelian ratio A ratio of *phenotypes* in the offspring of a particular union that does not reflect the simple operation of *Mendel's laws*. Non-Mendelian rations are often evidence of normal *recombinational* events such as *gene conversion*. (compare *Mendelian ratio*)

nonself Any cell or *antigen* that does not display the appropriate *MHC proteins* that mark it as belonging to the organism. (compare *self*)

nonsense codon (stop codon) Any of three different *nucleotide triplets*, none of which code for an *amino acid*. They serve as stop markers during *translation* of a *polypeptide*. (see *amber, ocher, opal, protein synthesis*)

nonsense mutation A *mutation* that alters the *nucleotide* sequence of a *gene* so as to produce a *nonsense codon*, which signals a premature stop to the growing *polypeptide* chain. (see *translation, protein synthesis*)

nonsense suppressor A *mutation* that alters the *nucleotide* sequence of a *tRNA* (transfer RNA) *gene*, so that is produces a *mutation* complementary to a *nonsense mutation*. When both mutations are present, an *amino acid* will be inserted during *protein synthesis*, and a full length, albeit *mutant*, protein will be synthesized. (see *translation*)

nonspecific immune response (immediate immune response) The immediate bodily response to any foreign invader identified as "nonself." Its effects are mediated by *macrophages, NK-cells* (natural killer cells), and a protein complex called *complement* that disrupts invading cells. The *inflammatory* response is a generalized response to infection and is characterized by the release of *histamines* and *prostaglandins*. These produce redness and swelling as a result of increased circulation and fluid leakage. An increase in body temperature, effected by *prosaglandins*, is also a general response and serves to inhibit microbial growth and to stimulate various aspects of the bodily defense systems. Elements of the specific *humoral* and *cell-mediated immune responses*, such as *interferon, interleukin-1,* and *T-cell* proliferation, are stimulated by the macrophages that first attack invading cells. The invading cells themselves release *pyrogens* in direct response to macrophage attack. (see *immune system*)

noradrenaline (see *norepinephrine*)

norepinephrine A *neurohormone* secreted by the *adrenal medulla* in response to stress, readying the body physiologically for a *fight-or-flight* response. The action throughout the body is identical to that

achieved initially by the *sympathetic nervous system* but works on a longer time scale. (see *neuromodulater*)

northern blot A laboratory technique used to analyze *RNA*. Unlike a *Southern blot* (*DNA*), RNA is not cut with *restriction enzymes*, the molecules having already been processed into manageable sizes by the cell. (*Restriction enzymes* are also specific to DNA.) The RNA molecules are separated by *electrophoresis*, often on an *agarose gel*. The *secondary structure* of the RNA is disrupted by a *denaturing* agent in the agarose gel such as formaldehyde. The molecules are then transferred out of the gel by capillary action to a solid support such as a nylon membrane. The immobilized RNA molecules are usually detected by *radioactively* or *chemiluminescently* labeled *probes* that bind to the membrane at the location of a *gene* or region of interest and are visualized on X-ray film.

notochord In chordates, a flexible supporting rod lying just below the dorsal *nerve*

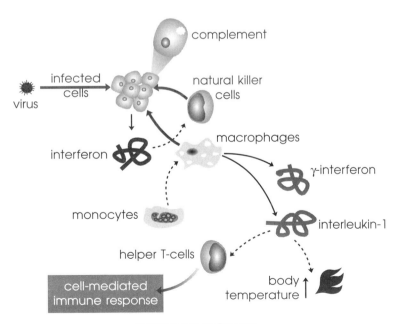

Immediate Immune Response
(Nonspecific)

cord and extending the length of the body. In vertebrates (craniates), it is present only as an *embryonic* structure and replaced wholly or partly by the *vertebral column* in the adult (except for *cartilaginous* fish such as sharks).

Notoryctemorphia The order of marsupials comprising the marsupial moles.

nucellus, nucelli (pl.) In plants, the mass of tissue in the developing *ovule* (seed) that contains the *embryo sac* enclosing the egg cell or *megaspore*.

nuclear chromosomes (see *nuclear DNA*)

nuclear division Division of the *nucleus* of a cell, including the equal distribution of replicated *chromosomes* to daughter cells. Normally, this occurs as the last step in *mitotic* or *meiotic cell division* before *cytokinesis*. It may also occur independently of cell division, such as in the generation of a new *macronucleus* after *conjugation* in *ciliates*. (compare *cytokinesis*)

nuclear DNA *DNA* contained within the *nucleus* of a cell and packaged into *chromosomes*. It is distinguished from *cytoplasmic DNA* elements, such as *episomes* and *plasmids*, and DNA contained in *subcellular organelles*, such as *mitochondria*. Nuclear chromosomes show *Mendelian inheritance*.

nuclear genome (see *nuclear DNA*)

nuclear membrane The specialized *phospholipid bilayer* membrane enclosing the *nucleus* of a *eukaryotic* cell and separating its components from the *cytoplasm*. It is a double membrane, perforated with pores that allow for the selective exchange of materials (such as *mRNA*) with the *cytoplasm*.

nuclear spindle The *microtubular* structure that forms at *metaphase* of both *mitotic* and *meiotic nuclear divisions* and upon which the *chromosomes* are organized and arrayed before they separate to opposite poles.

nuclease Any enzyme that acts on *nucleic acids*. (see *DNase, RNase, endonuclease, restriction endonuclease, exonuclease*)

nuclease protection assay A method for mapping the portions of a *gene* that are retained in a *transcribed* and processed *mRNA* molecule. *Single-stranded probe DNA* from the *cloned* gene of interest is allowed to *anneal* with *single-stranded* cellular *RNA* and exposed to *nucleases* that differentially degrade only single-stranded DNA. Any regions that do not pair exactly, such as where *introns* have been removed during RNA *splicing*, will be degraded by the nuclease and produce discrete double-stranded fragments. These can be further analyzed by *gel electrophoresis* and *DNA sequencing* to determine which portions of the *genomic DNA* are missing in the final mRNA molecule. (see *splicing, intron, exon*)

nucleated Any cell containing a *nucleus*. This defines all *eukaryotic* cells. Mature mammalian red blood cells (*erythrocytes*)

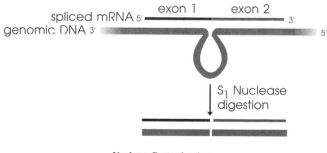

Nuclease Protection Assay

and plant *sieve cells* lose their nuclei early in development.

nucleic acid (polynucleotide) One of the four main classes of *macromolecules* (protein, *nucleic acid*, carbohydrate, lipid) found in biological systems. Nucleic acid forms the basic chemical structure of both *DNA* and *RNA*, the biological molecules in which genetic information is encoded. They are long-chain *polymers* consisting of a backbone of alternating sugar and phosphate units, with *nitrogenous bases* attached to the sugar units. In DNA, two chains interact via the bases to form a *double helix*. In RNA, a single chain often forms regions of *secondary structure* within itself. Also in RNA, the *deoxyribose* sugar is replaced by a *ribose*, and the base *thymine* is replaced by *uracil*.

nucleic acid hybridization A laboratory technique in which *nucleic acid* molecules (*DNA* or *RNA*) are immobilized on a solid support, often but not always after undergoing *electrophoresis* and challenged with a *probe* complementary to a specific *gene* or *locus*. Hybridization is based on the fundamental principle of *complementary base pairing*. Single-stranded complementary sequences of nucleic acid will reform a *duplex* (*double stranded*) under the correct conditions. A nucleic acid probe of known sequence is prepared, historically by isolating a known DNA fragment and more recently by synthesizing the desired sequence in the laboratory. The probe is labeled with a tag, usually *radioactive* or *chemiluminescent*, enabling it to be tracked. Both the nucleic acid to be queried and the probe must be single stranded, often accomplished by subjecting them to alkali and/or heat. Under controlled conditions, the probe binds exclusively to its complementary sequence, and its location is visualized by exposure to X-ray film. Nucleic acid hybridization and its kinetics may also be studied in solution ($C_o t$ curve). (see *double helix, DNA analysis*)

nucleic acid precipitation A technique used in the extraction and purification of *nucleic acids* (*DNA, RNA*) for laboratory analysis. Conditions for precipitation commonly include salts and alcohols that make nucleic acids less soluble in aqueous solution. The precipitate is removed by *centrifugation* or *filtration*.

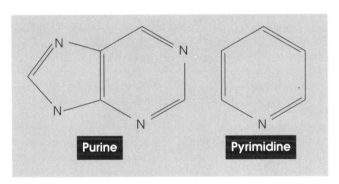

Nucleic Acid Base Structure

nucleoid The *DNA*-containing region of a *prokaryotic* cell where the main *chromosome* and any *plasmids* reside. In contrast to the *nucleus* of a *eukaryotic* cell, the nucleoid is not sequestered by a *nuclear membrane* but may be attached to the interior of the *plasma membrane*. It also refers to the *DNA* mass in any *endosymbiotically* derived *subcellular organelles* such as *mitochondria* and *chloroplasts*.

nucleolar organizer The region on *chromosomes* containing many copies of tandemly repeated *rRNA* (*ribosomal*

RNA) *genes* that is often *transcribed* at a high level, forming a *nucleolus*.

nucleolus, nucleoli (pl.) A region of intense *transcription* at the site of multiple tandem *rRNA* (*ribosomal RNA*) *genes* on *eukaryotic chromosomes*. Nucleoli are large enough to be visible under a light microscope. RNA-binding proteins and associated ribosomal proteins transported from the *rough endoplasmic reticulum* are also included. Nucleoli are especially conspicuous in cells that are manufacturing large quantities of protein, for which an increase in *ribosome* production is needed. (see *gene amplification, nuclear organizer*)

nucleoplasm The fluid contents of the *nucleus* of a cell.

nucleoprotein A complex of *nucleic acid* and protein. For instance, *eukaryotic chromosomes* are composed of *DNA*, some *RNA,* and *structural proteins* (*histones*). *Ribosomes* also consist of a nucleoprotein complex.

nucleoside A *purine* or *pyrimidine base* linked to a sugar, either *ribose* or *deoxyribose*. *Adenosine, cytidine, guanosine, thymidine,* and *uridine* are the most common nucleosides in biological systems. (see *nucleic acid;* compare *nucleotide*)

nucleosome The *tertiary structure* of *eukaryotic chromosomal DNA*. About every 200 *nucleotides*, the DNA *double helix* is coiled about a complex of eight *histone* proteins. This gives the appearance of beads on a string. The nucleosomes are themselves *supercoiled* into a *solenoid structure*.

nucleotide A *nucleoside* to which one phosphate group has been attached. Nucleotides are used in the synthesis of *nucleic acids* and other important biological molecules such as the respiratory *coenzymes NAD* and *FAD*. The terms *"base"* and "nucleotide" are often used informally and interchangeably in referring to the nucleotide residues that make up a *polynucleotide* chain such as DNA or RNA.

nucleotide pair (see *base pair*)

nucleotide triplet (codon) The unit of genetic information that specifies one *amino acid*. From the four different *nucleotides* that make up *DNA*, 64 different triplet combinations are possible, specifying 20 common *amino acids* with some redundancy. (see *genetic code*)

nucleus A *subcellular organelle* found in all *eukaryotic* cells (with the exception of mature mammalian *erythrocytes* and plant *sieve cells*) in which the majority of genetic information (*DNA*) is found organized into *chromosomes* and from which the cell's activities are thus directed. It is surrounded by a *nuclear membrane*.

nude mice Laboratory mice that lack a *thymus gland*, thus are unable to produce mature *T-cells* or coordinate immunological function. They are used in immunological studies.

null- A word element derived from Latin meaning "none." (for example, *nullisomic*)

null allele A genetic variant, usually a *mutation*, whose effect is either an absence of normal *gene product* at the molecular level or an absence of normal observed function.

nullisomic A cell or organism lacking both members of a particular *chromosome* pair. This situation is only viable where the *genome* is *polyploid* (contains several of each chromosome pair), such as in some agricultural crops.

numerical taxonomy (phenetics) The classification of organisms based on large numbers of measurable morphological traits. In modern classification schemes, phenetics is combined with *cladistics*, the *phylogenetic* classification of organisms, particularly protein data and *DNA analysis*.

nurse cells The sister cells of the developing egg cell (*oocyte*) in some insects that produce the bulk of the *cytoplasmic* contents of the mature oocyte.

nurse log In *old growth forests* particularly, fallen logs that begin to decay and provide an ideal environment for seeds to

Purines

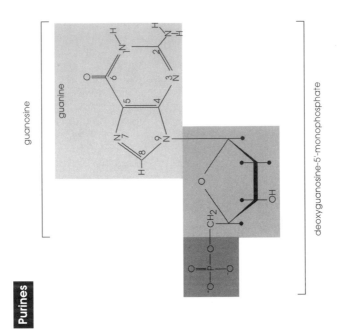

guanosine

guanine

deoxyguanosine-5'-monophosphate

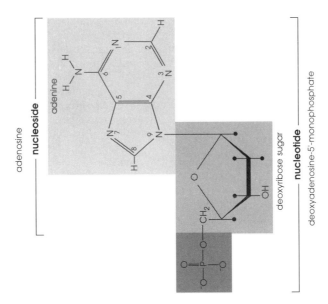

adenosine

nucleoside

adenine

deoxyribose sugar

nucleotide

deoxyadenosine-5'-monophosphate

259

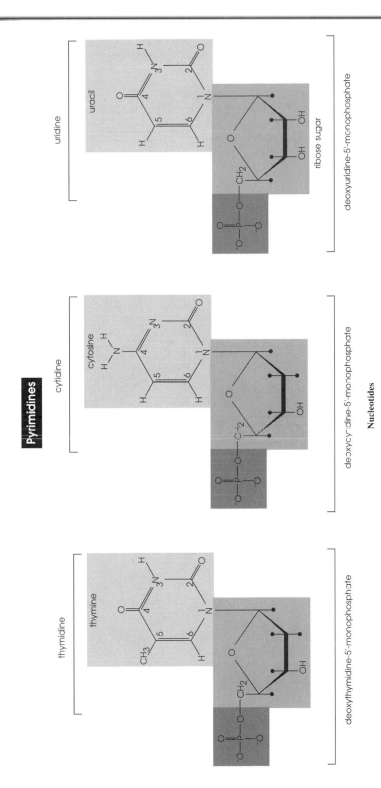

Pyrimidines

uridine

uracil

ribose sugar

deoxyuridine-5'-monophosphate

cytidine

cytosine

deoxycytidine-5'-monophosphate

Nucleotides

thymidine

thymine

deoxythymidine-5'-monophosphate

sprout and grow. They are an essential part of the *ecology* of both *tropical* and *temperate rainforests*.

nut A dry *indehiscent* fruit resembling an *achene* but derived from more than one *carpel* (female reproductive unit). It characteristically contains a single seed. The term "nut" is frequently misused in the common vernacular. For instance, Brazil nuts are seeds and walnuts are *drupes*. Beech, hazel, oak, and chestnut are examples of true nuts.

nutation (circumnutation) The spiral growth pattern observed in the shoot tips of climbing plants. The direction of rotation is often constant in a species. (see *thigmotropism*)

nutria (see *Rodentia*)

nutrition Any process whereby an organism obtains from its environment the material and energy required for growth, health, and reproduction. (see *autotrophic, heterotrophic*)

nyctinasty The opening and closing of flowers and leaves in response to daily cycles of temperature and light. (see *nastic movements, circadian rhythm*)

nylon membrane A paperlike material that is used extensively in the laboratory as a solid support for *nucleic acids* (*DNA, RNA*). It is positively charged, which helps to bind nucleic acids that are negatively charged because of their phosphate backbone. It has generally replaced the more fragile *nitrocellulose membrane* historically used for such purposes. (see *DNA analysis, Southern blot*)

nymph (see *instar*)

obligate An organism that requires a specific environmental condition. An obligate *aerobe* (most organisms) must have oxygen to survive, while an obligate *anaerobe* (such as methanogenic bacteria) may only survive in the absence of oxygen. The term is also used in conjunction with *parasitic* organisms. (compare *facultative*)

obtect In insects that undergo *metamorphosis*, the type of *pupa* in which the appendages adhere to the body by means of a secretion produced at the last *larval molt*. It is particularly characteristic of true flies, butterflies, and moths. (compare *coarctate, exarate*)

occipital condyle A protrusion at the back of the skull where it meets the first spinal *vertebra* (*atlas*). It is absent in fish and double in amphibians and mammals.

occipital lobe The rear portion of the mammalian *cerebrum*, anatomically separated from the three other lobes by deep grooves. Functionally, it contains the *visual cortex*. (see *cerebral cortex, telencephalon*; compare *parietal lobe, temporal lobe, frontal lobe*)

oceanic The marine environment beyond the continental shelf, usually deeper than 300 meters. It makes up about 99% of the total marine environment. *Pelagic* organisms (*plankton* and *nekton*) inhabit this zone. (compare *abyssal, neritic, photic zone*)

ocellus, ocelli (pl.) An eyelike *photoreceptor* found in some invertebrates, either alone or in conjunction with a pair of *compound eyes*. It contains light sensitive cells that detect the direction and intensity of light but generally do not form an image. Ocelli may function as horizon detectors and are important in the visual stabilization of flight in locusts and dragonflies. (see *median eye*)

ocher codon The *nucleotide triplet* UAA, which is a *nonsense*, or stop, *codon*.

octopod A cephalopod mollusk that has eight arms with suckers.

octopus, octopi (pl.) (see *Cephalopoda*)

oculomotor nerves (cranial nerves III). A pair of nerves in vertebrates that supply the muscles of the eyeballs. (see *cranial nerves*)

Odonata The order of *exopterygote* insects containing the dragonflies and damselflies. They are *carnivorous* insects whose fossil record dates back to the Permian period.

odontoblast A cell that secretes the *dentine* of a tooth.

Odontoceti A suborder of whales comprising the toothed whales, dolphins, and porpoises. They have many peglike teeth to feed on fish and other animals.

offset In plants, a short *runner* that grows aboveground and produces a new plant some distance from the parent. Offsets are a means of *vegetative reproduction* but store no food supply. (see *perennation, perennating structure*; compare *bulb, corm, rhizome, runner, stolon, sucker, tuber*)

oil gland (see *sebaceous gland*)

oil immersion A microscopic technique in which oil is used to reconcile the difference in refractive index between the objective lens and the specimen.

old growth forest A popular term that has yet to gain an agreed upon, precise scientific meaning but generally refers to any virgin forest that is at least approximately 175 years old. Mature stands frequently approach 500 years of age and may exceed 1,000. Old growth forests differ inherently from new, densely populated reforestations. Old growth forests have light gaps, dead and decomposing old trees, and an abundance of fallen logs and standing trees, all of which contribute to a diverse and complex *ecosystem*. Those of the Pacific Northwest have received much attention of late, and one of its denizens, the spotted owl, has become a flagship for the protection of *endangered species*. (compare *second growth forest*)

oleo- (see *elaio-*)

oleoplast (see *elaioplast*)

olfaction The sense of smell.

olfacto- A word element derived from Latin meaning "smell." (for example, *olfactory organ*)

olfactory Any structure or organ pertaining to the sense of smell.

olfactory nerve (cranial nerve I) The nerve that connects the *receptor* cells of the *olfactory organs* in the nose with the olfactory center in the forebrain of vertebrates. (see *cranial nerves*)

olfactory organ Any organ specialized in the detection of odors. It consists of a group of *sensory* receptors that are sensitive to air or water-borne chemicals, often in very dilute concentration. Olfactory organs are found on the *antennae* of insects, and in various other positions in other invertebrates. In terrestrial vertebrates, they are *neurons* whose cell bodies are embedded in the *mucous membrane* lining the nose and whose *dendrites* extend to the surface. A *nerve impulse* is initiated when the receptor is bound by particular molecules, often in very low concentration. (see *olfactory nerve*)

oligo- A word element derived from Greek meaning "few." (for example, *oligonucleotide*)

Oligocene An epoch of the Tertiary period, about 38 million to 25 million years ago. It is characterized by the disappearance of primitive mammal groups and their replacement by more modern forms.

Oligochaeta The class of annelid worms containing the terrestrial earthworms as well as many freshwater species. They have no head appendages as a result of living mostly underground. They are *hermaphroditic* but usually *cross-fertilize*. Earthworms are crucial to soil health as they literally eat their way through it, leaving it aerated and enriched.

Oligochetes (see *Oligochaeta*)

oligonucleotide A short chain of *nucleotides*, from a few to about twenty-five. Synthetic oligonucleotides are used in many molecular biology and *genetic* engineering techniques. (compare *dinucleotide, polynucleotide*)

oligopeptide A short chain of *amino acids*, from a few to about twenty-five. (compare *polypeptide*)

oligosaccharide Short chain sugar *polymers* made of various sugar monomers (about four to twenty) connected by *glycosidic bonds*. (compare *monosaccharide, disaccharide, polysaccharide*)

oligotrophic Lakes that are deficient in nutrients and consequently low in productivity. An oligotrophic lake may become *eutrophic* (rich in nutrients) by the accumulation of organic matter. This would initiate a *primary ecological succession*, where organisms gradually come to occupy the area.

omasum The third region of the specialized stomach of *ruminants* (for example, the cow), where water is absorbed.

ommatidium, ommatidia (pl.) One of the thousands of individual units that make up the *compound eye* of arthropods. Each ommatidium consists of a circular *lens*, below which is a crystalline core. Each core leads to a number of *retinal* cells that form a light sensitive *rhabdom*. This, in turn, stimulates a *nerve fiber* leading to the brain. In *apposition* eyes (for example, in bees), each ommatidium is surrounded by pigment cells that serve to isolate the individual inverted images that are integrated in the brain. *Superposition* eyes (for example, moths) lack screening pigment cells. The images from ommatidia are right side up and are combined on a *cornea* at the back of the eye.

Omnibacteria (see *Proteobacteria*)

omnivore An organism that eats a wide variety of foods, including both plants and animals. Man is an example. (compare *carnivore, herbivore*)

oncogene Any *gene* implicated in the initiation of *cancerous* growth. It is usually a gene involved in the regulation of *cell division* that is permanently or abnormally active because it has *mutated* and lost its normal *regulatory sequences*, sometimes due to transmission by a *retrovirus*.

Cancerous growth may be caused when such a gene is expressed inappropriately, causing uncontrolled proliferation and *tumor* formation. Oncogenic activity may also be caused by mutations in genes that normally inhibit cell proliferation. (see *proto-oncogene*)

oncogenic Denotes any cancer-causing agent, in particular viruses such as *retroviruses*.

one gene-one polypeptide The general dogma that each *gene* codes for one and only one *polypeptide* chain, often an enzyme. It was originally expressed as the *one gene-one enzyme hypothesis* in 1945 by George Beadle and Edward Tatum, who first recognized the principle in studying fungal *metabolic* mutants. The principle remains mostly true, although as with all scientific generalizations, exceptions have been noted. They shared the Nobel prize in physiology or medicine with Joshua Lederberg in 1958 for their combined work. (see *DNA, RNA, transcription, translation, central dogma*)

ontogeny The developmental course of an organism from *fertilized* egg to adult.

Onychophora A small invertebrate phylum of segmented worms, the velvet or walking worms, having some annelid and some arthropod features. They live primarily in tropical rainforests. Like annelids, they have an elongated body covered with a soft cuticle and non-jointed appendages. Like arthropods, they have a body cavity that is a *hemocoel* and tracheae for respiration. They secrete a sticky fluid that they throw relatively great distances to entangle their prey of wood lice and terrestrial mollusks, then secrete into the victim substances that kill, and partially liquefy its tissues. Different species may also be *herbivores* or *omnivores*. They have *nephridia* for excretion and simple eyes (*ocelli*) that sense light. In some species, *fertilization* is internal and the female nourishes her young by means of a true *placenta* and bears them alive. The best known example is *Peripatodes*.

oo- A word element derived from Greek meaning "egg." (for example, *oocyte*)

oocyst In spore-forming *parasitic* protists such as apicomplexans, a *zygote* that takes the form of a thick-walled cyst that is highly resistant to dessication and other environmental factors. Within it, *meiotic* division and subsequent *mitotic* divisions produce large numbers of infective *haploid sporozoites*. In the malarial parasite *Plasmodium*, this takes place in the gut of the mosquito.

oocyte A female reproductive cell (*germ cell*) that ultimately gives rise to a mature egg cell (*ovum*). During the process of *oogenesis*, the *diploid* primary oocyte undergoes the first *meiotic* division, producing the secondary oocyte and a small *polar body*. The second meiotic division produces the *ovum* and another *polar body*. Neither the first polar body, which may divide again, nor the second play any further role and eventually disintegrate. In some animals (including humans), the second meiotic division is not completed until *fertilization*. (see *estrus cycle, menstrual cycle*)

oogamous An extreme form of *anisogamy*, the *sexual* fusion of *gametes* differing in size and/or motility. The term is generally reserved for plants that produce large, nonmotile, female gametes in *oogonia* that are fertilized by minute, motile *spermatozoa*. (compare *isogamous*)

oogenesis The development of mature *ova* (mature egg cells). In female animals, precursor cells multiply *mitotically* during *fetal* development to form hundreds of thousands of *oogonia* that, at birth, are each surrounded by a *follicle*. From the onset of *sexual* maturity until senescence, an oogonium periodically develops into an ovum and is released from the *ovary*. The *diploid* oogonium increases in size, becoming an *oocyte* that undergoes *meiosis* (reductive division) and forms the *haploid* ovum. In some animals (including humans), the second meiotic division is not completed until after *fertilization*. (see *estrus cycle, menstrual cycle*)

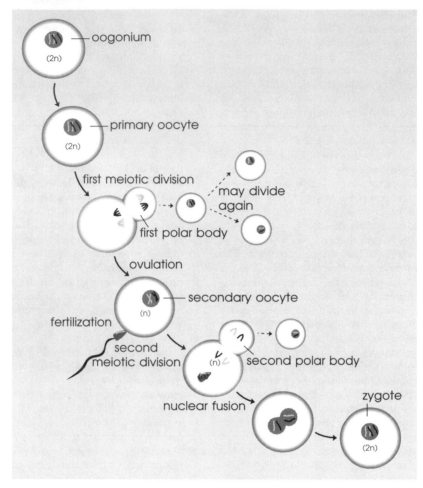

oogonium
(2n)

primary oocyte
(2n)

first meiotic division

may divide
again

first polar body

ovulation

secondary oocyte
(n)

fertilization

second
meiotic division
(n)

second polar body

nuclear fusion

zygote
(2n)

Oogenesis

oogonium, oogonia (pl.)
   1. In female animals, an immature reproductive cell (*germ cell*) that divides
   and grows to become an *oocyte* during
   the initial stages of *oogenesis*. (see
   estrus cycle, menstrual cycle)
   2. A *unicellular* female sex organ, or *gametangium*, that contains one or more
   nonmotile *haploid* egg cells, sometimes called *oospheres*. It is found in
   seedless plants, algae, and occasionally in fungi. (compare *antheridium*)
oomycetes (see *Oomycota*)
Oomycota A phylum of funguslike pro-

toctists, the oomycetes or water molds,
that include the white rusts and downy
mildews. All members of the group are
either *parasites* or *saprobes* (feed on
dead matter), and many are important
plant and animal *pathogens*. Their *cell
walls* are composed of *cellulose*. Their
sexual cycle is unique, complicated, and
distinctive. The female *gametangium*,
called an *oogonium*, and the male *antheridium* develop on the ends of specialized filaments called *hyphae*, similar to
those in fungi. The male and female gametangia fuse, permitting the male *nu-*

cleus to enter and *fertilize* the *oosphere*, or egg. The thick-walled *zygote* is called an *oospore* and divides *mitotically*, producing *diploid* motile *zoospores* that swim away and propagate the water mold. Oospores may also undergo *meiosis* to form *haploid nuclei* that then differentiate into haploid *zoospores*. Alternatively, haploid zoospores may be produced directly by mitotic division in a *sporangium*. Oomycetes include the plant *pathogens* associated with potato blight and grape mildew.

oosphere In certain plants, algae, and fungi, a large, nonmotile female *gamete* formed within an *oogonium*. *Fertilization* is followed by development of an *oospore*.

oospore In oomycete protoctists, the *diploid zygote* resulting from *fertilization*. It forms a thick-walled cell from which the phylum derives its name.

opal codon The *nucleotide triplet* UGA, which is a *nonsense*, or stop, *codon*.

open circulatory system A *circulatory system* in which blood is not maintained in closed vessels but rather bathes the tissues directly. It is typical of many invertebrates. (see *hemocoel, hemolymph*)

open reading frame (see *ORF*)

operator A region of *DNA* at one end of an *operon* that acts as the binding site for a *repressor* protein. When the repressor protein is bound to the operator site, *gene transcription* is blocked.

operculum, opercula (pl.) In general, a covering or closure to an open chamber. **1.** The circular lid of a moss capsule. **2.** The bony plate covering the *gill chambers* of bony fish. **3.** The disc that closes the aperture of the shell of some mollusks. **4.** The fold of skin that grows over to enclose the gills of a developing tadpole.

operon A cluster of *structural genes* coding for the enzymes in a particular biochemical pathway that are transcribed as a unit into a single *mRNA* molecule. It also includes the adjacent regulatory *nucleotide* sequences. *Transcription* of all

the genes in an operon is regulated coordinately by controlling the binding of *RNA polymerase* to the single *promoter* site, typically via an adjacent and overlapping *operator* site. Transcription of the operon is often repressed by the product of an independent gene, the *regulator*, that binds to the operator site, obstructing the binding of RNA polymerase. Expression of the operon is often induced by the presence of a suitable *substrate* (for example, β–galactoside for the *lac operon*) called the *inducer* that prevents the regulator protein from binding, allowing transcription to commence. Operons are a common mode of gene organization in bacteria but are rare in *eukaryotes*. The operon theory was first described by François Jacob and Jacques Monod in the 1950s. In 1965, they shared the Nobel prize in physiology or medicine with André Lwoff for their combined work.

Ophiuroidea (see *Stelleroidia*)

Opiliones The order of arachnids including the daddy longlegs, well known for their compact body and long legs. They are unusual among arachnids in that they engage in *copulation*. Most are predators of insects, other arachnids, snails, and worms. However, some live on plant juices, and a few are *saprobes*.

opossum rats (see *Paucituberculata*)

opossums (see *Didelphimorphia*)

opsin The protein component of the molecular *photoreceptor rhodopsin*.

optic chiasma The structure just beneath the vertebrate *forebrain* in which the *nerve fibers* from each eye cross to the opposite side of the brain.

optic nerves (cranial nerves II) The tracts of *sensory nerve fibers* that run from the *retina* of each eye to the brain. Each enters the *forebrain* through the *optic chiasma* where they cross, so that the nerve from the right eye connects to the left brain hemisphere and *vice versa*. (see *cranial nerves*)

oral groove (see *cytostome*)

orbit (eye socket) One of two cavities or depressions in the vertebrate skull that contain the eyeballs and their associated muscles, blood vessels, and nerves.

order In the *taxonomic* classification of organisms, a collection of similar *families.* Similar orders are grouped into a *class.*

Ordovician The second oldest period of the *Paleozoic* area, about 510 million to 440 million years ago. It is characterized by an abundance of marine invertebrates, many now extinct, and a lack of vertebrates other than jawless fish.

ORF (open reading frame) A section of *DNA* that begins with a *start codon,* ends with a *stop codon,* and contains uninterrupted *codons* for *amino acids* in between. It is usually used in describing a portion of DNA that has been *cloned* and *sequenced,* and is presumed to code for a *gene.*

organ A structure composed of several different types of *tissue* grouped together into a structural and functional unit. Examples in animals are heart, skin, kidney, and reproductive organs.

organ culture The maintenance or growth of living organs *in vitro.*

organelle A discrete *subcellular* structure specialized to perform a specific function. The largest organelle is the *nucleus,* other examples are *mitochondria* and *chloroplasts.* Organelles are characteristic of *eukaryotic* cells and probably all arose from evolutionary *endosymbioses.*

organic Any molecule containing carbon. The term was originally used to define compounds formed by a living organism but now includes those synthesized in the laboratory as well. Simple molecules such as carbon dioxide gas (CO_2) and carbon monoxide gas (CO) are not usually considered organic. But carbohydrates, for instance, derived from carbon dioxide during *photosynthesis* by plants, are included. Many industrial products and byproducts fall into the chemical category of organic but are in fact highly disruptive to living systems. Thus, organic molecules are at once the basis of life on earth and a potential cause of its demise.

organism Any individual living creature, either *unicellular* or *multicellular.*

organizer A region of the developing *embryo* that determines the future *differentiation* of adjacent tissue. (see *induction, notochord*)

organ of Corti In the *cochlea* of the terrestrial vertebrate *inner ear,* the arrangement of auditory receptors, the *hair cells,* sandwiched between two membranes, the *basilar* and the *tectorial.* Deflection of the *cilia* of these hair cells in one direction causes *depolarization,* initiating *neurotransmitter* release that results in a *nerve impulse* in the associated *afferent neurons.* Movements in the other direction cause *hyperpolarization,* inhibiting nerve activity.

organogenesis The *embryological* formation of body *organs.*

organotrophic Any organism that uses organic molecules as a source of hydrogen atoms (electron donors) in *cellular respiration.* In the case of animals, fungi, many protoctists, many prokaryotes, and a few achlorophyllous plants, these are the same molecules that provide a carbon source, typically carbohydrates ($C_X[H_2O]_Y$). Some *photosynthetic* bacteria use other small organic molecules, such as acetate or pyruvate, as both a source of carbon and hydrogen atoms. (compare *lithotrophic*)

organ system A group of *organs* that function together to carry out one the principal activities of the body. The major systems of the vertebrate body include the *circulatory, digestive, endocrine, urinary, immune, integumentary, muscular, nervous, reproductive, respiratory,* and *skeletal* systems.

origin of life Scientific evidence suggests that life on earth originated approximately 4,000 million years ago. Experiments have shown that water, methane, ammonia, and other related compounds in the presence of energy discharges may combine into increasingly complex organic molecules. Scientists theorize that particular combinations of these compounds eventually began to show the characteristics of life. This theory may be distinguished from *spontaneous generation* by both frequency of occurrence and time scale. (see *coacervates, biopoiesis*)

origin of replication A specific *nucleotide sequence* at which *DNA replication* is initiated.

ornithine An *amino acid.*

ornithine cycle (see *urea cycle*)

ornithology The study of birds.

ornithophilic Flowers that are *pollinated* by birds.

Orthonectida A newly autonomous phylum of parazoans, the orthonectids. They are small, wormlike organisms that exhibit bilateral symmetry but only have two body layers. They lack organs with the exception of a *gonad*. Orthonectids live within various marine invertebrates and show an *alternation of generations.*

Orthoptera The order of insects containing the grasshoppers, crickets, and locusts. They have characteristic *mandibulate* (chewing) mouthparts and hind legs modified for jumping. They undergo *incomplete metamorphosis* (*exopterygote*). *Stridulation* (sound production) is common, with varying mechanisms. Most species are *herbivorous*, some are *omnivorous*, and a few are *predators.*

orthotropism Growth directly toward or away from a stimulus, such as plants to light. (see *tropism*)

orthotropous The position of the *ovule* in flowering plants (*angiosperms*) when it is oriented so that the *micropyle* points away from the *placenta*. (compare *campylotropous, anatropous*)

osculum, oscula (pl.) The wide opening at the end of a sea sponge by which water leaves after it has passed through the internal structures.

osmium tetroxide A heavy metal compound used as a stain in *electron microscopy*. It is particularly useful in studying *phospholipid bilayer* membranes and the fatty *myelin sheaths* that surround *neurons.*

osmoconformer (see *osmoregulation*)

osmolyte A substance that maintains the salt balance in a cell.

osmoregulation The process by which animals regulate *osmotic pressure* by controlling the relative amounts of water and salts in their bodies. Most marine invertebrates are *osmoconformers*, they maintain their body fluids at about the same concentration as seawater as do sharks. Since all animals evolved from the ocean, their *physiology* still reflects the composition of seawater and must be maintained. Most vertebrates other than sharks are *osmoregulators*. They maintain an internal solute concentration that does not vary, regardless of the external environment. In freshwater animals, various methods are used to counteract the tendency of excess water to enter the body from the environment, including *contractile vacuoles* in protozoans, *nephridia* and *Malpighian tubules* in other invertebrates, and kidneys with well-developed *glomeruli* (important in water excretion) in freshwater fish. Marine vertebrates have kidneys with few *glomeruli* and short *uriniferous tubules* to prevent excess water loss and to excrete excess salts. Terrestrial vertebrates have evolved kidneys with long *renal tubules* in which reabsorbtion of both water and salt takes place. This is seen in its most extreme form in desert animals that excrete almost no water.

osmoregulator (see *osmoregulation*)

osmosis The diffusion of water across a selectively permeable membrane (*semipermeable*) that permits the free passage of water but prevents or retards the passage of a solute (*passive diffusion*). In the absence of a difference in pressure or volume, the net movement of water is from the side containing a lower concentration of solute to the side containing a higher concentration. Because *plasma membranes* act as semipermeable membranes, osmosis is an important feature in biological systems. (see *osmotic pressure, osmotic balance, osmoregulation, turgor,* *hypotonic, isotonic, hypertonic*; compare *facilitated diffusion, active transport*)

osmotic balance In biological systems, the delicate balance in *osmotic pressures* that must be maintained in order for the system to function properly and cells to remain intact. (see *osmosis, hypotonic, isotonic, hypertonic*)

osmotic pressure The pressure needed to prevent the passage of water or other pure solvent through a *semipermeable* membrane that separates the solvent from a solution containing dissolved solutes. Osmotic pressure rises with an increase in solute concentration. Where two solutions of different substances or concentrations are separated by a *semipermeable membrane*, the solvent will move to equalize osmotic pressure within the system. (see *osmosis, hypotonic, isotonic, hypertonic*)

osmotrophic A mode of nutrition in which nutrients are obtained by absorbtion directly across *cell membranes*. It is typical of fungi. (see *saprotropic*)

ospreys (see *Falconiformes*)

ossification The transformation of *connective tissue* into bone. *Osteoblast* cells deposit a matrix of *collagen* fibers impregnated with calcium salts in which they eventually become encased. (see *osteoclast, osteocyte, membrane bone*)

Osteichthyes The class of vertebrates containing the bony fish, characterized by a skeleton of bone. They possess a *swim bladder* that allows them to maintain and regulate their buoyancy. They have become the dominant class of fish in modern times and account for more than half of all living vertebrate species.

osteo-, oss-, ost- A word element derived from Greek meaning "bone." (for example, *Osteichthyes*)

osteoblast A cell that secretes the bone matrix of *collagen* fibers on which calcium salts are later deposited. They are found in the beginning stages of *ossification* on the outside of the cartilage or membrane undergoing transformation. Those that become part of the permanent structure become *osteocytes*. (compare *osteoclast*)

osteoclast A cell that erodes *calcified* cartilage or membrane formed in the early stages of *ossification* of bone in order to provide permanent channels for the *vascular* structure. (compare *osteoblast*)

osteocyte A mature *osteoblast* that has become embedded in the bone matrix it secreted.

ostium, ostia (pl.) In general, an opening. In sponges, it refers to the openings through which water enters the body and in arthropods, the openings from the heart into the surrounding space.

Ostracoda A class of minute aquatic crustaceans, the ostracods, that are found in both marine and freshwater habitats, They are abundant in *plankton*. At least one terrestrial species exists. *Parthenogenesis* is common among the freshwater forms.

ostracods (see *Ostracoda*)

ostrich (see Struthioneformes)

-otic, oto- A word element derived from Greek meaning "ear." (for example, *otolith*)

otic capsule (see *auditory capsule*)

otoliths (see statolith)

outbreeding Breeding between individuals that are not closely related. It tends to increase *genetic diversity* within a population and contributes to evolutionary *fitness*. (compare *inbreeding*)

outcrossing *Pollination* between genetically dissimilar individuals of the same plant species. (see *outbreeding*)

outer ear The region of the vertebrate ear that is external to the eardrum (*tympanic membrane*). It is present in birds, mammals, and some reptiles.

oval window (see *fenestra ovalis*)

ovary A *multicellular* reproductive organ that produces female *gametes*. In animals, it is an organ that produces egg cells (*ova*). In vertebrates, it is usually present as one of a pair. In plants, the ovary refers to the base of the *carpel* in the *gynoecium* (female reproductive structures), which contains at least one *ovule* (immature seed). (see *oogenesis,* *estrus cycle, menstrual cycle, pericarp*)

ovi-, ova-, ovu-, ov- A word element derived from Latin meaning "egg." (for example, *ovary*)

oviduct In animals, a duct that conveys *ova* (egg cells) from the *ovaries* to the uterus or the exterior. It may or may not be directly connected with the ovary. In vertebrates, it is called the *Mullarian duct* and in most mammals is separated into a pair of *Fallopian tubes.*

oviparity The type of reproductive mechanism in which undeveloped eggs are laid or spawned by the female. *Fertilization* may occur before their release, as in birds and some reptiles, or after, as in most invertebrates, fish, and amphibians. Large numbers of eggs are often produced because of their relatively poor chances of survival due to a lack of maternal protection (excepting birds). Each egg contains a large *yolk* store to nourish the developing *embryo.* (compare *ovoviviparity, viviparity*)

ovipositor An egg-laying structure located at the hind end of female insects. It is frequently long and needlelike to enable the piercing of animal and plant tissues to lay eggs. In wasps, bees, and ants, it is modified into a stinger.

ovitestis The reproductive organ of certain *hermaphrodites* (for example, snails) that contains both male and female reproductive organs, functioning as both an *ovary* and a *testis.*

ovoviviparity The type of reproductive mechanism in which eggs are produced, *fertilized,* and retained within the body of the female during *embryonic* development but derive nourishment from a *yolk* store. They only depend on the mother for physical protection. It is found in a few species of fish. (compare *oviparity, viviparity*)

ovulation The periodic release of an egg from a *Graafian follicle* at the surface of a vertebrate *ovary.* (see *estrus cycle, menstrual cycle, oogenesis*)

ovule In seed plants, the structure that will develop into the seed after *fertilization.* It contains an *integument*-covered *nucellus*

surrounding the *embryo sac* within which is situated the egg cell. In *angiosperms,* the ovule is contained within an *ovary* from which develops a fruit. In *gymnosperms,* ovules are borne naked, often on *ovuliferous scales* on cones.

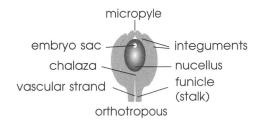

orthotropous

campylotropous

anatropous

Ovule Orientation

ovuliferous scales In conifers, the structures that bear the seeds, usually arranged spirally on a cone.

ovum, ova (pl.) The egg cell in animals. It is a large, immotile *gamete* produced in *ovaries.* A mature ovum consists of a *haploid nucleus* surrounded by *cytoplasm* containing a variable amount of *yolk* and a *vitelline membrane.* In some animals, particularly mammals, *fertilization* occurs before the ovum is fully developed, at the *oocyte* stage, and is the initiator for completion of *meiosis II.* (see *ovulation, oogenesis, oogonium, estrus* *cycle, menstrual cycle*; compare *spermatozoon*)

owls (see *Strigiformes*)

oxalic aid A substance occuring in rhubarb leaves, wood sorrel, garden oxalis, and spinach that results in a puckered feeling when eaten.

oxaloacetic acid An intermediate in the *Krebs cycle* in *aerobic respiration.*

oxi-, oxy- A word element derived from Greek meaning "oxygen." (for example, *oxyhemoglobin*)

oxic A habitat containing molecular oxygen (O_2).

oxidation The loss of one or more electrons from an atom or molecule. In biological systems, the electron(s) are usually part of a hydrogen atom. Oxidation takes place simultaneously with *reduction* (gain of an electron) of another atom, often as a result of combination with an oxygen atom. Reduction-oxidation (redox) reactions are an important means of energy transfer in biological metabolism.

oxidative metabolism The comprehensive group of biological reactions requiring molecular oxygen (O_2). It is believed that cyanobacteria first evolved the elements of oxidative metabolism as a mechanism for coping with an increasing atmospheric oxygen concentration. They were then incorporated into pre-*eukaryotic* cells as *endosymbionts*, eventually becoming integral as *mitochondria.* (see *oxidative phosphorylation, oxidative respiration, Krebs cycle, electron transport chain*)

oxidative phosphorylation The *chemiosmotic* production of *ATP* during *aerobic respiration.* It takes place in *mitochondria* of *eukaryotic* cells. Since it requires molecular oxygen (O_2) as a *terminal electron acceptor* to the *electron transport chain,* it produces *metabolic water* (H_2O) as a byproduct. (see *Krebs cycle, electron transport chain, NADH, FADH_2, proton pump, respiration*; compare *photophosphorylation*)

oxidative respiration *Respiration* in which the *terminal electron acceptor* is

molecular oxygen (O_2). (see *oxidative metabolism, oxydative phosphorylation, aerobic respiration*)

oxygen An element essential to living organisms both as a constituent of proteins, fats, and carbohydrates and for *aerobic respiration.* It enters biological organisms via the carbon dioxide (CO_2) and water (H_2O) molecules used by plants for *photosynthesis* and is released back into the atmosphere as gaseous oxygen (O_2), a byproduct of the same reactions. Oxygen is inhaled or absorbed by most organisms for use in aerobic respiration. The byproducts are CO_2 and H_2O, which are recycled back into the *environment.*

oxygenated A term usually referring to blood or other body fluids that are rich in oxygen and carry it to body tissues. (compare *deoxygenated*)

oxygen debt A *physiological* state that occurs when a normally *aerobic* (oxygen-using) animal is forced to respire *anaerobically* (without oxygen), for example during intense exercise. *Pyruvate,* an early intermediate of *cellular respiration,* is converted anaerobically to *lactic acid,* which is toxic. Lactic acid requires oxygen for its breakdown, thereby building up an oxygen debt, and is responsible for muscle soreness. The debt is repaid when oxygen is made available, allowing complete *metabolism* of lactic acid in the liver. (see *glycolysis*)

oxygenic The production of oxygen.

oxygen quotient The rate of oxygen consumption of an organism or tissue performing *aerobic respiration.* Small organisms tend to have higher oxygen quotients than larger ones.

oxyhemoglobin The *respiratory pigment hemoglobin* when it is bound with oxygen.

oxytocin A *peptide hormone* that promotes uterine contractions in pregnant women and stimulates the release of milk from the *mammary gland* after birth. It acts on *smooth muscle* and is secreted in large amounts during sexual activity in both sexes. (compare *vasopressin*)

oystercatchers (see *Charadriiformes*)

oysters (see *Bivalvia*)

ozone hole A region over the Antarctic where the natural layer of ozone in earth's stratosphere thins annually for a few months at the onset of the Antarctic spring (September). The major cause is believed to be *chlorofluorocarbons* (CFCs), once used in large amounts in cooling systems, fire extinguishers, and styrofoam containers. These molecules tend to percolate up through the atmosphere and reduce the ozone (O_3) molecules to gaseous oxygen (O_2). The thinning removes the natural stratospheric protection against harmful ultraviolet rays that induce damage in proteins and nucleic acids (*DNA* and *RNA*), increasing the *mutation* and *cancer* rates in humans. Other organisms, such as the *photosynthetic plankton* that form a main base of global productivity, are even more susceptible to damage. Ironically, CFCs were initially appropriated for use because they are extremely stable molecules and chemically inert under most conditions.

ozone layer The natural layer of ozone in earth's stratosphere (20 to 50 km above the earth's surface) that screens out harmful wavelengths of ultraviolet (UV) light. It is formed when some wavelengths of ultraviolet light split diatomic oxygen gas (O_2) and reform it as the triatomic (O_3) ozone molecule. Recently, "holes" in the ozone layer have been detected, particularly in the Antarctic region. They may be the consequence of atmospheric pollutants such as freon and aerosol propellants. Excessive amounts of short wavelength UV light are known to induce harmful *mutations* in *DNA*. Ozone itself, when present too low in the atmosphere, may contribute to the *greenhouse effect*. (see *pollution*)

P Shorthand for "parental generation." It refers to the parent organisms or strains in a particular genetic cross. (see F_1, F_2, *monohybrid, dihybrid, dominant, recessive, sex-linked*)

^{32}P An *isotope* of phosphorus that is commonly used in radioactive *labeling* of *nucleic acids.*

pacemaker (sinoatrial node, SA node) The point of origin of each heartbeat. It consists of a group of modified *cardiac muscle fibers* that have an innate tendency to *depolarize,* triggering a cascade effect leading to contraction of the entire cardiac muscle in a rhythmic fashion. The pacemaker is located where the *superior vena cava* enters the right *atrium* of the vertebrate heart. It is an evolutionary remnant of the *sinus venosus,* a major chamber in the fish heart. The pace is modified by input from the *sympathetic nervous system.* Mechanical pacemakers may replace this function in damaged hearts. (see *nerve impulse, Purkinje fibers*)

pachytene In the first *meiotic* division, the stage in mid*prophase* during which the replicated, paired *chromosomes* condense and *crossing-over* takes place. (see *meiosis; recombination, chromatid;* compare *leptotene, zygotene, diplotene*)

paed-, paedo-, ped-, pedo-, pedi- A word element derived from Greek meaning "child." (for example, *paedomorphosis*)

paedogenesis The accelerated maturation of *sexual* organs in relation to the rest of the body, enabling reproduction in an otherwise *larval* or immature organism. It is seen in the axolotl, the aquatic *larva* of a salamander, which is able to produce offspring similar to itself. (see *paedomorphosis, neotenic, heterochrony*)

paedomorphosis The evolutionary process whereby juvenile developmental stages of an animal become reproductive. (see *neotenic, paedogenesis, heterochrony*)

palate The roof of the mouth, separating the nasal passage from the *buccal cavity* in mammals. The front portion supported by bone is the *hard palate*, and the back portion is the *soft palate.*

Paleocene The oldest epoch of the Tertiary geological time period, about 65 million to 55 million years ago. It is characterized by the absence of dinosaurs and the appearance of primitive mammals.

paleoecology The use of *fossil* and sedimentary evidence to study the *ecology* of ancient times.

Paleognathae A superorder of modern birds containing the flightless ratites and tinamous. They are anatomically distinguished from the *Neognathae* by palate structure.

Paleolithic The beginning of the Stone Age, from 2 million to 10,000 years ago, when stone tools were first used.

paleontology The study of extinct organisms, including their *fossil* remains and impressions left by them. It may be further subdivided into paleozoology (animals) and paleobotany (plants).

Paleozoic The first and oldest geological era in which life became abundant, about 590 million to 230 million years ago.

palindrome A sequence of symbols that has a central axis of symmetry and reads identically in both directions. Most *restriction enzymes* recognize palindromic *DNA* sequences.

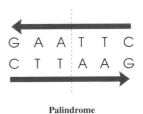

Palindrome

palisade parenchyma In a plant leaf, a layer of *columnar parenchyma* cells located just below the upper *epidermis*, or occasionally on both sides of the leaf, that contains numerous *chloroplasts* and is active in *photosynthesis*. (see *mesophyll*)

palium (see *cerebral cortex*)

palmella A stage in the *life cycle* of some *unicellular* algae in which cells remain within the parental envelope after *mitosis*. This type of *cell division* may continue producing a *multicellular* mass, or the cells may develop *undulipodia* and become independently mobile at any time. Cells of the genus *Palmella* typically exist in the multicellular form.

Palpigrada A minor order of minute, colorless arthropods, the palpigrades or microwhipscorpions. They inhabit damp spaces under stones and soil.

palps A pair of appendages situated near the mouth in many invertebrates including bivalves and some insects.

palynology (see *pollen analysis*)

pancreas The major digestive *gland* in vertebrates. It has both *exocrine* and *endocrine* components. It manufactures a host of *digestive enzymes* that are secreted through a duct into the *duodenum* as well as bicarbonate, which neutralizes stomach acid and allows the enzymes to function. Both *insulin* and *glucagon*, *hormones* regulating blood sugar levels, are made in the *islets of Langerhans*, while endocrine cells are distributed throughout the pancreas.

pangolins (see *Pholidota*)

panicle In plants, a multiple flowering structure (*inflorescence*) composed of several growing stalks (*racemes*). More loosely, it is applied to any complex, branched flowering structure.

panmixis The free interbreeding between members of a population such that there are no important barriers to *gene flow*. The population represents one *gene pool*.

pantothenic acid (vitamin B$_5$) A member of the *vitamin B complex* found in egg yolk, kidney, liver, and yeast. It is a constituent of *coenzyme A*, required by the *Krebs cycle* in *aerobic respiration*, thus is an essential nutrient.

paper chromatography A laboratory technique in which soluble small molecules are separated by differential migration along a solvent front proceeding by capillary action through absorbent filter paper. Particular substances have characteristic migration distances in different solvents (*Rf value*) that are determined by physical characteristics such as differential solubility, charge and mass.

papilla, papillae (pl.) Any projection from a cell, tissue, or organ. For instance, *dermal* papillae provide contact friction (and give fingerprint patterns), and tongue *papillae* increase the surface area for *taste buds* in mammals.

papilloma virus A group of small *DNA* viruses that cause warts and appear to be associated with cervical cancer.

pappus, pappi (pl.) In certain plants, a tuft or ring of hairs or bristles found at one end of the fruit. It is derived from the *calyx* and persists on the fruit, aiding in wind dispersal of seeds.

para- A word element derived from Greek meaning "beside," "along side," or "near" and sometimes implying modification. (for example, *parasexuality*)

parallel adaption (see *parallel evolution*)

parallel evolution (parallel adaption) The development of similar features in ancestrally related but geographically isolated organisms as a result of *natural selection* in similar environments. It is less common than *convergent evolution*. A good example is the similarities of some individual members of Australian *marsupial* mammals to *placental* mammals on other continents.

paralogous Two *genes* or gene clusters situated at different *chromosomal* locations in the same organism that have structural similarities indicating that they derive from a common ancestral gene.

Paramecium A common genus of *ciliate* protists found in freshwater containing decaying vegetable matter. They are covered with *cilia* and reproduce both *asexually* by *mitosis* and *sexually* by *conjugation*. They contain *contractile vacuoles* for *osmoregulation* and take in food through a specialized oral groove (*cytostome*). Food is then digested in food *vacuoles*. It has both a *macronucleus* and a *micronucleus*. (see *cytoproct, cytopharynx, peristome*)

Paramyxa A phylum of *parasitic unicellular* protists that are unique in their *spore* formation. They undergo an internal cell cleavage so that each spore is successively enclosed in the previous one. They are *parasitic* on various insects or mullusks and include *pathogens* that infect commercially important seafood such as oysters and clams. They were formerly classed in a larger grouping, the sprorozoans.

parapodium, parapodia (pl.) In polychaete worms, pairs of appendages occuring on most segments. They function as swimming paddles and provide a large respiratory surface.

parasexuality A *sexual reproductive* process that generates an offspring cell containing genetic material from more than a single parent but without standard *meiosis* or *fertilization*. It takes place in deuteromycetes and some other organisms, for instance cellular slime molds. Portions of *chromosomes* are exchanged between genetically distinct *nuclei* within a common *hypha* or *syncytium*. (see *dikaryon, heterokaryon*)

parasite The dependent organism of the two in a *parasitic* relationship. It lives on or in an organism of a different species and obtains nutrients and energy from it. An alternative description is a *symbiotroph* with *necrotrophic* tendencies. (see *host*)

parasitic flatworms (see *Trematoda, Cestoda*)

parasitism A *symbiotic* relationship between two organisms in which one, the *parasite*, benefits at the expense of the other, the *host*. The parasite lives on or in the host, on which it is *metabolically* dependent. The tolerance of the host varies from being almost unaffected to serious illness and death. (see *ectoparasite, endoparasite, symbiotroph, nectotroph, facultative, obligate*; compare *commensalism, mutualism*)

parasitoid An organism, often an insect, that lays its eggs in the body of another animal. Parasitoids grow and develop using the *host's* resources without killing it, although the host is often killed at hatching.

parasympathetic nervous system (PNS) One of the two divisions of the *autonomic* (involuntary) *nervous system*. It supplies *motor nerves* to the *smooth muscles* of the internal organs and *cardiac* muscles. The endings of parasympathetic nerves release the *neurotransmitter acetylcholine*. This lowers heart rate and blood pressure and promotes digestion, thereby antagonizing the effects of the *sympathetic nervous system*. (see *central nervous system, peripheral nervous system*)

parathyroid glands Four small *glands* embedded in the *thyroid gland* that produce *parathyroid hormone* and *calcitonin*. These hormones antagonistically control blood calcium levels through a *feedback regulation* mechanism.

parathyroid hormone (PTH) A *peptide hormone* produced by the *parathyroid glands* that regulates calcium levels in the blood. Since the release of calcium drives muscle contraction, absence of this hormone will result in cessation of *cardiac* (heart) muscle function and death. PTH also acts on *osteoclasts* to release calcium from bone when blood levels drop, promotes reabsorbtion from the kidneys, and activates *vitamin D* which is necessary for absorbtion of calcium by the intestine. It is one of two (the other being *aldosterone*) hormones absolutely essential to survival.

Parazoa A subkingdom comprising the most primitive animals, traditionally defined as organisms generally lacking definite symmetry or any distinct tissues or organs. The definition has become somewhat strained, as two phyla have recently been added (*Rhombozoa, Orthonectida*) that exhibit approximate bilateral symmetry and contain a *gonad*.

parenchyma
1. In animals, a spongy tissue made of loosely packed cells. In flatworms for example, it contains the organs, and oxygen and nutrients can diffuse through it.
2. In plants, an undifferentiated tissue in which many important *metabolic* processes are carried out. It also has the capacity to *differentiate* into various specialized plant tissues. The leaf *mesophyll* and stem *medulla* and *cortex* consist of parenchyma. (compare *collenchyma, sclerenchyma*)

parietal Any structure located in a peripheral position.

parietal cells Specialized cells that secrete *hydrochloric acid (HCl)*. They are contained in *exocrine glands* in the *gastric pits* in the stomach lining. (see *gastric juice, pepsinogen*)

parietal eye (see *median eye*)

parietal lobe The middle portion of the mammalian *cerebrum*, anatomically separated from the three other lobes by deep grooves. Functionally, it contains the *primary motor cortex* and the *primary somatosensory cortex*. (see *cerebral cortex, telencephalon*; compare *occipital lobe, temporal lobe, frontal lobe*)

parrots (see *Psittaciformes*)

parthenocarpous The development of fruit from un*fertilized* flowers, resulting in seedless fruit. It may occur naturally, as in bananas and pineapples, or may be induced artificially in various commercial applications.

parthenogenesis The development of an un*fertilized haploid gamete* (commonly an egg cell) into a complete organism. It occurs regularly in certain plants (for example, dandelion) and animals (for example, aphids). Animals resulting from parthenogenesis are always female and if *diploid*, are exact replicas of the parent. The *chromosomal* complement of the new individual is determined by the state of the *ovum* (egg cell) when development of the *embryo* begins. Genetic *recombination* is precluded by parthenogenesis.

In organisms where parthenogenesis is common, it is often interspersed with *sexual reproduction*. In the laboratory, artificial parthenogenesis can be induced by various physical or chemical treatments. (see *apomixis*)

partial digestion The incomplete cleavage of *DNA* by a *restriction endonuclease*. In other words, some of the sites specific to that enzyme are cut but not others. Partial digestion is sometimes used as a specialized technique in *cloning,* such as for *chromosome jumping*. At other times, it is an unwanted side effect caused by impure DNA or incorrect biochemical conditions that simply complicates an analysis.

partial dominance (see *codominance*)

partruition In animals bearing live young (*viviparous*), the process of giving birth at the termination of pregnancy.

parvovirus A group of *DNA* viruses that include some of the smallest and simplest viruses. They contain *single-stranded DNA* and are implicated as a factor in arthritis.

Passiformes The very large order of birds comprising all the perching birds, a few examples of which are flycatchers, thrushes, warblers, and creepers.

passive diffusion The passage of dissolved molecules across a *semipermeable membrane* as determined solely by pore sizes and concentration gradients. Molecules always move in the direction of lower ion concentration. (see *osmosis*; compare *facilitated diffusion, active transport*)

passive immunity A type of *immunity* in which *antibodies* against a particular disease are acquired from another individual who has produced them through *active immunity*. This is achieved artificially through injection with an *immune serum*. It occurs naturally through the *placenta* during pregnancy, through breast secretions (*colostrum*) just after birth, and through breast milk. (see *immune system*; compare *active immunity*)

pasteurization A method of preserving food products, such as milk, by heating to a temperature just below boiling, followed by rapid cooling. This serves to kill microorganisms without destroying flavor, theoretically. It is named after its discoverer, Louis Pasteur, who used it to prevent spoilage of wine and beer in the 1850s.

patch clamp A method of measuring electrical activity within a single *ion channel* in a *nerve cell*. It consists of isolating a small section of the *plasma membrane* by drawing it up into the narrow tip of a glass *pipette* and stimulating the trapped ion channel(s) with various ionic solutions introduced through the pipette. Erwin Neher and Bert Sakmann won the Nobel prize in physiology or medicine in 1991 for their development of the technique.

patella The small bone in front of the knee joint of most mammals, some birds, and some reptiles. In man, it forms the kneecap.

-path, -pathic A word element derived from Greek meaning "disease." (for example, *pathogen*)

pathogen Any organism that is capable of causing disease or a toxic response in another organism. Many bacteria, viruses, fungi and other microorganisms are pathogenic. (see *parasite, necrotrophic*)

Paucituberculata The order of marsupials comprising the opossum rats.

Pauropoda A class of uniramian arthropods, the pauropods. They are soft-bodied animals resembling centipedes and widely occur in temperate and tropical forest litter.

PC (plastocyanin) A copper-containing protein that functions as an electron carrier, linking *photosystems I* and *II*.

PCR (polymerase chain reaction) Perhaps the most important advance in molecular biology in the latter part of the 20th century, PCR allows millions of copies of a chosen segment of *nucleic acid* (*DNA, RNA*) to be replicated accurately and efficiently *in vitro*. Its inventor, Kerry Mullis, received the Nobel prize in chemistry in 1993 for his discovery. The ends of the segment to be copied are defined by small pieces of synthetic DNA called *primers*. The reaction is facilitated by a heat stable *DNA polymerase*, originally and most often *Taq*, that is able to withstand the many cycles of heating and cooling involved in PCR.

Pd (see *power of discrimination*)

peat Partially decomposed plant material that accumulates on the forest floor in *temperate rainforests* such as on the Northwest coast of the United States. It occurs when decomposition is slow due to *anaerobic* conditions imposed by water saturation. Peat is rich in nutrients, encouraging new growth. It is used directly as fuel and is also the first step in coal formation.

peccaries (see *Artiodactyla*)

pecking order (see *dominance hierarchy*)

pectins A heterogeneous group of *polysaccharides* that, together with *hemicellulose*, form the matrix within which *cellulose* fibers are embedded in plant *cell walls*.

pectoral Any shoulder structure.

pectoral fins In fish, the pair of fins that are situated just behind the *gills* with one on each side.

pectoral girdle (shoulder girdle) In vertebrates, a bony or *cartilaginous* skeletal structure consisting of several elements to which the forelimbs or fins are attached.

ped-, pedi
1. A word element derived from Greek meaning "foot." (for example, *pedipalps*; see also *pod-*)
2. (see *paed-*)

pedal Referring to the foot.

pedicel In plants, the stalk that attaches an individual flower to the main stalk of a multiflowered structure (*inflorescense*). (see *peduncle*)

pedigree A family tree showing inheritance patterns of specific genetic traits.

pedipalps In arachnids, a pair of segmented appendages attached at the back of the head behind the *chelicerae*. In spiders, they sometimes function in place of

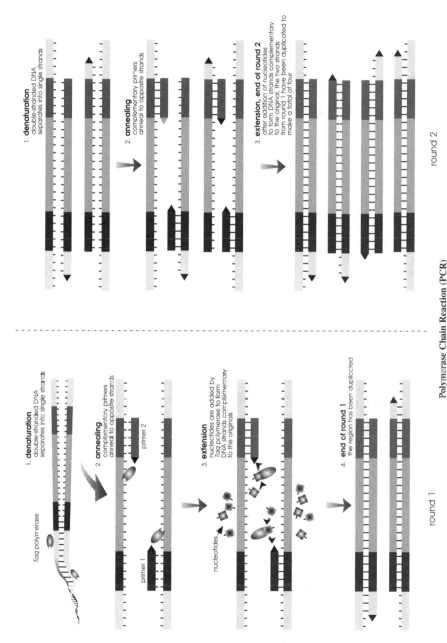

Polymerase Chain Reaction (PCR)

Modified from an illustration in *An Introduction to Forensic DNA Typing*, ©1997, by Inman and Rudin, CRC Press, Boca Raton, Florida.

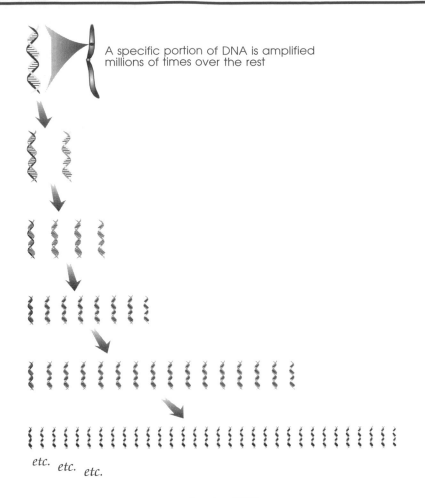

A specific portion of DNA is amplified millions of times over the rest

etc. etc. etc.

Overview of PCR

Modified from an illustration in *An Introduction to Forensic DNA Typing,* ©1997, by Inman and Rudin, CRC Press, Boca Raton, Florida.

antennae and in males, as specialized *copulatory* organs. They are often used for grasping prey, particularly in scorpions. The segment nearest the body functions in chewing food.

peduncle In plants, the main stalk of a multiflowered structure (*inflorescence*) to which individual stalks bearing single flowers (*pedicels*) are attached.

pelagic The organisms that inhabit the open waters of a sea or ocean (*oceanic*). They are synonymous with the *limnetic* organisms of a freshwater system. Pelagic organisms are divided into *plankton* and *nekton*. (compare *benthic*)

Pelecaniformes An order of birds containing a group of water-loving birds, the pelicans, tropicbirds, frigatebirds, boobies, gannets, cormorants, and anhingas.

P elements A family of *transposable genetic elements* found in *Drosophila*. Many are about 3 *kb* (kilobases) long, and the rest appear to be derived from these. They are able to mediate their own *transposition* to a distant location in the *genome* and are responsible for *hybrid dysgenesis*. P elements have been used extensively in the laboratory as a tool to insert *genes* in the *Drosophila* genome. (see *transposon*; compare *TY element, copialike element, FB element, LINEs*)

pelicans (see *Pelecaniformes*)

pellicle A transparent, flexible outer covering found in some protists, especially *ciliates* (for example, *Paramecium*), and also in many *parasitic* organisms. It is made mostly of protein and functions as an *exoskeleton* to protect and maintain body shape.

Pelmatozoa A subphylum of echinoderms containing a single class, the crinoids. They are distinguished by their attachment to a substrate on the surface opposite their mouth.

pelvic Any hip structure.

pelvic fins (ventral fins) In fish, the pair of fins positioned on the underside of the body.

pelvic girdle (hip girdle) In vertebrates, a bony or *cartilaginous* skeletal structure consisting of several elements to which the hind limbs or fins are attached.

pelvis (pelvic girdle) The hipbones in mammals. The term pelvis may also refer to the cavity formed by the bones of the *pelvic girdle*.

penetrance In genetics, the variable expression of particular *genes* in the population resulting in a range of *phenotypes* (physical manifestation). In complete (100%) penetrance, a particular *gene* is expressed as a predictable characteristic in all individuals in the population. Incomplete penetrance may result from the modifying action of other *genes* or environmental factors. (see *dominant, recessive*; compare *expressivity*)

penguins (see *Sphenisciformes*)

penis The male *copulatory* organ used by many animals during internal *fertilization* to introduce *spermatozoa* into the female reproductive tract. In mammals, both *urine* and *semen* pass via the *urethra* through the penis to the exterior. (see *intromittent organ*)

pennate A structure resembling a feather.

pentadactyl limb A limb having five digits at the end. It has a bone structure characteristic of all tetrapod (four-limbed) vertebrates. Numerous variations are seen that adapt the basic structure to running, swimming, digging, and flying. (see *digitigrade, plantigrade, unguligrade*)

Pentastoma A phylum of animals, the tongue worms, that are named for the five projections at the front of their bodies. They are *parasitic* on vertebrates, usually inhabiting the respiratory passages. Some *metamorphose* through three distinct stages. They mate within their *host*, and eggs are excreted in the feces.

pentose A sugar that has five carbon atoms in each molecule. (see *ribose, deoxyribose*)

pentose phosphate shunt An alternative pathway of glucose breakdown that is sometimes used in lieu of *glycolysis*. The principle functions of the pathway are the production of the *pentose* sugars *deoxyri-*

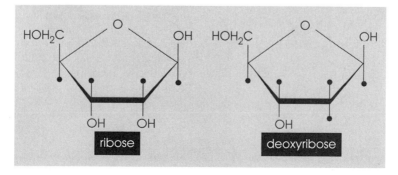

Pentose Sugars

bose and *ribose* for *nucleic acid* synthesis, the generation of reducing power in the form of *NADPH* for *fatty acid* and/or *steroid* synthesis, and the interconversion of carbohydrates. Intermediates in the pathway feed into the *Calvin cycle* in *photosynthesis*.

pepo The type of fruit seen in cucumber, squash, and pumpkin where the flesh is not easily separable from the rind.

pepsin An enzyme that catalyzes the partial *hydrolysis* of proteins to *polypeptides*. It is secreted in an inactive form by the *gastric glands* in the stomach wall as *pepsinogen* and is activated by stomach acid.

pepsinogen The inactive form of the gastric enzyme *pepsin*.

peptidases (see *proteases*)

peptide A *polymer* consisting of *amino acids* joined by *peptide bonds*. It is often used interchangeably with the term *polypeptide* when referring to biological *macromolecules*. (see *oligopeptide, polypeptide, protein*)

peptide bond The chemical bond linking two adjacent *amino acids* and formed by the reaction between the carboxyl and amino groups on each end. (see *peptide, oligopeptide, polypeptide, protein*)

Formation of a Peptide Bond

peptide hormone Any *hormone* that consists of a *polypeptide* chain. They typically work by binding to a *receptor* on the cell surface that triggers events within the cell through a *second messenger*. (see *transmembrane protein, transmembrane receptor*)

peptidoglycan A *glycoprotein polymer* found in the rigid layer of bacterial *cell walls*.

Peramelemorphia (Peramelida) The order of marsupials comprising the bandicoots.

Peramelida (see *Peramelemorphia*)

perching birds (see *Passiformes*)

perennating structure A specialized structure in which food is stored in many plants that use *dormancy* as a means of surviving seasonal changes in climactic conditions. In *biennial* and *perennial herbaceous* plants, this occurs in specialized underground structures such as *bulbs, tubers, corms,* and *rhizomes.* The aerial parts die back annually and the plant is restored the next season from the perennating structure. In *deciduous woody* plants, the leaves drop and buds on the stems are used to store food. Seeds are sometimes regarded as perennating structures. (see *vegetative propagation, perennation, perennating structure*)

perennation A method used by many plants to survive unfavorable conditions. The main body of the plant becomes *dormant*, and food is stored for the next growing season in various *perennating structures*.

perennial A plant that survives and continues growing from year to year. In *herbaceous* perennials, *vegetative* underground *perennating structures* such as *bulbs, tubers, corms,* and *rhizomes* store food from one growing season to the next. In woody plants, buds on the stems accomplish the same purpose. In larger woody perennials such as trees and shrubs, the body of the plant persists aboveground throughout the year but may show adaptions, such as leaf drop, to survive unfavorable conditions. (see *deciduous, evergreen, wood*; compare *annual, biennial, ephemeral*)

peri- A word element derived from Greek

meaning "about," "around," or "beyond." (for example, *pericycle*)

perianth The part of a flower that encircles the reproductive structures (*stamens* and *carpels*). It usually consists of two whorls of leaflike structures that, in *Dicotyledons*, are differentiated into the *sepals* (*calyx*) and *petals* (*corolla*). In flowers that are wind *pollinated*, the perianth tends to be reduced (for example, grasses) or absent (for example, willow).

pericardial cavity The cavity containing the heart.

pericardium The *epithelial* tissue layer lining the *pericardial cavity*.

pericarp In plants, the *ovary* wall that differentiates to become the fruit. It is separated into the outer layer (*exocarp, epicarp*) which is often a tough skin, the middle layer (*mesocarp*) which may range from juicy (for example, plum) to hard (for example, almond), and the innermost layer (*endocarp*) which forms the stony covering of the seed in a *drupe* (stone fruit) but in others is indistinguishable from the mesocarp.

periclinal In plants, the alignment of the plane of *cell division* approximately parallel to the outer surface of a structure. (compare *anticlinal*)

pericycle In plant roots, a layer of cells just inside the *endodermis*. It gives rise to lateral, or branch, roots.

periderm The outer protective covering of a mature stem or root in a *woody* plant. It consists of the *cork, cork cambium,* and *phelloderm.*

perigynous A flower in which the *calyx, corolla,* and *stamens* are located around the *ovary* rather than above or below it. The ovary is technically *superior*, although the floral parts may be inserted above it because of an extended *receptacle*. (compare *epigynous, perigynous*)

perilymph The fluid that surrounds the structures of the *inner ear* of vertebrates. (compare *endolymph*)

perineum The region between the anus and *urogenital* openings of *placental mammals.*

period (see *geological time scale*)

periosteum The sheath of *connective tissue* that surrounds vertebrate bones and to which tendons attach. It contains *osteoblasts* and bundles of *collagen* fibers.

peripheral nervous system The system of nerves in vertebrates that connects the *central nervous system* (*CNS*) to the organs and peripheral regions of the body. It includes both *efferent* (*motor*) and *afferent* (*sensory*) nerves. It is divided into the *autonomic nervous system* (involuntary) and the *voluntary nervous system.*

periphyton (see *aufwuchs*)

periplasm Cytoplasm found in the peripheral regions of a cell, specifically between the *plasma membrane* and *cell wall* of *prokaryotes.*

perisperm In plant seeds, an *embryonic* nutritive tissue that supplements or replaces the *endosperm* or *cotyledons* as a food source during *germination*. It is derived from the *nucellus.*

Perissodactyla The order of mammals that contains the odd-toed *ungulates*, including horses (one toe) and rhinoceroses (three toes). These *herbivorous* mammals typically have feet encased in a protective horny hoof, lips adapted for plucking, and teeth adapted for plucking and chewing. The stomach is simple, and *bacterial* digestion of *cellulose* occurs in the *cecum*. (see *ungulate*)

peristalsis Waves of muscular contraction that pass along tubular organs of the body, primarily the *alimentary canal*, where they serve to transport food through the *digestive system*. The rate and force of peristalsis is regulated by the *autonomic nervous system*, but the wave itself is an intrinsic property of the particular muscle tissue. (see *smooth muscle*)

peristome In general, a mouthlike opening.
1. The funnel-like region around the mouth of *ciliate* protists.
2. The edge of the opening in a gastropod shell.
3. An opening in the *spore* capsule of mosses. The teeth (*setae*) surrounding the moss peristome are *hygroscopic*

(absorb moisture from the air), causing the coiled fibers that comprise them to distort the capsule, aiding in *spore* dispersal.

perithecium, perithecia (pl.) A type of *fruiting body* found in some ascomycete fungi in which the *spores* are discharged through a pore. (see *ascocarp*)

peritoneal (see *coelemate*)

peritoneum The lining of the abdomen.

peritrichous Bacteria whose cell surfaces are covered with *flagella* (for example, *Proteus*).

permanent teeth The second set of teeth in most mammals, replacing the *deciduous teeth* (milk teeth).

permanent wilting point The point at which soil has dried to the extent that plants can no longer remove the remaining water held on the surface of soil particles and begin to wilt.

Permian The most recent period of the Paleozoic geologic era, about 280 million to 230 million years ago. Primitive seed plants replaced ferns and modern insect groups appeared. Amphibians declined, while a few types of reptiles thrived.

permissive The conditions under which a *conditional mutation* in *DNA* exhibits the normal (*wild-type*) *phenotype*. (compare *permissive*)

peroxisome (see *microbody*)

pest Any species whose existence in a particular area conflicts with human profit, convenience, or welfare.

pesticides Any compound that is intended to kill a wide variety of organisms including garden and crop pests. General chemical pesticides are usually detrimental, at least at some level, to all animal life. (see *food chain concentration, chlorinated hydrocarbons*)

petal In flowering plants (*angiosperms*), one of the inner floral leaves borne in a spiral or whorl peripheral to the reproductive structures. The petals collectively make up the *corolla*. They are often brightly colored and scented to attract insects and birds, and are usually reduced or absent in wind-*pollinated* flowers.

petiole In plants, the stalk by which a leaf blade is attached to the stem.

petrels (see *Procellariiformes*)

petri dish A shallow glass or plastic container with a removable lid that is used in the laboratory for growing microorganisms or culturing cells. Historically, petri dishes have been round, but newer plastic version are often square-shaped for more convenient stacking and storage. They are named after their inventor, the German bacteriologist Julias Petri.

P_{fr} (P_{730}) (see *phytochrome*)

Ph (pheophytin) A *chlorophyll* molecule lacking a magnesium (Mg^{++}) ion. It is an intermediate in *photosystem II*.

pH A measure of the acidity or alkalinity of a solution of a scale on 0 to 14. A neutral solution has a pH of 7. Acid solutions have a pH below 7, alkaline solutions above 7. Acids tend to give off a free hydrogen ion (H^+) in solution while bases release a hydroxide ion (OH^-). The pH value is calculated as the reciprocal of the hydrogen ion concentration expressed in moles per liter ($-\log[H]$).

Phaeophyta The phylum containing the largest *multicellular* protoctists, the brown algae, notably the giant kelp. They are found mostly in marine environments and contain various *photosynthetic pigments* including primary *chlorophylls* as well as the accessory *xanthophylls*, which confer their characteristic brown color. Like plants, the *diploid sporophyte* is manifest as the conspicuous seaweed, and sex organs in the form of *antheridia* and *oogonia* are produced on its surface. Male *gametes* are motile, and swim to the oogonium to *fertilize* the egg. They are not plants because they do not form *diploid embryos,* which are retained in the maternal tissue. Pheophytes are an important source of dyes, adhesives, vitamins, minerals, and various pharmaceuticals. (see *violaxanthin*)

phag-, -phage, -phagy, -phagic A word element derived from Greek meaning "eat" or "devour." (for example, *bacteriophage*)

phage (see *bacteriophage*)

phage library A collection of *genomic DNA* fragments, each contained in a *bacteriophage cloning vector*, usually bacteriophage λ, and propagated by infection of the *host* bacteria, usually *E. coli*.

phagocyte A cell that is specialized for engulfing and digesting particles by *phagocytosis*. They are found mainly in the vertebrate *immune system*, where they locate and destroy invading microorganisms. (see *macrophage, neutrophil*)

phagocytosis The mechanism by which a cell ingests solid particles. It is accomplished by extension of the *cell membrane* to engulf the substance and invagination to form a *vesicle* or *vacuole*, which is then pinched off inside the cell. Phagocytosis is used by both *multicellular* and *unicellular* organisms for transport, feeding, and defense. (see *endocytosis, lysosome*; compare *pinocytosis*)

phagotrophic A mode of nutrition in which a single cell ingests particles of food via *phagocytosis*. It is common among protoctists.

phalanges The small bones that form the *digits* (fingers and toes) of animal limbs. (see *pentadactyl, metacarpals, metatarsals*)

phanerophyte A *perennial* plant with persistent shoots and buds well above the soil level. (see *Raunkiaer's plant classification*)

Phanerozoic The more recent of the two eons of *geologic time*, during which there was an explosive evolution of diverse life forms.

pharyngeal slits (see *gill slits*)

pharynx The part of the *alimentary canal* between the mouth and the rest of the digestive tract. In mammals, it is usually considered the throat region and opens to the nasal passages, the *esophagus,* and the *trachea*. The *Eustachian tube* leading from the *middle ear* also opens into the pharynx.

phase contrast microscope A light microscope in which a portion of the emitted light is retarded, such that it is out of phase with the remainder. This produces a darkening of the field and increasing contrast, which is particularly useful for biological specimens. In addition, many biological substructures, such as cell *organelles*, retard light, further increasing the contrast and, consequently, the visibility of the specimen.

phasmids A pair of *sensory* organs in nematodes believed to be *chemoreceptors* and located at the rear end. (compare *amphids*)

pheasants (see *Galliformes*)

phelloderm In *woody* plants, a dense tissue composed mainly of *parenchyma* cells. It is located just inside the *cork cambium*.

phellogen (see *cork cambium*)

phellum (see *cork*)

phenetics (numerical taxonomy). The classification of organisms based on large numbers of measurable morphological traits. In modern classification schemes, phenetics is combined with *cladistics*, the *phylogenetic* classification of organisms, in particular protein data and, most recently, *DNA analysis*.

phenocopy A change in the appearance of an organism that resembles a *phenotype* caused by *gene mutation* but, in fact, is only a physiological adaption. Such changes, which are not inherited, are generally caused by environmental factors affecting the organism at an early stage of development.

phenol An organic chemical commonly used in the extraction and purification of *nucleic acids* (*DNA, RNA*), particularly DNA, for laboratory analysis. Contaminating molecules either partition to the phenol layer (for example, lipids) or to the interface between the phenol and aqueous layers (for example, carbohydrates), while nucleic acids partition to the aqueous layer.

phenology The study of periodicity in organisms, particularly in relation to the local climate and environment. It includes phenomena such as flowering, *migration,* and *metamorphosis*.

phenolphthalein An acid-base indicator

that is colorless in acids and red in alkali. It is also used, in combination with other reagents, to test for *hemoglobin* in order to identify blood.

phenotype The physical appearance of an organism. It is determined by the complex interactions of the totality of its *genes* with each other and with the environment. It is distinguished from the genetic makeup of an organism, or *genotype*, which is a description of all genes that are present in the *genome* regardless of their state of expression or modification. Genotype can be determined only by specific testing. The term phenotype may also be used to describe a discrete outward manifestation of a particular gene or genes, if known.

phenotypic plasticity The extent to which *gene expression* and the manifestation of a *phenotype* may be influenced by different environments. (see *ecotype, cline*)

phenylalanine (phe) An *amino acid.*

pheophytin (see *Ph*)

pheromone A chemical excreted by an animal in minute amounts that causes a response in others of the same species. Pheromones often function as *sexual* attractants, particularly in vertebrates, but also serve to communicate a variety of messages, for instance food availability and danger in insect *communities.*

philo-, -phil, -philic A word element derived from Greek meaning "loving." (for example, *hydrophilic*)

pH indicator (see *indicator*)

phloem In *vascular plants*, the internal tissue that distributes carbohydrates and other products of *photosynthesis* (*photosynthate*), sometimes over great distances. It is made of *sieve tubes*, which are columns of *enucleate* living cells, connected by perforations in their *cell walls*, along with *companion cells*, which supply their nutritional needs. (see *vascular cambium*, compare *xylem*)

-phobic, -phobe A word element derived from Greek meaning "aversion" or "hating." (for example, *hydrophobic*)

Phoenicopteriformes An order of birds including the flamingos and allies.

Pholidota An order of placental mammals comprising a single family, the pangolins or scaly anteaters.

-phore- A word element derived from Greek meaning "bear," or "carry," or the "part that bears or carries." (for example, *electrophoresis*)

Phoronida One of the three phyla of *lophophorates*, the phoronids. They are marine invertebrates superficially resembling tube worms and secrete a *chitinous* tube within which they live either singly or in groups. They extend their *lophophore*, a ring or U-shaped ridge of *tentacles* around their mouth, to capture food. They develop as *protostomes*, and some species show *spiral cleavage.*

phoronids (see *Phoronida*)

phosphagens The class of compounds in which cellular energy is stored in high-energy phosphate bonds. They include *ATP, ADP, creatine phosphate*, and in invertebrates, *argenine phosphate.*

phosphatase Any enzyme that catalyses the removal of a phosphate group from an organic compound. (compare *kinase, phosphorylase*)

phosphatidylcholine (see *lecithin*)

phosphodiesterases A group of enzymes that split (*hydrolyze*) the *phosphodiester bonds* found in *nucleic acids*. An important cellular function is the catalysis of cyclic AMP (*cAMP*) into noncyclic AMP (5'-AMP). Some also work on the *sugar-phosphate backbone* of linear nucleic acids (*DNA, RNA*). (see *second messenger*)

phosphodiester bond The chemical linkage found in *nucleic acids*. It is formed by a condensation reaction between the phosphate group (PO_4^{3-}) of one *nucleotide* with the hydroxyl group (OH-) of another, creating a diester (-O-) bond.

phospholipid A class of lipid known primarily for its importance in biological membranes. Each molecule contains one or more phosphate groups that are charged, therefore *hydrophilic*, and a nonpolar *fatty acid* tail that is *hydropho-*

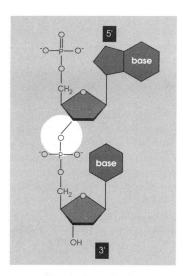

Phosphodiester Bond

bic. They are arranged tail-to-tail in the *phospholipid bilayer*, with their heads extending into the aqueous *intracellular* and *extracellular* environments. (see *amphipathic, plasma membrane*)

phospholipid bilayer The basic structure of all *plasma membranes* (cell membranes). It is composed of a double layer of *phospholipids* that orient spontaneously in an aqueous solution. Their *fatty acid* tails are oriented inward, toward one another, and their charged phosphate heads are oriented outward, into the aqueous environment. Other membrane constituents, such as proteins, are embedded in the phospholipid bi-

layer. The structure is fluid rather than rigid.

phosphoprotein A protein to which are conjugated one or more phosphate groups. *Casein* (milk protein) is an example.

phosphorescence The reemission of visible light as a result of the excitation of a substrate by low-energy radiation. The energy level of the reemitted light is always lower than the exciting radiation, and the wavelength (λ) concomitantly longer. The reemission is slightly delayed in comparison with *fluorescence*, thus the energy even lower and the wavelength even longer.

phosphorus One of the elements essential to living organisms. In vertebrates, calcium phosphate is the main constituent of bone and teeth, and *phospholipids* provide the basic structure of all *cell membranes*. Phosphate is part of the *sugar-phosphate backbone* of *nucleic acids*. It is essential for energy storage and transfer via high-energy bonds in molecules such as *ATP, creatine phosphate*, and *arginine phosphate*. The phosphate ion (PO_4^{3-}) is also an important biological buffer.

phosphorylase An enzyme that transfers a phosphate group, often from an inorganic source, to an organic compound that becomes *phosphorylated*. Phosphorylases are important in the mobilization of carbohydrate reserves. (compare *kinase, phosphatase*)

phosphorylation The addition of a phos-

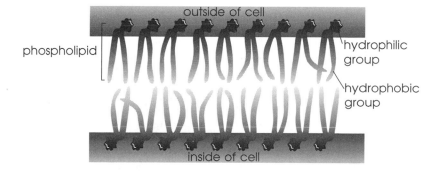

Phospholipid Bilayer

phate group to an organic compound. It is the most important energy transfer system in biological *metabolism*. (see *ATP, photophosphorylation, oxidative phosphorylation, phosphagen, kinase, phosphorylase*)

photic zone (surface zone) The surface layer of an ocean or lake (to about 200 meters) that is penetrated by enough sunlight to support *photosynthesis*. It is therefore rich in life, primarily *diatoms* and other *phytoplankton*. (compare *neritic, oceanic, abyssal*)

photo- A word element derived from Greek meaning "light." (see *photosynthesis*)

photoautotrophism The method of nutrition used by most *autotrophs*, in which light energy, via *photosynthesis*, drives the synthesis of organic molecules. Photoautotrophs manufacture all their organic requirements from inorganic carbon (CO_2) and are the *primary producers* of organic molecules for all *heterotrophs*. They include all plants as well as algae and photosynthetic bacteria. (see *trophism, autotrophism, phototrophism*)

photodynamic effect Any of the harmful effects on biological systems caused by light. They range from chromosomal mutations incurred as a consequence of exposure to UV light, to the burning of plant leaves caused by the focusing of light from the midday sun by water droplets. Although light is essential to most life, the amount and particular wavelengths (λ) absorbed is tightly regulated. Organisms have evolved a number of protection mechanisms, many of which utilize light-absorbing pigments such as *melanin, chlorophylls,* and *carotenoids*, to screen the interior of cells from inappropriate exposure to light.

photoheterotrophism The method of nutrition used by some bacteria in which energy is derived from light and carbon is derived from small organic molecules such as acetate, proprionate, and pyruvate.

photolysis The chemical breakdown, particularly of water (H_2O), by light. The process is important in the initial steps of *photosynthesis* as a source of electrons that are channeled into *photosystem II*. The hydroxyl (OH^-) radicals are reassembled into water and gaseous oxygen (O_2), while the H^+ ions (protons) augment the gradient established by the *proton pump* that drives the *chemiosmotic* synthesis of *ATP* in *chloroplasts*. They are are ultimately *fixed* into carbohydrate. (see *protein Z, Z scheme, photosystem II*)

photomicrograph (see *micrograph*)

photonastic In plants, a movement in response to a light stimulus. (see *nastic movement*)

photoperiodism The response of an organism to periodic changes in day length. Such responses include leaf drop and flowering in plants, and *molting* and *migration* in animals.

photophosphorylation The *chemiosmotic* production of *ATP* during *photosynthesis*. It is driven by the proton gradient created as high-energy electrons flow through the *electron transport chain* from *photosystem I* to *photosystem II* in *chloroplasts*. (see *cyclic photophosphorylation*; compare *oxidative phosphorylation*)

photopigment (see *photosynthetic pigments*)

photoplankton (see *plankton*)

photoreceptor In general, any light sensitive molecule, *organelle,* or organ that initiates a specific biological action in response to the absorbtion of light. This includes the *phytochromes* and *photosynthetic pigments* in green plants, the *eyespots* of some invertebrates, and the *compound* and simple eyes of higher animals. Often, a photoreceptor is sensitive to a particular wavelength (λ) or wavelengths of light.

photorespiration A process occuring in many plants in which organic carbon is combined with molecular oxygen (O_2) into carbon dioxide (CO_2) without the production of *ATP*, thus undoing the work of *photosynthesis*. It is initiated by a secondary activity of the enzyme *RuBP carboxylase*, whose primary job is to catalyze the key *carbon-fixing* reaction of photosynthesis. It is apparently an evolutionary relic, surviving from a time when not much oxygen was present in the earth's atmosphere. Some groups of plants have evolved secondary mechanisms to circumvent the process, thus conserving fixed carbon. (see *Calvin-Benson cycle, C_4 carbon-fixation, CAM carbon-fixation, bundle-sheath cell*)

photosynthate The combined products of *photosynthesis*.

photosynthesis The utilization of light energy to power biosynthesis. It is the process by which green plants, algae (photosynthetic protoctists), and some photosynthetic bacteria (cyanobacteria) capture energy from sunlight and combine it with carbon dioxide (CO_2) and water (H_2O) into carbohydrate, a medium for energy storage. In these organisms, oxygen (O_2) and water are characteristically released as byproducts of the process. In *eukaryotes*, the reaction occurs in two main parts. The first involves the *light reactions* where solar energy is captured by *chlorophylls*. Energy is generated in the form of *ATP* (*photophosphorylation*) and also reducing power, in the form of *NADPH*. ATP and NADPH are used in the *dark reactions* to fix carbon dioxide into carbohydrate. In green plants, photosynthesis occurs in specialized cellular *organelles* called *chloroplasts*, which are thought to have evolved from an *endosymbiosis* with cyanobacteria. Bacterial photosynthesis differs in that the enzymes are bound to the cell membrane rather than contained within a *plastid*, and oxygen is not requisitely evolved. In some variations of bacterial photosynthesis, the initial hydrogen

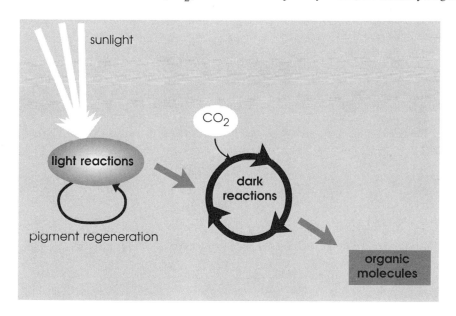

$$6CO_2 + 12H_2O \longrightarrow C_6H_{12}O_6 + 6H_2O + 6O_2$$

atmospheric   water vapor   sugar   water   oxygen gas
carbon dioxide   from water

Overview of Photosynthesis

donor can be another inorganic molecule such as hydrogen sulfide (H_2S). In those cases, molecular sulfur (S_2) or other sulfur compounds are released rather than O_2. Other sources include hydrogen gas (H_2) or even small organic molecules. The general photosynthetic equation is:

$$\overset{\text{(light energy)}}{2nH_2X + nCO_2 \longrightarrow nH_2O + nCH_2O + 2nX.}$$

(see *Calvin-Benson cycle, photosystem I, photosystem II, carbon fixation, photoautotrophism, photoheterotrohism*; compare *respiration, aerobic respiration*)

photosynthesizer Any organism able to carry out *photosynthesis* and thus convert solar energy into a form usable by the next level of the *food web*, the *herbivores*. Photosynthesizers include green plants, algae, and photosynthetic bacteria. (see *primary producer*)

photosynthetic pigments Molecules that are capable of absorbing light energy (photons) and using it to excite electrons that then initiate *photosynthetic* reactions. They are found in green plants, algae (photosynthetic protoctists), and photosynthetic bacteria. The main photosynthetic pigments are *chlorophylls*, with *carotenoids* and *phycobilins* generally acting as accessory pigments.

photosynthetic reaction center Any of the particular protein complexes found in green plants, algae, and photosynthetic bacteria that are excited by particular wavelengths (λ) of light and initiate the light dependent reactions of *photosynthesis*. (see *bacteriochlorophyll, chlorophyll, xanthophyll, photosystem I, photosystem II*)

photosystem I (PS I) One of the two reaction centers that are involved in *photosynthesis* in green plants, algae, and cyanobacteria. It utilizes the photopigment P_{700}, a type of *chlorophyll* whose electrons may be energized by relatively long wavelengths (λ) of light. Electrons are transferred to PS I from *PS II* through the *cytochrome bf complex*, concomitantly creating a *proton gradient* that is used to generate *ATP* by *chemiosmosis*.

The electrons are accepted by *plastocyanin (PC)*, and through a light-activated transfer, to P_{700}. *Ferrodoxin* is the penultimate acceptor, from which NADP reductase transfers the electrons to *NADP+*, generating *NADPH*. Both ATP and NADPH are funneled into the *Calvin cycle* where they fix atmospheric carbon (CO_2) into carbohydrates. An alternative pathway involves the flow of electrons back to PS I through the cytochrome *bf* complex. This *cyclic photophosphorylation* generates a proton gradient but without NADPH formation or carbon fixation. Similar cyclic photophosphorylation systems are seen in bacteria, where *bacteriochlorophylls* that absorb yet longer wavelengths, such as P_{840}, P_{870}, or P_{960}, are utilized. (see *Z scheme, photosystem II*)

photosystem II (PS II) One of the two reaction centers that are involved in *photosynthesis* in green plants, algae, and cyanobacteria. It begins with the photopigment P_{680}, a type of *chlorophyll* that requires relatively short wavelengths (λ) of light to energize its electrons. An electron is transferred from the excited P_{680} to *pheophytin* and then to *plastoquinone* (Q), generating the reduced form (QH_2). The reaction center regains electrons from water by the action of a manganese cluster that evolves oxygen (O_2) (*photolysis*). On its way from photosystem II to photosystem I, each energized electron drives a *proton pump*, generating *ATP* by *chemiosmotic* synthesis. (see *Z scheme, photosystem I*)

phototaxis A movement by an organism in response to light. Many motile algae (for example, *Volvox*) are positively phototactic, while cockroaches, for instance, are negatively phototactic. (see *taxis*)

phototrophism The nutritional mode in which organisms use energy derived from light to synthesize organic compounds by *photosynthesis*. Most phototrophic organisms are *autotrophic*, including green plants, algae, and photosynthetic bacteria. They are called *photoautotrophic.* A few bacteria are *photoheterotrophic,* using light as their energy source but relying on organic compounds synthesized by other organisms for primary nutrients. (see *trophism;* compare *chemotrophism*)

phototropism (heliotropism) The tendency for a plant to grow or bend toward a light source. It is mediated by the family of plant growth substances, *auxins.* Light patterns cause auxins to migrate to the shade side of a plant, which in turn stimulates the elongation of cells in that area, causing the stem to bend toward the light source. The roots of some plants are negatively phototropic, causing them to grow away from a light source. (see *tropism*)

phragmoplast The region that appears between the daughter *nuclei* of a dividing plant cell. It is initiated by *microtubules.* It develops into a barrel-shaped structure containing cell *organelles* associated with *protein synthesis* and secretion, including *rough endoplasmic reticulum* and a *Golgi apparatus.* It assembles the *cell plate* that partitions the *cytoplasm* into two cells at the end of *nuclear division.*

phyco- A word element derived from Greek meaning "seaweed" or "algae." (see *phycology*)

phycobilins A group of accessory *photosynthetic pigments,* found particularly in *cyanobacteria* and red algae, that allow growth at ocean depths and other regions where red and yellow light does not penetrate. They include the *phycocyanins* and *phycoerythrins,* which absorb orange and green light, respectively. Chemically, they are based on a *porphyrin* ring, as are the *chlorophylls* and *hemoglobin,* but have a linear, rather than circular, geometry. (see *phycobilisome, phycobiliprotein*)

phycobiliprotein A complex of *phycobilin* pigments with protein.

phycobilisome An *intracellular* structure, found particularly in cyanobacteria and red algae. It serves to concentrate and funnel the shorter, water-penetrating wavelengths (λ) of light to the *photosynthetic reaction centers.* They contain *phycobilin* pigments that are geometrically arranged as a protrusion on the surface of or within the *thylakoid* membrane of a plastid. Phycobilisomes allow these organisms to harvest the yellow and green light that penetrates more than a meter, enabling them to occupy an *ecological niche* unavailable to organisms relying only on *chlorophyll* a.

phycocyanin An accessory *photosynthetic pigment.* (see *phycobilins*)

phycoerythrin An accessory *photosynthetic pigment.* (see *phycobilins*)

phycology The study of algae.

-phyll- A word element derived from Greek meaning "leaf." (for example, *phyllopod*)

phylloclade (cladode) In plants, a specialized stem that resembles and functions like a leaf, including *photosynthesizing.* It is seen in butcher's broom and asparagus, and is an adaption to decrease surface area, preventing water loss. (compare *phyllode*)

phyllode An expanded, flattened leaf stalk (*petiole*) in which *photosynthesis* is carried out in plants where the leaf blades are reduced or absent. (compare *phylloclade*)

phyllopod A leaflike appendage found in some crustaceans.

phyllopodium, phyllopodia (pl.) **(leaf base)** The portion of the leaf attached to the stem. It is usually the expanded portion.

phylloquinone (see *vitamin K*)

phyllotaxis In plants, characterization by the arrangement of leaves on the stem, for instance whorled, spiral, opposite or alternate. (see *taxis*)

phylogenetics The classification of organisms based on their deduced evolutionary relationships. Currently, deductions are based on a comparison of the *DNA* sequences of particular *genes.* This has changed some previous classifications that were based on a comparison of protein sequences. (see *cladistics, biosystematics, biochemical taxonomy, molecular clock*)

phylogenetic tree A pictorial representation of the evolutionary divergences of organisms as determined by a comparison of the *DNA* sequences of particular *genes.*

phylogeny The evolutionary history of an organism or group of organisms.

phylum The major groups into which kingdoms are divided. Phyla may occasionally be divided into subphyla and then into classes. (compare *division*)

physiological saline A solution of sodium chloride and various other salts that mimics body fluids. It is similar in composition to the oceans from which all life evolved and is used in the laboratory to approximate physiological conditions.

physiological specialization The existence of physiologically distinct but morphologically identical *races* or subspecies within a species. The distinction is particularly important in *host-pathogen* relationships, particularly in agriculture.

physiology The study of the functional relationships within an organism. (compare *morphology, anatomy*)

phyto-, -phyte, -phyta A word element derived from Greek meaning "plant." (for example, *sporophyte*)

phytoalexin An antifungal agent produced by plants, usually in response to infection or injury.

phytochrome A plant pigment that acts as a *photoreceptor* in influencing a wide range of light-induced physiological processes. Examples include flowering, *dormancy*, leaf formation, and *germination*. It exists in two interconvertible forms: P_R (P_{660}), which absorbs red light at 660 nm, and P_{FR} (P_{730}), which absorbs far red light at 730 nm. It is used as an on-off switch and is linked to environmental light signals. (see *phototaxis, phototropism, photonastic*)

phytohemagglutinin A *mitogenic* plant *lectin* extracted from jack beans. It is used in the laboratory specifically to pro-

voke the division and *transformation* of *T-cells* in *culture.*

phytohormone (see *growth substance*)

phytol A component of the *chlorophyll* molecule

phytoplankton (see *plankton*)

pia mater The innermost membrane that surrounds and protects the vertebrate brain and spinal cord. (see *meninges*)

Piciformes The order of birds containing the woodpeckers.

pico- (p) A word element denoting 1/1,000,000,000,000, (one-trillionth) or 10^{-12}. For example, 1 picogram (pg) = 10^{-12} gram. The picogram is a common unit of measure for small amounts of *DNA.*

pigeons (see *Columbiformes*)

pigment Any molecule that absorbs light and therefore imparts a particular color to the tissue containing it. The wavelengths (λ) absorbed by a particular pigment depend on the available energy levels to which light-excited electrons can be boosted. (see *photoreceptor, phytochrome, rhodopsin, chlorophylls, xanthophylls, phycobilins, melanins, chromophore, chromatophore*)

pigs (see *Artiodactyla, Suina*)

pikas (see *Lagomorpha*)

pileus, pilei (pl.) The cap of a mature mushroom (*fruiting body*) in certain *basidiomycete* fungi.

pilus, pili (pl.) In certain bacteria (*Gram-negative*). Hollow protein filaments, found in *Gram-negative* bacteria, protrude through the *cell wall*. They may perform an adhesive function. Some (sex pili) are involved in *conjugation* and permit the passage of genetic material from one individual to another.

pineal eye (see *median eye*)

pineal gland An *endocrine gland* arising from the vertebrate *midbrain*. It secretes the *hormone melatonin* that influences pigmentation in amphibians and is released in response to darkness and promotes sleep in humans. In reptiles, it is located close to the surface and is structurally similar to the *retina*, responding directly to light (third eye). In humans, the pineal gland has lost all connection with the *central nervous system* but remains connected to the eyes via the *sympathetic nervous system*. It has been implicated in the establishment of daily biorhythms, but is not yet well understood.

pinna The external part of the *outer ear* in mammals, generally referred to as the ear.

Pinnipedia An order of aquatic mammals containing the seals, sea lions, and walruses. They are marine *carnivores* in which the digits are fully webbed and the limbs modified to form paddles.

pinocytosis The mechanism by which a cell ingests surrounding liquid. It is accomplished by a localized invagination of the *plasma membrane* that engulfs a minute drop of fluid forming a *vesicle* and is then pinched off to the inside of the cell. Pinocytosis is used by both *multicellular* and *unicellular* organisms for ingestion of dissolved *macromolecular* nutrients. (compare *phagocytosis*)

pipette A laboratory instrument used for measuring and delivering liquids. It is usually a long, thin, hollow tube of plastic or glass with a narrow tip. A micropipette is used for accurately measuring and delivering small amounts of liquid (in the range of *micro*liters) and requires a specialized suction device to which small tips are attached.

Pirellae A phylum of *Gram-positive* bacteria containing stalked bacteria with proteinaceous walls. It includes *Chlamydia* and *Planktomyces*. The reclassification has occurred based on DNA *homologies* in the small (16S) *ribosomal RNA* (*rRNA*) subunit.

pistil In female flowering plants (*angiosperms*), the traditional name for the central seed-containing structure. Technically, it may contain one or more separate or fused *carpels* and is better known as the *gynoecium*. (see *pistillate*)

pistillate Flowers in which the *stamens* (male reproductive structures) are missing or do not produce *pollen*. (see *dioecious, monoecious, outcrossing, cross-pollination*; compare *staminate*)

pith In plants, the central region of the stem that is composed of undifferentiated *parenchymal* tissue.

pituitary gland An *endocrine gland* in the vertebrate brain that acts as a control center for all other *endocrine systems* in the body. It is situated just behind the *optic chiasma*. Some of the *hormones* secreted by the pituitary include *growth hormone, vasopressin, adrenocorticotrophic hormone, oxytocin, and thyrotrophin*

placenta, placentae (pl).

1. A cluster of blood vessels in the uterus of a pregnant mammal where the maternal and *embryonic* systems meet. Both *metabolites* and nutrients are exchanged, but the *circulatory systems* remain essentially separate.

2. The part of the *ovary* wall of flowering plants (*angiosperms*) where the *ovules* (seeds) develop at the fused margins of the *carpels*.

placental mammals (see *Eutheria*)

placentation In flowering plant (*angiosperm*) seeds, the portion of the *placenta* (origin of seed development) within the *carpel*. It may be parietal (on the walls), axile (on the axis), basal (near the bottom), or central.

placoid scale (see *denticle*)

Placozoa A phylum of animals including probably the simplest animals alive today, the placozoans. They are *multicellular* but resemble large amebas in that they change shape as they creep along the sea bottom. Placozoans are classified as animals based on an *embryo* that develops through a *blastula* stage, characteristic of all animals, and results from *fertilization* of an egg with a sperm. They can also reproduce *asexually* by splitting into two multicellular organisms. Only one species is known.

placozoans (see *Placozoa*)

plagio- A word element derived from Greek meaning "oblique." (for example, *plagioclimax*)

plagioclimax (see *biotic climax*)

plagiosere A plant *succession* deflected into a new course through human intervention. (compare *prisere*)

plankton An ecological designation for various microscopic aquatic organisms that drift more or less freely in the upper regions of a sea or ocean (*oceanic*). They include *phytoplankton* (better termed *photoplankton* as *photosynthetic* protoctists (algae) and cyanobacteria are not plants). An important component of photoplankton are the *unicellular diatoms* upon which many *zooplankton* (animal plankton) feed. Zooplankton is made up of protists, small crustaceans, and aquatic *larvae*, some of which possess locomotory capability. Plankton forms the base of the *food web* in the sea. Phytoplankton (photoplankton), in particular, is responsible for regenerating much of the atmospheric oxygen (O_2) that many organisms, including humans, breathe. (see *pelagic*; compare *nekton*)

plant Any organism classified within the kingdom *Plantae*.

Plantae One of the *phylogenetic* kingdoms into which all life forms are divided for academic discussion. It contains *eukaryotic* organisms that are generally nonmobile and exhibit relatively slow response times. Although the vast majority of plants are *photoautotrophic* (make their own food by *photosynthesis*), a few lack *chlorophyll*, thus this is not a defining characteristic of the kingdom. The more definitive characteristics are found in their life cycle. Exchange of genetic material is accomplished by *fertilization* and *meiosis,* and new individuals develop from *diploid embryos* that are retained in the maternal tissue. Their cells are surrounded by *cell walls* in addition to the *cell membranes* also found in animals. A few more primitive members have motile *gametes* that are propelled by 9 + 2 *undulipodia*. Protists and fungi, which have sometimes been classified as plants, are now recognized as autonomous kingdoms.

plantaineaters (see *Muscophagiformes*)

plant hormone (see *growth substance*)

plantigrade The ambulatory mode of some mammals in which the entire sole of the foot is in contact with the ground. Humans use plantigrade locomotion. (see *pentadactyl limb*; compare *digitigrade, unguligrade*)

planula, planulae (pl.) The free-swimming, *ciliated larvae* of cnidarians and some ctenophores. They each develop from a *fertilized* egg and are common in *plankton.* (compare *ephyra*)

plaque A clear area on a bacterial *lawn* left by *lysis* of the bacteria through progressive infection by a *phage* and its descendants.

-plasm-, -plast- A word element derived from Greek meaning "formed" or "molded." (for example, *cytoplasm, chloroplast*)

plasma (see *blood plasma*)

plasma cell An *antibody*-producing cell resulting from the multiplication and differentiation of a *B-cell* that has interacted with an *antigen*. Plasma cells are derived from undifferentiated *lymphocytes* in *lymphoid tissue*, including *bone marrow*. A mature plasma cell can produce from 3,000 to 30,000 antibody molecules per second. (see *clonal selection*)

plasmagel The gel-like state of the outer layer of *cytoplasm* of some cells. It is important to *cytoplasmic streaming* and some cell locomotion, in particular *ameboid* movement. It can be converted to *plasmasol* (liquid cytoplasm) by hydrostatic pressure or the removal of calcium ions.

plasma membrane (cell membrane) The *phospholipid bilayer* membrane that surrounds all living cells and regulates exchange between the cell and the environment. In plants and some microorganisms, it is further surrounded by a rigid *cell wall.*

plasmasol The liquid state of *cytoplasm* found internal to *plasmagel*, if present. Plasmasol is present in all cells, comprises the bulk of the cytoplasm, and contains all cell *organelles*.

plasmid A small, generally circular, *extrachromosomal* genetic element that replicates independently of *nuclear DNA*. Plasmids occur naturally in bacteria and often carry genes encoding *antibiotic* resistance. They are occasionally encountered in *eukaryotes*, the best-known being the 2μ plasmid in yeast. They may occasionally *integrate* into and subsequently *excise* from the host *genome* by *homologous recombination* if similar sequences are present. Because of their accessibility and versatility, they were the first vehicles used as *cloning vectors* in *genetic engineering*. (see *episome, cytoplasmic inheritance, F factor, transduction*; compare *transposable genetic element*)

that can be converted to the active form by a variety of enzymes.

plasminogen The inactive precursor of *plasmin*.

plasmodesma, plasmodesmata (pl.) A *cytoplasmic* bridge lined by a *plasma membrane* that connects adjacent cells. At one time, they were thought to be confined to plant cells, extending through the *cell wall*, but have now been observed in some animal cells as well. They may provide pathways of communication and transport between cells. (see *transfer cell*)

plasmodial reticulum A *cytoplasmic* network of *proteinaceous* fibers by which *ameboid* cells achieve motility.

plasmodial slime molds (see *Myxomycota*)

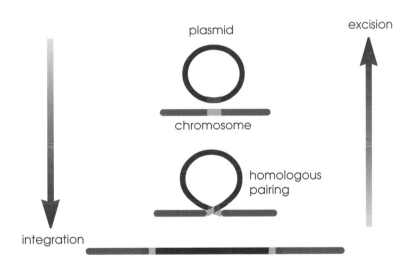

Plasmid Integration and Excision

plasmid library A collection of relatively small fragments of *genomic DNA* contained in a *plasmid cloning vector* and propagated in *E. coli*.

plasmin A *proteolytic* enzyme that degrades *fibrin* in blood clots, restoring the fluidity of the blood. It exists in the blood as an inactive precursor, *plasminogen*,

Plasmodiophora A phylum of protoctists, the *plasmodial parasites*. A few of their species are infamous for causing disease in commercial crops. The *plasmodial multinucleate* structure of the *trophic* stage lives and feeds inside the cells of the *host* organism (*necrotrophic*). It is known that they produce *zoospores*, but much remains to be discovered about their *life cycle*, genetics, and sexuality.

plasmodium, plasmodia (pl.) A membrane bound mass of *cytoplasm* containing multiple *nuclei*. It is found in various protoctists. The plasmodium is characteristic of a *life cycle* stage in plasmodial slime molds where the plasmodial mass is able to flow through a fine mesh and rejoin on the other side. (see *Myxomycota*)

plasmogamy The fusion of two cells without or preceding *nuclear* fusion. It occurs before nuclear fusion during *fertilization* of *gametes* and between *fungal mycelia* of different strains to form a *heterokaryon* (a cell containing genetically different nuclei). (compare *karyogamy*)

plasmolysis The result of water loss in a plant cell. This causes the *cytoplasm* to shrink, pulling the inner *plasma membrane* away from the *cell wall*. It occurs when the cell is surrounded by a *hypertonic* (more concentrated) solution, usually under experimental conditions.

plastid An *organelle* unique to *photosynthetic eukaryotes* (green plants and algae) that is surrounded by a double *plasma membrane* and contains its own *DNA* and protein-synthetic machinery (*ribosomes*). Various types exist, including *chloroplasts, xanthoplasts,* and *rhodoplasts,* in which *photosynthesis* occurs. Other plastids function as food storage and pigment-containing vesicles. (see *chromoplast, leukoplast, amyloplast, elaioplast, aleuroplast*)

plastocyanin (see *PC*)

plastoquinone (see *Q*).

plastquinol (see *QH₂*)

plastron
1. The flatter, bottom part of the *carapace* (shell) of a turtle or tortoise.
2. A semipermanent bubble of air through which respiratory gases are exchanged in some aquatic invertebrates.

plate
1. A flat glass or plastic dish used to culture microorganisms in the laboratory.

It is often filled with a solid *agar* growth medium
2. The technique of spreading cells, usually bacteria or yeast, over the surface of solid *agar* medium for the purpose of growing evenly spaced *colonies* or a *lawn.*

platelet (thrombocyte) In mammals, a white blood cell (*leukocyte*) fragment that circulates in the blood and initiates *blood clot* formation at the site of injury by the release of *thrombokinase*. Platelets also release *serotonin*, causing constriction of adjacent blood vessels to contain bleeding. They are not true cells and contain no *nuclei* or other *organelles*. As such, they are very short lived.

Platyhelminthes A phylum of primitive invertebrates, the flatworms, including free-living aquatic planarians, *parasitic* flukes, and tapeworms. They are the simplest animals that have organs other than gonads, are unsegmented, lack a *coelom* and *circulatory system*, and are *hermaphroditic*. A single opening serves as both mouth and anus. They have a simple nervous system including *eyespots*, a cluster of *nerve cells* in the head, and longitudinal *nerve cords*. An example is the common laboratory specimen *Planaria.*

Plecoptera The order of insects comprising the stoneflies.

pleiomorphism The occurrence of different *morphological* stages during the life of an organism. Examples are the *larval, pupal,* and *imago* forms of an insect. (see *metamorphosis, alternation of generations*)

pleiotropic A *gene* that may influence several physical (*phenotypic*) characteristics. This may occur if the gene codes for an enzyme or *regulatory protein* whose product is involved in several biochemical pathways. A *mutation* in such a gene may also have multiple effects.

Pleistocene The first epoch of the Quaternary geological time period, from about 2 million years ago until the last glaciation ended about 10,000 years ago. The four ice ages drove many organisms toward the equator, while others (for example, mammoth) became extinct. Modern man (*Homo sapiens*) evolved during this period.

pleural Related to the lungs.

pleural membrane A double membrane that surrounds the lungs in mammals. The space between the two membranes, the pleural cavity, is filled with air and cushions the lungs.

plexus Any central network of interconnecting nerves, blood vessels, or lymph vessels.

Pliocene The last epoch of the Tertiary geological period, about 7 million to 2 million years ago. During this period, *Hominids* such as *Australopithiecus* and *Homo* became clearly distinguishable from the apes.

-ploid A word ending used with an indicator of the number of *chromosomes*. (see *diploid*)

plovers (see *Charadriiformes*)

plumed worms (see *Polychaeta*)

plumule In early plant development, the *cotyledons* (first leaves) together with the portion of the shoot (*epicotyl*) that extends above them.

pluteus, plutei (pl.) The *larva* characteristic of some *echinoderms*, such as brittle stars and sea urchins. They are *ciliated* and free floating (*planktonic*).

pneumatophore A specialized "breathing root" that has developed in some plant species, such as mangroves, that grow in waterlogged or compacted soils. The aerial part of the root enables atmospheric gas exchange, while the internal system of *intercellular* air spaces allows diffusion of gases throughout the submerged portion of the root.

PNS (see *parasympathetic nervous system*)

pod A fruit that splits (*dehisces*) down both sides into separate halves. Pods are characteristic of the fruit of the family *Fabaceae* (peas and beans).

pode-, -pod-, -poda, -podium, -podia, -podial A word element derived from Greek meaning "foot." (for example,

pseudopodium, see *ped-*)

Podicipediformes The order of birds containing the grebes.

podium, podia (pl.) In general, *cytoplasmic* projections that are used by single cells in locomotion. In foram protists, they are additionally used for feeding and gathering of shell (*test*) materials. (see *pseudopodia, lamellipodia, ameboid*)

podsolic The type of soil found under coniferous forests in temperate climates. It is strongly acid and often deficient in nutrients as a result of leaching.

Pogonophora (Vestimenifera) A phylum of *coelomate* marine worms, the giant tube or beard worms. They live within *chitinous* tubes they secrete on bottom sediments or decaying wood and have long, beardlike *tentacles* behind their head. They reside among the sulfide and methane-rich *hydrothermal vents* on the bottom of the ocean. They obtain their energy and nutrition via the *chemoautotrophic* bacteria they harbor in their gut.

-poiesis A word element derived from Greek meaning "production." (for example, *hemopoiesis*)

poikilothermic (see *ectothermic*)

point mutation A *mutation* in DNA that is limited to one *base pair*. For instance, a **GC** base pair is replaced by an **AT** or *vice versa*. (see *complementary base pairing*; compare *deletion, insertion*)

Poisson distribution A mathematical expression giving the probability of observing the frequency of a particular random event in a given sample.

polar body A degenerate cell produced as a byproduct of *oogenesis* in animals. Only one cell from each of the two *meiotic* divisions continues to develop into a mature *ovum* (egg cell), leaving two or three meiotic products (if the first polar body continues to divide) as polar bodies.

polarizing microscope A *light microscope* that is specially modified to produce and analyze plane-polarized light. It is used in the visualization of some materials that are transparent to normal light but themselves have the ability to polarize light. It is also used analytically, to help determine the composition of a specimen. The polarizing element emits light in which the waves are oriented in a single plane. Light then passes through the specimen and its orientation is detected by an analyzer. Specimens often appear in vivid color and high contrast. Sometimes two sources of polarized light are used together in order to perform more complicated analyses.

polar nuclei In the seed development of flowering plants (*angiosperms*), the two *nuclei* in the *embryo sac* that fuse with one of two *spermatozoa* nuclei from the *pollen tube* to form the *triploid endosperm* (nutritive tissue that surrounds the *embryo*). (see *double fertilization*)

pollen (see *pollen grains*)

pollen analysis The study of *pollen grains* and *spores*, both fresh and *fossilized* specimens. The characteristics of the *exine* coat of a pollen grain can be distinctive to a family, genus, or even species of plant. It is also extremely resistant to decay, thus providing a good source of archeological information.

pollen culture A recently developed technique for regenerating a whole plant directly from a *pollen grain*. Since this gives rise to a *haploid sporophyte*, an entity not known in nature, all the *genes* are expressed directly, a useful feature for research and breeding.

pollen grains (pollen) The male *gametes* of seed plants that are derived from *microspores* and contain immature *microgametophytes*. Pollen is produced within the *anthers* of a male flowering structure (*stamen*) or the male cones of a nonflowering seed plant and are generally two or three-celled when shed. (see *angiosperm, gymnosperm, exine*)

pollen mother cell In seed plants, the precursor cell (*microsporocyte*) that eventually gives rise to four *haploid pollen grains* by *meiosis*.

pollen sac (microsporangium) A chamber in which *pollen grains* (male gametes) are formed in flowering plants

(*angiosperms*) and conifers (see *sporangium*)

pollen tube In plants, the tube extruded from a *pollen grain* (male *gamete*) that has begun to *germinate* on a female plant or flower *stigma*. In flowering plants (*angiosperms*), the tube penetrates the outer structures of the *gynoecium* and terminates at the *embryo sac* where it ruptures, depositing two male gametes. In nonflowering seed plants (*gymnosperms*), the tube temporarily halts at the *nucellus* and resumes growth the next year, when the female *gametophyte* is mature. In primitive plants such as cycads and *Ginkgo*, the pollen tube is poorly developed, and male gametes swim to the *ovule* by means of *undulipodia*. (see *fertilization, double fertilization*)

pollination In flowering plants (*angiosperms*), the transfer of *pollen* from the male reproductive structure (*stamen*) to the top of the female reproductive structures (*stigma*), facilitating *fertilization*. In nonflowering seed plants (*gymnosperms*), the *pollen grains* are transferred directly to the *ovule*. In *self-pollinating* plants, *fertilization* is virtually ensured, but the chance for genetic variation is bypassed. Thus in many plants, mechanisms exist to promote *cross-pollination*, even in *hermaphroditic* flowers. (see *self-incompatibility, self-sterility, monoecious, dioecious, dichogamous, protandry, protogyny*)

pollution The introduction into the environment of detrimental, man-made substances or of naturally occuring substances out of their normal proportion or place.

poly- A word element derived from Greek meaning "much" or "many." (for example, *polyploid*)

polyacrylamide A *polymer* of a synthetic molecule, acrylamide, used in the *electrophoretic* separation of *macromolecules* such as proteins and small *nucleic acids* (*RNA, DNA*). The *monomer* is a potent neurotoxin. (compare *agarose*)

polyadelphous Flowers in which the *sta-*

mens are formed into several separate groups around the *style*. (compare *diadelphous, polyadelphous*)

polyadenylation A *posttranscriptional modification* of *mRNA* that occurs in *eukaryotes* but not *prokaryotes*. A string of *adenosine* residues *(poly-A tail)* are added to the 3′ end of the RNA molecule, which helps in stabilizing it against *degradation* by endogenous *ribonucleases*.

polyamine A nitrogenous compound often found associated with *bacterial* and viral *nucleic acids (DNA, RNA)*. Polyamines are composed of two or more amino and/or imino groups. Examples include spermine, spermidine, cadavarine, and putrescine. (compare *histone*)

poly-A tail The string of *adenosine* residues added to the 3′ end of the RNA molecule in *eukaryotes* after *primary transcription*. (see *polyadenylation*)

polycarpic Plants that flower repeatedly. They are *perennials*. (compare *monocarpic*)

Polychaeta The class of annelids containing the marine representatives the bristle worms. Many are *carnivorous* and some burrow or build tubes in mud, sand, or mucous. Some are *commensal* with sponges or mollusks, and a few are *parasitic*. Each body segment bears a pair of locomatory *parapodia* in which stiff *setae* are embedded. The head is well defined and bears sense organs. The sexes are usually separate and *fertilization* is often external.

Polychetes (see *Polychaeta*)

polycistronic mRNA An *mRNA* molecule that codes for more than one protein. It is the exception and occurs mostly in bacterial *operons*. (see *central dogma*; compare *monocistronic*)

polycolpate pollen *Pollen grains* in which multiple furrows or pores are characteristic. They are found particularly in *dicotyledons*.

polygenic A trait that is determined by the action of several genes in concert. Polygenic traits, such as skin color and height, are seen in the population as a continuous set of variations.

polymer Any compound that is composed of multiple individual units that may or may not be identical. Examples of biological polymers include *nucleic acids,* proteins, and *polysaccharides*. (compare *monomer*)

polymerase Any enzyme that directs the synthesis of a *polymer* by linking individual *monomers*. In biological systems, it generally refers to *DNA polymerases* and *RNA polymerases*.

polymerase chain reaction (see *PCR*)

polymerization The process of *polymer* formation by chemical linkages between each adjacent *monomer*. In biological systems, this process may be but is not necessarily facilitated by specific enzymes. (see *polymerase, DNA polymerase, RNA polymerase*)

polymorph (see *granulocyte*)

polymorphism (genetic polymorphism) The occurrence in a population (or among populations) of at least several variations (*alleles*) of a *gene* or *genetic marker*. Many genes that code for functional proteins exist in two (biallelic) or three (triallelic) forms but are unlikely to exhibit high rates of polymorphism. *Loci* (genetic locations) that have not been under evolutionary pressure to produce or regulate functional proteins may be extremely polymorphic, having sometimes hundreds of different variations at a single locus. These types of genetic markers have recently been exploited for the purpose of individual identification. (see *DNA analysis, RFLP, PCR, tandem repeat, VNTR, conserved sequence*)

polymorphonuclear leukocyte (see *granulocyte*)

polynucleotide (see *nucleic acid*)

polyp The sedentary form of cnidarians. They remain anchored to a substrate by a stalk with the mouth and tentacles oriented upwards. Many species occur only as polyps, while others *alternate generations* with the *medusa* (free-floating) form.

polypeptide A single chain of *amino acids*

connected by *peptide bonds*. It is differentiated from a protein only by its smaller size and relative lack of complexity. (compare *oligopeptide*)

polypeptide hormone (see *peptide hormone*)

Polyplacophora A class of marine mollusks, the chitons, characterized by eight overlapping *calcareous* dorsal plates embedded in the *mantle*. They are grazing *herbivores*.

polyploid A cell or organism containing more than two sets of *chromosomes*. It is seen in plants, particularly in agricultural stocks. It is rare in animals because the *genes* governing reproduction generally only function properly in the correct number and proportion. (see *triploid;* compare *haploid*)

polyribosome (see *polysome*)

polysaccharide A long-chain *polymer* of *monosaccharide* sugar units joined by *glycosidic bonds*. Polysaccharides important in biological systems include *starch, glycogen,* and *cellulose*. (see *glucan, glycan*; compare *monosaccharide, oligosaccharide, disaccharide*)

polysome A group of *ribosomes* that are simultaneously involved in the *translation* of a single *mRNA* molecule into numerous, identical *polypeptides*. Each ribosome contains a polypeptide at a different stage of synthesis, depending on whether it is situated toward the beginning or the end of the mRNA. (see *protein synthesis*)

polysomy The abnormal condition in which multiple copies of one *chromosome* are present in a normally *diploid* organism. The organism is often abnormal or inviable. (see *aneuploidy, trisomy, nondisjunction*)

polyspermy During *fertilization*, the penetration of more than one *spermatozoa* into an *ovum* (egg cell). Generally, only one sperm actually fuses with the ovum *nucleus*. Polyspermy occurs only in a few animals where the eggs have large amounts of yolk, for example birds. In most other animals, a *fertilization mem-*

brane forms around the ovum in response to penetration by the first sperm, preventing others from entering.

polytene chromosomes (giant chromosomes) *Chromosomes* in which multiple rounds of *DNA replication* take place without separation of the *chromatids*. They are common in the *salivary glands* of some insects (*Diptera*), for example *Drosophila*, in which they have been used to great advantage in the study of *gene* activity and location. Specific regions (*puffs*) enlarge at particular developmental stages and have been correlated with a high rate of *RNA transcription*, evidencing the activity of the gene(s) at that location. When these regions are greatly distended, they are called *Balbiani rings*. (see *endomitosis*)

polyvoltine An organism that reproduces several times in a year.

pome A fruit in which the seed is protected by a tough *carpel* wall, the whole being embedded in a fleshy *receptacle*. For instance in an apple, the core is the actual fruit, the fleshy portion comprising the enlarged receptacle.

pons A thick tract of *nerve fibers* in the mammalian brain that links the two hemispheres of the *cerebellum*.

population A group of individuals of the same species occupying the same geographical area at a given time. They usually, but may not necessarily, interbreed.

population dynamics The study of changes in population sizes and the causes of these changes. (see *growth curve, carrying capacity, biotic potential*)

population genetics The study of inherited genetic variation and its modulation in space and time. (see *Hardy-Weinberg equilibrium*)

population risk The random variation in birth rates and death rates that might impact a small population enough to cause *extinction*. It is used in discussing *endangered species*.

population substructure The existence within a geographically defined population of smaller subgroups that tend to interbreed, thus potentially restricting *gene flow*. The existence of population substructure becomes an issue when applying various genetic principles and calculations that depend on large, randomly mating populations. (see *Hardy-Weinberg equilibrium*)

por-, -pore A word element derived from Greek meaning "small or minute opening." (for example, *blastopore*)

porcupine (see *Hystricognathi*)

Porifera A phylum of primitive *multicellular* animals, the sponges, defined by a lack of definite symmetry. All are *sessile*, and most are marine. The body is a loose aggregation of cells with little nervous coordination between them. It is lined with *undulipodiated* cells (*choanocytes*) that mobilize water currents in through apertures in the body wall and out through the top (*osculum*). Sponges have an internal skeleton of chalk, silica, or protein, the type used or copied as a bath sponge. They reproduce both *sexually* and *asexually*. (see *mesenchyme*)

porphyrins Cyclic organic molecules that form the basis for several important biological compounds. They form complexes with metal ions, iron in the case of *hemoglobin* and magnesium in *chlorophylls*, a property essential to their function. (see *photosynthesis, myoglobin, cytochromes, heme*)

porpoises (see *Cetacea*)

position effect A change in the control of *gene expression* as a result of embedding a *gene* in a foreign location on the *chromosome*. *Oncogenes* and *transposons* are examples of genetic elements whose state of *transcription*, therefore *phenotypic* effects, are often attributable to their immediate environment.

positive assortative mating (see *assortative mating*)

positive control Regulation of *RNA transcription* from a *gene* that is mediated by proteins required for the activation of transcription. (compare *negative control*)

positive feedback The perpetuation of a perturbation in a biological system.

Physiologically, positive feedback is important in the propagation of *nerve impulses*, resistance to disease, and childbirth labor. It also occurs within *ecosystems* and globally. In the case of processes leading to *global warming*, for instance, it may be detrimental. (compare *negative feedback*)

posterior The hind end of an animal. In *bilaterally symmetrical* animals, this is the end directed backwards in locomotion. In *bipedal* (two-legged) animals such as man, the posterior side corresponds to the *dorsal* side of animals that walk on all fours. (compare *anterior*)

postsynaptic In reference to *neural impulse* conduction, the nerve or muscle membrane on the far side of a *synapse*. Impulses travel from the *presynaptic* to the postsynaptic membrane via *neurotransmitters*.

posttranscriptional modification Various modifications to the *primary RNA transcript* that take place in *eukaryotes*. They include the addition of a *cap*, *polyadenylation* (poly-A tail), and *splicing*.

postzygotic isolating mechanisms. Factors that tend to help species keep their identity by preventing the propagation of any *interspecies zygotes* that might be formed. They may prevent proper development of an *embryo*, if it is formed at all, or render the adult *hybrid* sterile. Hybrid sterility is exemplified in the mule (the hybrid offspring of a horse and a donkey). (compare *prezygotic isolating mechanisms*)

potassium One of the essential elements for plants and animals. It is required for *protein synthesis*, is a *cofactor* for the enzyme pyruvate kinase in *glycolysis*, and is required for the transmission of *nerve impulses* in animals. It is absorbed by plant roots as the potassium ion K^+ and is obtained by animals from ingesting plant material.

potassium channel A transmembrane *ion channel* that is specific for potassium ions (K^+). They are particularly important in transmitting *nerve impulses*.

(see *channel, inhibitory synapse*)

potential difference A difference in electrical charge on two sides of a *cell membrane* caused by an unequal distribution of ions. (see *membrane potential, resting potential, equilibrium potential*)

potometer A laboratory instrument used to measure the rate of water uptake in a plant and indirectly to estimate *transpiration* (water loss) rates.

power of discrimination (Pd) In reference to a *genetic marker* or combination of markers, a numerical calculation that defines the potential power of a particular marker system to exclude portions of the population from having contributed the sample in question. It can be calculated from the *allele frequencies* in a defined population and is used in discussing the results of a *DNA profile* obtained from *genetic typing*.

poxviruses The largest and most complex animal viruses. They are *DNA viruses* and include the smallpox virus.

P_R (P_{660}) (see *phytochrome*)

prairie (temperate grassland) The temperate grasslands that tend to occur toward the interior of continents in regions with relatively long, cold winters and varying precipitation. They once covered large regions of North America and also include the steppes of Eurasia, the African plains, and the pampas of South America. The roots of *perennial* grasses penetrate deep into the soil, which is the most fertile in the world. Consequently, much of the original prairie has been converted to agricultural land. Native grasslands are populated by herds of grazing animals such as the North American bison. Both grazing and periodic fires clear the grassland of litter accumulation that ties up nutrients and blocks seedling growth, greatly limiting net productivity.

prebiotic Anything pertaining to the period of earth's history before life evolved.

Precambrian The geological eon, over 600 million years ago, preceding the first biological fossil-bearing period, the Cambrian period of the Paleozoic era. In the last part of the Precambrian, chemical fossils, which may be attributable to cyanobacterialike organisms, have been observed in sedimentary rock formations.

precocial Mammals and birds that are born alert and well developed. Although independent almost immediately from birth, precocial animals tend to reach full maturity slowly. (see *nidifugous*; compare *altricial*)

predation An interaction between individuals, usually of different species, in which one benefits and the other is harmed. Most typically, a predator catches, kills, and eats another animal for food. However, a continuous gradient of interaction can be extended to include *insectivorous* plants, grazing, and *parasitism*. It is a mode of *heterotrophic* nutrition that is *necrotrophic* but not *symbiotrophic*.

predator The animal in a predator-prey relationship that enacts the pursuing or consuming role. (see *predation, parasitism*)

preening Grooming behavior in birds that maintains the feathers. In addition to cleaning and straightening, the feathers are oiled with a substance secreted by a *gland* near the tail that renders them supple and water resistant. This is particularly important in birds such as the penguin that depend on their feather coat to protect them from icy ocean waters.

preformation The defunct theory that each *spermatozoa* contained a tiny, preformed human being (*homunculus*) in its head.

prehensile Any appendage, including a tail, adapted for grasping or holding.

pressure potential The pressure exerted on a column of water by the atmosphere. It, along with the *solute potential*, enables water to rise vertically within a plant stem. Their sum constitutes the *water potential*.

presumptive *Embryonic* tissues that have been shown by experimental means to exhibit a specific, predetermined developmental fate.

presynaptic In reference to *neural impulse* conduction, the nerve or muscle membrane on the near side of a *synapse*. Impulses travel from the presynaptic to the *postsynaptic* membrane via *neurotransmitters*.

prey The organism on the receiving end of a predator-prey relationship. (see *predation*)

prezygotic isolating mechanisms Factors that tend to help species keep their identity by preventing the formation of *hybrid zygotes*. The principle mechanisms are geographical (inhabitation of different regions), ecological (adaptation to different habitats, even within the same area), temporal (different breeding cycles), behavioral (different mating rituals and courtship behavior), mechanical (incompatible copulatory organs or flowering structures) and the prevention of *gamete* fusion (eggs and *spermatozoa* that are not attracted to each other or function poorly in *fertilization*). (compare *postzygotic isolating mechanisms*)

Priapulida A phylum of *coelomate* marine worms, the cucumber worms, that burrow in the sea bottom. Both their mouth and *proboscis* evert for food gathering. Their food includes other annelids and even other priapulids. Adults *molt* from time to time, shedding and replacing their *chitinous cuticle*.

Pribnow box An element of the *prokaryotic promoter* region defined by an invariant **T** residue located 10 bps (base pairs) upstream from the *transcription* start site. Its *consensus* sequence is **TATATT**. *Mutations* within it may have mild to severe effects on the level of *RNA transcription*. A similar site with a different consensus sequence is located 35 bps upstream from the *transcription* start site.

primary consumer The first organism in a *food chain* to consume material that has been generated by *photosynthesis*. They are generally *herbivores* although sometimes *omnivores*.

primary ecological succession The colonization of a newly bare area, for instance following glaciation or on a new lava flow. (see *ecological succession*; compare *secondary ecological succession*)

primary endosperm nucleus During *fertilization* in flowering plants (*angiosperms*), the *triploid nucleus* formed from the fusion of the second *pollen* nucleus with two *polar nuclei*. It develops into a nutritive tissue, the *endosperm*, that *metabolically* supports the seed during germination. (see *double fertilization*)

primary growth In plants, growth that results from *mitosis* at the tips of stems and roots (*apical meristem*). It causes an increase in length. This forms the *primary tissue* and the *primary plant body*. (compare *secondary growth*)

primary immune response (see *immune response*)

primary meristem In *vascular plants*, the three types of partially differentiated tissue derived from *apical meristem* cells, the *protoderm* (pre-*epidermis*), *procambium* (pre*vascular*) and *ground meristem* (pre*ground tissue*).

primary motor cortex The area of the *cerebral cortex* that controls voluntary motion. Each point on its surface is associated with the movement of a different body part. It is located at the rearmost portion of the *frontal lobe*.

primary phloem In plants, *phloem* (carbohydrate-conducting) tissue derived from the first wave of *procambium differentiation*. It combines with *primary xylem* into *vascular bundles*.

primary producers The organisms of an *ecosystem* that are able to synthesize organic molecules, usually by *photosynthesis*, and in some bacteria, by *chemosynthesis*. They form the bottom *trophic level* in a *food web* and include green plants, algae, and cyanobacteria.

primary production The total amount of organic material produced by *auxotrophs* from solar energy in a defined area during a given period of time. *Gross primary productivity* is the total amount produced, including that used by the *photosynthetic* organisms (*primary producer*) for their own needs. *Net primary productivity* is a measure of the amount left available for *heterotrophs*. (see *food web, trophic level, biomass*)

primary RNA transcript (see *primary transcript*)

primary sexual characteristics The reproductive organs that differ between the male and female of a species, and their associated *glands* and *hormones*. (compare *secondary sexual characteristics*)

primary somatosensory cortex The region at the leading edge of the *parietal lobe* of the brain that receives and integrates inputs from *sensory receptors*. Each body part, such as a fingertip or taste receptor, is allocated a specific region of the cortex.

primary structure The linear structure of a *polypeptide* or *nucleic acid* chain, determined solely by the number, sequence, and type of *amino acids* or *nucleotides*, respectively. (compare *secondary structure, tertiary structure, quaternary structure*)

primary transcript In *eukaryotes*, the RNA molecule as it is *transcribed* directly from *DNA*. It is usually modified before being released into the *cytoplasm* of the cell to be *translated* into protein. (see *hnRNA, splicing, cap, poly-A tail, intron, exon*)

primary xylem In plants, *xylem* (water-conducting) tissue derived from the first wave of *procambium differentiation*. It is combined with *primary phloem* into *vascular bundles*.

Primate The order of mammals that contains the chimpanzees, monkeys, apes, and man. Most primates are arboreal, with a highly developed brain, forward-facing eyes allowing binocular vision, and an opposable thumb (and usually big toe) for grasping.

primer A short piece of DNA or RNA (*oligonucleotide*) that forms a *duplex* with an otherwise *single-stranded DNA* molecule and permits synthesis of the remainder of the adjoining *complementary* strand. RNA primers are necessary for DNA synthesis *in vivo*. DNA primers are used extensively to define particular regions for DNA replication *in vitro*, in particular in *PCR* (polymerase chain reaction).

primer extension The sequential addition of *nucleotides* to the end of a short *nucleic acid* (*DNA, RNA*) fragment (*primer*) that has been *annealed* to a longer strand for which a *complement* is to be synthesized. The large single strand is used as a *template* to direct the order of nucleotides inserted by a *DNA polymerase* or *RNA polymerase*, ultimately resulting in the formation of a *duplex* molecule. It is a process that occurs *in vivo* in the *replication* of DNA and is utilized *in vitro* for various nucleic acid manipulations including *PCR* (polymerase chain reaction).

primitive streak In the early *embryo* of terrestrial vertebrates, a longitudinal invagination that is equivalent to *blastopore* formation in other animals. It forms a transitory opening between the interior and exterior of the embryo.

primordium, primordia (pl.) A collection of cells in their earliest stages of *differentiation* toward a specific structure or organ.

prion A class of infectious particles similar to viruses. They were initially thought to consist solely of protein and lack any *nucleic acid*, but the evidence is contradictory.

prisere A primary sere, in other words a plant *succession* allowed to proceed without human intervention. (compare *plagiosere*)

pro- A word element derived from Greek or Latin meaning "before," "outward," or "in front of" in biological words. (for example, *prosencephalon*)

probe In *DNA* technology, a small fragment of *nucleic acid*, either *cloned* or artificially synthesized, that is labeled with a *radioactive* or *chemiluminescent* tag. Under the appropriate conditions, it will *hybridize* by *complementary base pairing* to its partner sequence, identifying its location for the purpose of further analysis. (see *Southern blot, DNA analysis, RFLP*)

Probosidea The order containing the largest terrestrial mammals, the elephants. They are characterized by a trunk that is used for bathing, drinking, and collecting vegetation.

proboscis A tubular projection from the front of the head. It is common in sucking and nectar-feeding insects, but may refer to similarly shaped structures in other animals as well.

procambium In plants, the *primary meristematic* tissue that specifically develops into *vascular tissue*. (see *vascular bundle, phloem, xylem*; compare *ground meristem, protoderm*)

Procellariiformes The order of birds comprising the seabirds, including albatrosses, shearwaters, petrels, and fulmars.

process An extension of a cell, for example a *spicule, spine,* or *pseudopod.*

producer The first *trophic* (feeding) level in a *food chain*. Producers, such as green plants, algae, and *photosynthetic* and *chemosynthetic* bacteria, are capable of synthesizing complex organic molecules by harvesting solar, or sometimes inorganic, chemical energy. Producers are eaten by *herbivores*, which are primary *consumers*. (see *autotrophic, chemoautotrophic*)

product rule The assessment of the probability of two independent events occuring simultaneously by multiplying their individual known frequencies. It is used particularly when combining individual *allele frequencies* in order to estimate the rareness of an individual *genetic profile.*

profundal The deepest zone of a freshwater aquatic *ecosystem*. Little light penetrates this zone, thus the inhabitants tend to be *heterotrophic*, depending on *littoral* (surface) and *sublittoral* organisms for food. Inhabitants include particular bacteria, fungi, mollusks, and insect *larvae.* Organisms that are adapted to withstand the low oxygen concentration, low temperatures, and acidic conditions are

found in this region. The profundal zone is synonymous with the *benthic* zone of the ocean. (compare *limnetic*)

progesterone A *steroid hormone* secreted by the *corpus luteum* in the mammalian *ovary* after *ovulation* (egg release). It initiates preparation of the uterus for implantation of the *ovum* (egg), maintains it during pregnancy, and prepares the *mammary gland* for *lactation* (milk production). (see *prolactin;* compare *estrogen*)

progestogen Any *hormone* or chemical with progesteronelike effects on a female animal.

proglottid One of the many body segments of a tapeworm. As they mature, they develop reproductive organs in which *fertilization* and development of new organisms takes place. Proglottids containing young *embryos* (*hexacanths*) become detached sequentially from the end of the body and are passed out of the host in the *feces*, to be ingested by an *intermediate host*. (see *strobilization*)

programmed cell death (see *apoptosis*)

prohormone The inactive form of a *hormone*. Activation is usually enzymatic, often involving the removal of some portion of the prohormone.

Prokarya (prokaryotae, procaryotae) One of the two superkingdoms of life, the other being Eukarya. It contains a single kingdom, bacteria, comprising all the *prokaryotes*. These organisms are defined as lacking any membrane bound *organelles*, including a *nucleus.*

prokaryote A primitive type of *unicellular* organism lacking a *nucleus* and other cell *organelles*. Its *DNA* is usually a single, naked *chromosome*, generally circular, that lies free in the *cytoplasm*. Exchange of genetic material may be accomplished by *conjugation, transduction,* or *transformation*. Organisms may be *autotrophic* (*chemisynthetic* or *photosynthetic*) or *heterotrophic*. Movement, where present, is accomplished by bacterial *flagella*. It is generally agreed that prokaryotic cells evolved first and subsequently gave rise to the more complex

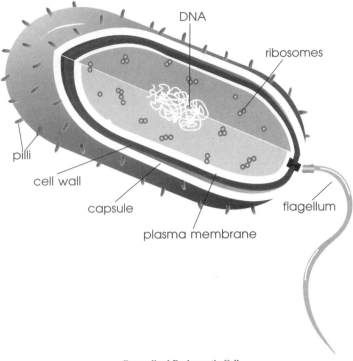

DNA

ribosomes

pilli

cell wall

capsule

flagellum

plasma membrane

Generalized Prokaryotic Cell
(One type of Bacterium)

and sophisticated *eukaryotic* cells by *symbiogenesis*. Modern prokaryotes, or bacteria, are divided into the archaebacteria and eubacteria.

prolactin (luteotrop(h)ic hormone) A *hormone* produced by the *pituitary gland*. In mammals, it promotes the secretion of *progesterone* by the *corpus luteum* and stimulates *lactation* (milk production) after birth. In some birds (for example, pigeons), it maintains egg incubation and stimulates *crop milk* secretion.

proline (pro) An *amino acid.*

promoter The region containing any specific *DNA* sequences that are essential for the initiation of *transcription* of *RNA* from DNA. They are located before *structural genes* and act as binding sites or facilitators of binding for *RNA polymerase*, although they are not themselves transcribed. In *prokaryotes*, the promoter is usually located directly adjacent to the *coding region* and is generally sufficient for at least a minimal level of *gene expression*. In *eukaryotes*, promoter sequences are located farther upstream and work in concert with other factors to promote RNA transcription. (see *operon, UAS, enhancer, Pribnow box,* **CCAAT** *box,* **TATA** *box,* **GC** *box*, compare *operator, regulator*)

propagule Any *unicellular* or *multicellular* structure produced by an organism and capable of survival, dissemination, and further growth. Propagules include all of the *sexual* and *asexual* methods of reproduction, survival, and propagation.

prophage A *bacteriophage genome* (*DNA*) that has temporarily *integrated* into the *host bacterial chromosome*. While it remains integrated, it causes no harm to the host and is replicated along with the bacterial *chromosome*. Excision of the phage DNA (*induction*) will cause an immediate resumption of the *lytic cycle*. (see *temperate phage, lysogeny*; compare *virulent, provirus*)

prophase The first phase of *nuclear division* in both *mitosis* and *meiosis*. During prophase, the *chromosomes* (which have been partially or completely replicated during *interphase*) condense and become visible. In addition, the *nuclear membrane* dissolves. The *microtubular spindle* forms along the future axis of *nuclear division*, and the chromosomes begin to move toward the equator of the spindle (*metaphase plate*). Prophase differs between mitosis and meiosis in that *homologous* chromosomes physically pair (form *bivalents*) during meiosis, whereas they remain separate during mitosis. Prophase may be subdivided into *leptotene, zygotene, pachytene, diplotene, and diakinesis*. (see *synapsis*; compare *metaphase, anaphase, telophase, interphase*)

proplastid A precursor *organelle* that differentiates into various types of *plastids* in plants, including *chloroplasts*.

propositus A member of a family who first comes to the attention of a geneticist.

proprioreceptor A *sensory receptor* that responds to physical or chemical stimuli originating from within the organism. In vertebrates, proprioreceptors supply the *cerebellum* with information about the position and movements of body parts and are important in maintaining balance and posture.

prop root An *adventitious* root arising from a plant stem above soil level and used for support. They are seen in plants that live in shallow waters, such as mangroves.

prosencephalon (see *forebrain*)

Prosimii (see *Scandentia*)

prosoma In spiders, the front portion of the body comprising the head and *thorax*.

prostaglandins A group of *hormone*like fatty acids that stimulate *smooth muscle* contraction (particularly in the uterus) and effect contraction and expansion of blood vessels, regulating blood pressure.

They are produced in virtually all cells of the body and do not circulate. Rather, they remain in regions of tissue disturbance or injury. Overproduction of prostaglandins causes blood vessels to swell so that their walls press against adjacent nerves causing pain (in the brain, causing a headache). Other actions include the induction of *endocrine* hormone secretion. Aspirin and other nonsteroidal analgesics such as ibuprofin counteract the pain, *inflammation,* and fever caused by prostaglandins, apparently by interfering with their synthesis.

prosthetic group The nonprotein component of a *conjugated protein*. Certain *cofactors* and *coenzymes* may be considered prosthetic groups, as well as incorporated metalloproteins such as the *heme* group of *hemoglobin*. The carbohydrate portion of *glycoproteins* and the lipid portion of *lipoproteins* are also prosthetic groups.

prostate gland A gland in male mammals surrounding the *urethra* in the region where it leaves the *urinary bladder*. It is under *androgenic* control and releases a fluid containing enzymes and anti*agglutination* factors that contribute to the production of *semen*.

protamines A group of simple *globular proteins*, rich in arginine and lysine and lacking in sulfur, that are found associated with *nucleic acids* (*DNA* and *RNA*), particularly in *spermatozoa*.

protandry In general, a sequence of events influencing *sexual reproduction* in which the male component occurs first. It is one mechanism to encourage *cross-* rather than *self-fertilization*.

1. In *hermaphroditic* (having both sex organs) animals, the production of *spermatozoa* before eggs.

2. The arrival of males before females at the breeding ground, for instance in some birds.

3. In certain coral reef fish that can change sex, the change from male to female.

4. In *dichogamous* (maturation of male

and female *gametes* at different times) flowering plants (*angiosperms*), the maturation of the *anthers* (male reproductive organs) before the *carpels* (female reproductive organs). It is more common than the opposite (*prodicals aretogyny*), though less effective.

protease (peptidase, proteinase) An enzyme that specifically catalyses the *hydrolysis* (breakage) of *peptide bonds* in proteins, producing *polypeptide chains* and, ultimately, *amino acids*. Individual proteases may be highly specific to the peptide bonds between particular amino acids. (compare *proteolytic enzyme*)

protein One of the four main classes of *macromolecules* (protein, *nucleic acid,* carbohydrate, lipid) found in biological systems. Proteins are composed of *amino acids* joined by *peptide bonds*. The *primary structure* of a protein is determined by the sequence of amino acids, the *secondary structure* by the interactions (hydrogen bonds) between the amino acids (for example, *alpha helix* or *beta pleated sheet*), and the *tertiary structure* by the way the chain is folded (for example, *disulfide bonds*). Sometimes yet another layer of infrastructure, *quaternary structure,* is added when multiple *polypeptide* chains interact to form a complex protein. (compare *polypeptide*)

proteinase (see *protease*)

protein channel (see *transmembrane protein channel*)

protein kinases A group of enzymes that regulate the activity of other enzymes by transferring a phosphate group from *ATP* to the protein. They are particularly important in *intracellular* communication and enzyme *cascades*. Their activity is often a function of the local concentration of *cAMP*. Particular kinases, such as tyrosine or serine kinases, target specific amino acid residues.

protein sequencing The laboratory determination of the *primary structure* of a protein, that is the type, number, and sequence of the *amino acids* in the *polypeptide chain*(s), by limited *proteolysis*. It has largely been supplanted by *DNA sequencing*, which is simpler, more reliable, and provides more information.

protein synthesis (see *translation*)

protein Z In the initial events of *photosynthesis* in *eukaryotes* (and cyanobacteria), the protein that facilitates a complex series of reactions in which high-energy electrons are removed from water (H_2O) and funneled to the *photopigment* P_{680}. It is driven by the photon-induced loss of an electron from P_{680}, which replaces it with an electron from protein Z. This is, in turn, rendered a powerful oxidant, able to replace its own lost electron by splitting water. One molecule of *ATP* is formed in the process. The remaining hydrogen (H^+) ions (protons) augment the proton gradient established as electrons move from *photosystem II* to *photosystem I*. The hydroxyl (OH^-) radicals are collected and reassembled as water and oxygen gas (O_2). (see *electron transport chain, proton pump, photosystem I, photosystem II, photophosphorylation*)

Proteobacteria (purple bacteria) A huge, diverse phylum of *Gram-negative* bacteria into which several previously separate groups have been combined including the omnibacteria, pseudomonads, *nitrogen-fixing aerobic bacteria*, chemoautotrophic bacteria, and *Desulfovibrio*. In fact, most organisms in the world are proteobacteria. They include the enteric bacteria (*coliform*), including the ubiquitous *E. coli* found in the gut of many vertebrates including man. Disease-causing organisms such as *Salmonella* and *Vibrio* (cholera) are also found in this group. The nitrogen-fixing aerobic bacteria are important *symbionts* of *leguminous* plants and are responsible for processing nitrogen into a form that can be used by all other organisms

proteoglycan A type of *glycoprotein* consisting of *glycosaminoglycan* molecules linked to a protein core of *amino acids* rich in serine. They characteristically have a greater carbohydrate content and higher molecular weight than typical glycoproteins and are found primarily in *extracellular* matrices. (see *mucin*)

proteolysis The enzymatic or less commonly chemical *hydrolysis* of proteins into their constituent *amino acids*. (see *protease, proteolytic enzyme*)

proteolytic enzyme Any enzyme involved in the breakdown of proteins. They may be less specific in their action than *proteases*.

proteoplast A colorless *plastid* (*leucoplast*) in which proteins are stored in plants. (see *aleuroplast*)

prothoracic gland A *gland* in arthropods that produces *ecdysone*, the *hormone* that initiates *molting*. (see *ecdysis*)

prothrombin The inactive form of the enzyme *thrombin* in the bloodstream. It is activated during *blood clotting* by the enzyme *thrombokinase*, which requires the presence of calcium ions (Ca^{++}).

protist A single- or few-celled protoctist.

protocooperation (facultative mutualism) A *symbiotic* relationship between organisms of different species where both benefit but neither depends on the relationship. (compare *mutualism*)

proto-oncogene A normal cellular *gene* that, when *mutated*, has the potential to become an active, cancer-causing *oncogene*. This may happen through transmission by a *retrovirus* or, more commonly, simply by mutation at the original *chromosomal* site. They tend to fall into the categories of *transcription factors* and their regulators, or *growth factors* and their receptors. (see *oncogene*)

protoctist Any organism classified into the kingdom Protoctista.

Protoctista One of the *phylogenetic* kingdoms into which all life forms are divided for academic discussion. It is defined mostly by exclusion and contains all the *eukaryotic nucleated* organisms that cannot be classified as animal, plant, or fungus, although different groups may exhibit characteristics normally ascribed to plants, animals, or fungi. Although many protoctists are primarily *unicellular* (*protists*), groups of *multicellular* organisms such as the kelps are also included. Protoctists include protozoans (single-celled animals), algae (single-celled plants and multicellular forms such as kelps), aggregate protists (slime molds and slime nets), as well as many other obscure *eukaryotes*. All contain *subcellular organelles* that are thought to have evolved from bacterial *endosymbioses*. Exchange of genetic material occurs by *fertilization* and *meiosis*. They may be *photosynthetic, heterotrophic,* or a combination. Movement may be *ameboid,* accomplished by 9 + 2 *undulipodia,* or by contractile fibrils.

protoderm In plants, the *primary meristematic* tissue that specifically develops into the *epidermis*. (compare *procambium, ground meristem*)

protofilament A *polymerized* filament of the *structural protein tubulin*. Protofilaments are organized into *microtubules* in the cell. (see *kinetosome, centriole, centrosome, MTOC*)

protogyny In general, a sequence of events influencing *sexual reproduction* in which the female component occurs first. It is one mechanism to encourage *cross* rather than *self-fertilization*.
1. In *hermaphroditic* (having both sex organs) animals, the production of eggs before *spermatozoa*.
2. The arrival of females before males at the breeding ground, for instance in some birds.
3. In certain coral reef fish that can change sex, the change from female to male.
4. In *dichogamous* (maturation of male and female *gametes* at different times) flowering plants (*angiosperms*), the maturation of the *carpels* (female reproductive organs) before the *anthers* (male reproductive organs). It is less common than the opposite (*protandry*), though more effective.

proton channel An *ion channel* associated with *ATP synthase*. They are found in *chloroplasts, mitochondria,* and bacteria. When protons (H^+) reenter a compartment from which they have been excluded by *proton pumps*, the energy released as they pass down the gradient is harvested by the ATP synthase for use in the *chemiosmotic* generation of *ATP*. (see *proton gradient, aerobic respiration, photosynthesis*)

protonema An early stage in the development of a moss or liverwort. A protonema is produced when the organism *germinates* and what is commonly known as the moss or liverwort arises from buds that develop on it.

protonephridium (see *nephridium*)

proton gradient A differential in the concentration of protons (H^+) from one side of a membrane to the other. It is established by a *proton pump* using the energy from excited electrons as they pass down an *electron transport chain* and is found in *mitochondria, chloroplasts*, and bacteria. Proton gradients are used in the *chemiosmotic* generation of *ATP*. (see *proton channel, ATP synthase, aerobic respiration, photosynthesis*)

proton pump An *active transport* system in the *plasma membrane* of a cell or cell *organelle* that moves protons (H^+) against a concentration gradient in an energy dependent fashion. When the protons diffuse back across the membrane through special channels associated with *ATP synthase*, their passage is coupled to the *chemiosmotic* production of *ATP*. Proton pumps are used in *mitochondria, chloroplasts*, and bacteria to generate ATP. (see *aerobic respiration, photosynthesis, photosystem I, photosystem II, electron transport chain*)

protoplasm The entire fluid contents of a cell including the *nucleoplasm* and the *cytoplasm*.

protoplasmic streaming (see *cytoplasmic streaming*)

protoplast The contents of a plant cell inside the *cell wall* including the *plasma membrane, cytoplasm,* and *organelles*.

protostomes One of the two major groups of *coelomate* animals. They show a distinct pattern of *embryological* develop-

ment. The mouth forms at or near the *blastopore*, and the anus, if present, forms subsequently on another part of the *blastula*. This pattern also occurs in all *acoelomate* animals. Additionally, the embryo shows a pattern of *spiral cleavage*, and each cell is *committed* to a particular developmental fate as soon as it appears. The group includes the mollusks, annelids, arthropods, and some smaller phyla. (compare *deuterostomes*)

Prototheria A mammalian subclass containing only one order, the monotremes, the only mammals that lay eggs. After hatching, the young are transferred to a pouch on the abdomen and nourished by milk secreted by primitive *mammary glands* whose ducts do not form nipples. Other primitive features include poor temperature control and possession of a *cloaca*. Monotremes include the duck-billed platypus and the spiny anteater.

prototrophic A laboratory strain, usually of a microoorganism, that will proliferate on a medium containing minimal nutrients. Most *wild-type* strains are prototrophs. (compare *auxotrophic*)

Protozoa A historical name for the kingdom *Protoctista*. The term has no current taxonomic significance but generally refers to *heterotrophic, unicellular* protoctists lacking *photosynthetic* ability, in other words single-celled animals.

Protura An order of minute primitive, wingless insects, the bark lice. They lack eyes and *antennae*, and have piercing mouthparts. They live under bark, stones, or among rotting vegetation. They are sometimes subterranean. Metamorphosis is *exopterygote*.

proventriculus The upper part of a bird stomach before the *gizzard,* where enzymes are secreted. In crustaceans and insects, it is synonymous with gizzard.

provirus A viral *chromosome* that has integrated into the *host genome* and replicates with it. In *RNA viruses*, this is accomplished by the enzyme *reverse transcriptase* that synthesizes *complementary DNA* from an RNA *template*.

The viral genome may either remain latent and innocuous, or may take over the host metabolic machinery, eventually killing the host cell and releasing many replicated virus particles. The *HIV* virus, which causes *AIDS*, works in this fashion and is specific for the T_4 white blood cell (*leukocyte*). It typically spends about 8 years as a provirus before resuming its infective *lytic* cycle. (see *retrovirus*; compare *prophage*)

proximal Denotes the part of a limb or structure that is nearest the origin or point of attachment. Thus, the shoulder is proximal to the fingers. (compare *distal*)

prymnesiophytes (see *Haptomonada*)

PS I (see *photosystem I*)

PS II (see *photosystem II*)

pseudocarp (false fruit) A fruit in which the ripened *ovary* and its contents are combined with another structure, often the *receptacle*.

pseudociliates (see *Discomitochondria*)

pseudocoelomate A group of bilaterally symmetrical invertebrates, including nematodes and rotifers, that possess a partially developed body cavity (pseudocoelom). The pseudocoelom serves as a *hydroskeleton*; fluids within it function as a loose *circulatory system*. They possess a complete *digestive system*, as do *coelomates*. (compare *acoelomate*)

pseudogene An ancestral *gene* that has become inactive through *mutation*. It usually has at least one actively *transcribed* counterpart and sometimes other pseudogenic counterparts elsewhere in the *genome*.

Pseudomonads (see *Proteobacteria*)

pseudoplasmodium, pseudoplasmodia (pl.)
1. An organismal state that occurs in some cellular slime molds. It resembles a *multinucleate plasmodium* but retains its *cell membrane* boundaries.
2. An aggregate of amebas.
3. A *uninucleate trophozoite* cell containing one to several *generative* cells.

pseudopod (see *pseudopodium*)

pseudopodium, pseudopodia (pl.)

(pseudopod) A temporary protrusion or projection on the body of an *ameboid protist*. It is formed by *cytoplasmic* movement and functions in locomotion and feeding. They are typical of the phylum Rhizopoda. (see *ameboid*)

Pseudoscorpiones An order of arthropods, the false scorpions. They resemble scorpions but lack a stinger.

Psilophyta The plant phylum comprising the whisk ferns. They are seedless *vascular plants* that lack both leaves and roots. As such, they resemble the earliest *vascular plants*. They *alternate generations*, the *sporophyte* being the prominent *diploid* form. *Haploid meiotic spores* develop in *sporangia*. Both generations develop *endomycorrhizal* associations.

Psittaciformes The order of birds containing parrots, macaws, cockatoos, and lories.

Psocoptera An order of *exopterygote* insects, the booklice and barklice. They are small, sometimes wingless, and have biting mouthparts. They feed on fragments of animal and vegetable matter and bookbinding paste.

psychrophilic Microorganisms whose optimum growth temperature is below approximately 20°C. (compare *mesophilic, thermophilic*)

Pteroclidiformes The order of birds including the sandgrouse.

Pterodatina (see *Filicinophyta*)

Pterophyta (see *Filicinophyta*)

PTH (see *parathyroid hormone*)

ptyalin One of the *amylase* group of enzymes. It is present in saliva and catalyses the conversion of *starch* to maltose.

ptyxis (see *vernation*)

puffins (see *Charadriiformes*)

pulmonary arch (see *aortic arches*)

pulmonary arteries The paired arteries that carry *deoxygenated* blood from the heart to the lungs in tetrapods and lungfish. They are derived from the sixth *aortic arch*.

pulmonary circulation The diversion of *deoxygenated* blood from the heart to the lungs and back again for bodily distribution. It is an evolutionary advance first seen in amphibians, improved in reptiles to achieve better separation of the *oxygenated* blood, and refined in birds and mammals to a true *double circulatory system.*

pulmonary respiration Respiration involving lungs.

pulmonary veins The paired veins that carries *oxygenated* blood from the lungs to the heart in tetrapods and lungfish.

pulp cavity The central core of a mammalian tooth, surrounded by *dentine.*

pulse The waves of pressure caused by each contraction of the heart. In humans, the rate ranges from approximately 50 to 80 beats per minute, increasing with activity or stress.

pulse-chase experiment A laboratory technique in which cells are grown in radioactive medium for a brief period (the pulse) then transferred to nonradioactive medium for a longer period (the chase). It is possible to measure the amount and location of any incorporated radioactive molecules. It was innovated in 1952 by Alfred Hershey and Martha Chase in their definitive experiment claiming *DNA* as the genetic material. Hershey received the Nobel prize in physiology or medicine for this work in 1969.

pulse-field gel electrophoresis A laboratory technique used to separate extremely large *DNA* molecules, including whole *chromosomes.* It is a type of *electrophoresis* in which the *agarose* gel is subject to electrical fields alternating between different angles, allowing large molecules to snake through the microscopic holes in the porous medium.

pulvinus, pulvini (pl.) A specialized group of cells located at the base of plant leaves involved in leaf bending. *Motor cells* on each side of the pulvinus absorb water differentially, such that the leaf moves away from the side in which more water has accumulated. (see *nastic movements, turgor pressure*)

Punnett square A classical genetic technique in which a grid is used to represent graphically the progeny *zygotes* resulting

from the fusion of different *gametes* in a specific *cross.* It was developed by Reginald Crundall Punnett working with William Bateson on sweet peas at the turn of the century.

pupa, pupae (pl.) The stage between *larva* and *imago* in the *life cycle* of insects that undergoes complete *metamorphosis.* During the pupal stage, the insect does not feed, is usually immobile, and undergoes a major internal reorganization. Pupae are frequently inactive, forming a hard shell (in butterflies, a *chrysalis*) or silken covering (*cocoon*) around themselves, an exception being mosquitoes in which the pupae are fully mobile.

puparium, puparia (pl.) A *pupa* formed from the *exoskeleton* of the final *larval instar.* (see *coarctate*)

pupil The hole in the center of the *iris* of vertebrate and cephalopod eyes through which light enters. The diameter of the pupil can be altered by the involuntary muscles of the iris.

purine A *nitrogenous* organic molecule with a double ring structure. Members of the purine group include *adenine* and *guanine*, which are constituent bases of nucleic acids (*DNA* and *RNA*) as well as certain plant *alkaloids* such as caffeine. (compare *pyrimidine*)

Purkinje fibers A bundle of specialized *cardiac muscle fibers* comprising the *bundle of His.* They are found in mammals and receive rhythmical nerve impulses from the *pacemaker* which spread as waves of contraction through both *ventricles.* They are named for Johannes Evangelista Purkinje who was one of the first to study nerve physiology in the 1830s.

Purple Bacteria (see *Proteobacteria*)

pycnidium, pycnidia (pl.) An *asexual multicellular* reproductive structure found in fungi. It is lined with *spore*-bearing *conidiophores.*

Pycnogonida A class of *chelicerate* arthropods, the sea spiders. They occur in coastal waters and are mostly external *parasites* or predators of other sea animals, in particular anemones. They completely lack an *excretory* or *respiratory system*. They appear to carry out these functions by direct diffusion of oxygen and waste products through the cells. Their eggs are carried on the legs of the males until they hatch.

pyloric sphincter (*pylorus*) A ring of *smooth muscle* in vertebrates that surrounds the opening from the stomach to the *duodenum* and regulates passage of food between them.

pylorus (see *pyloric sphincter*)

pyramid of biomass A diagrammatic description of the total amount of living material at each *trophic* (feeding) level in an *ecosystem*. The bottom of the pyramid consists of a large number of small organisms. At each level, the number of organisms becomes fewer and their relative size becomes larger. The total biomass at each level generally decreases. In some special systems (for example, *planktonic*), the pyramid of biomass may occasionally be inverted. (see *consumer, producer, decomposer*; compare *pyramid of energy, pyramid of numbers*)

pyramid of energy A diagrammatic description of the rate of energy flow at each *trophic* (feeding) level in an *ecosystem*. The loss of energy at each step is so great that very little of the original energy remains in the system as usable energy after it has been successively incorporated into the bodies of organisms at each trophic level. The pyramid of energy is never inverted. (see *consumer, producer, decomposer*; compare *pyramid of biomass, pyramid of numbers*)

pyramid of numbers A diagrammatic description of the total numbers of organisms at each *trophic* (feeding) level in an *ecosystem*, irrespective of size. The bottom of the pyramid is very broad, consisting of a large number of small organisms; at each level, the number of organisms becomes fewer and their relative size becomes larger. Larger animals are characteristically members of the higher levels; to some extent they must

be larger to capture enough prey to support themselves. In some special systems (for example, *planktonic*), the pyramid of numbers may occasionally be inverted. (see *consumer, producer, decomposer*; compare *pyramid of energy, pyramid of biomass*)

pyranose A sugar that has a six-membered ring structure with five carbons and one oxygen.

pyrenoid A region within the *chloroplasts* of many algae in which protein or starch is stored.

pyridoxine (vitamin B$_6$) A member of the *vitamin B complex* found in yeast, certain seeds, liver, and other foods. It is also obtained from *symbiotic* intestinal bacteria. It plays a role in forming a *coenzyme* required for the synthesis of *amino acids* from carbohydrate precursors.

pyrimidine A *nitrogenous* organic molecule with a six member, single ring structure. Members of the pyrimidine group include *cytosine, thymine,* and *uracil,* which are constituents of nucleic acids (*DNA* and *RNA*) and *thiamine* (vitamin B$_1$). (compare *purine*)

pyrimidine dimer A type of *DNA damage* resulting from exposure to UV (ultraviolet) light in which the absorbed energy results in a covalent bond being formed between two adjacent *pyrimidines* (*thymine, cytosine*). If left unrepaired, such a structure can block DNA *replication*, causing cell death.

pyrogens Chemical substances that boost the body's temperature by acting on the *neurons* in the *hypothalamus* that serves as a thermostat. They are released by invading microorganisms that have encountered *macrophages*. The fever inhibits microbial growth, stimulates *phagocytosis*, and reduces blood levels of iron, which bacteria need in large amounts.

pyruvate The three-carbon compound that links the *anaerobic* and *aerobic* portions of *cellular respiration*. It is the end product of *glycolysis* and the starting material of the *Krebs cycle*.

◆ Q ◆

Q (uniquinone, plastoquinone) Uniquinone and plastoquinone are both minor variations on the quinone molecule, and each functions as a molecular electron transporter. Plastoquinone is the oxidized form of the protein that accepts electrons from water in the *photosystem II* pathway. Uniquinone is an electron acceptor in the mitochondrial *electron transport chain*. Their reduced forms are, respectively, *plastoquinol* and *uniquinol,* both designated as QH_2. (see *aerobic respiration*)

QH_2 (uniquinol, plastoquinol) The reduced form of these similar electron transport proteins. Their oxidized forms are, respectively, *uniquinone* and *plastoquinone*, both designated as *Q*. (see *photosystem II, aerobic respiration, electron transport chain*)

quadrat A measured area of any shape and size that defines a sample area in a biological survey, particularly a survey of plants or *sessile* (sedentary) animals. (compare *transect*)

quail (see *Galliformes*)

qualitative variation (discontinuous variation) Traits or characteristics that are inherited as distinct entities. They are typically determined by different *allelic* forms of a single *gene*, for instance *ABO blood groups* in man or *mating types* in fungi or protoctists. Biological systems contain few clear examples of qualitative variation. (compare *quantitative variation*).

quantasomes Particles arrayed on the surface of *chloroplast thylakoid discs*. They include various molecular *photoreceptors* (*chlorophylls* and accessory *pig*-

ments) and electron carriers. They are considered to be photosynthetic units. They occur in two sizes, the smaller thought to represent the site of *photosystem I* and the larger, *photosystem II*.

quantitative variation (continuous variation) Traits or characteristics that vary on a continuum throughout the population. They are usually determined by more than one *gene* (*polygenic*) and may also be influenced by environmental factors. Examples include skin color, eye color, and height. Quantitative variation is more common than *qualitative variation* in biological systems, particularly in higher organisms.

Quaternary The most recent period of the *Cenozoic* geological era, from about 2 million years ago to the present day. It is characterized by glaciation in the northern hemisphere (*ice age*) and the emergence of modern man.

quaternary structure The fourth order of *macromolecular* structure. In proteins, it is characterized by the interaction of two or more individual *polypeptides*, often by *disulfide bonds*, to give larger functional molecules. In *nucleic acids*, the supercoiling of *DNA nucleosomes* into a *solenoid structure* represents a fourth order of structure that repeats several more times to form *eukaryotic chromosomes*. (compare *primary structure, secondary structure, tertiary structure*)

quetzals (see *Trogoniformes*)

quill The base of each bird feather that grows out of the feather *follicle* in the skin and continues into the *vane* of the feather as the *rachis*.

quillworts (see *Lycophyta*)

rabbits (see *Lagomorpha*)

race A collection of individuals of a species that have developed similar genetic characteristics as a result of living in a similar environment. Races were originally defined in plants and are called *ecotypes*. Particularly among humans, the delineations are ambiguous as well as somewhat arbitrary. In fact, genetic differences between individuals have been shown to be greater than differences between what may be defined as human races.

raceme An *inflorescence* (multiple flower) in which flowers are borne on a single main stalk (*peduncle*). An example is foxglove.

racemose inflorescence (indefinite inflorescence) An *inflorescence* (multiple flowering structure) in which apical (top) growth of the plant continues indefinitely while side branches continue to be produced. This results in older flowers being toward the base of the structure or in flat-topped clusters, to the outside of the flower head. (see *capitulum, corymb, raceme, spike, umbel, monopodial*; compare *cymose inflorescence*)

rachis In general, a main axis bearing smaller or divided structures.

1. In plants, the rachis is the main stalk bearing the flower or in compound leaves, the leaflets.
2. In feathers, the rachis is the central support of each, bearing the fibers.

racoons (see *Carnivora*)

radial cleavage The pattern of *cell cleavage* in *deuterostome embryos*. The cells divide parallel and at a right angle to the polar axis such that a line drawn through a sequence of dividing cells describes a radius outward from the polar axis. It results in a loosely packed array of cells. (compare *spiral cleavage*)

radial symmetry The arrangement of structures in an organism in such a way that an imaginary line across the diameter, in any plane, produces mirror images. It is characteristic of simple, *sessile* animals such as jellyfish and starfish, and of many flowers.

radicle The rudimentary root in plant *embryos*. It is the first structure to emerge from the seed at *germination*.

radioactive Any element that exists as an unstable *isotope* and emits high-energy radiation as it decays to a more stable form.

radioactive dating A general method of archeological dating that uses the decay rates of naturally occuring isotopes to assess the age of a specimen. Carbon-14 (^{14}C), which decays to carbon-12 (^{12}C), is commonly used to date organic matter by comparing the present atmospheric proportions of the isotopes in carbon dioxide (CO_2) to those found in the sample. After death, no more environmental carbon is assimilated, and the proportion of ^{14}C to ^{12}C continues to decrease at a predictable and measurable rate.

radioactive isotopes (see *radioisotope*)

radioactive labeling (see *labeling*)

radioimmunoassay (RIA) (see *immunoassay*)

radioisotope An alternative form of a chemical element (*isotope*) that is highly unstable and emits high energy radiation as it decays to a more stable form. Radioisotopes such as ^{32}P and ^{35}S have been widely used as biological tags, but such use is declining as safer ways of visualizing biological molecules are developed. (see *DNA analysis, Southern Blot, chemiluminescent*)

radius In tetrapods, one of the two long bones of the lower forelimb. In man, it extends from the elbow to the wrist. (see *ulna*)

radula A rough strip on the tongue of most mollusks that acts as a file to scrape the vegetation on which it feeds. Continual growth enables the radula to be replaced as it wears down.

rails (see *Gruiformes*)

rainforest Any coniferous forest receiving

a large amount of rainfall, at least 150 to 450 or more centimeters per year. True rainforests, those that receive rain regardless of season, are concentrated in a few equatorial regions. In most others including most *tropical rainforests* as well as the *temperate rainforests* of the Pacific Northwest, precipitation is concentrated in a wet season that alternates with a dry one. Both support giant trees and contain a large *biomass* (many organisms), but *biodiversity* (the number of different species) is much higher in tropical rainforests. The temperate rainforest is essentially synonymous with *old growth forest*. (see *biome*)

ram ventilation In bony fish, the process of continuously forcing water past the *gills* by swimming with their mouths open.

random assortment (see *independent assortment*)

random mating Mating between individuals where the choice of partner is not influenced by *genotype*, particularly with respect to the specific *genes* under study. It is one of the prerequisites for a population to reach *Hardy-Weinberg equilibrium*.

random primer labeling A technique for incorporating radioactive *nucleotides* into short fragments of *DNA*, usually for use as *probes*. Synthetic *oligonucleotides* of random nucleotide sequence are manufactured and used as *primers* for the synthesis of *double-stranded* DNA from a known *single-stranded template*. The oligonucleotides are short enough (usually 6 to 10 nucleotides) that they are not *gene* or *locus* specific and will be complementary to many short sequences within the chosen template. All four nucleotides, one or more of which may be radioactive, must be provided for a *DNA polymerase* to incorporate in the DNA synthesis. (see *DNA analysis, Southern blot, hybridization*)

raphides Needle-shaped crystals of calcium oxalate found in the *vacuoles* of certain plant cells.

rapid eye movement (see *REM*)

rare species A species that has a small total population size or is restricted to a small area. Although some species are naturally rare yet not necessarily *threatened* or *endangered*, rarity still raises questions about *extinction*.

ratites An informal grouping that contains the terrestrial flightless birds such as the emu and the ostrich. They are large, heavy, fast-running birds with long powerful hind limbs and reduced wings. Their sternum structure, palate, and feathers differ from other birds. Ratites may not be closely related to each other. (compare *carinates*)

rats (see *Rodentia*)

Raunkiaer's plant classification A classification of plant growth forms proposed by Danish botanist Christen Raunkiaer based on the position of *perennating* (resting) buds in relation to the soil surface. (see *chamaephyte, cryptophyte, epiphyte, helophyte, hemicryptophyte, hydrophyte, phanerophyte, therophyte*)

rays (see *Chondrichthyes*)

reaction As it pertains to the steps of an *ecological succession*, the return effect of vegetation on the local environment. The amount of organic material is increased, encouraging further growth and complexity.

reaction time The period of time between stimulation of an organism and a detectable response. *Reflex arcs* shorten this time considerably, while nonreflex responses are somewhat slower due to the delay introduced in *nerve impulse* transmission at each *synapse*.

reading frame The specific permutation of *nucleotide triplets* in *DNA* determined by the placement of a particular *start codon*. Proteins must be *translated* in the correct reading frame or the *codons* will either fail to specify the correct *amino acids* or induce premature termination by a *stop codon*. An addition or deletion or even one *nucleotide* anywhere in the *coding region* for a *gene* will disrupt the reading frame.

reannealing The reformation of a *double-stranded* molecule from two single strands of nucleic acid (*DNA* or *RNA*) that have previously been *denatured*. (see *renaturation*).

recapitulation (biogenetic law) The theory introduced by E. F. Haekel that *ontogeny* (*embryological* development) recapitulates the *phylogeny* (evolutionary history) of the individual. Although the study of embryological development is useful in making inferences about evolutionary relatedness, the theory itself has been rejected as an oversimplification.

receptacle In flowering plants (*angiosperms*), the tip of the stalk into which all other flower parts are inserted. (see *hypogenous, epigynous, perigynous*)

receptor

1. **(sensory receptor)** A cell, tissue, or organ that is specialized to receive and respond to stimuli from outside or inside the body of an organism. The eyes, ears, and nose are organ receptors, while *baroreceptors* and *proprioreceptors* are cell receptors.

2. Any of the various proteins and modified proteins that process molecular signals on the cell surface, convey them to the interior, and mediate their function at a molecular level. They often mediate or regulate *gene expression*. (see *second messenger, steroid receptor*)

receptor protein Any *transmembrane protein* that transmits information from the outside environment to the inside of the cell when bound by a specific chemical messenger molecule. The system is highly specific because each type of receptor is shaped to fit only one particular signal molecule. The act of binding initiates a cascade of actions that effect a biochemical change inside the cell. *Peptide hormones* work in this fashion. (see *second messenger, steroid receptor*)

hypogenous

epigynous

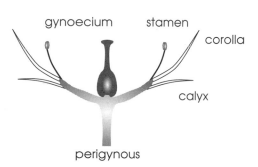
perigynous

Floral Receptacles

recessive A variant (*allele*) of a particular *gene* that fails to contribute to the *phenotype* (physical characteristic) of a *diploid* organism except when found in a *homozygous* (two of the same alleles) condition. In contrast to a *dominant* allele of the same gene which normally produces a functional protein, a *recessive* allele often produces a nonfunctional protein product or no protein at all. A *heterozygote* containing dominant and recessive alleles of a gene is usually phenotypically indistinguishable from a homozygote containing two dominant alleles.

reciprocal cross A pair of genetic crosses in which individuals with reciprocal *genotypes* for defined *loci* (gene locations) are mated in both combinations and the offspring analyzed. Such crosses are used to detect *sex linkage, maternal inheritance,* or *cytoplasmic inheritance* of particular *genes.* For instance CROSS I: genotype A (male) x genotype B (female); CROSS II: genotype A (female) x genotype B (male).

reciprocal recombination A type of genetic *recombination* in which two *chromosomes* trade equivalent segments that may, however, contain different *alleles* (variants) of the *genes* located there. This normally occurs during *meiotic prophase*, when the chromosomes have been duplicated (each into two *chromatids*) and are physically paired (*synapsis*). The result of reciprocal recombination is to produce *gametes* with different combinations of alleles than either of the parents, thus introducing new, possibly beneficial, variants into the population. Reciprocal recombination is important primarily in *eukaryotes*, which are typically *diploid* during some or all of their *life cycle.* (see *homologous recombination, unequal crossover*)

recombinant In classical genetics, a *sexually* derived individual or cell with a *gene* combination (*genotype*) different than either parent. It may also refer to a *DNA* molecule that has been altered by exchange with or incorporation of a foreign DNA segment. (see *recombination, recombinant DNA*)

recombinant DNA Any *DNA* molecule that has been altered by exchange with or incorporation of a foreign DNA segment. The process occurs naturally, particularly during *meiosis* in the normal development of organisms. More recently, the term has come to be synonymous with artificial DNA constructs combining DNA from different species. (see *recombination, genetic engineering, gene cloning*)

recombinant DNA technology (*see genetic engineering*)

recombination (genetic recombination) The formation of new *gene* combinations, thus introducing new, possibly beneficial, variants into the *population.* In *eukaryotes,* redistribution of *genes* occurs during *meiosis* as a result of *independent assortment* of *chromosomes* as well as the reciprocal or nonreciprocal exchange of chromosome segments. Certain types of recombination also occur at a much reduced rate (by several orders of magnitude) during normal *mitotic* division. In *prokaryotes,* recombination may occur when genetic information is transferred between cells by *transduction* or *conjugation.* (see *homologous recombination, reciprocal recombination, gene conversion, F-factor, episome, independent assortment, independent segregation*)

recombination frequency The measured frequency with which a particular genetic *recombination* event is observed in a population. Before the advent of physical *gene* mapping, genetic distances were estimated by the frequency of *reciprocal recombination* between the questioned *genes.* (see genetic map)

rectum The last part of the *alimentary canal* in which *feces* are stored and released at intervals to the exterior through an *anus* or *cloaca.* In mammals, it is closed by a *sphincter.*

red algae (see *Rhodophyta*)

red blood cell (see *erythrocyte*)

redia, rediae (pl.) An intermediate *larval*

stage of flukes (trematode worms) that develops *asexually* from a *sporocycst* or from other rediae inside a secondary *host*. It develops into a *cercaria* which is free swimming and emerges from the secondary host, often a snail, to penetrate the skin of its primary host (for example, man in *Schisotosoma*) or to *encyst* as a *metacercaria* awaiting ingestion by the primary host. (see *parasite*)

red nucleus A large group of *neurons* in the vertebrate *brainstem* that, in concert with the *cerebellum* and the *corpus striatum*, is able to generate complex patterns of activity in *motor neurons*.

redox reaction A chemical reaction that involves the simultaneous *reduction* and *oxidation* of two compounds by a transfer of electrons between them. In biological reactions, the electrons are usually transferred as part of hydrogen, oxygen, and sometimes other atoms.

red tide The name given to periodic population explosions of particular dinomastigote protists that take place in coastal waters. The water is colored red and contains high concentrations of a powerful respiratory toxin that poisons marine life and contaminates sea animals typically harvested for food, particularly *filter feeders* such as clams and oysters. Red tides have also been correlated with *blooms* of chrysophytes, euglenids, and a ciliate protist.

reducing power The ability of a compound to donate protons and accept electrons in a chemical reaction. In biological reactions, the electrons are usually part of a hydrogen atom. (see *reduction, NADH, FADH$_2$*)

reduction The addition of electrons to an element or compound in a chemical reaction. In biological reactions, the electrons are often part of a hydrogen atom. It takes place simultaneously with *oxidation* (loss of an electron) of another atom. (see *reducing power*)

reduction division A *nuclear division* in a *diploid* cell in which the progeny each retain half the number (*haploid*) of chro-

mosomes. The first division of *meiosis* is a reduction division. Each of the duplicated *chromatids* is counted as one chromosome since they are still attached to the same *centromere*. (compare *equational division*)

reflex In the *nervous system*, an immediate and involuntary reaction to a stimulus without conscious initiation or modification. It is the only type of reaction in the simplest animals (for example, cnidarians) but represents only a few responses, such as a knee-jerk reaction, in animals with more sophisticated nervous systems. (see *nerve net, reflex arc*)

reflex arc A nerve system in which the *sensory* and *motor* nerves are linked directly (for example, cnidarians, knee-jerk response in humans) or through only one or a few *interneurons* in more complex reactions. A reaction to a stimulus is extremely rapid, as the *nerve impulses* bypass any processing or integration in the brain. (see *reflex*)

refractory period The period following the passage of an impulse along a *neuron* when no stimulus can be evoked while the resting ion imbalance (*resting potential*) is restored by the *sodium-potassium pump*. During the absolute refractory period, initiation of another impulse is impossible. During the somewhat longer relative refractory period, an *action potential* can be initiated, but the stimulus must be stronger than normal. (see *depolarization*)

regeneration The regrowth of an organ or tissue that has been lost though injury or *autotomy* (deliberate disassociation). In plants and some lower animals, the complete organism may be regenerated from only a few cells. As organisms become more complex, regenerational ability is lost proportionally. In mammals, it is limited to wound healing, some organ regrowth, and minimal regrowth of peripheral *nerve fibers*. (see *totipotent*)

regulation In *embryonic* development, the ability to compensate for a disturbance that involves removal, addition, or rearrangement of material and still produce an apparently normal embryo. In a regulative egg, *cleavage* products are not *determined* (set in developmental fate) until after many divisions; any one *blastomere* (embryonic cell) can still reproduce a whole embryo. Embryonic regulation includes *twinning* (bisecting the embryo to produce identical animals) and *fusion* (combining more than one embryo to produce one large individual). (see *totipotent*)

regulator (see *operon*)

regulatory enzyme An enzyme that is involved in the regulation of *metabolic* pathways at the protein level. Regulation may involve *allosteric* changes modulated through molecules binding at sites other than the *active site*. Alternatively, it may involve modification of the enzyme by another enzyme, such as a *kinase*, by which active and inactive forms are alternated.

regulatory gene A *gene* that regulates the *transcription* of *structural genes*. Regulatory genes produce *regulatory proteins* that bind to *regulatory sites* upstream from structural genes. They may either inhibit (*negative control*) or permit, or facilitate (*positive control*) RNA transcription of the structural genes. (see *operon*)

regulatory protein (see *regulatory gene*)

regulatory site (see *regulatory gene*)

reinforcement The strengthening of a response during learning by introducing an associated action. In experiments, a reward for performing a desired action is a positive reinforcement to encourage repetition and learning of the action; punishment to teach avoidance of a behavior is negative reinforcement.

relative refractory period (see refractory period)

relaxin A *hormone* produced by the *corpus luteum* (site of *ovulation* in the ovary) in the last stages of pregnancy that causes dilation of the cervix, enables the *pelvic girdle* to widen, and prepares the uterus for labor.

releaser A standard external stimulus that evokes a stereotyped behavioral response. A social releaser emanates from a member of the same species and is commonly used in courtship and threat displays.

REM (rapid eye movement) The type of sleep during which dreams take place. It is essential to the restorative properties of sleep.

renal Any structure or function pertaining to the kidney.

renal tubule (uriniferous tubule) In the mammalian kidney, the tubule between each *Bowman's capsule* and a collecting duct. The middle portion of each renal tubule is folded into a hairpin loop called the *loop of Henle*, in which ionic *countercurrent exchange* takes place. This enables the reabsorbtion and conservation of water essential in terrestrial animals. (see *osmoregulation*)

renaturation The reestablishment of the *secondary* and *tertiary structure* of proteins or *nucleic acids* after it has been destroyed by heat, exposure to air, or extreme pH values. Once *denatured*, proteins usually cannot be renatured. *Complementary single strands* of *nucleic acid* (DNA, RNA) may often be reformed into a *double-stranded molecule* under the proper conditions. This property is exploited in the laboratory technique of *hybridization* in the course of a *Southern blot*, and in the *annealing* of *primers* in *PCR*.

renin An enzyme produced in the kidney that stimulates the conversion of a *plasma* protein into several forms of the *hormone angiotensin*. It is part of the *osmoregulatory* system and ultimately increases total sodium (Na^+) in the bloodstream, elevating *blood pressure*.

rennin An enzyme found in *gastric* secretions and responsible for the coagulation (*denaturation*) of milk protein. It is used commercially under the name rennet in the manufacture of cheese products.

reoviruses A group of *double-stranded RNA viruses* that infect mammals, arthropods, and plants. One is responsible for Colorado tick fever and others cause a form of diarrhea in young children.

repeated sequence In the *DNA* of *eukaryotes*, a *nucleotide* sequence that is repeated from two to thousands of times in a single *genome*. *Tandemly repeated* sequences are arranged in a head-to-tail fashion in one location, while dispersed repeats may be scattered throughout the genome. The number of nucleotides in the sequence itself may range from two to hundreds. (see *satellite DNA, rRNA, length polymorphism, VNTR*; compare *unique sequence*)

repetitive DNA (see *repeated sequence*)

replica plating A laboratory technique in which microorganisms are transferred between different solid media plates, usually for the purpose of selecting *mutants* or testing growth. Sterilized velvet, which contains numerous needlelike projections, is used as a transfer implement.

replication (DNA replication) The new synthesis of identical *DNA* strands from a parent molecule. The *double helix* unzips and each strand serves as a template for *DNA polymerase*, which adds complementary *nucleotides* in the order dictated by the existing strand. This mechanism is called *semiconservative replication* as each new molecule contains one original and one newly synthesized strand of DNA. (see *complementary base pairing*; compare *transcription, translation*)

replicon The region of DNA under the influence of one adjacent *replication* initiation sequence. *Prokaryotic* and viral *genomes* usually contain one replication origin per *genome*, while a *eukaryotic chromosome* usually contains several. (see *ARS*)

reporter gene A *gene* whose expression is easy to monitor either by visual inspection or by a simple test. They are often used in research to study the regulation of *RNA transcription* and to monitor *recombinant DNA* constructs that have been reintroduced into an organism. The bacterial *lacZ* gene, which breaks down the *lactose* analogue X-gal to yield a bright blue *metabolite*, and the *bioluminescent luciferin* gene are often used as reporter genes.

repressor A protein molecule that inhibits *RNA transcription*, hence *gene expression*, by binding to an *operator* site upstream of the *promoter* for a *structural gene*. Repressor proteins are produced by *regulatory genes*.

reproduction The genesis of new organisms by either *sexual* or *asexual* means from existing organisms. Reproductive capability is one of the criteria for life.

reproductive cell Any cell from which a new individual can be produced. This may occur either *sexually*, by fusion with another cell (*fertilization*), or *asexually*, by *mitotic* division. (see *gamete, spore, propagule*)

reproductive system The organs, cells, and *hormones* involved in the production of new individuals. In higher organisms, this includes the *gonads* of both sexes (*testes* and *ovaries*) and their associated *glands* and hormones.

reptiles (see *Reptilia*)

Reptilia The class of vertebrates that contains the first wholly terrestrial tetrapods, the reptiles, including lizards, snakes, crocodilians, and turtles. They are adapted to life on land by the possession of a dry skin with horny scales that serves to prevent evaporative water loss. *Fertilization* is internal and no *larval* stage exists; the young develop from a leathery, watertight egg laid on land (*amniote*). Respiration is by lungs only, and the heart has four chambers. Reptiles are unable to regulate their body temperature internally (*poikilothermic*). Like their probable descendants the birds, they contain *nucleated* red blood cells (*erythrocytes*).

reservoir The invagination in the outer *pellicle* of euglenoid protists from which the *undulipodia* project.

resin A liquid exuded from the wood or bark of some types of trees that becomes solid upon exposure to air. It is produced in special cells and is one of a number of derivatives of the terpene family. Some resins are important commercially, such as pine which is distilled into turpentine and others which are made into varnishes.

resistance transfer factor (R-plasmid) Bacterial *plasmids* that carry multiple *antibiotic* resistance *genes*. (see *gene mobilization*)

resolving power (resolution) The ability of an optical system to distinguish adjacent objects as separate. It is important in microscopy that resolving power increase along with magnification. (see *microscope*)

respiration
1. The inhalation of oxygen (and other atmospheric gases) and the exhalation of carbon dioxide and water vapor (*metabolic water*) in terrestrial vertebrates.
2. (see *cellular respiration*)

respiratory chain (see *electron transport chain*)

respiratory movement Muscular movement that results in the passage of water or air over a respiratory surface, allowing the exchange of oxygen and carbon dioxide. In mammals, it involves movements of the diaphragm and rib cage.

respiratory organ Any organ by which oxygen is absorbed into the body and carbon dioxide released. It normally has a large, often convoluted, surface area to facilitate the diffusion of gases. Examples include lungs, gills, and sometimes skin.

respiratory pigments Colored compounds that can combine reversibly with oxygen and are used to distribute it within the bodies of many animals. The color is conferred by a metal ion associated with the protein moiety, iron in *hemoglobin* (vertebrates) and *hemoerythrin*, copper in *hemocyanin*. (see *porphyrin, heme, leghemoglobin*)

respiratory quotient The ratio of the volume of carbon dioxide expired by an organism compared to the volume of oxygen consumed during the same period (V_{CO_2}/V_{O_2}).

respiratory system The organs that are involved in the gross exchange of respiratory gases. They include the lungs, *trachea, bronchi,* sometimes the skin in terrestrial organisms, and the *gills* and related structures in aquatic organisms.

response Any change in an organism or in part of an organism that is produced as a reaction to a stimulus.

resting cyst A *dormant* life cycle stage enabling the *genome* of an organism to survive unfavorable environmental conditions. It is typical of many protoctists, in particular *parasites*. Resting *cysts* often develop thick outer walls resistant to conditions inside a *host's* body, and *metabolic* activity is suspended.

resting potential The voltage difference that exists across the *cell membrane* of a nonconducting *neuron*. It is a balance between the concentration gradients of sodium (Na^+) and potassium (K^+) ions as determined by the selective permeability of the *voltage-gated ion channels* and the electrical potentials formed by the unequal distribution of ions. It is produced and maintained by the *sodium-potassium pump* that actively pumps K^+ into the cell and Na^+ out of the cell. The resting potential is much nearer to the *equilibrium potential* for K^+ than Na^+ because resting cells are more permeable to K^+ ions. It is only when the Na^+ permeability increases during a *nerve impulse* that the Na^+ concentration gradient dominates. (see *nerve impulse, action potential, depolarization, hyperpolarization*)

resting spore A cell enabling the *genome* of an organism to survive unfavorable environmental conditions. Resting *spores* often develop thick outer walls resistant to desiccation, and *metabolic* activity is suspended. They *germinate* when moisture levels and temperature are optimal. They may resume as *vegetative* single cells or divide to give rise to a *multicellular* organism, depending on the species. Spore formation is common among *prokaryotes*, protoctists, and fungi.

restriction endonuclease (restriction enzyme) A type of enzyme found in bacteria that cleaves *DNA* at internal sites. Its biological function is to fragment and destroy invading *bacteriophages*, thus restricting its target to foreign DNA. Class I restriction endonucleases cleave DNA sequences relatively randomly. Each class II restriction enzyme is highly specific for a defined *nucleotide* sequence that is protected in the host bacterium, often by methylation, so that it does not self-destruct. These enzymes, of which there exist at least hundreds, have been co-opted by molecular biologists for the purpose of manipulating DNA fragments in the laboratory. Using restriction endonucleases, particular fragments can be reproducibly isolated, identified, and inserted into other DNA molecules. When applied to *genomic* DNA, they produce a heterogeneous population of fragments with identical ends. Restriction enzymes form the basis of *recombinant DNA technology* and *genetic engineering*. They are named according to the organism from which they were originally isolated. For instance, *Eco*RI is obtained from *Escherichia coli* strain R and was the first isolate. (see *restriction-modification system, gene cloning, palindrome, sticky ends, blunt ends*)

restriction enzyme (see *restriction endonuclease*)

restriction map A compilation of the number and location of particular *restriction enzyme* sites within a *DNA* fragment, *chromosome,* or *genome.* These sites directly reflect the *nucleotide* sequence of the region in question. Fragments produced by restriction enzymes may be separated using *electrophoresis* and the

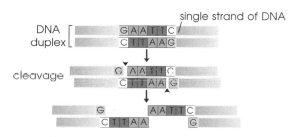

complementary single-stranded tails or "sticky ends"

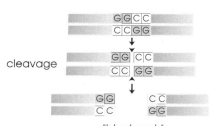

"blunt ends"

Restriction Enzyme Cleavage
(see *Restriction Endonulease*)

position of any gene or *locus* within a fragment determined by *Southern analysis*. The size, number, and location of particular fragments form a basis of comparison between species and between individuals. (see *DNA analysis, RFLP*)

restriction-modification system A mechanism present in virtually all bacterial species that restricts the entrance of foreign *DNA*, usually from a *bacteriophage*, into the cell. It consists of one or more types of site-specific *endonucleases*, called *restriction endonucleases*, that make *double strand* cuts only in the invading DNA. The bacteria's own *chromosome* is protected by marking the sensitive sites with a methyl group using a modifying enzyme (modification). If the phage DNA happens to be modified before being attacked by the restriction enzymes, it is able to complete the *lytic* cycle and escape to infect other cells. Restriction endonucleases have been commandeered by molecular biologists because of the plethora of different types that reliably cleave DNA at sites where a particular *nucleotide* sequence, often a *palindrome*, occurs.

restriction site The sequence-specific site in a *DNA* molecule where a particular *restriction endonuclease* cuts.

restrictive The conditions under which a *conditional mutation* in *DNA* exhibits the abnormal *phenotype*. (compare *permissive*)

reti-, reticul- A word element derived from Latin meaning "net" or "network." (for example, *reticulin*)

reticular activating system A portion of the *reticular formation* in the *brainstem* that controls consciousness and alertness. It monitors all of the *sensory* information that feeds into the brain and plays a role in regulating wake and sleep states. It also oversees the *cardiovascular*, *respiratory*, and *digestive systems*.

reticular formation A net of *neurons* in the *brainstem* that contains the *reticular activating system*. The pathways from these neurons to the *cerebral cortex* and other brain regions are depressed by anesthetics and barbiturates.

reticulin A protein that is closely related to and often converted into *collagen*. Reticular fibers participate in forming supportive networks between and around cells in vertebrate tissues.

reticulocyte An immature *erythrocyte* (red blood cell).

reticuloendothelial system The system of *macrophage* cells that engulf and digest foreign particles. They are important in bodily defense and also in the scavenging of dead cells, such as old *erythrocytes*. They may be attached or circulating.

reticulopods (see *Granuloreticulosa*)

reticulum The second region of the specialized stomach of *ruminants* (for example, cow). The *cud* passes back into the reticulum after it has been regurgitated and chewed.

retina The light sensitive region at the back of vertebrate and cephalopod eyes. It consists of two types of *photoreceptor* cells (*rods* and *cones*) that are connected by *synapses* to *ganglion cells* via *bipolar cells*. From the ganglion cells, *nerve fibers* pass over the inner surface of the retina and collectively become the *optic nerve* leading to the brain.

retinal A light sensitive *carotenoid* protein that is a constituent of *rhodopsin*. When struck by a photon of light, it undergoes an *allosteric* change from 11-cis-*retinal* to all-trans-*retinal*, initiating a series of events leading to light detection by the organism. Retinal is a derivative of all-trans-*retinol* (*vitamin A*). (see *eyespot, bacteriorhodopsin*)

retinol (see *vitamin A*)

retroposons (retrotransposons) Short, dispersed, repetitive *DNA* fragments that have spread throughout the *genome* of a species through *reverse transcription* of *RNA*. (see *retrovirus, transposon, Alu sequences*)

retrotransposons (see *retroposons*)

retrovirus A type of *RNA virus* that has the ability to *integrate* into the *host* cell *DNA*.

The enzyme *reverse transcriptase* catalyzes the synthesis of *double-stranded DNA* from the viral RNA (plus strand), which is then inserted into the host *chromosome*. Both viral RNA and *mRNA* for virus-specific proteins are *transcribed* from this segment by the host system. Retroviruses are known for instigating *cancer*. This may happen if they carry genetic material from a previous host that finds itself in an abnormal environment, causing inappropriate *gene expression*. *HIV*, the virus that causes *AIDS*, is a retrovirus. (see *arboviruses, cDNA, retroposon*)

reverse dot blot A laboratory technique for analyzing nucleic acids (*DNA, RNA*). In contrast to a standard *dot blot*, the known *probe* is attached to a solid support, and challenged with a *labeled* sample for the purpose of identifying particular *nucleotide* sequences that might be present. This technique determines the presence but not the size of a particular DNA sequence. The same information is determined as in a standard dot blot, only in a more convenient format. (compare *Southern blot*)

reverse genetics The process of taking a piece of *DNA* whose function is unknown and using *recombinant DNA* techniques to alter it in hopes of deducing its function. The *mutant gene* is reintroduced back into the organism, which is assayed for a mutant *phenotype* in hopes of correlating it to the gene of interest.

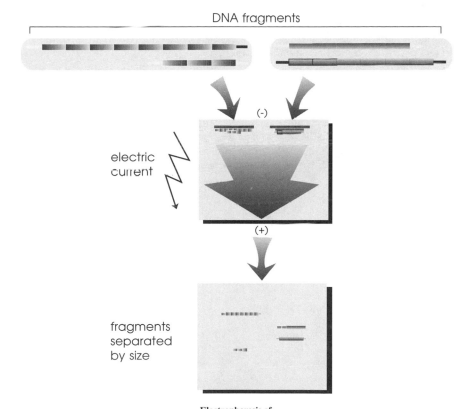

DNA fragments

electric current

(-)

(+)

fragments separated by size

Electrophoresis of
Restriction Fragment Length Polymorphisms (RFLPs)
Modified from an illustration in *An Introduction to Forensic DNA Typing,* ©1997, by Inman and Rudin, CRC Press, Boca Raton, Florida.

reverse transcriptase An enzyme carried by *retroviruses* that catalyzes the synthesis of *complementary DNA (cDNA)* from an *RNA* template. It also copies the first DNA strand to produce a *double-stranded DNA* molecule that may then be inserted into the host *chromosome*. Under most circumstances, the direction of *transcription* is from DNA to RNA, thus the name of the enzyme. Reverse transcriptase has been isolated and exploited by molecular biologists *in vitro*, in particular for studying *gene expression* and for *genetic engineering*. (see *central dogma*)

RFLP (restriction fragment length polymorphism) A type of *DNA* variation (*polymorphism*) based on differences in the length of fragments produced by *restriction enzyme cleavage*. The differences in length may be the manifestation of DNA *nucleotide* sequence changes resulting in the addition or deletion of enzyme cleavage sites or the addition or deletion of genetic material, often *tandem repeats,* between two defined sites. Uses for RFLP analysis include species comparison, analysis of *phylogenetic* relationships, and individual identification. (see *DNA analysis, Southern blot, VNTR*)

Rf value (relative front) In *chromatography*, the distance traveled by the solute (questioned molecule) divided by the distance traveled by the solvent front, the latter always being 1. The Rf value is characteristic of a particular molecule in a defined solvent. The procedure is used to categorize and identify *amino acids*, *chlorophylls,* and other relatively simple organic molecules.

R-group In *amino acids*, the side group that determines the identity and unique chemical properties of each.

rhabdom The group of light sensitive *retinal* cells extending from each individual *corneal lens* structure in a *compound eye*. Each rhabdom stimulates a *nerve fiber* leading to the brain. (see *ommaditium*)

rhabdoviruses A group of *single-stranded RNA viruses* that carry the minus strand. Thus, once it enters a *host cell, mRNA*

may be *transcribed* directly from the viral RNA strand. Rabies, measles, mumps, and distemper are all caused by rhabdoviruses.

rheas (see *Rheiformes*)

Rheiformes The order of birds containing the flightless rheas.

rheocrene A flowing spring with a substratum of sand and gravel.

rheotaxis A movement orientation to a water current, either positive or negative.

rhesus factor (see *Rh factor*)

Rh factor (rhesus factor) An *antigen* found on the outside of red blood cells (*erythrocytes*) in humans and also the rhesus monkey (hence the name). The antigen is present in most people (Rh-positive), and absent in a small percentage of the population (Rh-negative). Rh-negative women who have previously borne Rh-positive children may carry some Rh *antibodies* due to red blood cell (*erythrocyte*) leakage across the *placenta* during pregnancy. If the woman subsequently becomes pregnant with another Rh-positive fetus, the antibodies, which are relatively innocuous in the mother, may diffuse across the *placenta* and cause *hemolysis* in the developing fetus. If known, this condition can be prevented by destroying any Rh-positive *erythrocytes* in the mother's circulation immediately after the first birth.

rhinoceroses (see *Ceratomorpha*)

rhinoviruses A group of *single-stranded RNA viruses* that carry the plus strand and lack a lipid-rich outer envelope. The viral RNA acts directly as *mRNA*, attaching to the host *ribosomes* and being *translated* into viral proteins. Rhinoviruses account for about one third of all upper respiratory infections identified as colds.

rhiz-, -rhiza, -rrhiza A word element from Greek meaning "root." (for example, *mycorrhiza*)

Rhizobium A genus of bacteria that lives either freely in the soil or *symbiotically* in the *root nodules* of *leguminous* plants (peas and beans, *Fabaceae*). Rhizobium possesses the ability to "fix" atmospheric nitrogen (N_2) into nitrates, a form that can be utilized by plants. In return, the bacterium is supplied with carbohydrates. (see *nitrogen cycle, nitrogen fixation*)

rhizoid A single- or several-celled, hairlike structure that may serve as an anchor and/or root in nonvascular plants and ferns as well as some *multicellular* algae and fungi.

rhizome In plants, a stem that grows horizontally below ground and from which a new plant may arise at its tip or at *nodes* along its length. In may also function in food storage. (see *vegetative propagation, perennation, perennating structure*; compare *bulb, corm, offset, runner, stolon, sucker, tuber*)

rhizomorph A specialized rootlike structure found in certain fungi that transports nutrients from one part of the *mycelium* to another.

Rhizopoda A phylum of *unicellular heterotrophic* protists, the amebas. The hundreds of species are abundant in freshwater, saltwater, as well as soils, and many are animal *parasites*. Reproduction is mostly *asexual* by *mitosis*, and they lack *cell walls*. Some form *cysts* resistant to digestion, such as the species that causes amebic dysentery. They are known for their characteristic locomotion that occurs by *pseudopodia, cytoplasmic* extensions that extend and pull the ameba forward or engulf food particles. Rhizopods include the naked and shelled amebas and the acrasiomycotes, or cellular slime molds. The acrasiomycotes are an important class of *heterotrophic* protoctists that includes *Dictyostelium*, an organism often used as a model to study morphogenic differentiation cues. They were at one time regarded as fungi due to a lack of *chlorophyll* and *undulipodia*. They have a *cell wall* of *cellulose* and resemble amebas, including forming *amoeboid* cystlike *spores*. When their food supply becomes exhausted, the separate organisms display an elaborate form of *parasexuality* in which they aggregate into a *pseudoplasmodium*. As conditions become drier, this sluglike structure differentiates into a *sporophore* with a *spore*-containing *sorus* at its tip. *Meiotic sexual reproduction* is rare. They are distinct from plasmodial slime molds (*Myxomycota*).

rhizosphere The area of soil immediately surrounding plant roots (a few millimeters) in which the composition is affected by their biological processes. Various microorganisms and fungi live in this region, forming beneficial *symbiotic* relationships with the plant. The root surface itself is sometimes called the *rhizoplane*. (see *Rhizobium, Azobacter, root nodule, mycorrhizae, actinorrhize, mutualism*)

rho factor (ρ) A protein factor required to release both *RNA polymerase* and the newly synthesized *RNA* molecule at the end of a *transcription* unit in some *prokaryotic genes*. Rho factor recognizes certain transcription termination signals in the *nucleotide* sequence. (compare *sigma factor*)

Rhodophyta A phylum of some of the largest *multicellular* protoctists, the red algae, which are mostly marine. They contain *chlorophyll* and the *photosynthetic pigments phycoerythrin* and *phycocyanin* (*phycobiliproteins*), from which they derive their characteristic red color. Of the *eukaryotic* algae, they are the only ones lacking *undulipodia*, hence a motile form, at any stage of their *life cycle*. Because they do not form *diploid embryos* retained in maternal tissue, they are not plants. Many exhibit *sexual life cycle* stages, although some appear to reproduce only by *mitosis*. The genus *Gracilaria* is the source of laboratory *agar*. Other species produce carrageenan and other polysaccharides used in the manufacture of food products such as ice cream. (see *rhodoplast*)

rhodoplast A *plastid* containing the red *photopigment* characteristic of red algae.

rhodopsin (visual purple) A light sensitive pigment found in *rod cells* in the *retina* of vertebrate and cephalopod eyes. It consists of a protein component, *opsin*, linked to 11-cis-*retinal*, a derivative of *vitamin A*. When a photon of light strikes a rhodopsin molecule, the retinal portion undergoes a conformational (*allosteric*) change to all-trans-retinal, inducing an allosteric change in the opsin portion of the compound. This initiates a chain of events leading to generation of a *nerve impulse*. Variations of this scheme are ubiquitous in biological systems. Forms of rhodopsin are found in the *photoreceptors* of many light sensitive invertebrates and even in a light driven *proton pump* in halobacteria. (see *bacteriorhodopsin, eyespot*)

rhombencephalon (see hindbrain)

Rhombozoa A newly autonomous phylum of parazoans, the rhombozoans (dicyemids, heterocymids). They are small, wormlike organisms that exhibit bilateral symmetry but only have two body layers. They lack organs with the exception of a *gonad*. Rhombozoans live in the kidneys of cephalopod mollusks and show an *alternation of generations*.

Rhynchocoela (*Nemertina*). A phylum of *acoelomate* marine worms, the ribbon worms. They are characterized by a long *proboscis* that is thrust out from a sheath to capture prey. They are the simplest animals with a complete *digestive system* and with a *circulatory system* in which the blood flows in vessels. Some species are *hermaphroditic*.

rhytidome The dead outer zone of tree bark.

RIA (radioimmunoassay) (see *immunoassay*)

ribbon worms (see *Nemertina*)

riboflavin (vitamin B$_2$) A member of the *vitamin B complex* found in cereal grains, legumes, organ meats, and milk. It is a constituent of several *flavoproteins*, acting as a *coenzyme* in electron transfer (*redox*) reactions. (see *FMN, FAD*)

ribonuclease (see *RNase*)

ribonucleic acid (see *RNA*)

ribose A 5-carbon sugar ring (*pentose*). It is the sugar moiety found in *RNA*.

ribosomal RNA (see *rRNA*)

ribosome The molecular machine that carries out *protein synthesis* (*translation*) in all cells. Each ribosome contains several different types of *rRNA* transcribed from clusters of rRNA *genes* on the *chromosomes*. Along with ribosomal proteins, they form two main subunits that fit together and either float free in the *cytoplasm* or are bound to the *endoplasmic reticulum* in *eukaryotic* cells. Ribosomes actively engaged in protein synthesis are attached at different points along an *mRNA* molecule in a structure called a polyribosome, or *polysome*. Each ribosome facilitates the *polymerization* of *amino acids* introduced by *tRNA* molecules according to the order specified by the mRNA molecule. (see *rough endoplasmic reticulum, nucleolus*)

ribozyme A small group of recently discovered *RNA* molecules that surprisingly act as enzymes and able to self-splice. They are found in the *ciliate* protist *Tetrahymena*. (see *central dogma splicing*)

ribs A series of paired bones attached to the *vertebral column*, surrounding and protecting the heart and lungs in vertebrates.

ribulose bisphosphate (RuBP) A 5-carbon molecule that acts as an acceptor for carbon dioxide (CO$_2$) in the *Calvin-Benson cycle* (*dark reactions*) of *photosynthesis*. RuBP is regenerated when the carbon molecules are incorporated ("fixed") into carbohydrate at the end of the cycle by ribulose bisphosphate carboxylase (rubisco). (see C_4 *carbon fixation, carboxysome*)

ringtails (see *Carnivora*)

riparian The environment on the banks of lakes, ponds, and streams.

ritualization The process by which behavior patterns become stereotyped through evolution and become easily recognizable social signals. They are common in threat and courtship.

RNA (ribonucleic acid) A *nucleic acid*

that functions in various forms to *translate* the information contained in *DNA* into proteins. It also serves as the primary genetic material in some viruses. It is similar in composition to DNA, with a *ribose* sugar exchanged for the *deoxyribose* and the replacement of the *pyrimidine base thymine* with *uracil*. RNA is generally *single stranded*, with the exception of *double-stranded RNA viruses*, but often forms regions of *secondary structure* between complementary regions on the same molecule. This particularly occurs in *ribosomal RNA (rRNA)* and *transfer RNA (tRNA)*. RNA is generally *transcribed* from DNA, with the exception of *retroviruses* where the reverse occurs. Self-replicating RNA systems have begun to be explored in the laboratory. *Messenger RNA (mRNA)* functions as a direct link between *structural genes* and *protein synthesis*. rRNA, transcribed from complexes of rRNA *genes* on the chromosome, forms *ribosomes*. These are the physical site of *protein synthesis* in the *cytoplasm*. tRNA molecules contain *anticodon nucleotide* sequences on one end and the matching *amino acid* on the other. They bring amino acids to ribosomes then insert them in the order specified by the mRNA. (see *genetic code, reverse transcriptase, central dogma, hnRNA, ribozyme*)

RNA polymerase An enzyme that *polymerizes ribonucleoside triphosphates* into *RNA*, in the order dictated by a *DNA* or, less commonly, an RNA *template*. They are ubiquitous to all cells. Only one type is found in prokaryotes while three types occur in *eukaryotic* cells, *polymerase I* specific to *rRNA*, *polymerase II* specific to *mRNA* transcribed from *structural genes*, and *polymerase III* specific to *tRNA* and *rRNA*. (compare *DNA polymerase, reverse transcriptase*)

RNA processing (see *splicing*)

RNase (ribonuclease) Any enzyme that degrades *RNA*. RNases work by catalyzing the hydrolysis of sugar-phosphate bonds in the *RNA* backbone. There are several types, each acting in a slightly different fashion. RNases are ubiquitous in living organisms and are normally carefully sequestered in cells, acting only at specified times and places. (see *sugar-phosphate backbone*)

RNA splicing (see *splicing*)

RNA virus A virus that employs *RNA*, either *single stranded* or *double stranded,* as its primary genetic material. In plus strand *RNA* viruses, the viral *genome* serves directly as *mRNA* in the *host* cell as well as the template for a minus strand RNA intermediate. In minus strand RNA viruses, the viral genome serves as a template for mRNA *transcription*, similar to the normal function of DNA. In *retroviruses*, the plus strand also serves as a *template* for double-stranded DNA synthesis by *reverse transcriptase*. (see *enveloped viruses*; compare *DNA virus*)

rod cells One of the two main types of light sensitive cells found in the *retina* of the vertebrate eye. Rods provide monochromatic (black and white) vision in dim light and are found chiefly in the periphery of the retina. Several rods connect with the same *bipolar cell* leading to the *optic nerve*, resulting in high sensitivity but low visual acuity (fuzzy images). Rod cells contain the molecular *photoreceptor rhodopsin*. Each molecule of rhodopsin is altered by a single photon of light, initiating a series of reactions leading to a *nerve impulse*. (compare *cone cell*)

Rodentia The largest and most successful order of mammals, including rats, mice, squirrels, and beavers. They are *herbivorous* or *omnivorous*, possessing chisel-like incisor teeth specialized for gnawing. They are mostly *nocturnal* and terrestrial, and are noted for their rapid breeding.

rollers (see *Coraciiformes*)

rolling circle replication The process by which a bacterial *F* (fertility) *plasmid* is duplicated and transferred to another cell. It may also perform this process while *integrated* into the bacterial *chromosome*, carrying a copy of the *host DNA* along with it. The F plasmid binds to a site in the donor F⁺ cell just beneath the *pilus*, or *conjugation* bridge. It begins DNA *replication* at that point, passing the copy across the bridge and into the recipient F⁻ cell as it rolls off the circular molecule. A *complementary* strand is synthesized in the new host cell, restoring it to a *double-stranded* DNA molecule.

root The structure by which a plant is anchored to its substrate and by which the majority of water and nutrients are absorbed. Roots differ from shoots in that they lack *chlorophylls*, do not produce buds or leaves, and have a solid rather than hollow central core. They also function as storage organs for *biennial* and *perennial* plants. (see *perennating structure, bulb, corm, tuber, rhizome, mycorrhizae, root cap, root nodule*)

root cap A structure at the tip of a root that protects it as it grows through soil.

root hairs Specialized cells that occur near the tips of plant roots. Each hair is a single cell. They occur in masses, helping to keep the root in intimate contact with soil particles. Virtually all the absorbtion of water and minerals occurs by way of root hairs in *herbaceous* plants.

root nodule Small growths that develop on the roots of certain plants, particularly legumes (*Fabaceae*), as a result of the normal infection by *nitrogen-fixing aerobic bacteria*. The relationship between the plant and the bacteria is *mutualistic*, the bacteria "fixing" atmospheric nitrogen in a form that can be used by the plant and in return obtaining nutrients, particularly carbohydrate, from the plant. (see *Rhizobium, Azobacter, nitrogen fixation, nitrogen cycle*)

root pressure A phenomenon in which water pressure in plant roots increases, in particular at night. It is due to the continued *active transport* of ions into the roots, causing water uptake by *osmosis* with a concurrent lack of *transpiration* driven water movement away from the roots. When root pressure is very high, it may cause *guttation*, in which water is lost in a liquid form through special cells at the edges of leaves.

rosette A tuft of leaves that forms in *biennial* plants in their first year, when they do not flower. In the second year, energy stored in the rosette and underground parts of the plant is used to produce flowering stems.

rostrum, rostra (pl.) Any extended part of the head end of an invertebrate.

Rotifera A phylum of microscopic aquatic invertebrates, the wheel animals. They are characterized by a *ciliated* crown on the head that resembles a rotating wheel when beating. The body is divided into head, trunk, and tail. The jaws are well developed. The body is covered with a *chitinous lorica*. They possess *eyespots* and *protonephridia*. Males are often degenerate and *parthenogenesis* is common. Rotifers possess the unusual trait of *cell constancy*, the number of cells comprising their body is constant over their life. They have an extreme tolerance for both freezing (*cryptobiosis*) and desiccation.

rotifers (see *Rotifera*)

rough endoplasmic reticulum (see *endoplasmic reticulum*)

round window (see *fenestra rotunda*)

roundworms (see *Nematoda*)

R-plasmid (see *resistance transfer factor*)

rRNA (ribosomal RNA) A class of *RNA* molecules that have an integral role in *ribosome* structure. rRNA molecules are *transcribed* from clusters of thousands of copies of rRNA *genes* located in tandem at one location on a *chromosome*. Each neighboring copy is separated by a short, nontranscribed spacer region. (see *translation, nucleolus*; compare *mRNA, hnRNA, tRNA*)

r strategist Organisms whose populations tend to exhibit rapid exponential growth followed by sudden crashes; thus they are somewhat self-limiting. They are generally small, fast-growing organisms with short generation times and live in unpredictable, rapidly changing environments. They mature early in their *life cycle* and produce many offspring that receive little or no parental care. Some of the more extreme examples include bacteria, aphids, cockroaches, and dandelions. (compare *k strategist*)

rubisco (see *ribulose bisphosphate*)

ruderal A plant growing in disturbed soil, often waste ground near human habitation. They often have high nutrient demands and are intolerant of competition.

ruffled membrane (see *lamellipodium*)

rumen The first region of the specialized stomach of *ruminants* (for example, the cow). It contains *symbiotic* bacteria that produce the enzyme *cellulase*, facilitating the breakdown of *cellulose* in the plant material that is the mainstay of a ruminant's diet. The partly digested food is then regurgitated as the *cud* and chewed again before being swallowed and passed into the second stomach region (*reticulum*).

ruminant An animal that chews its cud. (see *Ruminantia*)

Ruminantia A suborder of even-toed ungulates including deer, antelope, giraffe, cattle, sheep, and goats. They are characterized by the possession of complex stomachs in which *cellulase*-producing *symbiotic* bacteria are employed to digest the *cellulose* in plant material. The partially digested material is then regurgitated as the *cud* and rechewed. (see *rumen, Artiodactyla*)

runner In plants, a stem growing horizontally along the ground that roots at its tip. Eventually, a new plant forms, and the runner decays. (see *vegetative propagation*; *perennation, perennating structure*; compare *bulb, corm, offset, rhizome, stolon, sucker, tuber*)

rusts *Parasitic* basidiomycete fungi of the order Uredinales that are plant *pathogens*. They are characterized by the formation of yellow-brown streaks or spots on the leaves and have a complicated *life cycle* involving the formation of a series of *spore* types. Some need two *hosts* to complete their *life cycle*.

^{35}S An *isotope* of sulfur that is commonly used to radioactively *label* proteins.

sacchar- A word element derived from Greek meaning "sugar." (for example, *Saccharomyces*)

saccharide An alternative term for sugar.

Saccharomyces A genus of *unicellular* ascomycete fungi that falls into the subgroup of yeasts. They can *respire* either *aerobically* or *anaerobically*. They are important in brewing and baking for the alcohol and carbon dioxide, respectively, that they produce under anaerobic conditions. *S. cerevisiae*, in particular, has become increasingly important as a laboratory tool for the investigation of genetic and biochemical processes, and the *cloning* and propagation of *eukaryotic* (including human) genes. (see *Ascomycota*)

sacculus The lower of two chambers in the *labyrinth* of the *inner ear* of most vertebrates. It is part of the *macula* and contains a patch of *sensory epithelium*, or *statocyst*, that detects changes in head tilt with respect to gravity and acceleration. (see *utriculus*)

sacral Several specific *vertebrae* in the lower back of tetrapods.

sacrum One or more fused vertebrae in the lower back region that support the *pelvic girdle* in tetrapods.

sagenogen (see *bothrosome*)

salamanders (see *Caudata*)

saliva A secretion produced by the *salivary glands* of animals consisting mainly of *mucus*. It is used to moisten and lubricate food. In some animals, it contains *anticoagulants* (for example, mosquito) or *digestive enzymes*, particularly for carbohydrates (*amylases*).

salivary glands *Glands* that secrete a solution, sometimes containing *digestive enzymes* or *anticoagulants*, into the mouth (*buccal cavity*). (see *saliva*)

saltatory conduction The mode of transmission of a *nerve impulse* along a *myelinated nerve fiber* whereby it jumps between breaks in the insulation (*nodes of Ranvier*). This results in a considerably faster rate of propagation than in a non-myelinated fiber. An *action potential* can be initiated only in the absence of the insulating *myelin sheath* at that point. (compare *giant fiber*)

samara A one-seeded *nut* or *achene* characterized by membranous wings that aid in dispersal. An example is the fruit of the ash tree.

sand A soil type high in mineral content and with relatively large particles facilitating good drainage. Sandy soils are often lacking in organic nutrients, because they wash away with the water flow.

sand fleas (see *Branchiopoda*)

sandgrouse (see *Pteroclidiformes*)

sandpipers (see *Charadriiformes*)

sanguiverous An organism that feeds on blood.

SA node (see *pacemaker*)

sapr-, sapro- A word element derived from Greek meaning "rotten." (for example, *saprophyte*)

saprobes *Heterotrophic* bacteria that obtain organic nutrients by digesting the products or remains of other organisms. They are important *decomposers* in the *food web*. Mutants that are able to break down *pollutants* such as petroleum products and pesticides are being further refined and exploited. (see *saprotroph, saprophage*)

saprophage Any organism that obtains organic nutrients by digesting and breaking down the products or remains of other, dead organisms. Many bacteria and fungi and some insect *larvae* are important saprophages. They act as *decomposers*, recycling nutrients in the *food web*. (see *heterotrophism, saprobe, histophage*; compare *saprotrophism*)

saprophyte A plant that obtains soluble organic nutrients from the inanimate products or remains of other organisms. (see *saprotrophism*)

Saprospirae A phylum of *Gram-negative* bacteria containing the fermenting gliders, including *Bacteroides* (previously classified fermenting bacteria) and *Cytophaga* (previously classified myxobacteria). The reclassification has occurred based on DNA *homologies* in the small (16S) *ribosomal RNA* (*rRNA*) subunit. They are structurally complex, and some contain *carotenoid* pigments.

saprotrophism A method of obtaining soluble organic nutrients by absorbing them directly from the inanimate products or remains of other organisms. This often involves the secretion of extracellular enzymes followed by absorbtion of the digestion products. Many bacteria and fungi and some insect *larvae* are saprotrophs. They act as *decomposers*, recycling nutrients in the *food web*. (see *trophism, osmotrophism, histophage*; compare *saprophage*)

saprozoite An animal that obtains soluble organic nutrients from the inanimate products of other organisms, such as *parasites* in the *gut* of a *host*. (see *saprotrophism*)

sapwood The outer living cells (*xylem*) of a tree trunk that are actively involved in water transport and food storage. (compare *heartwood*)

SAR (scaffold attachment regions) Positions along a *DNA* molecule where it is anchored to the central protein *scaffold* of the *chromosome*.

sarcoma A *cancer* of the *connective tissue*. (compare *carcinoma, leukemia, lymphoma*)

sarcomere The basic contractile unit in *striated muscle*. The two major contractile proteins, *actin* and *myosin*, are arranged as overlapping filaments in repeating units called sarcomeres. Each are bounded by and anchored at *Z-lines* made of the protein α-*actinin*. During *muscle contraction*, the filaments interact such that the degree of overlap increases up to a maximum, shortening each sarcomere, hence the muscle. (see *muscle contraction*)

sarcoplasm The *cytoplasm* in *striated muscle*.

sarcoplasmic reticulum The *endoplasmic reticulum* of a *striated muscle* cell. It wraps around each individual *myofilament* like a sleeve and contains specialized *ion channels* for calcium (Ca^{++}), the entity that initiates *muscle contraction*. The sarcoplasmic reticulum mediates the coordinated contraction of the muscle in response to a *nerve impulse*. (see *muscle contraction*)

Sarcopterygii The subclass of bony fish that contains the lobe-finned fish. There are only four surviving genera, three of which are lungfish. They are characterized by paired swim bladders that serve as functional lungs, and internal as well as external nostrils that allow air-breathing at the water surface without opening the mouth. The young breathe by temporary external gills. The heart and *circulatory system* are adapted for *pulmonary* respiration and resemble those of amphibians. The lungfishes are found in freshwater, often in areas of seasonal drought, where they lie dormant in burrows at the bottom of ponds awaiting the wet season.

satellite DNA Originally defined as a portion of *genomic DNA* separable by high-speed centrifugation, it is now known to consist of thousands of highly repeated DNA sequences, such as those found flanking the *centromere* in *eukaryotic* cells. Satellite DNA has a distinctive *base* composition and is not *transcribed* into protein. Different classes can be characterized by particular repeated *nucleotide* sequences. (see *repeated sequence, tandem repeat, microsatellite, minisatellite*)

saturated fatty acid A *fatty acid* in which hydrogen atoms have filled all possible slots, thus leaving them relatively inert. At room temperature, fats consisting of saturated fatty acids tend to solidify. There is general agreement that the overconsumption of fats containing saturated fatty acids contribute to *atherosclerosis*. (compare *unsaturated fatty acid*)

saturation mutagenesis The deliberate

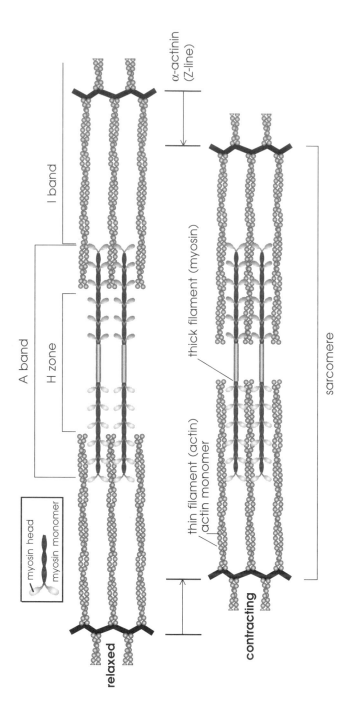

α-actinin
(Z-line)

I band

A band

H zone

thick filament (myosin)

thin filament (actin)
actin monomer

sarcomere

myosin head

myosin monomer

relaxed

contracting

Sarcomere

induction and recovery of large numbers of *mutations* in one area of the *genome* or in one function. The purpose is to identify all the *genes* in that area or that affect that function. (see *genetic dissection*)

savanna An area of open grassland with widely spaced trees. It receives little annual rain, about 100 to 150 centimeters per year, and usually experiences seasonal droughts. The soils are nutrient-poor but rich in aluminum, a substance toxic to many plants. The African savanna is characterized by herds of large grazing animals. Both grazing and periodic fires clear the grassland of litter accumulation, which ties up nutrients and blocks seedling growth, greatly limiting net productivity. (see *biome*)

scaffold The central protein framework of a *chromosome* to which the *DNA solenoid* is attached in loops.

scaffold attachment regions (see *SAR*)

scale In general, a rigid, platelike outgrowth of the *epidermis*. Various forms are ubiquitous among plants and animals, providing protection. They often contribute to characteristic patterns and coloring of species or individuals.

scaleworms (see *Polychaeta*)

scallops (see *Bivalvia*)

scaly anteaters (see *Pholidota*)

Scandentia An order of placental mammals comprising the tree shrews and lemurs. They have sometimes been classified as Prosimii, a suborder of primitive primates.

scanning electron microscope (see *electron microscope*)

scapula, scapulae (pl.) One of two large, flat, triangular bones forming the back of the *pectoral* (shoulder) girdle in vertebrates. In man, they form the shoulder blades.

schizocarp A dry fruit, intermediate between *dehiscent* (bursting to release seeds) and *indehiscent*, that is formed from two or more *carpels* (female reproductive structures), each of which matures into a single-seeded unit.

Schwann cell One of the many cells that are spirally wrapped around *myelinated*

nerve fibers, forming the insu. myelin sheath. (see *glial cells, neurog. saltatory conduction, nerve impulse*)

scientific method A directed approach to problem solving. It includes the proposal of a hypothesis and systematic testing, analysis, and formulation of conclusions regarding the hypothesis. The hypothesis may be proved or disproved, or the results declared inconclusive. (see *independent variable, dependent variable*)

scion In a plant *graft*, the bud or shoot that is grafted onto the rootstock.

scler-, sclero- A word element derived from Greek meaning "hard." (for example, *atherosclerosis*)

sclera The fibrous outer protective coat of the vertebrate and cephalopod eye. The opaque sclera is continuous with the transparent *cornea* in the front of the eye.

sclereid (stone cell) Specialized woody (*sclerenchyma*) plant cells that are found in seed coats and some fruits (for example, pears).

sclerenchyma The woody tissues in some plants that provide mechanical support. They are formed from cells that are thickened and strengthened by the deposition of *cellulose* and *lignin*, and when mature, die and lose their *cytoplasm*. Sclerenchyma cells include fibers and *sclereids*. (see *secondary growth*; compare *parenchyma, collenchyma*)

sclerite A hardened area of an insect body bounded by flexible *sutures* or membranes.

scleroproteins A group of fibrous *structural proteins* found in animal bodies. They include *keratin* (hair) and *collagen* (connective tissue).

sclerosis Any hardening of tissues. Examples are the formation of scar tissue at the site of a healed injury and the deposition of *cholesterol* and *triglycerides* in artery walls producing the stiffness called *atherosclerosis*.

sclerotium, sclerotia (pl.) The resting body of certain ascomycete fungi (for example, *ergot*) that forms a thickened outer covering, rendering it resistant to unfavorable environmental conditions. It is different from a *spore* in that it is a *vegetative* mass of fungal tissue (*hyphae*). It can remain *dormant* for long periods of time, and when nutrients and water become available, gives rise to *fruiting bodies*.

scolex The head of a tapeworm. It is spherical, with a narrow neck leading to the region where *proglottids* (body segments) are produced. It has a crown of hooks and four suckers for attaching to the lining of the gut of the final *host*. (see *Cestoda*)

Scorpiones The order of arachnids containing the scorpions. Their *pedipalps* are modified into pincers that they use to manipulate and tear their food. Their well-known venomous stings are used primarily to stun prey and only secondarily in self-defense. The species are all terrestrial, and respiration is by means of *book lungs*. Courtship rituals are elaborate, and the young are born alive.

scorpionflies (see *Mecoptera*)

scorpions (see *Scorpiones*)

screening The testing of a library of *genomic DNA clones* for a desired fragment. The test usually involves a physical, biochemical, or genetic assay.

scrotum (scrotal sac) A pouch of skin that hangs external to the body directly behind the penis in most male mammals. It is divided into two compartments, each containing a *testis*. The external location helps to maintain the testes at a lower temperature than that of the body in order to ensure optimum development of *spermatozoa*.

scutellum, scutella (pl.) The single *cotyledon* (first leaf) of a grass *embryo*, specialized for absorbtion of the *endosperm* (nutritive tissue).

scyphistoma The *polyp* stage in the *life cycle* of some jellyfish (scyphozoans) during which *larvae* are released by transverse splitting (*strobilization*).

Scyphozoa A class of cnidarians, the jellyfish, in which the *medusa* (free-floating form) is the only or dominant form and the *polyp* (sedentary form) is absent or

restricted to a small *larval* stage (*scyphistoma*).

sea anemone (see *Anthozoa*)

seabirds (see *Procellariiformes, Charadriiformes*)

sea cucumbers (see *Holothuroidea*)

sea lilies (see *Crinoidea*)

sea lions (sea *Pinnipedia*)

seals (sea *Pinnipedia*)

sea moss (see *Ectoproca*)

sea spiders (see *Pycnogonida*)

sea stars (see *Stelleroidea*)

sea urchins (see *Echinoidea*)

sea walnuts (see *Ctenophora*)

sea wasps (see *Cubozoa*)

sebaceous gland (oil gland) A *gland* situated at the upper end of a *hair follicle* near the skin surface that secretes an oily substance, *sebum*, into the follicle.

sebum An oily secretion produced by the *sebaceous glands* in mammalian skin. It prevents desiccation of the skin, repels water from hair and skin, and contains *antiseptic* ingredients to kill bacteria.

secondary consumer An organism that consumes other *heterotrophs*, often *herbivores* and sometimes other *carnivores*. They are at least twice removed from a *producer* that has generated organic material by *photosynthesis* or *chemosynthesis*. They are generally *carnivores* or *parasites*, and considered at the top of the *food web*. (see *trophic level*)

secondary ecological succession The recolonization of an area in which major environmental disruption has destroyed a previously existing *climax community*, leading to a marked change in the definitive vegetation. Examples of initiators include fire, removal of grazing pressure, or abandonment of previously cultivated areas. Biological remnants of the previous *ecological community* remain, such as organic matter and seeds.

secondary growth (secondary thickening) The increase in girth of a stem or root by lateral division of cells in the *cambium* layer of plants (*lateral meristems*). Both secondary *phloem* and secondary *xylem* are produced in this fashion. In a woody plant, much of the woody part is a result of secondary growth and can be seen as *annual rings.* The capacity for secondary growth leading to a mechanism for physical support, was important in the evolution of *vascular plants.* (see *vascular cambium*; compare *primary growth*)

secondary immune response (see *immune response*)

secondary sexual characteristics The sex-associated manifestations in the *somatic* tissues of *sexually dimorphic* animals. In primates, they develop in response to male (*androgens*) or female (*estrogens*) *hormones* at the onset of puberty. For example, males develop facial hair and the voice deepens, females develop breasts and the hips enlarge. (compare *primary sexual characteristics*)

secondary structure The second order of structure of a *polypeptide* or *nucleic acid chain.* It is determined by interactions between the sequential units defining the *primary structure.* In proteins, *hydrogen bonding* between particular *amino acids* results most often in an *alpha helix* or *beta pleated sheet* configuration. In nucleic acids, the *double helix* consisting of two single nucleic acid chains twined around each other is held together by specific hydrogen bond interactions between *complementary base pairs.* (compare *primary structure, tertiary structure, quaternary structure*)

secondary thickening (see *secondary growth*)

second growth forest A forest that has been cut and regrown. (compare *old growth forest, virgin forest*)

second messenger An intermediary compound, often cyclic AMP (*cAMP*) that couples *extracellular* signals to *intracellular* processes and also amplifies the signal. *Peptide hormones* (for example, *adrenaline, insulin,* and *luteinizing hormone*), in particular, act by second messengers. When the hormone binds to a *receptor* on the cell surface, it initiates a series of reactions leading to the activation of *adenyl cyclase* (an enzyme embedded in the *cell membrane*). Adenyl cyclase produces *cAMP* from *ATP*, in turn activating another specific cascade of enzymatic reactions leading to the final action in the cell. (see *transmembrane protein, transmembrane receptor, protein kinases*)

secretin A *peptide hormone* secreted by the *duodenum* in response to *bile* release when the stomach empties its contents into the intestine. It stimulates alkaline pancreatic secretions that neutralize stomach acids, enabling the *digestive enzymes* to function.

secretion The release of substances or fluids produced in cells and released into the surrounding medium. The term also refers to the substance itself.

sedge flies (see *Trichoptera*)

sediment
1. Matter that settles to the bottom of a liquid, including bodies of water.
2. Material that is deposited by wind, water or glaciers.

seed In the *sexual reproduction* of seed plants, the structure from which a new plant develops. It is formed from a *fertilized ovule.* The seed contains an *embryo* and discrete food store (*cotyledon* or *endosperm*) surrounded by an outer covering (*testa*). In flowering plants (*angiosperms*), seeds are contained within a *fruit*, but in other seed plants (*gymnosperms*), they are shed naked from the plant. In *annual* plants, seeds provide the only mechanism for propagation from season to season. The development of the seed habit, which makes water unnecessary for *fertilization*, is one of the most significant advances in plant evolution, enabling plants to colonize terrestrial habitats.

seed pool The seeds of a plant species that do not *germinate* during the first possible season but remain as a genetic reservoir.

segment One of a series of repeated body parts. (see *metamere, segmentation*)

segmentation (metamerism) The repetition of body parts of an animal along the longitudinal axis of the body to produce a series of similar units (segments or *metameres*). It is most clearly seen in annelids such as the earthworm. In arthropods, it is obscured by *cephalization* (development of a head). In chordates, exterior segmentation in the adult has been completely lost, but evolutionary remnants may be seen in the *embryo*.

segregation In a *diploid* organism, the separation of each *homologous chromosome* of a pair, and consequently the *genes* and other *loci* on them, into different cells at *meiosis*. (see *Mendel's laws, independent segregation*)

seismonasty In plants, a movement in response to sudden touch. (see *nastic movements*)

selection pressure Environmental influences on the evolutionary process of *natural selection*. Weak selection pressure results in little evolutionary change, while strong selection pressure has a greater effect. (see *Darwinism*)

selective permeability The mechanism by which the *plasma membrane* of a cell limits the passage of particular molecules both into and out of the cell. It is accomplished by *transmembrane proteins* that form *membrane channels*, allowing the transport under selective conditions of non-lipid-soluble materials. (see *facilitated diffusion, coupled channel*)

self A shorthand term for *self-fertilization*.

self-assembly The ability of some cellular structures to *polymerize* from *monomers* without the help of enzymes or other non-functional components. An example is the formation of *microtubules* from *tubulin*.

self-fertilization Fusion of male and female *gametes* from the same individual to produce a *zygote*, leading to a new individual. It is more common in plants than animals, though many plants have mechanisms to prevent it. (see *self-pollination*; compare *cross-fertilization*)

self-incompatibility In flowering plants (*angiosperms*), a mechanism to prevent *self-fertilization*, thus promoting *cross-*

fertilization. It is due to the inability of the *pollen tube* to penetrate the *style* of a *stigma* from the same individual. Self-incompatibility results in *self-sterility*, thus preventing inbreeding. (see *dichogamous, protandry, protogyny, heterothallism*)

self-pollination In plants, the transfer of *pollen* (male *gametes*) from the *anther* (male reproductive structure) to the *stigma* (female reproductive structure) of the same plant, resulting in *self-fertilization*. It may occur between different flowers or within the same flower, but it is always within the same individual. Self-pollination is frequently found where there is a strong selective pressure to produce large numbers of genetically uniform individuals adapted to particular, relatively uniform habitats or where insects and other animals are scarce. An example of both would be the Arctic tundra. (compare *cross-pollination, self-sterility, self-incompatibility*)

self-sterility In *hermaphroditic* (producing both male and female *gametes*) organisms, the situation in which male gametes cannot *fertilize* female gametes from the same individual. (see *self-incompatibility*)

semelparous An organism that breeds only once during its life history.

semen The fluid produced by male mammals and reptiles containing *spermatozoa* and various nutritive substances. Spermatozoa are produced by the *testes*. The other constituents of semen are produced by the *prostate gland* and the *seminal vesicles*.

semicircular canals The three, looped canals that form part of the *labyrinth* of the vertebrate *inner ear* and function in maintaining equilibrium. They detect changes in the rate of movement (acceleration) of the head in all three planes. (see *cupula, ampulla*)

semiconservative replication The mechanism by which *DNA* molecules are able to replicate themselves faithfully in order to propagate genetic information. It is called semi-conservative because each

parent molecule

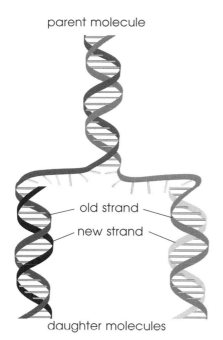

old strand

new strand

daughter molecules

Semiconservative Replication

new molecule contains one original and one newly synthesized strand of DNA. (see *replication, complementary base pairing*)

seminal receptacle (spermatheca) A sac-like organ in some female or *hermaphrodite* invertebrates, for example the earthworm, that acts as a store for *spermatozoa* received during *copulation* until they are required for *fertilization* of *ova* (egg cells).

seminal vesicles In most male mammals, a pair of *glands* that secrete an alkaline nutritive fluid that contributes to *semen* and counteracts *vaginal/uterine* acidity. They open into the *vas deferens* (the tube carrying semen from the *testes* to the exterior). In lower vertebrates and invertebrates, the seminal vesicles are the organs that store *spermatozoa*.

seminiferous tubules The tubules within the vertebrate *testis* in which *spermatozoa* are produced.

semipermeable membrane A selectively

permeable membrane, that permits the free passage of water but prevents or retards the passage of solute. (see *osmosis*)

semivoltine An organism that reproduces once every two years.

senescence The advanced phase of aging of a cell, organ, or organism, prior to death. It manifests as a progressive deterioration in function, the mechanisms of which are not well understood. (see *apoptosis*)

sense organ A group of *sensory receptors* and associated tissues specialized for detection of one specific sense such as touch, taste, smell, or light. They may be concentrated in certain regions (for example, ear, eye, nose), or distributed throughout the body (for example, touch, temperature). The stimulus is transmitted via *sensory neurons* to the brain for interpretation and response.

sensitivity (irritability) The responsiveness of an organism or part of an organism to a change in its *environment*. It is a characteristic feature of living organisms.

sensitization

1. The increase in reactivity of a cell or organism to an *antigen* to which it has previously been exposed. This occurs through the production of *antibodies* for which *B-cells* have already been induced. (see *memory cell*)

2. The increase in likelihood that a stimulus will produce a response in an animal repeatedly exposed to it. (compare *habituation*)

sensory In general, refers to *receptors, neurons, tissues,* and *organs* involved in receiving *environmental* or *physiological* information. Signals are then transmitted via *afferent* neurons to the *central nervous system* or central *ganglion* of an organism. (see *sensory neuron, receptor;* compare *motor*)

sensory epithelium Any *epithelium* (surface layer of cells) in which *sensory receptors* are embedded.

sensory hair (see *hair cell*)

sensory neuron (sensory nerve) A *neuron* that transmits *nerve impulses* from a sensory *receptor* to the *central nervous system* or central *ganglion* in an organism (*afferent*). *Stimulus-gated ion channels* are used to convert sensory inputs into a neural signal. (compare *motor neuron*)

sensory receptor (see *receptor*)

sepal In a flower one of the outer floral structures situated below the *petals*. Collectively, the whorl or spiral of sepals is called the *calyx*. They serve to protect the flower in the bud stage and may sometimes be brightly colored to attract insects for *pollination* (for example, orchids).

separation layer (see *abscission*)

septum, septa (pl.) Any wall or partition between two cavities. For instance, the nasal septum separates the two nasal cavities in the nose. In annelids, septa separate each body segment. In fungi, incomplete septa divide the filamentous *hyphae* into cellular compartments.

sequence polymorphism In *DNA*, any variation in the sequence of *nucleotides* between different *alleles* of the same *gene*. It usually refers to a difference of only one or a few *base pairs*. (compare *length polymorphism*)

seral stage Each successive phase in the sequential development of a plant *climax community*. (see *sere*)

sere The characteristic sequence of developmental stages occurring in *ecological succession*. Examples are *hydroseres* (starting in water) and *xeroseres* (starting in dry conditions).

serial endosymbiosis (**SET**) The theory put forth by Lynn Margulis that all *eukaryotic* cells evolved from mergers among *prokaryotes*. According to the theory, *undulipodia*, *mitochondria*, and *plastids* began as swimming, *respiring*, and *photosynthetic* free-living bacteria that, in this order, established *symbioses* with archaebacteria. Although intially met with resistance, it is generally accepted in the biology community today.

serine (ser) An *amino acid.*

sero- A word element derived from Latin meaning "blood serum." Its etymological

ancestor is "whey." (for example, *serology*)

serology The *in vitro* study of blood components. It was initially based on *antigen-antibody* reactions. However, it currently encompasses any blood component or group of components that can be analyzed in the laboratory and classified based on its chemical structure and/or biological function. This includes functional enzyme tests as well as various forms of *chromatography* and *electrophoresis*. Serology was once the main system used for individual and species comparison and identification, but it is being replaced by *DNA analysis*. (see *agglutination, blood groups*)

serosa. (see *mesothelium*)

serotonin (hydroxytryptamine) A *neurohormone* that acts on muscles and nerves, affecting a variety of involuntary functions. Within the brain, it controls sleep and evokes a variety of generally pleasurable emotional responses. Hallucinogenic compounds (for example, LSD) appear to antagonize the effects of serotonin in the brain, simulating one of the possible causes of clinical depression. Some antidepressants, such as Prozac and its relatives, block reuptake of seratonin into the *presynaptic* terminals at the ends of the *neurons* that produce it. Research is ongoing into the function and effect of serotonin.

serous membrane (see *mesothelium*)

Sertoli cells Large cells in the vertebrate *testis* that support and nourish developing *spermatozoa*. (see *spermatogenesis*)

serum (see *blood serum*)

sessile
1. An animal that lives permanently attached to a substrate, for example sea sponges which are attached to rocks (compare *vagile*).
2. A plant structure that is directly attached to the main body of the plant rather than to a secondary stalk.

SET (see *serial endosymbiotic theory*)

seta, setae (pl.) (**chaeta**)
1. The stalk of a moss or liverwort capsule.

2. The teeth surrounding a moss *peristome.*

3. One of many stiff bristlelike structures found in some animals (for example, the earthworm) that are used for gripping the soil during locomotion.

sex The formation of a new organism from more than a single genetic source, usually from two parents. (see *fertilization, meiosis, sexual reproduction*)

sex cell (see *gamete*)

sex chromosomes The *chromosome* pair that determines the *sexual* characteristics of an individual in species that exhibit sexual differentiation (*bisexual, dioecious*). This includes most animals and some higher plants. In humans, the **X** chromosome determines female genetic traits, and the **Y** determines male traits. **XX** (*homogametic*) individuals are female, while **XY** (*heterogametic*) individuals are male. The two chromosomes also look physically different. The **Y** is much smaller and contains less information than the **X**. In birds, butterflies, and some fish, where females are the heterogametic sex, their chromosomes are named as **Z** and **W**. The homogametic males have a *genotype* of **ZZ**. (see *sex determination, sex-linked*; compare *autosome*)

sex determination The mechanism by which the sex or *mating type* of an individual is determined in species that exhibit sexual differentiation (*bisexual, dioecious*). In humans, the **X** *chromosome* determines female genetic traits, and the **Y** determines male traits. **XX** (*homogametic*) individuals are female, while **XY** (*heterogametic*) individuals are male. In birds, reptiles, butterflies, and some fish where females are the heterogametic sex, their *sex chromosomes* are named **Z** and **W**. The homogametic males have a *genotype* of **ZZ**. In some fish, amphibians, and insects, the sex of the individual is determined by the number of **X** chromosomes in relation to the number of *autosomes* (non-sex chromosomes) and may also involve nutritional

and environmental factors (for example, bees). In many reptiles, sex is determined by the incubation temperature of the egg. The mating type of most fungi is determined by a single main *gene* that may switch back and forth by physical replacement within the *chromosome*. (see *dosage compensation, haplodiploidy, sexual dimorphism*)

sexduction The *sexual* transmission, via *conjugation*, of *E. coli chromosomal* genes that have been incorporated into an *F' factor*.

sex hormone Any *hormone* responsible for the development and functioning of the reproductive organs, and the development of *secondary sexual characteristics*. (see *androgen, testosterone, estrogen, progesterone*)

sex-linked A genetic trait that is determined by a *gene* located on a *sex chromosome*. It shows a characteristic pattern of inheritance different than *autosomal* (non-sex-linked) genes. In humans, sex-linked traits are usually located on the **X** chromosome, such that a *recessive mutation* is exhibited in all males (because there is no paired gene on the **Y**) but is *phenotypically* manifested only in females *homozygous* (having two identical *alleles*) for that particular gene. Examples in humans are *color blindness* and *hemophilia*. (see *X-linked, Y-linked*)

sex pilus (see *pilus*)

sex ratio The proportion of males to females in a population.

sexual Any of the processes, organs or structures involved in the ultimate formation of a new individual by the transfer and mixing of genetic material.

sexual dimorphism The occurrence of morphological differences, other than *primary sexual characteristics* (sex organs), that distinguish males from females within a species. For instance, male birds may have more brightly colored plumage than females, and male deer have large antlers. In humans, *secondary sexual characteristics* such as breasts in women and facial hair in men are examples of sexual dimorphism.

sexual reproduction The formation of new individuals by the transfer and mixing of genetic material. In *eukaryotes*, it involves the fusion of two *haploid nuclei,* usually *gametes,* to form a *diploid zygote.* The haploid cells are products of *meiosis,* in which *genetic recombination* takes place at a relatively high frequency. The definition of *sexual* reproduction is sometimes limited to a cycle of reproduction that involves *meiosis* and *fertilization.* (see *conjugation*; compare *asexual reproduction*)

sexual spore In most plants and fungi, a *spore* that can engage directly in or is a product of *fertilization,* or is one of the *haploid* products of *meiosis.* They may become new *unicellular* adults or divide *mitotically* to produce a *multicellular* individual. (see *spore, megaspore, microspore, sporophyte, gametophyte, alternation of generations*; compare *asexual spore*)

sharks (see *Chondrichthyes*)

shearwaters (see *Procellariiformes*)

sheep (see *Ruminantia*)

shellfish A commercial term for the many freshwater and marine mollusks that are farmed and harvested for food.

shoot The aerial portion of a plant that generally consists of a stem upon which leaves, buds, and flowers may be borne. It is the portion of the plant in which *photosynthesis* takes place. (compare *root*)

shorebirds (see *Charadriiformes*)

short-day plant A plant in which flowering is favored by short days and correspondingly long nights. The critical factor is actually the length of the dark period. This phenomenon is exploited commercially by briefly flashing chrysanthemums with a bright light during the night to delay flowering until Christmas. (see *photoperiodism*; compare *long day plant*)

short interspersed elements (see *SINEs*)

short tandem repeats (see *STRs*)

shotgun cloning The capture of a large number of different *genomic DNA* fragments in *cloning vectors* as a prelude to selecting one particular fragment for study or use. (see *genetic engineering*)

shoulder girdle (see *pectoral girdle*)

shrews (see *Insectivora*)

shrimps (see *Crustacea, Malacostraca*)

shuttle vector A *DNA cloning vector,* such as a *plasmid,* constructed in such a way that it can *replicate* in a least two different *host* species. This allows a DNA segment to be tested or manipulated in several cellular settings. (see *genetic engineering*)

siblings Offspring sharing at least one common parent.

sieve elements Elongated cells that form the *sieve tubes* in all *vascular plants.* They conduct carbohydrates away from the areas where they are manufactured, generally the leaves. Sieve elements have lost most cell *organelles,* including their *nuclei.* They retain only their *cytoplasm,* which is connected from cell to cell through the perforated ends of each *cell wall.* They are *metabolically* supported by *companion cells* that provide them with nutrients. In most vascular plants, the sieve elements are called *sieve cells,* having perforations of relatively even diameter. In *angiosperms,* specialization has produced sieve elements with areas containing larger and more efficient pores. These sieve elements are called *sieve tube members.* (see *phloem*; compare *tracheary element*)

sieve plate Any anatomical element that contains pores, allowing liquid to be transferred between cells or body compartments.
  1. The regions at the ends of *sieve elements* in *vascular plants* containing perforations that allow *cytoplasm* to flow through them. (see *phloem*)
  2. (see *water vascular system*)

sieve tube In *vascular plants,* a column of cells (*sieve elements*) connected by perforations in their *cell walls* (*sieve elements*) in which nutrients, in particular carbohydrates, are transported. (see *phloem*)

sigma factor (σ) A component of the *RNA*

polymerase holoenzyme in *prokaryotes.* It is responsible for initiation of *transcription* at the correct *promoter* region and dissociates from the complex soon after, leaving behind the *core enzyme.* (compare *rho factor*)

sigmoid growth curve (see *growth curve*)

signal sequence An *amino acid* sequence located at the amino terminal end of a secreted protein that is required for transport through the *plasma membrane* of the cell.

silaceous Any substance containing silicon or silicates.

silcula (see *siliqua*)

silent mutation A *DNA mutation* in which the normal function of the *gene product* is unaltered.

silicon A trace element found in many organisms, although not essential for growth in most. It is found in large quantities in the *cell walls* of certain protoctists and plants, such as *diatoms* and horsetails, and in smaller amounts in the cell walls of some higher plants. It forms the skeleton of certain marine animals such as sponges.

siliqua An elongated, capsular fruit found in *cruciferous* plants such as cabbage and mustard. A similar type of fruit, but shorter and broader, called a *silcula* is found in honesty and shepherd's purse.

Silurian A period of the Paleozoic geologic era, about 440 million to 405 years ago, characterized by early land plants, primitive jawless fish, and abundant invertebrates.

simple cuboidal epithelium *Epithelial* cells that have a cuboidal shape. They tend to be *metabolically* active and are found in the interior linings of various organs, such as the walls of kidney tubules. Individual cells often bear *cilia* or have secretory functions. (see *simple epithelium*)

simple columnar epithelium *Epithelial* cells that have a columnar shape. They tend to be *metabolically* active and are found in the interior linings of various organs, such as the walls of the intes-

tine. Individual cells often bear *cilia* or have secretory functions. (see *simple epithelium*)

simple epithelium Epithelial tissues that are a single cell layer thick. The cells come in three different shapes, *simple squamous, simple cuboidal,* and *simple columnar.*

simple squamous epithelium *Epithelial* cells that have a flattened shape. They are relatively *metabolically* inactive and are found in the interior linings of major body cavities, the lining of the lungs and the interior of blood vessels. (see *simple epithelium*)

SINEs (short interspersed elements) A collective term for the families of relatively short *repeated sequences* that are dispersed throughout mammalian *genomes.* They are on the order of hundreds of *base pairs* long, and are present in hundreds of thousands of whole and partial copies. SINEs are thought to spread as *retroposons.* An example is the *Alu sequence.* (compare *LINEs*)

single locus probe A *DNA probe* that detects genetic variation at only one particular *locus* in the *genome* of an organism. This type of system is commonly used in *RFLP genetic typing.* (see *DNA analysis, Southern blot, DNA profile*)

single-stranded DNA *DNA* unwound from its native *duplex* form and separated (*denatured*) into two single strands. In this form, it is able to interact with other *nucleic acid* molecules and enzymes. Single-stranded DNA is formed during natural processes, such as *replication* and *transcription.* It is made and used extensively in the laboratory for techniques such as *nucleic acid hybridization.* (compare *double-stranded DNA*)

single strand nick A form of *DNA* damage in which the *phosphodiester bond* on only one strand of *duplex* DNA is split. This type of damage may be easily repaired by a *DNA ligase* but may also sometimes initiate a *genetic recombination* event at a *homologous chromosome* or region. (see *DNA* *repair*; compare *double strand break*)

sink The region in a *vascular plant*, often the roots, where carbohydrates that have been distributed in the *sieve tubes* are actively withdrawn for use in roots, fruits, and stems. (see *mass flow*; compare *source*)

sinoatrial node (see *pacemaker*)

sinus Any *anatomical* cavity or space, for instance the nasal sinuses.

sinusoid A specialized type of blood *capillary* with large gaps between the cells, allowing high permeability. They occur in the liver and spleen.

sinus venosus In fish and amphibians, the first chamber of the heart that receives *deoxygenated* blood from the body and contains the *pacemaker*, the initiator of each heartbeat. In higher vertebrates, it has degenerated to the *sinoatrial (SA) node* on the wall of the right atrium and remains only as the *pacemaker*.

Siphonaptera The order of *endopterygote* insects containing the fleas. All are *ectoparasites* of mammals and birds, and some transmit serious diseases. The species tend to be *host* specific. Fleas are small, wingless insects with compressed bodies, legs adapted for jumping and clinging, and mouth parts modified for piercing and sucking. The *larvae* feed on organic *detritus* of the host.

Sipuncula A phylum of unsegmented *coelomate* sedentary marine worms, the sipunculans or peanut worms. They have *metanephrida* but lack a circulatory or respiratory system. They extrude a contractile organ, the *introvert*, to feed. Like annelids, they burrow in their substrate, extruding digested *detritus* as they go. Some species develop through *trocophore* larvae, and others develop directly or go through several larval stages. A few reproduce *asexually* by *binary fission*.

Sirenia An order of large, wholly aquatic mammals, the manatees (sea cows) and dugongs. They are unrelated to whales. They have front flippers, no hind legs, and a broad, horizontally flattened,

rounded tail. They inhabit warm, shallow, brackish water such as lagoons and river mouths and feed on aquatic vegetation. Because they are slow and sluggish, they are easily injured or killed by speedboats and water skiers. The name derives from the sirens of Greek mythology, the female sea creatures that lured sailors astray with their songs.

sister chromatid exchange A relatively rare *recombination* event between duplicated copies of a single *chromosome* (*chromatids*) before they have separated during *meiosis*. Unequal sister chromatid exchange in regions of *tandem duplications* results in a change in the number of unit copies. This is a common form of *mutation* in these non*transcribed* chromosomal regions.

site-directed mutagenesis A technique for creating specific *mutations* at preselected sites in *cloned DNA*. A small, synthetic *oligonucleotide* containing the desired *nucleotide* sequence is *annealed* to its *single-stranded complement* under per-

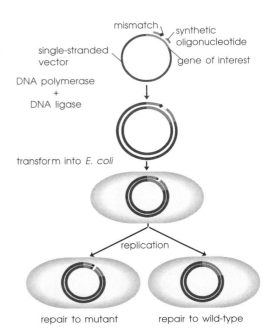

Site-Directed Mutagenesis

missive *hybridization* conditions that allow for some *base pair mismatching*. The oligonucleotide serves as a *primer* for *in vivo* DNA synthesis so that a *double-stranded* DNA molecule is reformed containing a small region of *heteroduplex*. In further replications, half the molecules will be propagated as the original sequence, while the introduced mutation will be propagated in the other half.

site-specific recombination Genetic *recombination* that occurs between specific *nucleotide* sequences that need not be *homologous*. It is always mediated by a specific recombination system, including enzymes that recognize only these sites. Some examples are the *integration* and *excision* of bacteriophage λ in *E. coli*, control of *gene expression* in the *flagellin* gene of the bacterium *Salmonella*, and the expression of various surface proteins that define the host ranges in *bacteriophages* μ and P1

skates (see *Chondrichthyes*)

skeletal muscle (see *striated muscle*)

skeletal system The structures that supply protection and support for the body, and provide an anchor for muscular movement. In vertebrates, they include the bones, cartilage, and ligaments. (see *exoskeleton, endoskeleton, hydroskeleton*)

skeleton A hard structure that supports and maintains the shape of an animal. It may be internal to the body (*endoskeleton*) or external (*endoskeleton*). (see *hydroskeleton*)

skin The largest animal *organ*, covering the entire surface of the body. It guards against excessive water loss, helps in *thermoregulation* in *homeothermic* animals, acts as a barrier against the entry of microorganisms, and protects against damage by UV light and from mechanical injury. It contains numerous nerve endings, thus acting as a peripheral *sense organ*. Outgrowths of the skin include

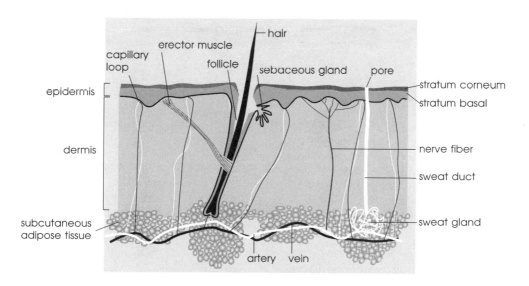

Cross-Section Through Skin

hair, nails, scales, and feathers. It contains *sweat glands* and *sebaceous glands*. The *epidermis* is the outermost layer, lying above the *dermis*, which is supported by *adipose* and *connective tissue*.

skin gills In many echinoderms, fingerlike *epidermal* projections that act as respiratory organs and through which waste removal takes place.

skuas (see *Charadriiformes*)

skull The bony plates that surround and protect the vertebrate brain.

skunks (see *Carnivora*)

sleep A normal recurrent state of reduced responsiveness to external stimuli. It is not a loss of consciousness. It occurs in all vertebrates. In man and other mammals, it is characterized by typical brain wave patterns (EEG). It is less well defined in amphibians and fish. Dreams take place during *REM* (rapid eye movement) sleep, which is essential for the maximum restorative properties of sleep. (see *serotonin, melatonin*)

sliding filament mechanism A unifying mechanism explaining many diverse forms of *intra-* and *intercellular* movement including *muscle contraction*, cell locomotion, *cyclosis,* and *cytoplasmic streaming*. It involves the energy dependent sliding of *actin*-containing *thin filaments* over *myosin*-containing *thick filaments* that generally produces a localized shortening.

slime net (see *Labyrinthulata*)

sloths (see *Xenarthra*)

slugs (see *Gastropoda*)

small intestine The part of the mammalian gut (*alimentary canal*) between the stomach and the large intestine. Most of the digestion and absorbtion of fluid and solid nutrients, as well as water, takes place over its long length and large surface area. The internal surface of the small intestine is covered with microscopic projections (*villi*) that are coated with a surface layer of specialized *epithelial* cells from which even smaller *cytoplasmic* projections (*microvilli*) extend. (see *duodenum, jejunum, ileum*)

small ribonucleoprotein particles (see *snRNPs*)

smooth muscle (involuntary muscle) The type of muscle found in all internal organs and arteries. It is found as long tubes or sheets made of long, narrow *muscle cells* containing the contractile proteins *actin* and *myosin*. In contrast to *striated muscle*, each cell has only one *nucleus*, and the *myofibrils* are not aligned into organized assemblies. Parallel arrangements of *thick filaments* (*myosin*) and *thin (actin) filaments* cross diagonally from one side of the cell to the other. Myosin molecules are attached to *dense bodies*, the functional equivalent of *Z-lines* in *striated muscle*, or to the muscle membrane. It is generally not under voluntary control and is supplied by the *autonomic nervous system* in vertebrates. Smooth muscle is specialized for slow, sustained contraction with minimal energy utilization. It may have spontaneous rhythmic contractions, such as *peristalsis* in the gut. All invertebrates except arthropods have only smooth muscle. (see *calmodulin, muscle contraction*)

snails (see *Gastropoda*)

snakes (see *Squamata*)

snRNPs (small ribonucleoprotein particles) A species of *RNA* that remains in the *nucleus* and combines with protein into particles thought to facilitate *RNA splicing*.

sociobiology The study of the biological, particularly the genetic, basis of social behavior. It was developed by E. O. Wilson in the 1970s.

socioecology The study of the influence of environmental conditions on vertebrate social organization.

sodium An element essential in animal tissues and often found in plants although apparently not essential in most. It acts with *potassium* to maintain the *osmotic balance* of animal tissues, a less important process in plants because of their *cell walls*. It is essential to nerve and muscle function, although an excess can produce high blood pressure. (see *sodium-potassium pump*)

sodium channel A transmembrane *ion channel* specific for sodium ions (Na⁺). They are particularly important in transmitting *nerve impulses.*

sodium inactivation The spontaneous closing of *voltage-gated* sodium (Na⁺) *ion channels* after an *action potential* passes through a *neuron.* Along with the opening of voltage-gated potassium (K⁺) channels, it promotes a return to the *resting potential* of the cell.

sodium-potassium pump (Na⁺-K⁺ pump) An *active transport* system, that maintains the *resting potential* of *neurons* by conveying three sodium ions (Na⁺) out of the cell for every two potassium ions (K⁺) transported to the interior. It is a *transmembrane protein* with a distinct pathway for each ion and works by conformational changes fueled by *ATP.* (see *membrane pump, nerve impulse, hyperpolarization, depolarization, action potential*)

phospholipid bilayer membrane

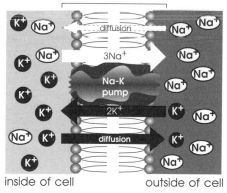

The Sodium-Potassium Pump

soft palate (see *palate*)

softwood A commercial designation of the wood of conifers. It has little to do with the actual hardness of the wood.

soil The mineral and organic material that forms a superficial layer of much of the earth's surface. It provides a growth medium and physical support for plants and is inhabited by numerous microorganisms.

solenocyte (see *flame cell*)

solenoid The *quaternary structure* of *chromosomes* into which *nucleosomes* are further arranged. It resembles a giant coil.

Solifuga (Solipugida). A minor order of arthropods, the solifugids or windscorpions. Most are desert dwellers.

Solipugida (see *Solifuga*)

solute potential The concentration of ions and *metabolites* in plant roots that tends to draw water in by *osmosis.* Along with the *pressure potential,* it enables water to rise vertically within a plant stem, constituting the *water potential.*

somat-, somato-, -some, -soma A word element derived from Greek meaning "body." (for example, *chromosome*)

somatic All the cells of a *multicellular* organism other than the *germ cells* (*gametes*). Changes in somatic cells are not heritable. They divide only by *mitosis,* producing progeny cells identical to the parent cell.

somatic cell genetics The study of *genes* in the nongermline cells of an organism. It is made possible by advances in the *in vitro culture* and manipulation of both plant and animal cells, including human cells.

somatic cell hybridization A laboratory technique in which human *fibroblast* cells are fused with mouse *tumor* cells. The final hybrid cell is immortal and typically contains a full mouse *genome* plus a few random human *chromosomes.* These constructs are used as a simplified system to map the location of specific *genes* on the human chromosomes.

somatic cell line The cell line that, early in the development of many animals, becomes differentiated (*determined*) from the few specialized cells that will become the *germ cell line.* Somatic cells lose the potential to undergo *meiosis* and form *gametes.* They form the bulk of any organismal body.

somatic genetic disease The novel appearance of an abnormal *gene* form in an organism, *cancers* being the most prominent example. Although the actual changes are not passed on, it is now appreciated that various predispositions to cancer are inherited as abnormal genes. (compare *inherited genetic disease*)

somatic mutation A *mutation* in the *DNA* of a *somatic cell*. Because somatic cells do not contribute to the formation of *gametes*, any changes are not inherited by the next generation. Certain diseases, such as *cancers* however, are caused by somatic mutations.

somatic nervous system (see *voluntary nervous system*)

somatic rearrangement (see *somatic recombination*)

somatic recombination (somatic rearrangement) The rearrangement of genetic material in the *somatic* cells of an organism. It occurs in all cells at a very low rate, but is particularly important in the generation of the millions of different *stem cells* in the *bone marrow*. Each stem cell is specific for a different *antigen*. (see *clonal selection, antibody, B-cell, plasma cell, T-cell*)

somatomedin A *peptide hormone* produced by the action of *growth hormone* on the liver and kidneys. It mediates the action of growth hormone on cartilage.

somatostatin A human *growth hormone*.

somatotrophin (see *growth hormone*)

somites Blocks of *mesodermal* tissue in the vertebrate *embryo* that differentiate into muscle blocks (*myotomes*) or kidney portions, or contribute to the skeleton or skin.

soredium, soredia (pl.) The *vegetative propagule* of *lichens*. It consists of minute clusters of *algal* cells surrounded by fungal *hyphae*. Soredia are formed in huge numbers over the lichen surface, resembling a fine powder. They may be dispersed by various agents including wind, water, and insects.

sorocarp A *multicellular, sexual reproductive* structure in which *haploid meiotic*

spores are formed. It is a stalked structure characteristic of some cellular slime molds.

sorosis A type of fruit incorporating a spike, such as the pineapple.

sorus, sori (pl.) A fruiting structure (*sporocarp*) consisting of a mass of *spores* or *sporangia*. It is found in various nonseed plants, fungi, and protoctists.

source The region in a *vascular plant*, usually the leaves, where carbohydrates manufactured by *photosynthesis* are actively loaded into the *sieve tubes* of the *phloem* for distribution to the rest of the plant. (see *mass flow*; compare *sink*)

Southern blot A laboratory technique developed by E. M. Southern in 1975 for detecting and analyzing specific fragments of *DNA*. It is one of the mainstays of modern molecular biology. Isolated *genomic* DNA is usually cleaved with *restriction enzymes* and the fragments separated by *electrophoresis* on an *agarose* gel. The DNA is *denatured* (made *single stranded*) and then transferred to a solid support (a cellulose acetate or nylon membrane) by capillary action. The membrane is then exposed to specific, labeled *probes*, usually *radioactive* or *chemiluminescent*, and the results visualized on X-ray film. The specifics of the technique vary with its particular application. It has been adapted for *RNA* (*northern blot*) and proteins (*western blot*).

spacer DNA The *DNA* found in the vast regions between *genes* and other recognizable units that has no known function.

specialization (see *adaption*)

speciation The evolution of populations of organisms within a species into distinct species themselves that can no longer interbreed. This occurs as a result of accumulated genetic differences and *natural selection*. (see *adaptive radiation*)

species A group of organisms that resemble each other and are generally able to interbreed and produce fertile offspring. (see *allopatric, sympatric*)

species biomass (see *biomass*)

species diversity A term loosely used to

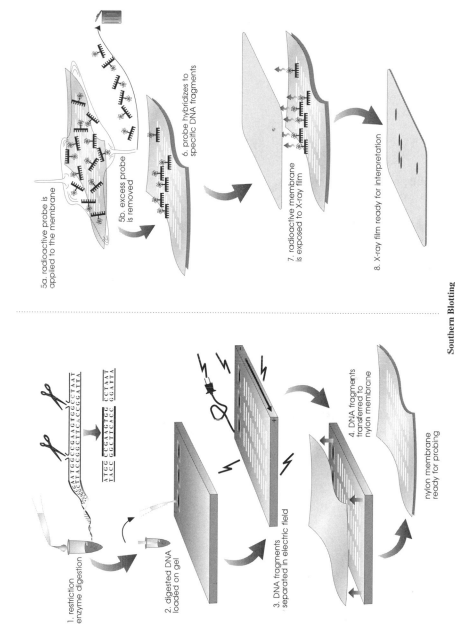

1. restriction enzyme digestion

2. digested DNA loaded on gel

3. DNA fragments separated in electric field

4. DNA fragments transferred to nylon membrane

nylon membrane ready for probing

5a. radioactive probe is applied to the membrane

5b. excess probe is removed

6. probe hybridizes to specific DNA fragments

7. radioactive membrane is exposed to X-ray film

8. X-ray film ready for interpretation

Southern Blotting

Modified from an illustration in *An Introduction to Forensic DNA Typing*, ©1997, by Inman and Rudin, CRC Press, Boca Raton, Florida.

mean the variety of species in an area or on the earth. Technically, it is composed of three components: species richness (the total number of species), species eveness (the relative abundance of species), and species dominance (the most abundant species). (see *biological diversity*)

spectrin A protein found in *eukaryotic* cells that helps determine cell shape. It forms a scaffold beneath the *cell membrane* and is anchored to both it and the *cytoskeleton*. It is responsible for the biconcave shape of red blood cells (*erythrocytes*).

spectrophotometer A laboratory instrument used for measuring the amount of light of different wavelengths (λ) absorbed by a solution. It is used in identification, analysis, and quantitation.

sperm (see *spermatozoon*)

spermatheca (see *seminal receptacle*)

spermatid In animals, each of the four *haploid* products that result from the *meiotic* division of a *spermatocyte* (male *germ cell*). (see *spermatogenesis*)

spermatocyte In animals, the *diploid* reproductive cells formed by enlargement of the *spermatogonia* (unspecialized male *germ cells*). The first *meiotic* division of the *primary spermatocyte* produces two *secondary spermatocytes* that complete the second *meiotic* division to produce four *haploid spermatids*. (see *spermatogenesis*)

spermatogenesis In male animals, the development of *spermatozoa* (male *gametes*) from undifferentiated *germ cells*. Even during *embryonic* development, the *diploid* precursor cells in the *germinal epithelium* lining the *seminiferous tubules* in the *testis* multiply and enlarge, forming *spermatogonia*. This process increases at the onset of *sexual* maturity. A spermatogonium undergoes further development and growth to become a still diploid *primary spermatocyte* that then undergoes *meiosis*. The first meiotic division results in two *secondary spermatocytes* that complete the second *meiotic* division to produce four *haploid spermatids*. The spermatids undergo further development to become spermatozoa. When mature, the spermatozoa pass from the seminiferous tubules into the *epididymis* for tem-

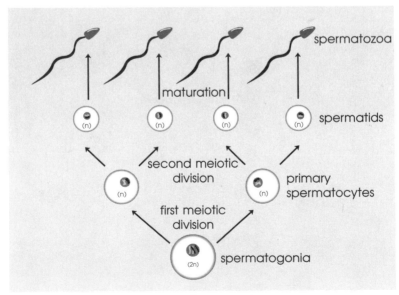

Spermatogenesis

porary storage. (see *capacitation*)

spermatogonium, spermatogonia (pl.) In male animals, the *diploid* precursor cells that develop in the *testis* from the *germinal epithelium* into *spermatocytes* that subsequently undergo *meiosis* and further development into mature *spermatozoa*. (see *spermatogenesis*)

spermatophore A gelatinous packet of *spermatozoa* produced by some species of animals with *internal fertilization*. It may be transferred directly to the female (cephalopods and insects) or deposited in water or moist soil to be taken up by the female (salamanders and newts).

spermatozoid The *undulipodiated*, motile male *gamete* found in some primitive plants. It is produced in an *antheridium*, except in certain *gymnosperms* (for example, *Ginkgo*) in which they develop from generative cells of the *pollen tube*.

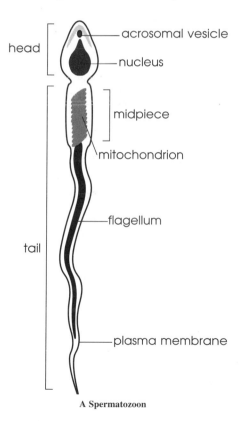

A Spermatozoon

spermatozoon, spermatozoa (pl.) **(sperm)** In animals, the mature male reproductive cell (*gamete*) formed in the *testis*. It is motile and generally much smaller than the *ova* (egg cell) that it is to *fertilize*. The head region, containing the *haploid nucleus*, is capped with an *acrosome* that assists in locally dissolving the egg cell coating so the sperm can penetrate. The midsection contains *mitochondria* that supply the energy for swimming. The long tail is an *undulipodium* that propels the sperm to its destination. During *fertilization*, only the head actually enters, leaving the tail and the midsection, including the *mitochondria* with their *DNA*, on the outside.

S phase (see *cell cycle*)

Sphenisciformes The order of birds containing the penguins.

Sphenodonta An order of reptiles containing only one known species, the tuatara. It is a *carnivorous*, primitive, lizardlike creature with a *pineal eye* and lacking a *copulatory* organ. The tuatara survives only on a few islands in the Bay of Plenty, New Zealand.

Sphenophyta A plant phylum containing only one living order (Equisetales), which itself contains one genus, horsetails (*Equisetum*). They are nonseed, *vascular plants* with characteristic ribbed, jointed stems containing silica embedded in the tissue. The visible *diploid sporophyte* bears cones (*strobili*) with *sporangia* in which *meiotic haploid spores* are produced. They fall to the ground and develop into insignificant *gametophytes*. They may also propagate by underground stems (*rhizomes*). Horsetails are common in wasteland areas and on silica-rich soils.

spheroplast A yeast cell in which the outer *cell wall* has been enzymatically removed in order to render the cell temporarily permeable to *exogenous DNA*. After the DNA has been introduced, the cell wall is allowed to regenerate. Yeast is a common *host* organism for *cloned* DNA in various constructs.

sphincter Any muscle that surrounds an opening and upon contraction completely closes it. Examples are the *pyloric sphincter* between the stomach and the *duodenum* and the muscle around the *urethra* where it leaves the *urinary bladder.*

spicule Any hardened, pointed structure, for instance the hard internal structures in sponges composed primarily of silica or calcium salts.

spiders (see *Araneae*)

spike In plants, a type of multiple flowering structure (*inflorescence*) in which flowers are directly attached to the main body of the plant (*sessile*) and borne on an elongated stem (for example, wheat).

spinal column (see *vertebral column*)

spinal cord The main *nerve* tract of the vertebrate *central nervous system* (CNS) which is contained in and protected by the bony *vertebral column*. It connects the brain to the *neurons* of the *peripheral nervous system* (PNS). It also contains *reflex arcs* that bypass the CNS. (see *spinal nerves*)

spinal nerves The paired *nerves* that arise at intervals along the length of the *spinal cord* to supply each segment of the body. Each nerve contains both *sensory* (incoming, *afferent*) and *motor* (outgoing, *efferent) neurons.*

spindle (spindle apparatus) The structure formed during *prophase* of *nuclear division* (both *meiosis* and *mitosis*) upon which the *chromatids* or *chromosomes* are arrayed and physically attached by *kinetochores*. It directs their separation and movement to opposite poles of the cell at *anaphase*. The spindle consists of *polymerized tubulin monomers* (*spindle fibers*) that generate movement by removing *monomers* from one end (attached to a pole by an *aster*) while remaining anchored at both ends. An attached object will then continue to move toward the end from which units are being lost. In animal cells, the microtubules are organized by *centrioles*.

spindle plaques In higher plant and fungal *cell division*, a somewhat amorphous structure that directs *microtubule* formation into a *spindle* during *prophase*. It is not visible in the light microscope. (compare *centriole*)

spine
1. In animals, the *vertebral column*.
2. In plants, a modified leaf structure reduced to a sharp point as protection against predators.

spinneret In spiders, a paired appendage modified for spinning silk, used in making webs and *cocoons* and in binding prey. The liquid silk *polymer (fibroin)* secreted by the silk *glands* hardens in response to tension (rather than evaporation or exposure to air). It may also refer to a pore at the hind end of nematodes out of which certain substances are secreted.

spiny anteater (see *Monotremata*)

spiny-headed worms (see *Acanthocephala*)

spiral cleavage The pattern of *cell cleavage* in *protostome embryos*. The cells divide at an oblique angle such that each new cell fits into the space between the adjacent older ones, resulting in a closely packed array. A line drawn though a sequence of dividing cells spirals outward from the polar axis. It results in a tightly packed array of cells. (compare *radial cleavage*)

spiral valve A fold of *mucous membrane* projecting into the intestine of many fish. It provides an increased surface area for secretion and absorption.

spiracle
1. One of the pores on the body of an insect at which a *trachea* (windpipe) opens.
2. A vestigial *gill slit* located behind the eye in *cartilaginous* fish (sharks and rays) and some primitive bony fish.

spirillum, spirilla (pl.) Any coiled or spiral-shaped bacterium. An example is the genus *Spirillum*.

Spirochaetae A phylum of *Gram-negative,* spirally coiled bacteria, the spirochaetes, in which 2 to 100 *flagella* are wound beneath a flexible outer *cell wall*. The spirochete moves by flexing the entire cell through mud or water. They include the *pathogens* causing syphilis and lyme disease as well as a group of *symbiotic* spirochaetes that live in the intestines of termites.

spleen A *lymphoid* organ that is one site of *leukocyte* (white blood cell) production and that destroys both worn *erythrocytes* (red blood cells) and *platelets*. It also stores blood and produces red cells before birth. It is part of the *reticuloendothelial system*.

splicing The process of removing *introns* (intervening sequences not necessary for protein synthesis) from *primary RNA transcripts* in *eukaryotes*. The remaining blocks of RNA are called *exons* and are linked together to form *mRNA* molecules that are *translated* into protein. Occasionally, different mRNAs may be generated from the same primary transcript by altering the splicing pattern.

sponges (see *Porifera*)

spongy parenchyma The interior tissue of plant leaves consisting of spherical cells

ous *chloroplasts* for *photosynthesis*. (see *mesophyll, stoma*)

spontaneous generation A disproved hypothesis that was once offered to explain the formation of living organisms. It stated that complete organisms can spontaneously arise from inorganic starting materials under the right conditions. It was debunked by the experiments of Redi in the 17th century and Pasteur in the 19th century. (see *abiogenesis*; compare *origin of life*)

spontaneous mutation Changes in *DNA* and *chromosome* structure that occur naturally with relatively low frequency. They arise from a variety of sources including errors in DNA *replication*, *transposable genetic elements*, and spontaneous lesions. (see *transition, transversion, frameshift mutation, base pair mismatch, base substitution, deamination, depurination, deletion, tandem duplication, free radical*; compare *induced mutations*)

spoonbills (see *Ciconiiformes*)

spor-, spori-, sporo-, -sporic, -sporous A

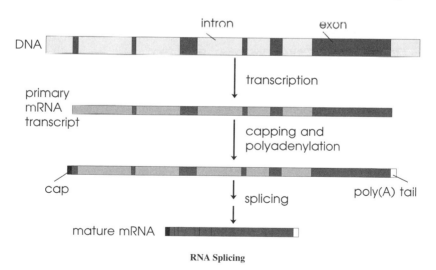

RNA Splicing

between which are located large *intercellular* spaces. They function in gas exchange and the passage of water vapor from the leaves. The cells contain numer-

word element derived from Greek meaning "seed." (for example, *heterosporous*)

sporangiophore In zygomycete fungi, the *asexual reproductive* structure consisting of a stalk on which *spores* are borne.

sporangium, sporangia (pl.) In nonseed plants and most fungi, the reproductive organ within which *haploid spores* are formed by *meiosis*. In plants with *morphologically* differentiated spores, *microspores* form in a *microsporangium* while *megaspores* form in a *megasporangium*. (compare *gametangium*)

spore A general term that encompasses a wide variety of reproductive and regenerative *propagules*. Most simply, it is a single cell that contains at least one *chromosome* complement that is given off or formed by an organism and can develop, either directly or indirectly, into a new individual. Spores are ubiquitously found in fungi, bacteria, protoctists, and nonseed plants. They may be produced in great numbers and dispersed widely, serving to increase the population rapidly. Microorganismal spores, in particular, often develop a thick protective *cell wall* in order to survive adverse environmental conditions. *Sexual* spores can directly engage in *fertilization*. They are produced by *meiosis* or fertilization. (see *megaspore, microspore, ascospore, basidiospore, conidiospore, zygospore, zoospore, sporocyst, endospore, meiospore, mitospore, oospore, sphorophyte;* compare *gamete*)

spore mother cell In plants, a *diploid* cell in the *sporophyte* generation that undergoes *meiosis* to produce four *haploid* spores. (see *megasporocyte, microsporocyte*)

sporocarp Any structure, usually stalked, with a primary function of producing *spores* from one initial cell. Sporocarps occur particularly in nonseed plants and some protoctists. (see *sorus, sporangium*)

sporocyst
1. The tough covering formed around some protist *spores*.
2. In *parasitic* flukes, a resting structure formed in the *intermediate host* containing *embryonic germ cells* that develop into *redia larvae*. (see *miracidium, cercaria*)

sporocyte (see *spore mother cell*).

sporogonium, sporogonia (pl.) The *spore*-producing structure in the *sporophyte* generation of mosses and liverworts.

sporogony The multiple fission process of protoctists, in particular apicomplexan protists, in which a *spore* or *zygote* undergoes multiple rounds of *mitosis* with no increase in cell size. This results in the generation of *sporozoites*.

sporophore A *spore*-bearing stalk found in some protoctists.

sporophyll A plant leaf that bears *sporangia*. It is typical of ferns.

sporophyte In plants, the *life cycle* generation that is *diploid* and in which some cells undergo *meiosis* to form *haploid spores*. These continue to divide *mitotically* and form the body of the *gametophyte* generation. The sporophyte arises from the union of *gametes* produced mitotically by a haploid *gametophyte*. In vascular plants, the sporophyte is the dominant generation, while in nonvascular plants (mosses, hornworts and liverworts), it is *parasitic* on the *gametophyte* generation. (see *alternation of generations, meiosis*)

Sporozoa A group of *spore*-forming *parasitic* protoctists once lumped into a single group and now separated into different phyla. (see *Apicomplexa, Myxozoa, Microspora, Paramyxea*)

sporozoan An informal name for the various *spore*-forming, usually *parasitic* protoctists now divided into several different phyla. (see *Apicomplexa, Myxozoa, Microspora, Paramyxea*)

sporozoite The product of multiple *mitotic* divisions (*sporogony*) of a *zygote* or spore. It is motile and usually infective. Sporozoites are characteristic of a *life cycle* stage of apicomplexans. In the malarial *parasite Plasmodium*, it is the form injected into humans by mosquitoes.

spring overturn In environments in which the temperature dips below freezing in the winter, the surface of freshwater lakes and ponds often freezes. Below that layer, the water remains liquid and life survives. In the spring, melted surface ice becomes warmed, and a thermal inversion occurs, mixing the oxygen dissolved in the surface layer with the nutrients held in the lower layers. (see *upwelling*; compare *fall overturn*)

springtails (see *Collembola)*

Spumellaria (see *Actinopoda)*

Squamata The order that contains the most successful living reptiles, the lizards and snakes. Lizards typically have a long tail, four limbs, an eardrum, and movable eyelids, although some are limbless. Snakes lack an eardrum and their eyes are covered by transparent eyelids. They have an elongated body lacking limbs, a deeply forked protrusible tongue used in *sensory* perception, and a wide jaw gape facilitating the swallowing of whole prey. Some suffocate their prey by strangulation, while others inject poison through fangs.

squids (see *Cephalopoda)*

squirrels (see *Rodentia)*

stable population A population whose numbers remain constant over a relatively long time period. In other words, births plus immigration balances deaths plus emigration. The age structure also tends to remain stable.

stamen The male reproductive organ in flowering plants (*angiosperms*) consisting of a fine stalk, or *filament*, bearing the *pollen*-producing *anther*. Collectively, the anthers make up the *androecium*. (see *microsporophyll*; compare *carpel*)

staminate Flowers in which the *pistils* (*carpels*) (female reproductive structures) are missing. (see *dioecious, monoecious, outcrossing, cross-pollination*; compare *pistillate*)

standard In a test or experiment, the criteria used to measure a result. Often, standards are a set of reagents for which the results are known and reliable. (see *molecular weight size marker*)

stapes (stirrup) The stirrup-shaped bone that forms one of the *ear ossicles* in the *inner ear* of terrestrial vertebrates. It is attached to and transmits signals to the *fenestra ovalis*. The stapes is derived from the *hyomandibular* bone in fish.

Staphylococcus A genus of *Gram-positive coccus* bacteria, many of which are *pathogenic* in animals. They cause wound infections and food poisoning. They are killed by *pasteurization* and common disinfectants.

starch A *polysaccharide* found exclusively in green plants that stores the energy captured by *photosynthesis*. It is actually a mixture of two types of molecules, *amylose* (10% to 20%), which contains only straight chain carbohydrates and is water soluble, and *amylopectin* (80% to 90%), which contains branched carbohydrate molecules and is water insoluble. It is formed into grains, which may be characteristic, and laid down in a series of concentric circles. Starch is stored in special *leukoplasts* (*amyloplasts*) and the outer layers (*stroma*) of *chloroplasts*. It is easily broken down into its component *glucose* subunits by the salivary enzyme *amylase*.

starch sheath In young flowering plant (*angiosperm*) stems, the innermost layer (*endodermis*) of cells that contains *starch* grains. The starch component may be lost at a later stage.

starfish (see *Asteroidea)*

start codon The *nucleotide triplet* **AUG** on an *mRNA* molecule that signals initiation of a *polypeptide* chain. In eukaryotes it directs the insertion of the *amino acid methionine* and in prokaryotes, *N-formylmethionine*. **AUG** also codes for the insertion of methionine at internal positions.

stat-, -static A word element derived from Greek meaning "standing," "set," or "stationary." (for example, *bacteriostatic*)

statoblast An internal bud produced in marine moss animals (ectoprocs) by which the animals reproduce *asexually*. It consists of a mass of cells containing stored food that is enclosed in a *chitinous* envelope. The statoblast can withstand

354

desiccation and freezing, and may remain *dormant* for prolonged periods. The structure and shape of statoblasts are important taxonomic features of fresh-water ectoprocs.

statocyst
1. Any organ involved in the perception of gravity. It usually contains mineral crystals (*statoliths*) in a vesicle lined with *sensory cilia* (*hair cells*).
2. In vertebrates, a *receptor* in the vertebrate inner ear that detects changes with respect to gravity. It consists of two chambers, the *utriculus* and the *sacculus*. Each contains crystals of calcium carbonate (*statoliths*) that rest in *hair cells*. The hair cells are *sensory neurons* and perceive any displacement by the statoliths. (compare *macula*)

statocyte A specialized plant cell thought to be involved in the perception of gravity. They contain *statoliths* (starch grains) that settle at the lowest surface, enabling the plant to detect the direction of gravity. (see *gravitropism*)

statoliths
1. Calcium carbonate crystals found in various organs, including the vertebrate *inner ear*. They are involved in the perception of gravity. (see *macula, statocycst*)
2. Starch grains found in specialized plant cells (*statocytes*) that are involved in the detection of gravity. (see *gravitropism*)

stele (see *vascular cylinder*)

Stelleroidia A class of echinoderms that were once classed into two: Asteroidea, the brittle stars, and Ophiuroidea, the sea stars.

stem In *vascular plants*, the structure that bears the buds, leaves, and flowers. It forms the central axis of the plant and often provides mechanical support. It is usually found aboveground but may also lie below ground (for example, *rhizome, bulb, or corm*)

stem cells The undifferentiated cells in *bone marrow* that give rise to all blood cells. They develop into red blood cells

(*erythrocytes*) as well as various white blood cells (*leukocytes*) of the *immune system* including *B-cells, T-cells,* and *macrophages.* (see *hemopoiesis*)

stereoscopic vision The type of vision found in vertebrates where both eyes are located in the front of the head so that their fields of vision overlap. By utilizing the slight parallax in perceiving the same image with each eye, the brain is able to interpolate distance and, consequently, achieve depth perception. Most predators tend to have forward-facing eyes, while prey animals have side-facing eyes, enlarging the field of vision at the expense of sensitive depth perception.

sterigma, sterigmata (pl.) In many fungi particularly, basidiomycetes, a projection upon which *spores* are formed in the *basidia.* (compare *ascus*)

sternum In tetrapods, the bone to which the ribs are attached in front of the chest. In birds and bats, it is the point of attachment for wing muscles.

steroid (steroid hormone) Any member of a group of lipids having a particular structure of four fused carbon rings. The *sex hormones, adrenal cortical hormones, vitamin D,* and *bile* acids are steroids and all derived from *cholesterol.* Certain *carcinogens* are also steroids.

steroid receptors A class of *hormone* dependent *transcription factors* that are em-

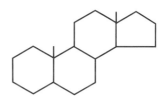

Steroid Ring Structure

ployed not only by all *steroids* but also by *thyroid hormones, retinoic acid,* and others. These hormones are lipid soluble and therefore able to pass through the *cell membrane.* They complex with their specific protein *receptors,* which float free in the *nucleus* or *cytoplasm* of the cell. The hormone-receptor complex produces an *allosteric* change in the receptor's conformation, allowing it to bind to a specific *enhancer* on the *chromosome* in order to maximize *RNA transcription* of an adjacent *gene*.

sterol A type of *steroid* compound with long side chains and at least one hydroxyl group. They frequently occur in *cell membranes*, an example is *cholesterol*.

sticky ends (cohesive ends) A type of cut produced by certain *restriction endonucleases* in which the *DNA* molecule is cut in a staggered pattern such that single-*stranded* tails of a few *nucleotides* extend from each end. It is an invaluable characteristic in manipulating *recombinant DNA*, because the sticky ends readily permit splicing of non*homologous* DNA fragments.

stigma, stigmata (pl.)
   1. In flowering plants (*angiosperms*), the tip region of the *carpel* (female reproductive organ) that serves as a receptive surface for *pollen grains*.
   2. (see *eyespot*)

stimulus A change in the external or internal environment of an organism to which it responds. (see *sensitivity, adaption*)

stimulus-gated ion channels The *ion channels* responsible for both the transmission of *nerve impulses* from one cell to another and the conversion of *sensory* inputs into a neural signal. They provide for the specific and directional passage of particular ions, usually sodium (Na^+). They open in response to chemical signals such as *neurotransmitters* or physical signals such as pressure. (compare *chemically gated ion channel*; *voltage-gated ion channel*)

stinging cell (see *cnidocyte*)

stipe
   1. The stalk of certain algae (for example, kelp) between the *holdfast* and the blade.
   2. The stalk of the *fruiting body* in certain basidiomycete fungi.

stipule A modified leaf, usually found in pairs near or along the base of a leaf attachment (*petiole*). They are found in the garden pea, where they serve as additional *photosynthetic* structures, and in the rose, where they protect the *axillary* bud.

stirrup (see *stapes*)

stolon
   1. In *colonial* invertebrates, a stalklike structure from which individuals are produced by *budding*.
   2. In plants, a stem that grows horizontally above the ground and puts out roots at one or more *nodes* where new plants may eventually develop. An example is the strawberry. (see *vegetative propagation, perennation, perennating structure*; compare *bulb, corm, offset, rhizome, sucker, tuber*)

stom-, -stome, -stomata A word element derived from Greek meaning "mouth." (for example, *cytostome*)

stoma, stomata (pl.)
   1. The portion of the *alimentary canal* between the mouth and the *pharynx*.
   2. In plants, small pores in the outer layer (*epidermis*) through which carbon dioxide (CO_2) gas enters and water exits. In most plants, they are located on the bottom surface of the leaf. Each stoma is surrounded by two *guard cells* that regulate the opening and closing of the pore by changes in their *turgidity*.

stomach The part of the *alimentary canal* in vertebrates that acts as a storage organ, permitting food to be eaten at intervals rather than continuously. Food enters the stomach from the *esophagus* through the *cardiac sphincter*. Within the stomach, hydrochloric acid, various enzymes (for example, *pepsin, rennin*), and muscular contractions reduce the food to a semiliquid state (*chyme*) allowing it to pass gradually through the *pyloric sphincter* to the *duodenum*. In birds, the stomach is divided into the *crop* and the *gizzard*. Some mammals (ruminants) possess complex stomachs harboring *symbiotic* bacteria. (see *gastric juice, gastrin, goblet cell, cecum, rumen, reticulum, abomasum, omasum, cud*)

stomium, stomia (pl.) In ferns, the location in the wall of the *sporangium* (*spore*-producing organ) where rupture occurs at maturity, releasing the spores.

stone cell (see *sclereid*)

stoneflies (see *Plecoptera*)

stone fruit (see *drupe*)

stop codon Any of the three *nonsense codons*, UAA, UAG, and UGA, that signal termination of a *polypeptide* chain. (see *amber, ochre, opal*)

storks (see *Ciconiiformes*)

strain A pure-breeding lineage of microorganisms such as yeast or bacteria. It may also refer to viruses.

stratification (after ripening) The requirement for the seeds of some plants to be held at low temperatures for a period of time before *germination* can take place. This prevents germination and seedling growth until winter has passed in cold regions.

stratified epithelium *Epithelial* tissues that are several cell layers thick. The most prominent is found in the *epidermis* of skin, which consists of *stratified squamous* (flattened) *epithelium*. They characteristically accumulate large amounts of *keratin*, a strong fibrous protein.

stratified squamous epithelium (see *stratified epithelium*)

stratum basal A layer of dividing cells at the base of the *epidermis* of vertebrates. Its cells contain granules of the pigment *melanin*, which protects the body against ultraviolet radiation. Pockets of this basal layer form *hair follicles* and *sebaceous* (oil) *glands* in mammals. The pattern of *papillae* in this skin layer is reflected in the dermal ridges that make up fingerprints.

stratum corneum The outer layer of dead, *keratinized* cells in the *epidermis* of vertebrate skin.

Strepsiptera An order of small *endopterygote* insects, the twisted-winged parasites. The *larvae* are almost all *endoparasitic* on other insects. The female adults are *endoparasites*, while the males are free living. It usually causes its *host* to become infertile.

Streptococcus A genus of *Gram-positive coccus* bacteria, many of which are *pathogenic* in animals, causing respiratory and alimentary tract infections. Some infect red blood cells (scarlet fever and rheumatic fever). They are killed by *pasteurization* and common disinfectants. Many strains are sensitive to *antibiotics*.

striated muscle The type of muscle organization found in voluntary skeletal muscles and *cardiac muscle*. It derives its name from its microscopic striped appearance, reflecting the arrangement of *myofibrils* into *sarcomeres* that are lined up in register with each other. In contrast to *smooth muscle*, each cell is *multinucleate* and contains numerous *mitochondria* for energy production. Striated muscles are specialized for rapid contractions and large forces, possible because the cells act in concert. (see *sarcoplasm, sarcoplasmic reticulum, Z-line, muscle contraction*)

stridulation The production of sound by the rubbing of body parts. It is common in insects and is used in courtship rituals.

Strigiformes The order of birds comprising the owls.

stringency As applied to *nucleic acid hybridization*, the specific conditions under which it is performed. This can be varied to specify the degree of *base pair mismatching* that will be tolerated. (see *DNA analysis, Southern blot*)

strobila, strobilae (pl.) An animal structure in which *asexual reproduction* occurs by transverse splitting (*strobilization*). Examples include the *scyphistoma* stage in jellyfish and the *proglottids* in tapeworms.

strobilization A method of *asexual reproduction* in animals that involves the transverse splitting of a particular body structure. (see *strobila, scyphistoma, proglottis*)

strobilus, strobili (pl.) A plant reproductive structure found in nonflowering seed plants (*gymnosperms*) and some ferns. It consists of an organized collection of *ovule*-bearing scales grouped terminally on a stem.

stroma In plant *chloroplasts*, the internal fluid matrix in which the *thylakoids*, organized into *grana*, are embedded. It contains the enzymes involved in the *Calvin cycle* (*dark reactions*) of *photosynthesis* in which carbon is "fixed" into carbohydrates.

stroma lamellae (lamellae) The membranes in plant *chloroplasts* that connect the stacks of *thykaloid discs*, grana, in which *photosynthetic* enzymes are housed.

stromatolites Massive limestone deposits produced by cyanobacteria that were abundant in the fossil record from about 2.8 billion to 1.6 million years ago. They continue to be produced today but under conditions of unusually high salinity, acidity, and light intensities, as in warm lagoons. (see *microbial mat*)

STRs (short tandem repeats) *Tandem repeats* in *DNA* that are very short, approximately 2 to about 6 *nucleotides*. They are used as *genetic markers* for disease genes and, under certain conditions, as individual genetic markers. (see *DNA analysis, PCR*)

structural gene A *gene* that encodes a protein. (compare *regulatory gene*)

structural protein A class of proteins, mostly fibrous, that play structural roles in the body. Examples include *keratin, collagen, actin,* and *myosin.* (compare *globular protein*)

Struthioneformes The order of birds containing the flightless ostriches.

styl-, -stylic, -stylous A word derived from Latin meaning "stylus," a long, thin, pointed structure. (for example, *heterostylic*)

style In plants, the stalklike portion of a *carpel* (female reproductive organ) joining the *ovary* and the *stigma*. In may be elongated in plants relying on wind, insect, or animal *pollination*. During *fertilization*, the *pollen tube* grows through the style. (see *double fertilization*)

subcellular A biological level of organization within a single cell.

subclavian artery In mammals, a large, paired artery that carries *oxygenated* blood to the forelimbs.

subcloning The subdivision and propagation of desired portions of *DNA* fragments that have already been *cloned*.

subcutaneous tissue The layer of tissue just beneath the *dermis* of the skin in vertebrates, composed primarily of fat cells. It restricts heat loss, particularly in aquatic mammals, and acts a shock absorber. In *hibernating* mammals, it also acts as an essential food store. (see *adipose tissue*)

suberin A fatty substance produced by the inner layer of *cork cambium* in woody plants. It is similar to *cutin* in that it makes the cork layer nearly impenetrable to water.

sublittoral
1. (see *limnetic*)
2. In a marine *ecosystem*, the zone extending from low tide to a depth of about 200 meters. The more shallow end is inhabited by kelp (brown algae) while red algae are characteristic of the deeper end. Marine animals found in this zone include various crustaceans (for example, mollusks, arthropods) and higher invertebrates such as echinoderms and cnidarians. (compare *littoral, benthic*)

subspecies The taxonomic grouping below the species level. It is a *race* for which a Latin name is allocated. Subspecies are generally geographically distinct populations that bear a physical distinction from other populations of a species.

substrate
1. In biochemistry, the substance upon which an enzyme acts. For instance, the substrate for *DNA polymerase* is DNA.
2. In ecology, the surface upon which an organism grows. This is generally nonliving, such as soil or rock, but may also be the body of another animal, such as barnacles on whales.

substrate level phosphorylation The generation of *ATP* from *ADP* in which the energy is obtained from breaking the bonds of the same molecule that donates the terminal phosphate group. It is essentially a reshuffling of chemical bonds that couples the formation of ATP to another, highly exergonic reaction. Many bacteria use this system exclusively. (compare *chemiosmosis, cyclic phosphorylation, photophosphorylation, oxidative phosphorylation*)

succession (see *ecological succession*)

succinic acid An intermediate in the *Krebs cycle.*

succulent Any plant that has leaves specially adapted for water storage. They are characteristic of desert environments.

sucker In plants, an underground shoot that emerges to give rise to a new plant that is nourished by the parent until it becomes established. Suckers are troublesome in certain ornamentals (for example, roses) in which they grow from rootstock at the expense of the graft. (see *vegetative propagation, perennation, perennating structure*; compare *bulb, corm, offset, rhizome, stolon, tuber*)

sucrose A common plant sugar, a *disaccharide* formed from one *glucose* and one *fructose* unit. It is hydrolyzed into its two component *monosaccharides* by the enzyme *invertase*. Sucrose is extracted commercially from sugar cane and sugar beet, and commonly appears as table sugar.

sugar (saccharide) Any of a group of simple sugars that are water soluble. (see *monosaccharide, disaccharide, polysaccharide, glucose, fructose, sucrose, pentose, hexose, pyranose*)

sugar-phosphate backbone The alternating chain of *ribose* or *deoxyribose* sugar units that are linked together through *phosphodiester bonds* in a single chain of nucleic acid (*DNA, RNA*). The *bases* extend at right angles from each sugar *monomer.*

Suina A suborder of even-toed ungulates including the pigs and hippopotamuses.

sulfur An essential element in living tissues. It is contained in the *amino acids cysteine* and *methionine*, thus in nearly all proteins. It is integral to *ferrodoxins*, iron-containing proteins that participate in the *electron transport chain* in *photosynthesis*. It is also integral to the *iron-sulfur centers* in photosynthesis and *aerobic respiration*. Like nitrogen and carbon, there is a cycling of sulfur in nature.

sulfuretum A marine habitat in which the sand is permeated with the odor of hydrogen sulfide (H_2S, rotten eggs) from the bacterial reduction of sulfate.

summation The additive effect of several impulses arriving at the *synapse* of a nerve or muscle cell when none of the impulses can individually invoke a response. This may be accomplished by the impulses arriving simultaneously from a number of *neurons* synapsing at the same cell (*special summation*) or a rapid succession of impulses from the same neuron (*temporal summation*). Summation is a major mechanism of integration in the *nervous system*. (compare *facilitation*)

superior The position of a structure or organ that is above relative to another. In botany, it is applied to the position of the *ovary* in *hypogynous* flowers, where the *sepals, petals,* and *stamens* are inserted at the base. (compare *inferior*)

supermutagen An agent that induces genetic *mutations* at an extremely high rate. They are used in the laboratory to stimulate the production of mutations for study by *mutational analysis*. Examples include *Aflatoxin* and benzyl chloride.

superoxide dismutase An enzyme that constitutes a front line defense against *DNA* damage. It deactivates potentially *carcinogenic free radicals* (superoxide radicals) by catalyzing their conversion to hydrogen peroxide that is, in turn, converted to water by *catalase*.

superoxide radical (see *free radical*)

superposition eyes (see *ommatidium*)

suppressor mutation A *mutation* that counteracts the effects of another mutation. It is located at another physical location (*locus*) but may be within the same *gene* or in another gene. Different suppressors act in various ways.

suppressor T-cell (see *T-cell*)

surface zone (see *photic zone*)

survivorship The percentage of an original population that survives to a given age. A survivorship curve is a way of expressing the characteristics of populations with respect to age distribution. In type I, a large number of individuals reach their theoretical maximum age; in type II, mortality is largely independent of age; in type III, mortality is highest during infancy. As humans approach a type I survivorship curve more and more closely, particularly in industrialized countries, and the birth rate remains stable or increases, overpopulation results.

suspension feeding (see *filter feeding*)

suspensor The column of cells that attaches the *embryo* to the *embryo sac* in *angiosperm* seeds. It is derived from the *basal cell*. (see *double fertilization*)

sustainable harvest The rate at which any particular resource can be harvested without decreasing the capacity of the environment to regenerate a similar level in the future.

suture In general, an area of fusion between adjacent structures. Examples are the line of fusion between adjacent *carpels* in flowering plants (*angiosperms*) and the joints between adjacent *sclerites* of an arthropod *exoskeleton*.

Svedberg unit (S) A unit of sedimentation velocity commonly used to describe molecular units.

swallowing (see *deglutition*)

swans (see *Anseriformes*)

swarmer cell (see *zoospore*)

sweat The fluid secreted by *sweat glands*. It contains small amounts of ions as well as *bactericidal* and *fungicidal* substances (urea and lactate). The evaporation of sweat is part of *thermoregulation* to help keep the body cool. Thermoregulation takes place over the entire body surface. In contrast, emotional sweating is confined to the palms, soles, and armpits. Excessive sweating is accompanied by the release of *bradykinin*, which stimulates dilation of the blood capillaries in the skin (*inflammation*), providing another avenue for heat loss.

sweat glands The many coiled, tubular *glands* in the skin of mammals that produce *sweat*. They are *exocrine glands*. Modified sweat glands produce ear wax.

swifts (see *Apodiformes*)

swim bladder In bony fish, a thin-walled sac in the abdominal cavity by which the fish is able to adjust its buoyancy. The volume of gas in the bladder may be adjusted by exchange with the extensive network of blood vessels surrounding it. The swim bladder evolved from an accessory respiratory organ. In lungfish it functions as a lung. It is also adapted in some fish for sound production and reception.

swimmerets In lobsters and crayfish, appendages that are used for swimming and in reproduction. They occur in lines along the sides of the abdomen.

syconus A succulent fruit that develops from a hollow *capitulum*, for example the fig.

symbiogenesis The appearance in *evolution* of new forms of life resulting from *symbioses*, such as lichens from associations between algae and fungi.

symbiont Each member of a *symbiotic* relationship.

symbiosis The intimate and protracted association of individuals of different species. It includes interactions of mutual benefit (*mutualism*), of benefit to one and of no significance to the other (*commensalism*), and of harm to one and benefit to the other (*parasitism*). Symbiosis is sometimes used as a synonym for mutualism.

symbiotroph A mode of nutrition in which a *heterotrophic symbiont* derives both its carbon and its energy from a living partner. (see *parasitism*)

sympathetic nervous system One of the two divisions of the *autonomic* (involuntary) *nervous system*. It prepares the body for stress (*fight-or-flight response*) by increasing heart rate, respiratory rate, and blood pressure and slowing digestive processes. It supplies *motor nerves,* which release the *neurotransmitter noradrenaline*, to the smooth muscles of internal organs and to heart muscle, antagonizing the effects of the *parasympathetic nervous system.* The *medulla* of the *adrenal gland* is supplied only by sympathetic *neurons*, which trigger the release of *adrenaline* into the bloodstream, further enhancing the effects of the sympathetic nervous system.

sympatric Populations that have diverged into separate species without being geographically isolated. They become species whose habitats temporarily overlap. (compare *allopatric*)

Symphyla A class of small, uniramian arthropods that superficially resemble centipedes. Their mouthparts are closer to that of insects. Most feed on decayed vegetation, but some are crop pests. They are found in the soil in warm climates.

symphysis A type of joint in which two bones are connected by fibrous cartilage.

symplast In plants, the continuum of *cytoplasm* through a plant that results from the interconnection of different cells. It is made possible by cytoplasmic bridges (*plasmodesmata*) between adjacent cells.

sympodial (definite growth, cymose branching) In plants, a type of branching in which the terminal bud of the main stem axis stops growing, and growth is taken over by lateral buds. This process may repeat itself a number of times, leading to a multibranched structure. It is typical of the formation of a *cymose inflorescence.* (compare *monopodial*)

synapse The junction between two nerve cells (*neurons*) or a nerve and a muscle cell. Between nerves, the *axon* of one nerve transmits the signal to the *dendrites* or *cell body* of another neuron. *Nerve impulses* are relayed across the gap (*synaptic cleft*) by small *peptides* called *neurotransmitters.* (see *presynaptic, postsynaptic, facilitation, summation*)

synapsis The physical pairing of replicated *homologous chromosomes* during *prophase* of *meiosis I.* The majority of genetic *recombination* takes place at this time. (see *chiasma, homologous recombination*)

synaptic cleft The space between two adjacent *neurons* or between a neuron and a

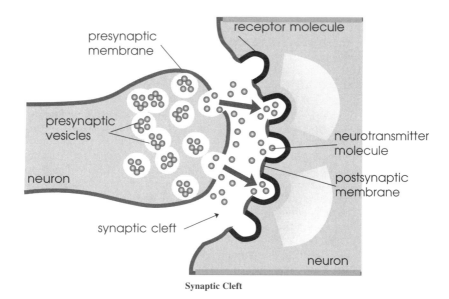

Synaptic Cleft

muscle cell across which a *nerve impulse* is chemically transmitted. A *neurotransmitter* is released from the *presynaptic membrane* and diffuses across the synaptic cleft to bind to receptors in the *postsynaptic membrane. This* causes *chemically gated ion channels* to open.

synaptonemal complex The lattice of protein that holds replicated *homologous* chromosomes exactly in register (*synapsis*) during *prophase* of *meiosis I*. Within it, the DNA *double helix* unwinds, and single strands of DNA may pair and exchange portions with single strands from *chromatids* of the other homologue.

syncarpous A plant *ovary* made of fused *carpels* (female reproductive structure), as in the primrose. (compare *apocarpous*)

syncytium, syncitia (pl.) A *multinucleate* cellular structure. At one extreme is *striated muscle*, which is essentially one large cell containing many *nuclei*. At the other is the *symplast* in plants, which consists of *cytoplasmic* bridges between adjacent cells. (see *acellular*)

synecology The *ecology* of *communities*, including the interactions of living and nonliving elements. (compare *autecology*)

synergid cells Two short-lived cells in the *embryo sac* of flowering plants (*angiosperms*) that nourish the egg cell and degenerate soon afterwards. They do not participate in *fertilization.*

synergism The cooperativity of two or more processes or systems.
1. Drugs or *hormones* have similar effects such that their individual activities are additive.
2. The coordinated actions of muscles produce a particular movement.

syngamy (see *fertilization*)

synovial membrane The membrane that forms the capsule surrounding a joint. It primarily contains *collagen* fibers and secretes a lubricating *synovial fluid.*

syrinx The sound-producing organ in birds. It is similar in structure to the *vocal cords* in mammals but located in the trachea as opposed to the *larynx.*

systemic arch A major vessel or vessels in tetrapods that carries *oxygenated* blood from the heart to the *dorsal aorta* that serves the rest of the body. It is derived from the fourth *aortic arch*. In amphibians, both left and right arches persist in the adult. In adult birds, only the right persists, and in adult mammals, only the left persists.

systemic circulation The blood that travels from the heart to the rest of the body and returns back to the heart. The separation of *pulmonary* and systemic circulation is an evolutionary advance first seen in amphibians, improved in reptiles to achieve better separation of the *oxygenated* blood, and refined in birds and mammals to a true *double circulatory system.*

systole The phase of the heartbeat cycle when the heart muscle is contracted, propelling blood from the *atria* into the *ventricles* that contract to propel it into the *aorta* and *pulmonary artery*. The measurement of systolic pressure is reflected in the upper, higher number of a blood pressure measurement. (compare *diastole*)

tactic movement (see *taxis*)

tadpole. The aquatic *larval* form of amphibians. It *metamorphoses* into the adult.

tagma, tagmata (pl.) In arthropods, body parts that have become fused into functional groups. Tagmatization is of central importance in the evolution of arthropods. For instance in some arthropods, the head is fused with the *thorax* to form the *cephalothorax.*

taiga (boreal forest) The northern coniferous forest, consisting primarily of spruce, hemlock, and fir, that extends across Eurasia and North America. It is characterized by long cold winters, wet summers, and extremes in day length according to the seasons. All life must either adapt to being covered by a thick blanket of snow for many months or migrate to warmer regions. Many large mammals and smaller rodents live in the Taiga, but *biodiversity* is relatively poor. (see *biome, endothermic, ectothermic*)

tandem duplication Adjacent identical *chromosome* segments.

tandem repeat A type of *chromosome* organization common in non*transcribed* regions of *eukaryotic genomes*. A particular *nucleotide* sequence ranging from two through at most about fifty bases is sequentially repeated in a head-to-tail fashion as many as hundreds of times. An exception is *rRNA genes*, which are arranged on the *chromosome* in tandem repeats; each transcribed copy is separated by a non*transcribed* spacer region. (see *repeated sequence, satellite DNA, microsatellite, ministatellite, VNTR;* compare *inverted repeat*)

tannin A group of complex astringent substances occurring widely in the bark, unripe fruit, and other parts of plants as well as in some insect *cuticles*. Their natural function may be to render plant tissues unpalatable to *herbivores*. Commercially, they are employed in the tanning of hide into leather.

tapetum, tapeta (pl.) A layer of cells surrounding the *sporocytes* (spore mother cells) developing in the *anthers* in the *microsporangia* of nonseed *vascular plants*. They disintegrate during development, liberating nutrients that are then absorbed by the developing *spores*. (see *sporangium*)

tapetum lucidum A light-reflecting layer at the back of the eyeball of *nocturnal* vertebrates and deepwater fish. It may consist of reflective connective tissue or *guanine* crystals situated within the *choroid* layer. Some of the reflected light passes out through the *pupil* so that the eyes seem to glow in the dark (for example, when caught in the headlights of a car).

tapiors (see *Ceratomorpha*)

***Taq* polymerase** A temperature resistant *DNA polymerase* isolated from the *Thermus aquaticus* bacterium. It was the first and remains the predominant enzyme used in *PCR* (polymerase chain reaction).

Tardigrada A phylum of small invertebrate animals, the tardigrades or water bears, named for their gait. They are mostly *pseudocoelomate* but have a small *coelom* surrounding their *gonad*. They live in the moisture that covers lichens and mosses, and eat nematodes and rotifers. Some species reproduce *sexually* with either internal or external *fertilization*, and some reproduce *parthogenetically*. After hatching, tardigrades grow by enlargement of existing cells rather than by *mitotic* division to increase the number of cells. They may desiccate into a *dormant* structure that will revive decades later when rehydrated. Tardigrades can also withstand over 100 times more exposure to X-rays than can humans. (see *cryptobiotic*)

targeted gene knockout (see *knockout*)

tarsal bones Bones in the distal regions of the hind limb of tetrapods. In man, they form the heel and ankle (*tarsus*). (compare *carpal bones*)

tarsus (see *tarsal bones*)

taste bud The vertebrate taste *receptor* consisting of a group of *sensory* cells located on a *papilla* of the tongue in terrestrial animals but scattered over the body in aquatic organisms. On the *primate* tongue, they are localized in regions recognized as salty, sweet, bitter, and sour (acid). Other taste modalities are detected in fish.

TATA box An integral *DNA* sequence element of *eukaryotic gene promoters*. It directs *RNA polymerase* to begin *transcribing* at a specified point *downstream* from the **TATA** box. Unlike *prokaryotic* promoters, it requires additional sequence elements, located yet farther *upstream* for efficient transcription *in vivo*. (see *UAS, enhancer,* **CCAAT box, GC box**)

tautomer One of the different isomeric forms that each of the *nucleic acid bases* can take. The rare presence of the wrong tautomer during *DNA replication* can cause a *base pair mismatch* and subsequent *mutation*. The *keto* form is normally present, whereas the *imino* or *enol* forms are rare. (see *tautomeric shift*)

tautomeric shift The chemical change from the normal *keto* form of a *nucleic acid base* to the rare *imino* or *enol* form. (see *tautomer*)

tax-, taxi-, taxo-, -taxis A word element derived from Greek meaning "to order" or " to arrange." (for example, *photo taxis*)

taxis A directional movement by a cell or animal in response to an *environmental* stimulus. Movement toward the stimulus is *positive taxis* and away is *negative taxis*. (see *aerotaxis, chemotaxis, phototaxis*; compare *nastic movements, tropism*)

taxon, taxa (pl.) The members of a particular taxonomic group such as a *class, family,* or *genus*. The members of the class *Mammalia* form a taxon.

taxonomy The study of the classification of living organisms. Modern taxonomy takes into account both the order and

magnitude of *phylogenetic* divergence of organisms. Molecular genetic techniques are increasingly being employed in order to refine and reclassify the data. Consequently, the field is currently in a state of flux while this information is being gathered. (see *phenetics, cladistics*)

TCA cycle (see *Krebs cycle*)

T-cell (T-lymphocyte) The type of *lymphocyte* (white blood cell) responsible for initiating the specific *immune response* in vertebrates. T-cells are also essential participants in the *cell-mediated immune response*. They arise from *stem cells* in the *bone marrow* and migrate to the *thymus* (hence the designation "T"), where they develop the ability to detect and recognize invading microorganisms as well as *cancerous* cells. Through *somatic recombination* in the *immune receptor genes* in the *genomes* of individual T-cells, a large library of *T-receptors* are developed. These detect and initiate the destruc-tion of bacteria and infected body cells. This interaction initiates the release of *lymphokines* including *interleukin*-2. Interleukin-2 stimulates T-cells to divide, establishing large *clonal* populations of *antigen* specific T-cells. Specific types of T-cells include *helper T-cells* (T_4) that initiate the immune response, *inducer T-cells* (T_4) that mediate the maturation of T-cells in the thymus, *cytotoxic T-cells* (T_8) that lyse virus-infected cells, and *suppresser T-cells* (T_8), that terminate the immune response. (see *cell-mediated immune response*, K-cell, NK-cell, *hemopoiesis, CD3, CD4, CD8*; compare *phagocyte, B-cell*)

tectorial membrane In the *inner ear* of terrestrial vertebrates, the gelatinous structure into which the *cilia* of the auditory receptors, the *hair cells*, project. They sandwich the hair cells with the *basilar membrane*, forming the *organ of Corti*.

teeth Hard structures growing on the jaws of vertebrates and used for seizing, biting and chewing. In mammals, each tooth consists of *dentine* covered by hard *enamel* and enclosing a pulp cavity. It has a crown above the gum (*gingiva*) and a root embedded in a socket of the jaw bone. (see *denticle*)

tel- A word element derived from Greek meaning "distant" or "end." (for example, *telocentric*)

telencephalon The front and largest part of the *forebrain* in terrestrial vertebrates. It is largely devoted to associative activity and includes the *frontal lobe*. (see *cerebrum, cerebral cortex, basal ganglia, corpus callosum, hippocampus*)

telocentric A *chromosome* having the *centromere* at one end.

telolecithal Some animal eggs, including most fish, amphibians, reptiles, birds, and some other aquatic invertebrates, that exhibit extreme asymmetric yolk distribution. (see *animal pole, vegetal pole*; compare *centrolecithal*).

telomerase The enzyme responsible for maintaining the proper length and structure of *telomeres*, or linear chromosome ends.

telomere The defined end of a *chromosome* containing specific, highly *repeated DNA sequences*. Telomeres participate in chromosome pairing during *meiosis*, and their structure is crucial to normal chromosome behavior over the life of a cell. They solve a functional problem inherent in the replication of linear chromosomes (found in most *eukaryotic* organisms) by turning back on themselves in an unusual hairpin structure. This provides a free 3′ end to finish *replication* of the chromosome end. The length of telomeres is maintained from generation to generation by the enzyme *telomerase*. The aging process of cells may be controlled by a molecular clock consisting of telomeres. Telomeres shorten each time a cell divides, and when they reach a certain critical length, they signal the cell to age, or senesce, thus the age of a cell can be inferred from the lengths of its telomeres. Variations in telomere length, perhaps resulting from mutant telomerase activity,

are also being investigated for their link to cancer.

telophase The final stage in either *meiosis* or *mitosis*. The *nuclear spindle* degenerates and new *nuclear membranes* form. The *chromosomes* become organized into two separate *nuclei* and return to their extended state in which they are no longer visible as distinct entities. (compare *prophase, metaphase, anaphase, interphase*)

telson The hindmost segment of the arthropod abdomen, forming the sting of scorpions and part of the tail fan of lobsters. In insects, it is present only in the *embryo*.

temperate A virus or *bacteriophage* that, after cell penetration, is able to integrate its *nucleic acid* into the *host genome*. In this state (*lysogeny*), most viral genes are repressed, the *host* cell remains intact (not *lysed*), and the viral or phage *genome* (*prophage*) is *replicated* along with the host *DNA*. Certain factors can *induce* the virus or phage to excise itself from the host *chromosome* and resume the infective *lytic* cycle. (see *lysogenic*; compare *virulent*)

temperate deciduous forest The forests of the Northern hemisphere containing mostly *deciduous* trees. They are characterized by cold winters and warm, rainy summers. Precipitation varies (75 to 250 centimeters per year) and is distributed throughout the year, but is locked in ice during the winter. *Perennial* herbs are common. They flower early in the season, before the deciduous trees have regained their leaves. Much of the development of civilization took place within the temperate deciduous forests. They are the remnants of a richer, more continuous forest that once stretched across the northern hemisphere. (see *biome*)

temperate grasslands (see *prairie*)

temperate rainforest A coniferous forest receiving a large amount of rainfall, at least 150 to 450 or more centimeters per year, and located in a temperate climate. Precipitation is concentrated in a wet season that alternates with a dry one. The temperate rainforest is essentially synonymous with *old growth*. Thus, it has a unique ecology attainable only with the complex natural aging process of an *ecosystem*, making its preservation important in preserving *biodiversity*. The main ones occur in the Pacific Northwest, along the coast of Chile, and on the southern coasts of Australia and New Zealand. (see *biome*)

temperate shrub lands (see *chaparral*)

temperate woodlands Woodlands existing in a slightly drier climate than *temperate deciduous forests*. They are dominated by open stands of small, mostly evergreen trees. As with *chaparral*, the vegetation is adapted to and even depends on periodic fires. (see *biome*)

temperature-sensitive mutation A *mutation* that is expressed only at one temperature or temperature range. At other temperatures it appears normal (*wild type*). It is a type of *conditional mutation*.

template A term used in describing the role of a single strand of *DNA* in specifying the synthesis of a complementary strand during *DNA replication* and repair or during *RNA transcription*. (see *complementary base pairing, PCR*)

temporal lobe The side portion of the mammalian *cerebrum*, anatomically separated from the three other lobes by deep grooves. Functionally, it contains the *auditory cortex*. (see *cerebral cortex, telencephalon*; compare *occipital lobe, parietal lobe, frontal lobe*)

tendon The tough nonelastic *connective tissue* that joins muscle to bone. It is composed of *collagen* fibers that are continuous with those of the muscle sheath. When the muscle contracts, the tendon pulls on the bone, causing movement at the joint.

tendril Part of a plant structure that is modified as a threadlike appendage and used as an aid in climbing and support. The concave surface of each tendril is *thigmotropic* (responds to touch). The cells respond by losing water, inducing a bend and resulting in a twining of the tendril around the support.

tentacles Long, slender structures found in many marine invertebrates and cephalopods. In cnidarians (corals, sponges, anemones, jellyfish), they contain *cnidocytes* (stinging cells) and are used in feeding. In cephalopods (octopi, squids), they are used in locomotion and feeding, and are studded with sucker discs.

teratogen Any factor that interferes with the normal development of an *embryo*, primarily by distorting normal growth patterns early in development. Examples included drugs (for example, thalidomide), diseases (for example, measles), and radiation (for example, X-rays). (compare *carcinogen, mutagen*)

terminal bud A bud that gives rise to and lengthens the main stem of a plant. It is located at the tip of the main stem.

terminal chiasmata An important feature in *meiosis I*, the *reduction division*. By *metaphase,* the *chiasmata* (*cross-overs*) have migrated to the ends of the *bivalent* arms and hold the replicated, paired *chromosomes* together while *microtubules* from the *spindle* attach to the *kinetochores*. This ensures that only one face of each *centromere* can be bound so that the *replicated chromatids* of each *homologue* migrate together and each to the opposite pole. In the *nuclear division* of *mitosis* and *meiosis II*, both faces of the centromere are bound, and the chromatids are split, migrating separately to opposite poles.

terminal electron acceptor The atom or molecule that is reduced in the final *redox reaction* of *cellular respiration*. It is always an oxidizing agent. In *aerobic respiration*, it is most commonly molecular oxygen (O_2), which is reduced by a hydrogen atom to form water (H_2O). However in some bacteria, molecular sulfur (S_2) may be reduced to hydrogen sulfide (H_2O) or nitrates (NO_3^-), sulphates (SO_4^{2-}), or even phosphates (PO_4^{3-}) reduced to ammonia (NH_3), hydrogen sulfide (H_2S), and phosphine (PH_3), respectively. In *anaerobic* *respiration,* the terminal electron acceptor is usually a small organic molecule. (see *fermentation*)

terminal transferase An enzyme that catalyzes the addition of *single-stranded nucleotide* tails to the 3′ ends of *DNA* chains. It can generate *sticky ends* on DNA fragments for the purpose of joining them *in vitro* when the use of a *restriction enzyme* is inconvenient or impossible.

termites (see *Isoptera*)

terns (see *Charadriiformes*)

terpenes A group of compounds including *vitamins A, E,* and *K, carotenoids, phytol, gibberellic acid, rubber,* essential oils, and other lipids. They are all based on an isoprene molecule (C_5H_8) and may be linear or cyclic.

terrestrial Any organism that lives on land. (compare *aquatic, marine*)

territory An area or space occupied and defended by an animal or group of animals.

Tertiary The oldest period of the Cenozoic geologic era, about 65 million to 2 million years ago. It is divided into the *Paleocene, Eocene, Oligocene, Miocene,* and *Pliocene* epochs. It is characterized by the emergence of *primates* and eventually *hominoids* and *hominids.*

tertiary structure The third order of structure of a *polypeptide* or *nucleic acid chain,* resulting in a three-dimensional configuration. In proteins, it is a folding or coiling of the molecule. This is primarily determined by hydrophobic interactions and, to a lesser extent, by hydrogen bonding between the *amino acid* side groups. In *nucleic acids,* the *double helix* is organized by *histone* proteins into a supercoiled structure consisting of *nucleosomes.* (compare *primary structure, secondary structure, quaternary structure*)

test The loose-fitting, pore-studded calcium carbonate shells that are characteristic of foram protists. They have contributed to limestone deposits all over the world, the most famous being the

White Cliffs of Dover in Southern England.

testa, testae (pl.) The outer coat of a plant seed. It develops from the *integument* of the *ovule.*

test cross (see *back cross*)

testicle *(see testis)*

testis, testes (pl.) **(testicle)** The reproductive organ in animals that produces male *gametes (spermatozoa).* In vertebrates it also produces male sex *hormones (androgens).* (see *spermatogenesis*)

testosterone The male sex *hormone* in mammals. It is a *steroid* secreted by the *testes.* It stimulates the production of *spermatozoa* and the development of *secondary sexual characteristics.* (see *androgen*)

tetanus The continuous maintenance of muscle force resulting from the *summation* of *motor neuron impulses.* At high stimulation frequencies, the firing of individual *muscle fibers* is no longer evident.

tetra- A word element derived from Greek meaning "four." (for example, *tetrapod*)

tetrad A group of four *spores* produced as a result of *meiosis* of a single parent cell. Tetrads are commonly seen in certain ascomycete fungi such as *yeasts* and in *sporocyte* (spore mother cell) development in nonseed plants.

tetraploid A cell or organism containing four times the *haploid* (one of each) number of *chromosomes.* (see *polyploid*)

tetrapod Any animal belonging to the superclass Tetrapoda.

Tetrapoda A superclass used by some scientists to consolidate all animals with four limbs including those that are vestigial, such as in whales. It includes amphibians *(Amphibia),* reptiles *(Reptilia),* birds *(Aves),* and mammals *(Mammalia).*

thalamus The part of the vertebrate brain that governs the flow of information from all other parts of the *nervous system* to the *cerebral hemispheres.* It is located in the *diencephalon.*

thallus, thalli (pl.) A primitive type of plant body that is not differentiated into stem, leaves, and roots although *analogous* structures may be present. The term is used mainly in reference to nonvascular plants and the *gametophyte* generation of some ferns and *lichens*.

theca, thecae (pl.) (see *test*)

theory Similar to a *hypothesis* but usually wider in scope. Explanatory theories for sets of phenomena are developed by observation and experimentation. They are proved by making predictions and testing their veracity.

Theria. The subclass of mammals containing those that bear live young.

thermo-, -therm, -thermic A word element derived from both Greek and Latin meaning "heat." (for example, *endothermic*)

thermoacidophile (see *Crenarchaeota*)

thermocline The vertical temperature gradient in an ocean or lake. (see *cline*)

thermonastic In plants, a movement in response to a change in temperature, such as the opening of flowers. (see *nastic movement*)

thermoperiodism The response of an organism to periodic changes in temperature. Such responses include determining the time of flowering in certain plants.

thermophilic Microorganisms whose optimum growth temperature is about 60°C, including bacteria that thrive in hot springs, deep vents in the ocean, and compost piles. (compare *mesophilic, psychrophilic*)

thermoregulation The processes by which *homeothermic* animals regulate their body temperatures. Examples include sweating to increase evaporation and lower body temperature, and shivering to raise body temperature.

Thermotogae The phylum of bacteria comprising the *thermophilic fermenters*. This is a new group based on heat resistance and DNA *homologies* in the small (16S) *ribosomal RNA* (*rRNA*) subunit.

therophyte A plant that completes its *life cycle* rapidly during periods when conditions are favorable and survives unfavorable conditions as a seed (that is *annual*

and *ephemeral* plants). Therophytes are typical of desert environments. (see *Raunkiaer's plant classification*)

thiamine (vitamin B$_1$) A member of the *vitamin B complex*. It is found in unrefined cereal grains, liver, heart, and kidney. Lack of thiamine results in the disease beriberi. (see *Krebs cycle*)

thick filament The *myosin* filament in the *actomyosin* complex in vertebrate muscle. Myosin *monomers* are arranged in an overlapping fashion into long filaments with the head portion of the molecule protruding so that it can interact with the *thin filaments* of actin during *muscle contraction*.

thigmotropism (haptotropism) A *tropism* (directional response) in which the stimulus is touch. It is responsible for the twining of tendrils in many climbing plants. (see *tropism, nutation*)

thin filament The *actin* filament in the *actomyosin* complex in vertebrate muscle. Actin *monomers* are arranged into long twisted filaments that interact with the *thick filaments* of *myosin* during *muscle contraction*.

thin layer chromatography A laboratory technique in which soluble materials are separated by differential migration in a particular solvent through a thin layer of a porous *polymer* such as silica gel or aluminum oxide. The chromatography material is usually supported on a glass plate, and separated materials may be further tested using specific enzymes or *antibodies*. (compare *electrophoresis*)

Thiopneutes (see *Proteobacteria*)

third eye (see *pineal gland*)

thoracic cavity The body cavity in terrestrial vertebrates that contains the lungs and heart.

thoracic vertebrae The *vertebrae* of the upper back region.

thorax

1. In vertebrates, the portion of the body containing the heart and lungs.
2. In arthropods, the fused, leg-bearing segments between the head and the abdomen.

thorn In *vascular plants*, a stiff, sharp process found on stems. One function is to deter *herbivores*.

thread cell (see *cnidoblast*)

threatened species (vulnerable species) A species that is experiencing a decline in population but is not yet considered *endangered*. (compare *rare species*)

three prime (3′) The end of a strand of *nucleic acid* (*DNA*, *RNA*) at which a hydroxyl group (OH⁻) is attached to the 3′ carbon of the sugar residue. (compare *five prime* [5′])

threonine (thr) An *amino acid*.

threshold potential In a *neuron*, the minimum change in *membrane potential* (the charge difference across the *cell membrane*) necessary to produce an *action potential*. (see *equilibrium potential, resting potential*)

threshold value The minimum stimulus intensity necessary to provoke a response in a nerve or muscle cell. (see *action potential*)

thrips (see *Thysanoptera*)

thrombin The enzyme that converts the soluble protein *fibrinogen* into *fibrin* fibers during *blood clotting*. It is itself derived from *prothrombin* by the action of *thrombokinase*.

thrombocyte (see *platelet*)

thrombokinase A *lipoprotein* liberated from *platelets* and damaged tissues that initiates the cascade of reactions resulting in *blood clotting*.

thromboplastin (see *thrombokinase*)

thrushes (see *Passiformes*)

thylakoid A membranous disc containing *chlorophylls, carotenoids, and their associated proteins*. It is the basic *photosynthetic* unit in plant and algal *chloroplasts* and in cyanobacteria. In chloroplasts, they are organized into stacks called *grana* that are situated along the inner of the two membranes.

thymidine A *nucleoside* of *thymine* linked to D-ribose with a β-glycosidic bond. Thymidine triphosphate (*TTP*) is derived from thymidine.

thymine A nitrogenous base found in *nucleic acids* and *nucleosides* It is based on a *pyrimidine* ring structure.

thymine dimer Two adjacent *thymine* bases in a *DNA* molecule that have become joined by a covalent bond. This can occur by the absorbtion of UV radiation. If left unrepaired, this *mutation* would block *DNA replication*, proving lethal to that cell. Normally, *DNA repair* mechanisms excise and replace the thymine dimer. These processes may also be error-prone and lead to further *mutation* at that site. (see *pyrimidine dimer*)

thymus gland The *lymphoid gland* in which *T-cells* mature and that coordinates immunological functions in the body. Its size and relative function decrease after *sexual* maturity.

thyroid gland The *endocrine gland* that is the main regulator of *metabolic* rate in the body. It produces two *hormones, thyroxine* and *calcitonin.*

thyrotropin (thyrotrophin, thyroid-stimulating hormone) A *hormone* produced by the *pituitary gland* that stimulates the *thyroid gland* to release *thyroxine*. The level of thyroxine controls thyrotrophin release by a *negative feedback* mechanism.

thyrotropin-releasing hormone (see *TRH*)

thyroxine (thyroid hormone) An iodine-containing *peptide hormone* secreted by the *thyroid gland*. It stimulates *metabolic* rate and is essential for normal growth and development.

Thysanoptera An order of minute *exopterygote* insects, the thrips. They have sucking mouthparts. Most are *herbivorous* and many are agricultural pests.

tibia The larger of the two long bones in the lower hind limb of tetrapods. In man, it forms the shin bone.

ticks (see *Acari*)

tight junction (zona occludens) A type of *intercellular junction* in which the membranes of two adjacent animal cells are fused. It prevents materials from leaking through the tissue, for example in the intestinal *epithelium*. (see *intercellular junction*)

Tinamiformes The order of birds containing the flightless tinamous.

tinamous (*see Tinamiformes*)

tissue A group of similar cells that are organized into a structural and functional unit. (compare *organ*)

tissue culture The growth of cells, tissue, or organs outside of the body (*in vitro*).

titer A measure of the concentration of biologically active entities in a solution. It is commonly applied to *antibodies* in *serum* as well as *microorganisms* in laboratory culture.

T-lymphocyte (see *T-cell*)

toads (see *Anura*)

tocopherol (see *vitamin E*)

ton-, -tonic A word element derived from Greek meaning "stretched." (for example, *isotonic*)

tongue The organ of taste in terrestrial vertebrates. It is situated in the mouth (*buccal* cavity) and is covered with *papillae* containing *taste buds*. It is also used to facilitate swallowing (its only function in fish) and in amphibians, to catch insects.

tonsils Pairs of *lymphoid* organs in vertebrates that are sites of *lymphocyte* production, thus involved in bodily defense and the *immune response*.

tonus The state of continuous partial muscle contraction that enables an animal to maintain its posture.

toothed whales (see *Odontoceti*)

top carnivore An animal at the highest trophic level of the *food web*. They are *secondary consumers* and feed off other *carnivores*.

tornaria The free-floating *planktonic larvae* of acorn worms (Hemichordata). Their *ciliated, radially symmetric* form resembles that of sea stars and suggests an evolutionary relationship.

torsion A twisting and development of asymmetry of the body of gastropods during *embryological* development. It is due to more rapid growth on one side and is independent of any twisting of the shell.

tortoises (see *Cheloni*)

totipotent The ability of a cell to differentiate into any of the different types of tissue that make up an organism. It reflects the fact that each cell contains the full genetic potential of the species. Initially, all cells are totipotent but, particularly in animals, become restricted by their developmental environment. Plant cells may sometimes be restored to totipotency by altering their environment and providing the appropriate biological factors. Restrictions on the expression of the *genome* of most animal cells are less easily removed. (see *regulation, differentiation, determination*)

tox-, -toxic A word element derived from Greek meaning "poison." (for example, *cytotoxic*)

toxin Any substance produced by an organism that is injurious to another. Toxins are commonly produced by *pathogenic* microorganisms such as bacteria and fungi and are responsible for the symptoms of many diseases. They may also be produced by plants and animals and are often extremely potent. They usually have very specific modes of action that include *competitive inhibition* of host enzymes and the *blockage of ion channels*.

trace element A chemical element required in minute amounts by a biological organism for health, growth, or survival. Trace elements include boron, cobalt, copper, fluorine, magnesium, manganese, zinc, iron, molybdenum, iodine, and selenium.

trachea

1. In terrestrial animals, a tube through which oxygen enters the body. In vertebrates, it leads from the throat to the *bronchi*. Its walls are stiffened with incomplete rings of *cartilage*, preventing collapse yet retaining flexibility. In terrestrial arthropods, multiple tracheae lead from *spiracles* (pores) on the body, dividing into finer and finer branches (*tracheoles*). These penetrate muscles and organs, where oxygen is distributed by diffusion.

2. In *vascular plants*, a tubelike series of dead cells (*tracheids* or *vessel elements*) through which water and salts are transported and that also provides mechanical support.

tracheary element (see *tracheid*)

tracheid (tracheary element) In plants, an elongated dead cell found in the *xylem* of some *vascular plants*. The walls are fortified with bands of *lignin*. A column of such cells forms a *trachea* through which water and dissolved salts are transported. Fluid moves from cell to cell through paired pits in the *cell walls* of adjacent tracheids. (compare *vessel element*)

tracheophyte (vascular plant) Any plant containing a *vascular system*, which transports water and nutrients throughout the plant and also provides mechanical support. The system is made of differentiated tissues forming the *xylem*, which is responsible primarily for water (and dissolved solute) transport, and the *phloem*, which transports carbohydrates. Vascular plants are able to colonize drier habitats inaccessible to more primitive plants. (see *vascular bundle, vascular tissue, vascular system*)

trama The sterile *gill* tissue in *basidiomycete* fungi.

trans-acting A factor, usually a soluble protein, produced at a *gene* distant from the *chromosome* region that it affects. (compare *cis-acting*)

transamination The enzymatic transfer of an amino group from an *amino acid* to a keto acid, producing in the process a new amino acid and a new keto acid. It is catalyzed by a *transaminase*.

transcript An *RNA* molecule that is synthesized using a *DNA template*. (see *primary transcript*)

transcription The process by which the genetic information in *DNA* is converted into *RNA*. In *eukaryotic* cells, the process is carried out in the *nucleus* and produces a complementary copy of all the information encoded on a particular stretch of *chromosome*. This molecule (*hnRNA*) is then processed (*spliced*) to remove all non-protein-coding portions (*introns*), and the protein-coding portions (*exons*) are rejoined into an *mRNA* molecule. The mRNA molecule then exits the nucleus to be *translated* into protein. In *prokaryotes*, which contain no introns in their DNA and in which no nucleus is present, transcription and translation are directly coupled.

transcription factors A class of *DNA*-binding proteins that regulate *RNA* transcription in *eukaryotes*. (see *enhancer, steroid receptor*)

transcript localization The use of *cDNA complementary* to an *mRNA* species of interest as an *in situ probe* to localize the time and location of *gene expression* in development.

transduction The transfer of genetic infor-

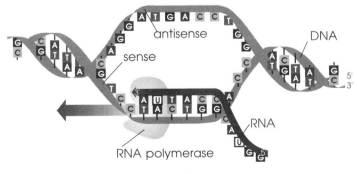

Transcription

mation from one bacterium to another by a *bacteriophage*. It occurs when a bacteriophage *integrates* into the bacterial genome (*lysogeny*) then *excises*, sometimes carrying with it a piece of the host *genome*. When it infects another cell, the foreign genes are carried along. (compare *conjugation*)

transect A measured area in the shape of a line or belt that is used as a sample area in a biological survey. Transects are most applicable to surveys of plants, or *sessile* or sedentary animals. (compare *quadrat*)

transfection The uptake of *exogenous DNA* fragments in solution directly into animal cells in laboratory culture. It is one way of introducing foreign *genes*. It has been used as a technique to determine the *cancer*-causing potential of different pieces of *nuclear* DNA isolated from *tumor* cells. (compare *transformation*)

transferase Any enzyme that catalyzes biochemical reactions in which entire chemical groups (functional groups) are transferred from one molecule to another. *Transamination* (catalyzed by a *transaminase*) is an example.

transfer cell A specialized type of cell found in some vascular plants that plays a role in the transfer of solutes over relatively short distances. They posses a convoluted inner *cell wall*, increasing the absorptive surface area, and many *plasmodesmata* (*cytoplasmic* bridges), facilitating transfer between adjacent cells.

transfer RNA (see *tRNA*)

transformation

1. The uptake and incorporation of *DNA* fragments or *plasmids* by a cell. It was first observed in pneumococcal bacteria by Oswald Avery in 1944, implicating DNA as the genetic material. It is now a common laboratory technique used in *genetic engineering*. (see *gene cloning, competent bacteria*)

2. The conversion of normal cells growing in laboratory culture into cells with *cancerous* properties, such as de*differentiation* and uncontrolled growth. Transformation is often induced by viral infection or *transfection*.

transgenic Any organism whose *genome* has been altered by the incorporation of foreign genetic material. *DNA* is introduced early in development into the cells of the organism by *microinjection* or *transformation*. Organisms that permanently incorporate the foreign genes into their own genome become transgenic.

transition A type of *DNA mutation* in which a *base pair* is replaced with one of like structure, **GC** to **AT** for instance. *Adenine (A)* and *guanine (G)* are both *purines. Cytosine (C)* and *thymine (T)* are both *pyrimidines*. (see *base pair substitution*)

translation (protein synthesis) The process whereby *polypeptide chains* are synthesized on *ribosomes* in the *cytoplasm* of a cell. An *mRNA* (messenger RNA) molecule, which has previously been *transcribed* from *DNA*, dictates the sequence of *amino acids* to be added. Usually, a number of ribosomes are attached along the length of an mRNA, each in a different stage of synthesis. The whole complex is called a *polysome*. As each ribosome moves along the mRNA, *tRNA* (transfer RNA) molecules, each bearing a specific amino acid on one end and an *anticodon* (complementary *nucleotide triplet*) on the other, match up to the series of nucleotide triplets specified by the mRNA. In this way, the next amino acid in the sequence is incorporated. The amino acids are attached by *peptide bonds*. (see *genetic code*)

translocation

1. In plants, the movement of carbohydrates by *mass flow*, usually from a site of uptake or synthesis to sites of growth or storage. (see *phloem*)

2. A *chromosomal mutation* in which a fragment is inserted in a foreign location in the *genome*. Translocations are most commonly *reciprocal*, resulting from the exchange of parts between two non*homologous* chromosomes but may occasionally be non*reciprocal*. Removal of genes from their normal

genetic environment often results in inappropriate *gene expression* and the manifestation of physical defects. Translocations often produce at least partial sterility, as *homologous* chromosomes fail to pair correctly at *meiosis*.

transmembrane carrier protein A protein complex extending through the *plasma membrane* of a cell that transports large uncharged polar molecules, such as sugars, via a flip-flop mechanism. At no time is a continuous pore open through the plasma membrane. (see *facilitated diffusion*)

transmembrane protein channel (protein channel) A protein extending through the *plasma membrane* of a cell that forms a continuous tunnel allowing passive transport of small charged molecules. (see *channel*)

transmembrane proteins A collection of proteins that are suspended in the *phospholipid bilayer* that forms the *plasma membrane* of a cell. They provide a means through which polar, ionic, or large molecules and other information pass from the exterior environment to the interior of the cell. These proteins are oriented so that their nonpolar portions (*hydrophobic*) are oriented facing the lipid bilayer of the membrane and their polar (*hydrophilic*) portions forming internal channels through which other hydrophilic molecules may enter or exit. (see *transmembrane carrier protein, membrane channel, transmembrane receptor, cell surface marker, membrane pump, second messenger*)

transmembrane receptor (cell surface receptor) Any protein that extends across the *plasma membrane* and whose protruding end is able to bind specific *hormones* or other signal molecules on the exterior surface of the cell. In many cases, the binding alters the shape of the interior end of the receptor protein, inducing an enzyme cascade or opening an *ion channel*. All known *neurotransmitters*, protein *hormones,* and *growth fac-*

tors mediate their effect in this fashion. (see *second messenger, cAMP, adenylate cyclase, B-receptor, T-receptor*)

transmission electron microscope (see *electron microscope*)

transpiration The loss of water vapor from a plant surface. It is the main route of moisture loss and takes place, concomitant with gaseous exchange, through *stomata* in the leaves and, to a lesser degree, through *lenticels* in the stems.

transplant A *tissue* or *organ* that has been transferred from one individual to another or one part of an animal to another part. (see *transplantation, MHC complex, MHC proteins, HLA complex*; compare *graft*)

transplantation The transfer of a *tissue* or *organ* from one individual to another or one part of an animal to another part. Transplantation may occasionally involve trans-species transfers. (see *transplant, MHC complex, MHC proteins, HLA complex*)

transplantation antigens (see *MHC proteins*)

transposable genetic element (jumping gene) Any of the numerous types of mobile pieces of *DNA*, which are found in both *prokaryotes* and *eukaryotes*. A transposable genetic element, once incorporated into the *genome*, may at any time *excise* and re*integrate* at a new random location, usually by the action of a specific enzyme. Most organisms contain their own characteristic elements and matching enzymes. Mobile elements that have incorporated *genes* (*transposons*) have been exploited in the laboratory to study their response to different genetic environments. (see *retrovirus, retroposon, resistance transfer factor, IS element, TY element, copia element, FB element, Alu sequences, SINEs, LINEs*)

transposition The one-way transfer of a *gene* to a random location on a *chromosome*. Genes move in this fashion when they are associated with a *transposable genetic element* or *retrovirus*.

transposon A large class of *transposable*

genetic elements found in virtually all organisms. They tend to be characteristic of an organism. They typically include a specific expressed *gene* or other genetic element along with the minimal *nucleotide* sequences needed for *transposition*. They are often flanked by direct or *inverted repeat* sequences that are an integral part of the transposition mechanism. (see *retroposon, resistance transfer factor, IS element, TY element, copia element, FB element, Alu sequences, SINEs, LINEs*)

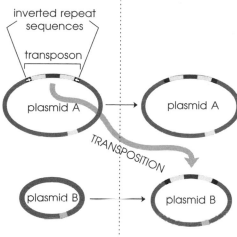

Transposition of a Bacterial Transposon

transversion A type of *DNA mutation* in which one *base* of a *base pair* is replaced with one of differing structure, for instance a **GC** to **TA**. In other words, a *purine* (*adenine* [**A**] or *guanine* [**G**]) is substituted for a *pyrimidine* (*cytosine* [**C**] or *thymine* [**T**]) and *vice versa*.

T-receptor A protein found exclusively on the surface of *T-cells*. T-receptors perform the complicated process of separating body cells carrying *MHC* markers, which identify them as "self," from "nonself" entities or infected body cells presenting foreign *antigens*. The structure of a T-receptor resembles one arm of a *B-receptor* or *antibody*, and the two component chains, α and β, resemble the *light* and *heavy chains*. The gross protein

structure is also similar, exhibiting *variable, hypervariable,* and *constant regions* that are formed by *somatic rearrangement* in the *immune receptor genes*. (see *immune system, immune response*; compare *MHC proteins*)

tree ring (see *annual ring*)

tree shrews (see *scandentia*)

Trematoda A class of *parasitic* flatworms, the flukes, including the human blood fluke *Schistosoma* and the human liver fluke *Clonorchis*. They have a thick *cuticle* to prevent digestion by the *host* and suckers, anchors, or hooks for attachment. Although they are *hermaphroditic* (contain both male and female reproductive organs), *cross-fertilization* is usual. They often have complex *life cycle*s with more than one host. The intermediate host is frequently a snail and the final host a vertebrate. (see *cercaria, metacercaria, miracidium, redia, sporocyst*)

TRH (thyrotropin-releasing hormone) A *peptide hormone* secreted by the *hypothalamus* that triggers the release of *thyrotropin* from the *anterior pituitary*.

tri- A word element derived from both Latin and Greek meaning "three." (for example, *triploblastic*)

Triassic The oldest period of the Mesozoic geological era, about 230 million to 195 million years ago. It is marked by a decrease in the number and variety of *cartilaginous* fish and an increase in primitive amphibians and reptiles.

tribe In plant classification, a group of closely related genera. Tribes are usually only employed in very large families such as grasses. A suffix of -*eae* is usually attached. Tribes may be grouped to form subfamilies.

tricarboxylic acid cycle (see *Krebs cycle*)

trichogyne In ascomycete fungi, an outgrowth on the female *ascogonia* that fuses with the *antheridium* so that male *nuclei* of the opposite *mating type* can migrate through. This gives rise to *heterokaryotic hyphae* in which some of the *haploid* nuclei will eventually fuse to form *zygotes*.

trichome In plants, any of a variety of small outgrowths of the *epidermis* (outer layer). They result in a woolly or fuzzy leaf appearance and play a role in maintaining heat and water balance. They may also excrete sticky or *toxic* substances for the purpose of detering *herbivores*.

Trichoptera An order of *endopterygote* insects including the caddis flies and sedge flies. They are aquatic insects that spend most of their lives as *larvae* inhabiting characteristic casings that they construct from mineral grains or plant material. Most adults do not feed, their *mandibles* being reduced or absent. Those that do feed on nectar. The wings are covered with bristles or hairs rather than scales. They are widespread, and their absence is considered an important indicator of polluted waters.

tricolpate pollen *Pollen grains* in which three furrows or pores are characteristic. They are found particularly in *dicots*.

tricuspid valve A valve in the mammalian and *avian* heart that prevents the backflow of blood into the right *atrium* when the *right ventricle* contracts.

trigeminal nerves (cranial nerves V) A pair of nerves that supply the vertebrate mouth and jaws. (see *cranial nerves*)

triglyceride The principal lipid in mammalian blood consisting of a *fatty acid* and a glycerol group. Triglycerides normally function as building blocks in *cell membranes*, as a transport mechanism for nonpolar material, and as a food reserve. The amount and proportion of different types of triglycerides in the bloodstream may be diagnostic in heart disease and *diabetes*.

tripeptide A *peptide* consisting of three *amino acids*.

triploblastic Animals with a body organization derived from three *embryonic* germ layers: *endoderm, mesoderm,* and *ectoderm*. Most animals are triploblastic with the exception of sponges, cnidarians, and ctenophores. (compare *diploblastic*)

triploid A cell or organism containing three times the *haploid* (one of each) number of *chromosomes*. Triploid animals are usually sterile because one set of chromosomes remains unpaired at *meiosis*, which disrupts *gamete* formation. In flowering plants (*angiosperms*), the *endosperm* (nutritive) tissue is usually triploid. (see *polyploid, double fertilization*)

trisomic A *diploid* organism that contains three copies of one *chromosome*. Although the situation may lead to a viable organism, physical defects are usually manifest. Down's syndrome in man is caused by a trisomy in all or part of chromosome 21.

tRNA (transfer RNA) A class of small *RNA* molecules that transfer individual *amino acids* to a growing *polypeptide* chain during *translation* (protein synthesis). Each tRNA molecule has two functional sites, one end bears a specific amino acid while the other carries the *nucleotide triplet* (*anticodon*) that specifies that amino acid. During translation as each *ribosome* moves along the *mRNA* (messenger RNA), *tRNA* molecules sequentially match up to

Structure of a tRNA

the series of nucleotide triplets specified by the mRNA, thus incorporating the next amino acid in the sequence. (see *genetic code*)

trochlear nerves (cranial nerves IV) A pair of nerves that supply the main muscle of the vertebrate eye. They contain mostly motor *nerve fibers*. (see *cranial nerves*)

trochocyst An *organelle* underlying the surface of many cilate protoctists and some mastigote protoctists that is discharged to sting prey.

trochophore A *planktonic, ciliated larva* characteristic of some mollusks, annelids, and various other *marine* invertebrates. (see *veliger*)

trophic A description related to feeding. It often refers to a feeding form in the *life history* of an organism.

trophozoite A growing, feeding stage in the life history of *parasitic* protists.

troglobite An obligate cave dweller. These organisms are characterized by a reduction in eye structure and a lack of *photosynthetic* or *visual pigments*.

troglophile A nonobligate cave dwelling organism.

Trogoniformes An order of birds containing the trogons and quetzals.

trogons (see *Trogoniformes*)

troph-, -trophic, -trophism A word element derived from Greek meaning "feed." (for example, *heterotroph*)

trophic level One of a succession of steps in the movement of energy and matter through a *food chain* in an *ecosystem*. Organisms are considered to occupy the same trophic level when the matter and energy they contain have passed through the same number of steps since their entrance by way of *photosynthesis* or *chemosynthesis*.

trophic level efficiency The ratio at which stored energy from one *trophic level* in a *food chain* is converted to usable energy in the next. Usually about 95% to 99% of the stored energy is lost to the external environment in such conversions. (see *pyramid of energy*)

trophism The method an organism uses to obtain nutrients. The two basic ways are *autotrophism* and *heterotrophism*. (see *photoautotrophism, chemoautotrophism, photoheterotrophism, chemoheterotrophism*)

trophoblast In the early vertebrate *embryo*, the outer *ectodermal* cell layer of the *blastocyst*. In mammals, it attaches to the uterus and forms the *placenta*. In reptiles, it forms the membrane that functions as a watertight outer covering.

trop-, tropism, -tropic A word element derived from Greek meaning "turn." (for example, *phototropism*)

tropical rainforest An evergreen forest receiving a large amount of rainfall, at least 200 to 450 or more centimeters per year, and located in a tropical climate. True rainforests, those that receive rain regardless of season, are concentrated in a few equatorial regions. In most others, precipitation is concentrated in a wet season that alternates with a dry one. The tropical rainforest is the *biome* richest in terms of *biodiversity* (the number of different species), containing perhaps half the world's species of terrestrial organisms. Because they occur on relatively infertile soils, and most of the nutrients are held in the plant material itself, the conversion of rainforests to agricultural land is unsuccessful in the long-term, and recovery of the former complex *ecology* is virtually impossible.

tropicbirds (see *Pelecaniformes*)

tropism A directional growth movement of part of a plant in response to an external stimulus. Tropisms are named according to stimulus (for example, *phototropism, gravitropism*) and may be negative or positive. (compare *nastic movements, taxis*)

tropomyosin An accessory protein in *striated muscle*. It forms filaments that wrap spirally around the *thin filaments*. When the muscle is at rest, it blocks the *myosin*-binding sites on the *actin* molecules. During *muscle contraction*, it is displaced by a complex of calcium ions

(Ca^{++}) with another protein, *troponin*, allowing the myosin heads to contact the actin filament.

troponin An accessory protein in *striated muscle*. It is positioned periodically along *thin filaments* and binds to both the *actin* molecules and to *tropomyosin*. At rest, it permits tropomyosin to block the *mysoin*-binding sites on the *thin filaments*. During *muscle contraction*, calcium ions (Ca^{++}) enter the *sarcoplasm*, forming a complex with troponin that displaces the tropomyosin filaments, allowing the myosin heads to contact the actin filament.

true bug (see *Hemiptera*)

trypsin A *protease* enzyme that catalyzes the partial *hydrolysis* of proteins to *polypeptides*. It is specific for the bond between lysine and arginine residues. Trypsin is secreted into the small intestine by the pancreas in the inactive form *trypsinogen* and is activated by *enteropeptidase*.

trypsinogen The inactive form of the enzyme *trypsin*.

tryptophan (trp) An *amino acid*.

tube feet The means by which echinoderms move and attach themselves to a substrate. In some, they are located near the mouth and are specialized for feeding. They are part of the *water vascular system*. (see *ampulla, radial canals*)

tuber An enlarged stem or root that functions as an underground storage organ and from which new plants may be propagated vegetatively. It is essentially the swollen end of a *rhizome*. Stem tubers (for example, potato) often produce buds from which new stems arise the following season. Root tubers may produce buds only at the junction with the stem. (see *vegetative propagation, perennation, perennating structure*; compare *bulb, corm, offset, rhizome, stolon, sucker*)

tube worms (see *Pogonophora*)

Tubulidentata An order of placental mammal containing only one species, *Orycteropus afer*, the aardvark or ant bear.

tubulin A *globular protein* that *polymerizes* to form long, hollow cylinders of *microtubules* that are essential to cellular and *intracellular* movement. (see *protofilament, cilia, flagella, undulipodia*)

tumor An overgrown group of cells caused by uncontrolled *cell division*. (see *cancer, oncogene*)

tumor virus A virus that can transform normal cells in certain animals into *cancerous* cells. They usually carry a single *oncogene*, derived by mutation from a normal cellular gene present in the virus's *host* species.

tundra The regions farthest north, just before permanent ice. The Arctic tundra covers a fifth of the earth's land surface and is dominated by scattered patches of grasses, heathers, and lichens, with small trees at the margins of streams and lakes. Precipitation is low, less than 25 centimeters per year, but the water is often trapped near the surface by the widespread permafrost, creating boggy conditions. Perennial herbs grow rapidly during the brief summers, and large grazing animals and *carnivores* are characteristic.

tunic The characteristic *cellulose* sac that surrounds tunicates and gives the class its name.

Tunicata (see *Urochordata*)

tunicates (see *Urochordata*)

turacos (see *Muscophagiformes*)

Turbellaria A small class of flatworms, the only one of the four classes that is free living. They are mostly aquatic, but some may inhabit moist places on land. They have a flat body covered with *cilia* and often have *tentacles* and eyes on the head. An example is the common planaria used for laboratory exercises.

turgor pressure The pressure exerted on the inside of a plant *cell wall* by the fluid contents of the cell. The interior of the cell is *hypertonic* in relation to the *extracellular* environment, so it tends toward the uptake of water by *osmosis*. The cell wall is strong enough to contain the excess pressure and to prevent water from entering to the point of bursting. Turgidity is the main source of mechanical support in *herbaceous* plants. (compare *wall pressure*)

turion An enlarged, detached winter bud that contains food reserves and is protected by an outer layer of leaf scales and *mucilage*. It is characteristic of various aquatic plants and functions in *perennation* and *vegetative propagation*.

turkeys (see *Galliformes*)

turtles (see *Cheloni*)

twisted-winged parasites (see *Strepsiptera*)

two-dimensional electrophoresis A laboratory technique in which *macromolecules* are separated through a porous solid medium (an *agarose* or *polyacrylamide* gel) in an electric current, then the gel turned 90° and run again. Proteins digested with certain *proteolytic* enzymes will produce a distinctive "fingerprint" when analyzed in this fashion, and various forms of *DNA* can be further separated. (see *electrophoresis*)

TY elements A family of *mobile genetic elements* found in yeast. They are about six *kb* (kilobases) long and are able to mediate their own *transposition* to a distant location in the *genome* through an *RNA* intermediate. (see *transposon*; compare *P element, copia-like element, FB element, LINEs*)

Tylopoda A suborder of even-toed ungulates including the camel and relatives.

tympanic cavity (see *middle ear*)

tympanic membrane (tympanum, ear drum)
1. The thin membrane that separates the outer from the *middle ear* of vertebrates. Incoming sound waves cause vibrations of the membrane that are then transmitted to the *inner ear* via the *ear ossicles*.
2. In insects, a thin membrane associated with the *tracheal* air sacs that is used to detect sound.

tympanum (see *tympanic membrane*)

tyrosine (tyr) An *amino acid*.

tyrosine kinases (see *protein kinases*)

UAS (upstream activating sequence) A *regulatory sequence* on the *chromosomes* of yeast cells that serves to maximize levels of *mRNA transcription* from a particular *promoter*. They are analogous to *enhancers* in mammalian cells. (see *CCAAT box, GC box*)

ubiquitin A highly *conserved protein* present in all *eukaryotic* cells. It binds to a variety of *nuclear* and *cytoplasmic* proteins, and is used to tag proteins for selective degradation.

ubiquitous species A species that is found almost anywhere on the earth. Humans are an example as is the bacterium *E. coli*.

ulna One of the two long bones of the lower forelimb in tetrapods. In man, it is the bone in the back of the forearm.

ultracentrifuge A laboratory apparatus used to separate *macromolecules* such as proteins and *nucleic acids* suspended in solution. This is accomplished by spinning tubes in a rotor at the appropriate speed, angle, and length of time for a specific desired separation. The material spun to the bottom of the tube or suspended at some point within the tube is then collected. It differs from a conventional low to medium speed centrifuge chiefly in the speeds that are used. Rotors and other components may be made from titanium, which can withstand the high G-forces applied. The process is carried out under vacuum. (see *centrifuge*)

ultrastructure The detailed structure of biological material as revealed at the level of *electron microscopy*.

ultraviolet radiation (se *UV radiation*)

umbel An *inflorescence* (multiple flowered structure) in which individually stalked flowers arise from the same point on the stem, usually the apex, giving the appearance of an umbrella.

umbilical cord In placental mammals, the cord of tissue connecting the *embryo* with the mother. It contains two arteries and one vein through which blood containing nutrients, wastes, and other small molecules is cycled between the mother's circulation and the embryo's.

unassigned reading frame (see *URF*)

undulipodium, undulipodia (pl.) The *eukaryotic flagellum*. It has a characteristic structure, the *axoneme*, consisting of nine pairs of *microtubules* surrounding an inner pair. It is anchored to the cell by a *basal body* (*kinetosome*). It has been proposed by Lynn Margulis that the eukaryotic flagellum be renamed an *undulipodium*, a nomenclature that eliminates confusion. *Cilia* are shortened undulipodia.

unequal crossover A *reciprocal recombination* event between the *chromatids* of *chromosome homologues* at *meiosis* in which the chromosomes are not perfectly aligned. Thus, a *duplication* of the region is produced on one chromosome and a *deletion* on the other.

unequal sister chromatid exchange (see *sister chromatid exchange*)

uneven-aged stands A forest area with at least three distinct age classes. It is typical of *virgin* or *old growth forests*. (compare *even-aged stands*)

ungulate An informal term for a hoofed, grazing animal. (see *Artiodactyla, Perissodactyla*)

unguligrade The ambulatory mode of some mammals in which only the tips of the *digits* are in contact with the ground. It is typical of hoofed mammals (*ungulates*) such as the horse. (see *pentadactyl limb*; compare *digitigrade, plantigrade*)

unicellular Any organism consisting of only one cell. It is characteristic of all *prokaryotes* and primitive *eukaryotes* including some protoctists (the protists) and one group of fungi, the yeasts. (compare *multicellular*)

uniformitarianism The principle first suggested in 1785 by James Hutton that the physical and biological processes presently active in forming and modifying the Earth can help explain its geologic and evolutionary history.

uninucleate The presence of only a single *nucleus* in a cell.

uniparental inheritance The transmission of certain traits from only one parent to all the progeny. It is seen in the *genes* carried by cellular *organelles* such as *mitochondria* and *chloroplasts*.

unique sequence A specific *DNA* sequence in an organism's *genome* that occurs only once. It is characteristic of most *structural* and *regulatory genes*, that is those that specify the functioning of the organism. (compare *repeated sequence*)

uniquinol (see QH_2)

uniquinone (see Q)

Uniramia (see *Mandibulata*)

uniramous appendage The single-branched appendages of terrestrial arthropods, including insects, centipedes and millipedes. (compare *biramous appendage*)

unisexual An organism bearing only male or female reproductive organs. It is the most common situation in higher animals. (compare *hermaphrodite*)

universal indicator (see *indicator*)

univoltine An organism that reproduces once every year.

unsaturated fatty acid A *fatty acid* in which hydrogen atoms have not filled all possible slots, thus leaving them potentially reactive with other compounds. *Monounsaturated* compounds (for example, in olive oil) have one slot left unoccupied; *polyunsaturated* compounds (for example, in safflower oil) have several free slots. At room temperature, fats consisting of unsaturated fatty acids tend to remain liquid. Considerable controversy remains over the relative health benefits of various types of unsaturated oils. (compare *saturated fatty acid*)

upstream The direction opposite to that of *mRNA transcription* in a particular region of a *chromosome*. It is used in reference to particular sequence elements such as *promoters*, *coding sequences*, and *regulatory sequences*. (compare *downstream*)

upstream activating sequence (see *UAS*)

upwellings The upward flow of waters that brings nutrients to the surface, allowing the growth of algae and, subsequently, animals that depend on the algae. They are found near the Arctic and Antarctic ice sheets and under certain conditions along the shores of continents. They take place on a smaller scale in ponds and lakes. Upwellings are an integral part of the seasonal *life cycle* typical of any particular area in which they occur. (see *spring overturn, fall overturn*)

uracil A nitrogenous base found in *RNA* in positions analogous to the *thymine* in *DNA*. It is based on a *pyrimidine* ring structure.

urea The main excretory product of amphibians and mammals. It is a *nitrogenous* compound that results from the *catabolism* of the *amino acids* in proteins. (see *urea cycle, ureotelic*; compare *uric acid*)

urea cycle A cyclic series of reactions in which the ammonia derived from *amino acid catabolism* is combined with carbon dioxide to produce *urea* for excretion. Part of the cycle takes place in the *mitochondria*, which involves a specific membrane transport system for an intermediate compound, *ornithine*.

ureotelic Animals that excrete excess nitrogen in the form of *urea*. This includes amphibians and mammals. (compare *uricotelic*)

ureters A pair of ducts in reptiles, birds, and mammals that transports *urine* from the kidneys to the *cloaca* (reptiles, birds) or urinary bladder (mammals). (compare *Wolffian duct*)

urethra A duct in mammals that conveys *urine* from the urinary bladder to the exterior. In males, it passes through the penis and also transports *spermatozoa*.

URF (unassigned reading frame) A region in a *cloned gene* that apparently codes for a protein (open reading frame [*ORF*]) but whose function has not yet been determined.

uric acid The main excretory product of reptiles and birds. It is a nitrogenous compound that results from the *catabolism* of the *amino acids* in proteins. (see *uricotelic*; compare *urea*)

uricotelic Animals that excrete excess nitrogen in the form of *uric acid* This includes reptiles and birds. (compare *ureotelic*)

uridine A *nucleoside* of *uracil* linked to D-ribose with a β-glycosidic bond. Uracil triphosphate (*UTP*) is a *nucleotide* derived from uridine.

urinary bladder (see *bladder*)

urinary system The body system in vertebrates that removes soluble *metabolic* wastes from the bloodstream and excretes them to the exterior. It includes the kidneys, bladder, and associated ducts. (see *excretion, ureotelic, uricotelic*)

urine The liquid waste product excreted through the *urethra* or *cloaca* of vertebrates. It is produced in the kidneys and contains *urea* or *uric acid* as well as other substances in smaller amounts. (see *ureotelic, uricotelic*)

uriniferous tubule (see *renal tubule*)

Urochordata A newly autonomous phylum of marine chordates, the tunicates, in which characteristic chordate features are seen in *larval* forms but disappear in the adult. These include the presence of a *notochord, dorsal nerve cord,* and *gill slits,* which are clearly present in the free swimming larva. In the *sessile*, saclike adults, these features are usually reduced or absent; they also lack *segmentation* and a body cavity, as well as the brain and *vertebral column* diagnostic of craniate chordates.

urochordates (see *Urochordata*)

Urodela (Caudata) The order of amphib-ians that possess tails, including the salamanders, newts, and mud puppies. Young caudates are *carnivorous* like adults and look like small versions of them. Although they return to the water to breed, most adults live in moist terrestrial habitats. A few remain aquatic and essentially *larval* throughout their lifespan and breathe through *gills*. (see *axolotl, paedomorphosis, neotenic*)

urogenital Any structure or function pertaining to the sometimes intermingled *reproductive* and *excretory systems* in animals.

uterus, uteri (pl.) **(womb)** A muscular cavity(s) in female placental and marsupial mammals (but not monotremes) in which the *embryo* develops. It is usually paired although single in humans. Eggs are discharged into the uterus through the *Fallopian tubes*, where they implant if *fertilized*. The uterus opens into the vagina, through which the fully developed *fetus* is delivered after *gestation*. (see *endometrium, oogenesis, estrus cycle, menstrual cycle*)

utriculus The upper chamber of the *labyrinth* in the vertebrate inner ear. It is part of the *macula* and contains a patch of *sensory epithelium*, or *statocyst*, that detects changes in head tilt with respect to gravity and acceleration. (see *sacculus*)

UV radiation (*ultraviolet radiation*) Short wavelengths of light that are invisible to the human eye but perceptible by some other organisms, notably many insects. UV radiation can cause damage to *DNA* and is therefore *mutagenic* and *carcinogenic*. (see *ozone layer, ozone hole, pyrimidine dimer*)

vaccination The introduction of an *antigenic* substance into the body to induce *immunity* to a particular disease by stimulating the production of specific *antibodies*. The antigens may be carried on dead or *attenuated* (weakened) *pathogenic* microorganisms or extracted from them. (see *active immunity, vaccine;* compare *antiserum, immune serum, passive immunity*)

vaccine A preparation of dead or *attenuated* (weakened) *pathogenic* microorganisms or *antigenic* substances extracted from them. It is injected or ingested to produce *active immunity* to a particular disease. (see *vaccination;* compare *antiserum, immune serum, passive immunity*)

vacuole A membrane bound sac found in the *cytoplasm* of many types of cells that serves to isolate its contents from the general *intracellular* environment. Vacuoles may be used to store a variety of substances including liquids, food, and pigments. In many mature plant cells, a single vacuole occupies most of the cell volume and contains a solution of sugars and salts called the *cell sap*. Vacuoles are notably absent from *prokaryotes*. (see *contractile vacuole*)

vagile The ability of an organism to move freely. (compare *sessile*)

vagility The distance that individuals in a population tend to move from their birthsite, particularly at the time of mating. Populations that have low vagility tend to breed amongst themselves, thus have higher rates of *homozygosity*. Conversely, populations with high vagility tend to crossbreed with other populations, thus increasing their rate of *heterozygosity*.

vagina A duct in female mammals that extends from the uterus (or uteri) to the exterior. It receives the male penis during mating and also channels the menstrual flow to the exterior.

vagus nerves (cranial nerves X) The pair of nerves that arise from the *medulla oblongata* in the vertebrate brain and supply many internal organs. They also function in swallowing, speech, and breathing. They are the major nerves of the *parasympathetic nervous system* and contain both *motor* and *sensory neurons*. (see *cranial nerve*)

valine (val) An *amino acid*.

valve A *test*, shell, or other hard covering.

vane The body of a feather, extending to either side of the *rachis* or shaft. It consists of rows of *barbs* held together by *barbules*.

variable number tandem repeat (see *VNTR*)

variable regions In *immune receptor genes*, the V segments that exhibit small differences in *nucleotide* sequence between the *tandemly repeated* copies. Within them are the *hypervariable regions* that show extreme variation and determine *antibody* specificity. V segments are incorporated into both *B-receptors* (which are also released as *antibodies*) and *T-receptors*. In the region coding for the B-receptor *light chain*, as many as 300 different copies of the V segment may be seen. (see *antibody diversity;* compare *constant regions, joining regions, diversity regions*)

variegation The occurrence of sectors with differing appearances within one tissue of an organism. It is well known in some plants in which patches of two or more different colors occur on the leaves or flowers. It may be caused genetically, by the random movement of *transposons* (*transposable genetic elements*), or by viral disruption of the *genome*.

variety A taxonomic term used mostly in botany to describe variation below the subspecies level. Some disagreement exists as to whether it is an equivalent ranking to subspecies in animals or one taxon lower.

vas-, vaso- A word element derived from Latin meaning "vessel." (for example, *vasoactive*)

vascular Any aspect of a system of tubes

and/or vessels that transport and distribute liquids, along with any dissolved *solutes*, throughout an organism. Vascular systems are found in plants and animals.

vascular bundle In *vascular plants*, a discrete strand of conducting tissue containing both primary *xylem* and primary *phloem*. Groups of vascular bundles form a continuous conductive system throughout the plant for the distribution of water and nutrients. The pattern of *veins* seen in a leaf is due to the distribution of vascular bundles, which is also reflected in their arrangement in the stem.

vascular cambium In *vascular plants*, a cylindrical sheath of *meristamatic* cells capable of dividing indefinitely. They increase stem or root diameter by producing secondary *phloem* outwardly and secondary *xylem* inwardly. (see *secondary thickening*)

vascular cylinder (stele) In *vascular plants*, the cylinder of primary *vascular tissue* in the center of stems and roots. It contains both *xylem* and *phloem*, and is surrounded by the *endodermis*, which contains the *Casperian strip*. The arrangement of tissues varies considerably and is characteristic of different plant types.

vascular plant (see *tracheophyte*)

vascular system A system of vessels and or tubes that enables the efficient distribution of water and nutrients throughout an organism and also facilitates waste disposal. In *vascular plants,* it also contributes to mechanical support. (see *blood vascular system, water vascular system, circulatory system*)

vascular tissue Generally, any tissue containing vessels that conduct fluid. In plants specifically, the types of tissue that conduct and distribute water along with any dissolved salts (*phloem*) and carbohydrates (*xylem*) throughout the plant. It is one of the three main types of tissue in plants. (compare *dermal tissue, ground tissue*)

vas deferens In male animals, the tube carrying *spermatozoa* from the *testes* to the exterior. In male *amniotes*, it leads from the *epididymis* to the *cloaca* or *urethra*. (see *spermatogenesis*)

vas efferens In male vertebrates, one of a number of small ducts that convey *spermatozoa* from the *seminiferous tubules* of the *testis* to the *epididymis*. In invertebrates, they lead directly from the testis to the *vas deferens*. (see *spermatogenesis*)

vasoactive Any factor or substance that has an effect on blood vessels. Examples include *autonomic motor neurons*, various *hormones* (for example, *adrenaline*), and certain drugs. (see *vasoconstriction, vasodilation, vasomotor nerves*)

vasoconstriction The *reduction* in diameter of blood vessels due to contraction of the *smooth muscle* in their walls. It is controlled by *vasomotor nerves* and the secretion (or injection) of *adrenaline*. It occurs in response to decreased *blood pressure*, low external temperature, and pain. Vasoconstrictor pharmaceuticals perform this action. (see *vasoactive*; compare *vasodilation*)

vasodilation The increase in diameter of blood vessels due to the relaxation of the *smooth muscle* in their walls. It is controlled by *vasomotor nerves* and occurs in response to increased *blood pressure*, exercise, and high external temperature. (see *vasoactive*; compare *vasoconstriction*)

vasomotor nerves *Nerve fibers* of the *autonomic nervous system* that control the diameter of blood vessels. They transmit impulses from the *medulla oblongata* of the brain to the *smooth muscle* in the vessel walls, causing them to either constrict (*vasoconstrictor nerve fibers*) or dilate (*vasodilator nerve fibers*).

vasopressin (see *antidiuretic hormone*)

vector

1. An agent that transmits *pathogenic* organisms from one *host* to another. It can be a person, animal, or microorganism. An example is the mosquito that carries malarial *parasites* and

transmits them to man when it takes a blood meal. (compare *carrier*)

2. In *gene cloning,* a *DNA* element such as a *plasmid* or *bacteriophage* that is used to introduce foreign DNA into a *host* cell. The *gene* or DNA fragment of interest is inserted into the vector *in vitro* and then *transformed* or injected into a host cell for propagation or experimentation. The vector, including the inserted DNA, may either replicate independently as an *episome* or *integrate* into the host *genome,* depending on its specific properties. (see *genetic engineering*)

vegetal pole In fish and other aquatic invertebrates with asymmetric yolk distribution in their eggs, the pole of the *blastula* in which the yolk is concentrated. It is situated opposite the *animal pole.*

vegetative A stage in the development or *life cycle* of an organism in which growth rather than *sexual reproduction* occurs. *Asexual reproduction* is not excluded. (see *vegetative cell*)

vegetative cell A cell that is actively growing rather than resting, dormant (for example, *spores*), or engaged in *sexual reproduction.* They are characteristic of large portions of the *life cycle* in *prokaryotes,* protoctists, fungi, and lower plants.

vegetative nucleus In flowering plants (*angiosperms*), one of the three *pollen nuclei* that migrate down the *pollen tube* after *germination* on the *stigma* surface. The vegetative nucleus disintegrates as the pollen tube grows down the *style* leading to the egg cell. The remaining two *nuclei* participate in *double fertilization.*

vegetative propagation A term basically synonymous with *asexual reproduction* but used in particular to describe propagation in plants from *bulbs, corms, offsets, rhizomes, stolons, suckers,* and *tubers.* Many of these structures also serve as *perennating structures,* that is they can produce a new plant after laying dormant through inclement conditions. (see *graft*)

vegetative reproduction (see *asexual reproduction*)

vein

1. The thin-walled type of blood vessel that returns blood to the heart from the rest of the body. All veins except the *pulmonary vein,* which leads from the lungs to the heart, carry *deoxygenated* blood. (see *venule*; compare *artery*)

2. One of the *vascular bundles* in a plant leaf.

3. One of the *chitinous* tubes that support and strengthen an insect wing.

velamen The multiple layers of dead, spirally thickened cells covering the aerial roots of *epiphytic* plants (for example, orchids). It is spongelike, thus able to absorb surface water, and also translucent, so that light is able to penetrate to the *photosynthetic* tissue below.

veliger The second *larval* stage of aquatic mollusks (except cephalopods) that develop from the *trocophore.* During this stage, the foot, *mantle,* and shell develop. In gastropods, the body becomes torsionally rotated.

velvet worms (see *Onychophora*)

vena cava In tetrapods, either of the two main veins through which *deoxygenated* blood is returned to the heart. The *inferior vena cava* receives blood from the parts of the body below the diaphragm; the *superior vena cava* draws blood from the head, neck, chest, and arms.

venation

1. The distribution of *veins* (*vascular bundles*) in a *vascular plant* leaf. *Dicotyledons* usually show a netlike arrangement whereas *monocotyledons* tend to show a parallel pattern.

2. The distribution of *veins* within an insect's wing, which is often characteristic and may be used in identification.

venter In non*vascular plants,* the swollen base of an *archegonium* (female reproductive structure) that contains the egg cell.

ventilation The anatomical movements facilitating the movement and exchange of gases across various respiratory surfaces in an animal. (see *respiration, lungs, gills, spiracle, ram ventilation, diaphram*)

ventral The underside of an organism. In four-legged animals, ventral refers to the belly region. This is continued when describing bipedal animals such as man, whereas the front side would be described as *anterior* in other animals. (compare *dorsal*)

ventral fins (see *pelvic fins*)

ventricle
1. A thick-walled muscular chamber of the vertebrate heart that receives blood from an *atrium* and pumps it out, either to the lungs to exchange waste gases for oxygen or to the body to distribute freshly *oxygenated* blood. Fish and amphibians have one ventricle, while reptiles, birds, and mammals have two.
2. One of the four fluid-filled, interconnected internal cavities found in the mammalian brain.

venule The type of blood vessel forming the smallest branch of a vein leading away from *capillaries.*

vermiform Any structure resembling a worm in shape.

vermiform appendix (see *appendix*)

vernal The late spring portion of the year.

vernalization. The exposure of *germinating* plant seeds to low temperatures to induce flowering at a particular preferred time. It is a technique used in agriculture and horticulture.

vernal pool Small pools of water that collect from spring rains and induce the flowering of specialized annual wildflowers and provide a reproductive *habitat* for certain invertebrates. The pools dry up over the summer and reappear annually. Several of the species that depend on vernal pools are *endangered.*

vernation The relative arrangement of leaves in a bud or young plant shoot.

vertebra, vertebrae (pl.) One of the series of bones or cartilage segments forming the *vertebral column* of vertebrates. (see *caudal vertebrae, cervical vertebrae, lumbar vertebrae*).

vertebral column (backbone, spinal column) A series of bones or *cartilage* segments (*vertebrae*) that run along the back of the vertebrate body from head to tail and form a protective channel for the *spinal cord.*

Vertebrata (see *Craniata*)

vertebrate eye (see *eye*)

vesicle Any small *vacuole* in an animal cell used for various storage and transport functions.

vessel element In plants, a specialized dead cell found in the *xylem* of flowering plants (*angiosperms*) and one phylum of nonflowering seed plants (*gymnosperms*), Gnetophyta. The walls are fortified with bands of *lignin.* A column of such cells forms a *trachea* through which water and dissolved salts are transported. They have evolved from *tracheids* by the fusion of multiperforate end walls to form a single perforation with a large surface area and are also shorter and wider. This presents a more efficient arrangement through which fluid can be exchanged between cells. (compare *tracheid*)

vestibular nerves (see *vestibulocochlear nerves*)

vestibulocochlear nerves (acoustic nerve, auditory nerve, cranial nerves VIII) A pair of major *sensory* nerves that carry impulses from the *inner ear* to the brain for interpretation. They contain *neurons* specific for hearing, as well as motion and equilibrium detection and adjustment. (see *cranial nerves*)

vestigial structure A structure or organ that serves no known function but resembles the structure of presumed ancestors. For instance, whales contains reduced pelvic bones, which serve only to suggest a common evolutionary ancestry with tetrapod mammals. The appendix is a vestigial organ in humans.

Vestimenifera (see *Pogonophora*)

viability The probability that a *fertilized* egg will survive and develop into an adult organism.

villus, villi (pl.) The minute, fingerlike projections lining the vertebrate small intestine that serve to increase the absorptive surface area. (see *microvilli*)

vimentin The protein subunit that is most commonly *polymerized* into *intermediate filaments* in the *eukaryotic* cell.

virgin forest A forest that has never been cut. (see *old growth forest*; compare *second growth forest*)

virion The complete virus particle, it exists outside of a *host*. It consists, at a minimum, of a highly organized protein coat (*capsid*) surrounding the viral *nucleic acid* (*DNA* or *RNA*). The capsid shape may be either helical (for example, tobacco mosaic virus) or isometric. The structural pattern in isometric viruses is the icosahedron, a figure with 20 equilateral triangular facets. It is the same basic design as the geodesic dome and is the most efficient symmetrical arrangement that subunits can take to form an external shell with maximum internal capacity. In some groups of viruses, an outer lipid-rich envelope may be acquired through the genetic machinery of the *host*.

viroid A circular, *single-stranded RNA virus* lacking the protein coat characteristic of most viruses and *bacteriophages*. Viroids cause infectious diseases in plants.

virulence The relative *pathogenicity* of an organism.

virulent A virus or *bacteriophage* that, after cell penetration, immediately begins to subvert the *host* metabolic machinery and multiply. It soon causes destruction of the host, releasing many mature virus or phage particles. (see *lytic*; compare *lysogenic, temperate*)

virus An infectious agent that is not considered alive. Viruses consist of *nucleic acid* (*DNA* or *RNA*) encased in a protein coat. They do not have the ability to grow or replicate on their own and may do so only when they invade a *host* cell and subvert its cellular machinery. Viruses that attack bacteria are called *bacteriophages*. It is thought that viruses originate as particles of organismal *genomes*, generating their great diversity. (see *virion, temperate, virulent, lysogeny, lytic, reverse transcriptase, RNA virus, DNA virus, viroid*)

visceral Pertaining to the internal organs of an animal.

visceral arches The series of bony or *cartilaginous* skeletal arches in the *pharyngeal* region of fish and other vertebrate *embryos* that may differentiate to serve various specific functions.

visceral cleft (see *gill cleft*)

visceral mass The main body portion of mollusks containing the organs of the *digestive, excretory,* and *reproductive* systems.

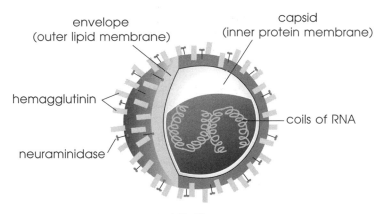

A Flu Virus

visual cortex The region on the *occipital lobe* of the *cerebral cortex* that receives and integrates information from the *optic nerve*. Different sites correspond to positions on the *retina* equivalent to particular points in the visual fields of the eyes.

visual pigment (see *rhodopsin*)

visual purple (see *rhodopsin*)

vitalism A doctrine that life embodied a force extrinsic to even the most complex chemical and physical explanations. It was first proposed in the 17th century and popularized during various periods stretching into the 19th century. The *mechanistic* view, in which all living processes can be explained by chemical reactions, began to gain general acceptance around the turn of the century.

vitamin A (retinol) A fat-soluble vitamin that, among other things, is converted to *retinal*, a component of the visual pigment *rhodopsin*. Vitamin A is a derivative of *carotene*, an accessory *photosynthetic pigment* found in large amounts in red, yellow, and dark green plants. It is found as *retinol* in milk products and organ meats, including cod-liver oil. Mammals cannot synthesize vitamin A and must obtain it from outside sources. Vitamin A deficiency leads to night blindness.

vitamin B complex A group of water-soluble vitamins, many of which act as *coenzymes* in *respiratory* pathways. They tend to occur together in unprocessed grains, meat, in particular organ meats, and brewer's yeast. They included *thiamin* (B_1), *riboflavin* (B_2), *nicotinic acid* (niacin), *pantothenic acid* (vitamin B_5), *pyridoxine* (vitamin B_6), *cyanocobalamin* (vitamin B_{12}), *biotin*, *lipoic acid*, and *folic acid*.

vitamin C (ascorbic acid) A water-soluble vitamin required for many metabolic functions, among others as a precursor to *hyaluronic acid*, an important component in *connective tissue*. It is found particularly in citrus fruits. A deficiency in vitamin C may lead to scurvy.

vitamin D (calciferol) A group of closely related *sterol* compounds. They are fat-soluble and important in the absorbtion of calcium and phosphorus from the intestine as well as their uptake by bone. They are synthesized from precursors in the skin by the action of ultraviolet light. They are also found in animal products such as milk, milk products, eggs, and organ meats including fish-liver oil. A deficiency may result in rickets in children and osteomalacia in adults.

vitamin E (tocopherol) A fat-soluble vitamin necessary for normal reproduction and other functions. It has antioxidant properties and is found in wheat germ and meat.

vitamin H (see *biotin*)

vitamin K (menaquinone, phylloquinone) A fat-soluble vitamin required for the synthesis of *prothrombin*, a *blood-clotting* factor. Intestinal bacteria provide most of the vitamin K needed in higher animals, although it also occurs in spinach, cabbage, liver, and egg yolk.

vitamins Organic compounds that are essential in small quantities for the growth and health of an organism. Many act as *coenzymes* for *metabolic* pathways. Deficiencies often result in specific disease symptoms. Plants have the capacity to synthesize vitamins, but animals generally must ingest them in their diet (with the exception of vitamins K and D). (see *vitamin A, vitamin B complex, vitamin C, vitamin D, vitamin E, vitamin K*)

vitelline membrane (egg membrane) The primary protective coating of an animal egg (*ovum*) lying just outside the *plasma membrane*. In most animals, it is altered by the first *spermatozoon* that penetrates it, forming the *fertilization membrane* that prevents any other sperm from entering the egg. In a few animals such as birds, it completely surrounds the yolk and does not prevent the entrance of subsequent sperm. (see *polyspermy*)

vitreous humor The gel-like substance that fills the space behind the *lens* in that vertebrate and cephalopod eye and helps maintain the shape of the eyeball.

viviparity The type of reproductive mechanism in which the *embryo* develops within the mother's body and derives nourishment from the mother via a *placenta*. *Fertilization* is internal, and young are born live. (compare *oviparity, ovoviviparity*)

VNTR (variable number tandem repeat) A motif in the *genomes* of many higher organisms in which a particular *nucleotide* sequence of *DNA* is *tandemly repeated*, the number of repeat units being highly variable between individuals. It is usually different even between *homologous chromosomes* in the same individual (*heterozygous*). The length of each repeat unit varies from about 9 to 50 nucleotides, the number of repeat units at a single locus may reach the hundreds. This type of *polymorphism* is used in individual identification by *DNA analysis*. (see *minisatellite, RFLP, PCR*)

vocal cords A pair of membranous folds stretched across the opening of the *larynx* in mammals and some other tetrapods and across the *syrinx* in birds. When air is forced across them, sound is produced by the vibrations. The pitch of the sound varies according to the length and tension of the cords, which may be controlled by muscles.

voice box (see *larynx*)

voles (see *Rodentia*)

Volkmann's canals A system of channels in bone that provide for the passage of blood vessels from the surface to the interior. They connect with the *Haversion canals*.

voltage-gated ion channel A specialized type of *ion channel* found, in particular, in the *cell membrane* of *neurons* and *muscle cells*. It opens and closes in response to the voltage across the membrane and allows for the specific and directional passage of ions. The channels are ion specific and include those for potassium (K^+), sodium (Na^+), calcium (Ca^+), and occasionally chloride (Cl^-). (see *action potential, resting potential, nerve impulse*; compare *chemically gated ion channel, stimulus-gated ion channel*)

voltine The number of generations produced by an organism per year or reproductive season.

voluntary muscle The *striated muscles* of the *skeletal system*. They are generally under conscious control of the individual.

voluntary nervous system The part of the *peripheral nervous system* in which *motor* pathways, mostly to the *skeletal muscles*, are under conscious control. (compare *autonomic nervous system*)

-vore, -vora A word element derived from Latin meaning "eating." (for example, *insectivore*)

vulnerable species (see *threatened species*)

vultures (see *Falconiformes*)

vulva The exterior opening to the *vagina*.

waders (see *Charadriiformes*)

wall pressure In plants, the pressure exerted against the inner *protoplast* by the outer *cell wall*. It opposes *turgor pressure*.

walruses (sea *Pinnipedia*)

warblers (see *Passiformes*)

warm-blooded (see *homeothermic*)

warning coloration (see *aposematic*)

water bears (see *Tardigrada*)

water cycle The earth-wide circulation of water molecules, powered by the sun. All life depends on water since it forms the basis for most biochemical reactions, and all organisms contain large amounts of it. Among other things, it is the source of the hydrogen ions (H$^+$, protons) whose movements generate *ATP* in many organisms. It is also the source of the hydrogen incorporated into carbohydrates during *photosynthesis*.

waterfowl (see *Anseriformes*)

water fleas (see *Branchiopoda*)

water molds (see *Chytridomycota*)

water potential In plants, a measure of the potential energy of water. There is a net movement from a region of high water potential to one of low potential. Water thus rises in a plant due to the positive pressure of the atmosphere and the negative pressure caused by evaporation of water from leaves. Technically, it is the algebraic sum of the *pressure potential* (the downward atmospheric pressure that forces liquid up a narrow tube in a reservoir) plus the *solute potential* (osmotic-driven diffusion).

water table The portion of *groundwater* that exists in the upper, unconfined regions of *aquifers*. It flows into streams and is accessible by plants.

water vascular system In echinoderms, a system of canals filled with seawater that lead to the *tube feet*, which operate by hydrostatic pressure. Water enters through the *madreporite* (sieve plate) on the

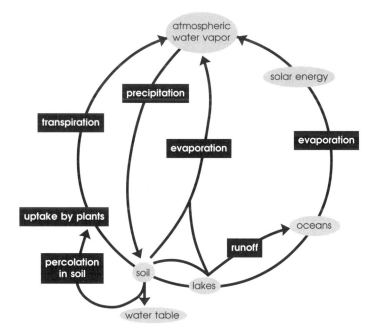

The Water Cycle

upper surface and passes down through the *stone canal* to the *ring canal*, which empties into five *radial canals* leading outward (into the arms in starfish). (see *ampulla*)

waxes A group of water-insoluble fatty substances. They form protective waterproof coverings on plant surfaces, insects, feathers, and animal fur and are also found in beeswax. They are composed of high molecular weight molecules, generally esters of long-chain alcohols with *fatty acids*. Commercially, they are harvested or synthesized for use as varnishes, polishes, and candles.

W chromosome In birds, reptiles, and butterflies, the chromosome found only in the *heterogametic* sex (**WZ**). (see *sex determination*; compare *Y chromosome*)

weasels (see *Carnivora*)

Weberian ossicles In certain fish (for example, carp and catfish), a few modified *vertebrae* that convey pressure changes from the *air bladder* to the *auditory capsule*. They are comparable in function with the *ear ossicles* in higher vertebrates.

web spinners (see *Embyoptera*)

weeds Plants growing where they are not wanted. (see *herbicide*)

weevils (see *Coleoptera*)

western blot A laboratory technique for identifying proteins based on their reaction with specific *antibodies*. It is a modification of the *Southern blot* (used for *DNA*). Proteins are *denatured* and separated by *electrophoresis* then transferred to a solid support and exposed to *radioactively* or *chemiluminescently* labeled antibodies. The locations of the antibody bound proteins are revealed by exposure to X-ray film and compared to standards for identification.

wetlands The *biome* consisting of freshwater swamps, marshes, bogs, ephemeral ponds, and saltwater marshes. They are characterized by continual or seasonal standing water, which creates a specialized soil environment with very little oxygen, retarding decay. Although wet-

lands occupy only a small portion of Earth's land area, the organisms that have adapted to this environment are very specialized and perform important functions in the environment. Most coal is derived from what was once wetland vegetation. *Anaerobic* bacteria and a large diversity of higher animals flourish in the wetland *habitat*.

whalebone (see *baleen*)

whales (see *Cetacea*)

wheel animals (see *Rotifera*)

whisk ferns (see *Psilophyta*)

white blood cell (see *leukocyte*)

white matter The *myelinated* nerves found in parts of the *spinal cord* and brain. They appear white because of the insulating *myelin sheath* surrounding them. (compare *gray matter*)

whorl Three or more leaves, flowers, or other plant structures originating from one point on a plant.

wild type (+) The *allele* (variant) of a particular *gene* most common in wild populations. Wild-type alleles (designated **+**) are often but not always *dominant* and are considered the norm for *phenotypic* comparison.

wilting The condition in which plants droop as a result of a loss of *turgidity* (water pressure).

windscorpion (see *Solifuga*)

wobble In the translation of *mRNA* into protein, a degeneracy of the *base pairing* rules that occurs at the third *nucleotide* of the *anticodon* on a *tRNA*. It may allow a second tRNA species to insert its attached *amino acid* in response to a particular *codon*. Wobble itself follows particular rules.

Wolffian ducts A pair of ducts in fish and amphibians that transport urine from the kidneys to the *cloaca*. In males, they also transport *spermatozoa* from the *testes*. (compare *ureter*)

wolves (see *Carnivora*)

womb (see *uterus*)

wood The hard, fibrous material found in woody *perennials* such as trees and shrubs. It is formed from secondary *xylem*, thus found only in *gymnosperms* and *dicotyledons*. (see *annual ring, cambium, cork cambium, bark, secondary growth*)

woodcocks (see *Charadriiformes*)

woodpeckers (see *Piciformes*)

woody Plants that produce *secondary growth* in the form of *wood*. Among other things, the mechanical support provided by wood provides them the ability to grow taller and effectively compete for available sunlight. They are always *perennial*, as wood is laid down over the lifetime of a plant. (compare *herbaceous*)

woody perennial (see *perennial*)

◆ X ◆

xanthin A yellow or orange *carotenoid*.

xanthophyll One of a class of yellow, orange, and red accessory *photosynthetic pigments* derived from *carotene*. Xanthophylls are found particularly in some algae and confer on them a characteristic yellow-brown or yellow-green color because they mask the green pigment of *chlorophylls*. (see *carotenoids*)

Xanthophyta A phylum of *photosynthetic unicellular* protoctists, the yellow-green algae. They are among the many organisms that make up pond scum. Their photosynthetic pigments include *xanthins*, which color them a distinctive yellow-green. Some are *colonial*, and all overwinter as *cysts*. In spring, they *germinate* as motile *zoospores* that either *differentiate* into an algal body or reproduce *asexually*.

xanthoplast A *plastid* containing *xanthophylls*. They are characteristic of certain groups of algae.

X chromosome In organisms with dissimilar *sex chromosomes* where the male is the *heterogametic* sex, the larger of the pair. It carries *sex-linked* genes, in other words, *loci* that have no counterpart on the *Y chromosome*. (see *sex determination*)

xen-, xeno- A word element derived from Greek meaning "strange" or "alien." (for example, *Xenarthra*)

Xenarthra The order of mammals containing the sloths, armadillos, and anteaters.

xenoma A *symbiotic aggregate* formed by intracellular *parasites* multiplying within growing tissue cells.

Xenophyophora A phylum of rarely seen protoctists, the xenophyophores, that inhabit the oceanic *abyss*. They are *heterotrophic plasmodial* organisms that sequester themselves inside a cemented, branched tube system called a *granellare*, which includes bits of foreign material. They have not yet been well characterized. It is known, however, that some

species contain crystals of barite (barium sulfite) within their *cytoplasm*, and that others can grow up to 23 centimeters while maintaining only a 1 millmeter thickness.

xer-, xero- A word element derived from Greek meaning "dry." (for example, *xerophyte*)

xerarch (see *xerosere*)

xeromorph Any organism that is structurally adapted to withstand dry conditions. For instance, desert organisms often have specialized water storage organs.

xerophyte Any terrestrial plant that is adapted to grow in dry conditions such as deserts.

xerosere The characteristic sequence of *communities* reflecting the developmental stages of an *ecological succession* that begins on dry rock. (see *sere*; compare *hydrosere*)

X-hyperactivation In male *Drosophila* (**XY**), the elevated *expression* of genes on the *X chromosome* necessary to equalize their levels with those produced by the **XX** females. (see *dosage compensation, Barr body*; compare *X-inactivation*)

X-inactivation In female mammals, the early inactivation of one of the two **X** chromosomes necessary to equalize their expression in both males (**XY**) and females (**XX**). (see *dosage compensation, Barr body*; compare *X-hyperactivation*)

X-linkage The inheritance pattern of *genes* found on the *X chromosome* but not on the *Y chromosome*. (see *sex-linked, heterogametic*)

X-ray crystallography A laboratory technique that uses *X-ray diffraction* to obtain information about the atomic structure of crystal molecules. It has been important in determining the structure of biological molecules such as *nucleic acids* (*DNA* or *RNA*), proteins (for example, *hemoglobin, lysozyme*), and viruses. (see *X-ray diffraction*)

X-ray diffraction A laboratory technique

in which X-rays are used to probe the atomic structure of crystal molecules. Characteristic diffraction patterns are obtained, much like light passing through a diffraction grating. (see *X-ray crystallography*)

X-rays (see *ionizing radiation*)

xylem The water-conducting tissue of *vascular plants*. It consists of dead, hollow cells (*tracheids* or *vessel elements*) that form columns within which water and dissolved salts are transported. The *cell walls* are strengthened and fortified with *lignin*. Xylem also contributes additional supporting tissue in the form of fibers and *sclereids*. (see *tracheary element*; compare *phloem*)

XY-linkage The extremely rare inheritance pattern of *genes* found on both the *X* and *Y chromosomes*. (see *sex-linked, heterogametic*)

Y chromosome In organisms with dissimilar *sex chromosomes* (except for birds, reptiles, and butterflies) where the male is the *heterogametic* sex, the smaller of the two different chromosomes. It is found only in the heterogametic sex (**XY**) and carries mostly male-specific genes, lacking counterparts for many *loci* found on the *X chromosome*. (see *sex determination*; compare *W chromosome*)

yeasts A subgroup of *ascomycete* fungi. They are *unicellular* and most of their reproduction is *asexual*, taking place by *budding* (the formation of a smaller cell from a larger one). However, cells of opposite *mating types* may fuse, forming a *zygote* that immediately undergoes *meiosis* and *ascospore* formation. The *spores* may survive adverse conditions and *ger-minate* under more optimal ones. Of particular importance are *Saccharomyces cerevisiae*, used in bread and beer making and as a tool in genetic research, and *Candida albicans*, the cause of yeast infections in humans.

yolk The *embryonic* food store found in many animals, consisting mostly of proteins and fats.

yolk sac One of the *extraembryonic* membranes present in vertebrates. It is particularly conspicuous in reptiles, birds, and sharks, where it contains a large amount of *yolk* that nourishes the *embryo* throughout its development and sometimes for the first few days after hatching. In mammals, it is usually devoid of yolk, forming part of the *chorion*. In *marsupials* it forms part of the *placenta*.

Z chromosome In birds, reptiles, and butterflies, the *chromosome* found in both *genders*, and in two copies (**ZZ**) in the *homogametic* sex, males. (see *sex determination*; compare *X chromosome*)

Z DNA A form of DNA that occurs in crystals of synthetic *oligonucleotides* containing alternating **GC** residues. It creates a left-handed zigzag structure compared to the right-handed helices of *A DNA* and *B DNA*. It is unclear what, if any, role stretches of Z DNA may play *in vivo*. (see *guanine, cytosine*)

zinc (see *trace elements*)

zinc finger A structural motif in some *DNA binding proteins* that occurs in *cysteine*- and *histidine*-rich regions of the protein and complexes the element zinc. Zinc fingers have been found in *steroid receptor* proteins and among *transcription factors*. (compare *leucine zipper, helix-turn-helix, helix-loop-helix*)

Z-line In *striated muscle*, the delineation between individual *sarcomeres*. It contains the protein *alpha-actinin* and serves as an anchor for the *thin filaments* (*actin*) in each *sarcomere*.

zo-, zoo-, -zoa, -zoid, zooid, -zoite A word element derived from Greek meaning "animal" or "living being." (for example, *protozoa*)

zoecium, zoecia (pl.) A *chitinous* chamber secreted by individual ectoprocts through which they attach to rocks and other members of the *colony*.

zoid An animal-like organism.

zona occludens (see *tight junction*)

zona pellucida The thick, clear membrane surrounding the mammalian egg (*ovum*). It lays outside the *vitelline membrane* and beneath a layer of *follicle* cells that disperse before *implantation*. (see *oogenesis*)

zoo blot A *Southern blot* containing *genomic DNA* from a range of animals. It is used in detecting and analyzing the evolutionary relationship between *genes* and *gene families* of variously related organisms.

zoology The study of animals.

Zoomastigina A phylum of protoctists, the zoomastigotes, that comprise a miscellany of unicellular motile *heterotrophs*. Many derive their food by absorbtion (*osmotrophy*) and some are *parasites*. The free-living choanomonads (choanoflagellates) are believed to be the ancestral cells of animals.

zoomastigotes (see *Zoomastigina*)

zooplankton (see *plankton*)

zoosporangium, zoosporangia (pl.) A reproductive structure, found in particular in some protoctists, in which *asexual zoospores* develop.

zoospore A motile *spore* produced, in particular, by some protoctists. They are also called swarmers. They bear two unequal *undulipodia* facing opposite directions and are produced *asexually* in a *zoosporangium*. Zoospores are capable of *differentiating* into a different developmental stage without being *fertilized*.

Zoraptera A small order (containing only one genus) of *exopterygote* insects, the zorapterans. They occur under bark or in decaying plant material and superficially resemble termites. The adults may be winged and pigmented with eyes and *ocelli* or nonpigmented, wingless, and blind.

zorapteran (see *Zoraptera*)

Z protein (see *protein Z*)

Z scheme The combination of *photosystem II* and *photosystem I* in the *photosynthetic* process of green plants, algae, and cyanobacteria. It is named for the shape generated by a charting of electron energy levels, which resemble a Z turned on its side.

zwitterion An ionic molecule bearing both positively and negatively charged functional groups. *Amino acids* are biologically important zwitterions. The amino group forms $-NH_3^+$ and the acid group ionizes to $-COO^-$ in a solution of the appropriate pH, such that the molecule as a whole is electrically neutral. (see *amphoteric*)

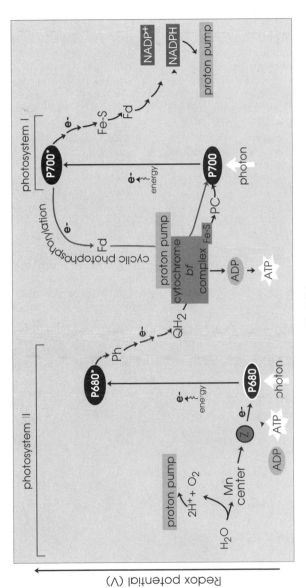

The Z-Scheme
(The Photosystems in Higher Plants)

zygo-, -zyg- A word element derived from Greek meaning "yolk" or "yoked." (for example *zygote*)

zygomorphic *Bilateral symmetry* as applied to flowers. For instance, snapdragons are zygomorphic and can be divided into equal halves along one plane.

zygomycete (see *Zygomycota*)

Zygomycota A phylum of fungi, the mating molds, characterized by the lack of divisions (*septae*) in their *hyphae* except to wall off reproductive regions. The vegetative fungal filaments (*hyphae*) clump together as a *mycelium* in which *sexual reproduction* takes place. They form characteristic resting structures, called *zygospores*, around each cell in which one or more *zygotes* have formed by the fusion of *nuclei* of opposite *mating types*. Collectively, these *multinucleate* structures are called *zygosporangia*. *Asexual sporangia* may also be produced on raised stalks called *sporangiophores* and are manifest as some of the more common bread molds. A few species of widely distributed zygomycetes participate in the formation of *endomycorrhizae*.

zygosporangium A *sporangium* in which *zygospores* are produced.

zygospore A *diploid* resting cell found in zygomycetes. A thick wall is formed around a cell in which two *nuclei* from cells of opposite *mating types* have fused, forming a large, *multinucleate zygote*. *Meiosis* occurs during *germination*, producing *haploid* products that continue to grow *vegetatively*.

zygote The *diploid* cell resulting from the fusion of male and female *gametes* (*fertilization*).

zygotene In *meiosis*, the stage in mid-*prophase I* in which *homologous* (paired) *chromosomes* physically pair (*synapsis*), forming *bivalents*.

zygotic meiosis *Meiosis* occurring during maturation or *germination* of a *zygote*.

zymo-, -zyme A word element derived from Greek meaning "ferment" or "leaven." (for example, *enzyme*)

zymogen Any inactive enzyme precursor, such as *pepsinogen, prothrombin, or trypsinogen*. Activation generally involves enzymatic cleavage of the *polypeptide*. (see *zymogen granule*)

zymogen granule A vesicle found in large numbers in *secretory* cells that sequesters the inactive precursor of an enzyme (*zymogen*). They are usually derived from the *Golgi apparatus*, where the enzyme is processed after *protein synthesis*.

◆ APPENDIX I ◆

THE METRIC SYSTEM*

Metric Length

1 meter ×	10 = decameter (10 m)
	100 = hectometer (10^2 m)
	1000 = kilometer (10^3 m)
	1,000,000 = megameter (10^6 m)
1 meter ÷	10 = decimeter (10^{-1} m)
	100 = centimeter (10^{-2} m)
	1000 = millimeter (10^{-3} m)
	1,000,000 = micrometer (10^{-6} m)
	1,000,000,000 = nanometer (10^{-9} m)
	10,000,000,000 = Angstrom (Å) (10^{-10} m)

Metric Weight/Mass

1 gram ×	1000 = kilogram
1 gram ÷	1000 = milligram (mg) (10^{-3} g)
	1,000,000 = microgram (μg) (10^{-6} g)
	1,000,000,000 = nanogram (ng) (10^{-9} g)
	1,000,000,000,000 = picogram (pg) (10^{-12} g)
	1,000,000,000,000,000 = femtogram (fg) (10^{-15} g)
	1,000,000,000,000,000,000 = attogram (ag) (10^{-18} g)

* Standard International (SI) units

♦ APPENDIX II ♦

THE METRIC-ENGLISH CONVERSIONS

LENGTH

English	Metric
inch	2.54 cm
foot	0.30 m
yard	0.91 m
mile (statute) (5280 ft)	1.61 km
mile (nautical) (6077 ft,	1.85 km
1.15 statute mile)	

Metric	English
millimeter	0.039 in
centimeter	0.39 in
meter	3.28 ft
kilometer	0.62 mi

WEIGHT

English	Metric
grain	64.80 mg
ounce	28.35 g
pound	0.45 kg
ton (2000 lb)	0.91 metric ton

Metric	English
milligram	0.02 grain
gram	0.04 oz
kilogram	2.20 lb
metric ton (1000 kg)	1.10 tons

VOLUME

English	Metric
cubic inch	16.39 cc
cubic foot	0.03 m^3
cubic yard	0.765 m^3
ounce	3 ml (cc)
pint	0.47 liter
quart	0.95 liter
gallon	3.79 liters

Metric	English
milliliter*	0.03 oz
liter	1.06 qt

* cubic centimeter

FARENHEIT-CELSIUS CONVERSION

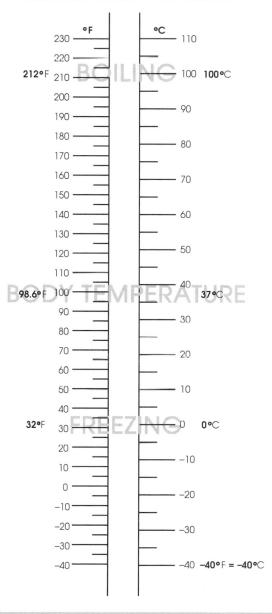

To convert temperature scales:

Farenheit to Celsius $°C = 5/9 \ (°F - 32)$

Celsius to Farenheit $°F = 9/5 \ (°C + 32)$

pH SCALE

H⁺ ion concentration (H⁺)	pH value −log(H)	Example of solution
10	−1	Nitric acid
1	0	Hydrochloric acid
10^{-1}	1	Stomach acid
10^{-2}	2	Lemon juice
10^{-3}	3	Vinegar, cola, beer
10^{-4}	4	Tomatoes
10^{-5}	5	Black coffee, Normal rainwater
10^{-6}	6	Urine, Saliva
10^{-7}	7	Blood
10^{-8}	8	Sea water
10^{-9}	9	Baking Soda
10^{-10}	10	Great Salt Lake
10^{-11}	11	Household ammonia
10^{-12}	12	Bicarbonate of soda
10^{-13}	13	Oven cleaner
10^{-14}	14	Sodium hydroxide (NaOH)
10^{-15}	15	Drain opener

GEOLOGIC TIME

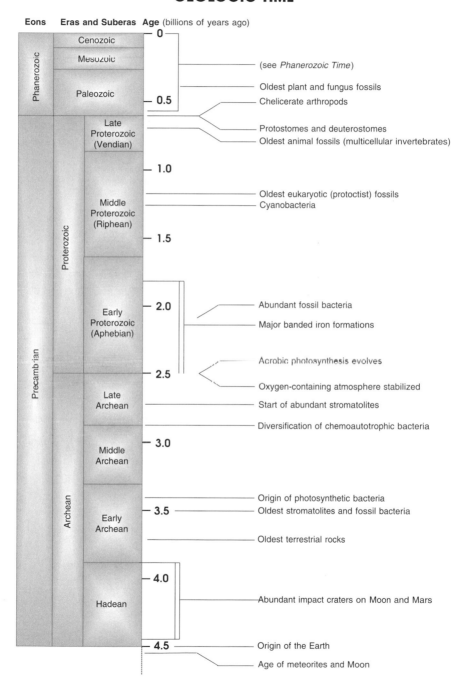

Eons	Eras and Suberas	Age (billions of years ago)	

Phanerozoic
Cenozoic
Mesozoic
Paleozoic

Precambrian
Proterozoic
Archean

Late Proterozoic (Vendian)
Middle Proterozoic (Riphean)
Early Proterozoic (Aphebian)
Late Archean
Middle Archean
Early Archean
Hadean

0
0.5
1.0
1.5
2.0
2.5
3.0
3.5
4.0
4.5

(see *Phanerozoic Time*)
Oldest plant and fungus fossils
Chelicerate arthropods
Protostomes and deuterostomes
Oldest animal fossils (multicellular invertebrates)
Oldest eukaryotic (protoctist) fossils
Cyanobacteria
Abundant fossil bacteria
Major banded iron formations
Aerobic photosynthesis evolves
Oxygen-containing atmosphere stabilized
Start of abundant stromatolites
Diversification of chemoautotrophic bacteria
Origin of photosynthetic bacteria
Oldest stromatolites and fossil bacteria
Oldest terrestrial rocks
Abundant impact craters on Moon and Mars
Origin of the Earth
Age of meteorites and Moon

◆ APPENDIX VI ◆

PHANEROZOIC TIME

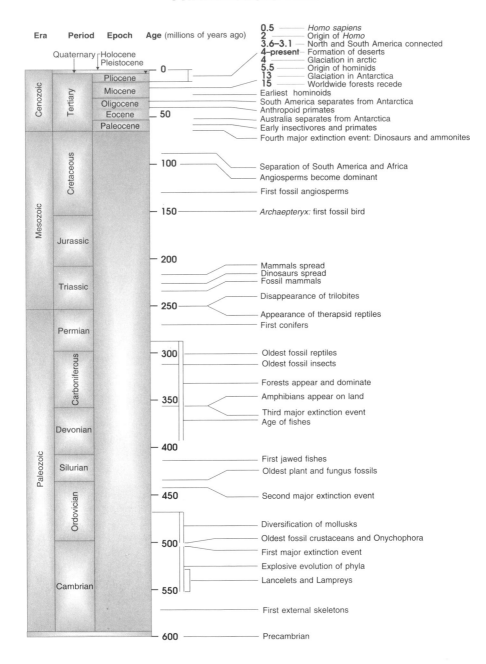

Era	Period	Epoch	Age (millions of years ago)

0.5	*Homo sapiens*
2	Origin of *Homo*
3.6–3.1	North and South America connected
4–present	Formation of deserts
4	Glaciation in arctic
5.5	Origin of hominids
13	Glaciation in Antarctica
15	Worldwide forests recede

Quaternary — Holocene / Pleistocene

Cenozoic

Tertiary — Pliocene, Miocene, Oligocene, Eocene, Paleocene

— 0

Earliest hominoids
South America separates from Antarctica
Anthropoid primates
Australia separates from Antarctica
Early insectivores and primates

— 50

Fourth major extinction event: Dinosaurs and ammonites

Mesozoic

Cretaceous

— 100

Separation of South America and Africa
Angiosperms become dominant

First fossil angiosperms

— 150

Archaepteryx: first fossil bird

Jurassic

— 200

Mammals spread
Dinosaurs spread
Fossil mammals

Triassic

Disappearance of trilobites

— 250

Appearance of therapsid reptiles
First conifers

Paleozoic

Permian

— 300

Oldest fossil reptiles
Oldest fossil insects

Carboniferous

Forests appear and dominate

— 350

Amphibians appear on land

Third major extinction event
Age of fishes

Devonian

— 400

First jawed fishes
Oldest plant and fungus fossils

Silurian

— 450

Second major extinction event

Ordovician

Diversification of mollusks

— 500

Oldest fossil crustaceans and Onychophora
First major extinction event
Explosive evolution of phyla
Lancelets and Lampreys

Cambrian

— 550

First external skeletons

— 600

Precambrian

402

◆ APPENDIX VII ◆

THE GENETIC CODE*

SECOND LETTER

FIRST LETTER	U	C	A	G	THIRD LETTER
U	Phenylalanine	Serine	Tyrosine	Cysteine	U
	Phenylalanine	Serine	Tyrosine	Cysteine	C
	Leucine	Serine	STOP	STOP	A
	Leucine	Serine	STOP	Tryptophan	G
C	Leucine	Proline	Histidine	Arginine	U
	Leucine	Proline	Histidine	Arginine	C
	Leucine	Proline	Glutamine	Arginine	A
	Leucine	Proline	Glutamine	Arginine	G
A	Isoleucine	Threonine	Asparagine	Serine	U
	Isoleucine	Threonine	Asparagine	Serine	C
	Isoleucine	Threonine	Lysine	Arginine	A
	Methionine (Start)	Threonine	Lysine	Arginine	G
G	Valine	Alanine	Aspartate	Glycine	U
	Valine	Alanine	Aspartate	Glycine	C
	Valine	Alanine	Glutamate	Glycine	A
	Valine	Alanine	Glutamate	Glycine	G

* The amino acids encoded by mRNA triplets (codons). Although the code depicted above is used by most organisms, subcellular organelles and some ciliate protists show minor variations. The code is also said to be degenerate because more than one triplet may code for a single amino acid.

Codon-Anticodon Pairing Allowed by the Wobble Rules

5' end of anticodon	3' end of codon
G	U or C
C	G only
A	U only
U	A or G
I	U,C, or A

403

AMINO ACIDS COMMONLY FOUND IN PROTEINS

Amino Acid	3-letter code	1-letter code	Essential (for humans)
alanine	Ala	A	
arginine	Arg	R	in early life
asparagine	Asn	N	
aspartic acid	Asp	D	
cysteine	Cys	C	
glutamic acid	Glu	E	
glutamine	Gln	Q	
glycine	Gly	G	
histidine	His	H	in children
isoleucine	Ile	I	+
leucine	Leu	L	+
lysine	Lys	K	+
methionine	Met	M	+
phenylalanine	Phe	F	+
proline	Pro	P	
serine	Ser	S	
threonine	Thr	T	+
tryptophan	Trp	W	+
tyrosine	Tyr	Y	
valine	Val	V	+

20 amino acids are commonly found in proteins. Essential amino acids (gray) cannot be synthesized by the organism and must be obtained from outside sources. Humans require the 10 essential amino acids listed above.

◆ APPENDIX IX ◆

THE HIERARCHY OF BIOLOGICAL CLASSIFICATION

	Nominal Designations
Most Comprehensive	**Kingdom**
	Phylum/Division
	Class
	Order
	Family
	Genus
Most Specific	**Species**

The following modifiers are sometimes used with the nominal designations:

Super—(e.g., a Superclass is a slightly more comprehensive grouping than a Class)

Infra—(e.g., an Infraclass is more comprehensive than a Class but less than a Superclass)

Sub—(e.g., a Subclass is a less comprehensive grouping than a Class but greater than an Order)

SOME GENERAL CHARACTERISTICS OF THE FIVE KINGDOMS

	Prokaryotes	Protoctists (Protists)	Fungi	Plants	Animals
cell size	small ~1 - 10 μm	large ~10 - 100 μm	large ~10 - 100 μm	large ~10 - 100 μm	large ~10 - 100 μm
cell type	prokaryotic	eukaryotic	eukaryotic	eukaryotic	eukaryotic
nucleus	absent	present	present	present	present
chromosomal structure	nucleoid—single naked circle of DNA	chromosomes made of DNA, RNA, and protein	chromosomes made of DNA, RNA, and protein	chromosomes made of DNA, RNA, and protein	chromosomes made of DNA, RNA, and protein
mitochondria	absent	sometimes present	sometimes present	present	present
respiration	range from obligate anaerobes to obligate aerobes, included facultative forms of both	most obligate aerobes	most obligate aerobes	most obligate aerobes	most obligate aerobes
photosynthesis	if present, enzymes bound to cell membrane, large variation in metabolic raw materials and end-products	if present, enzymes packaged into plastids, all generate oxygen	absent	generally present, enzymes packaged into chloroplasts, all generate oxygen	absent
metabolism	many variations	oxidative metabolism: glycolysis, Krebs cycle, electron transport	oxidative metabolism: glycolysis, Krebs cycle, electron transport	oxidative metabolism: glycolysis, Krebs cycle, electron transport	oxidative metabolism: glycolysis, Krebs cycle, electron transport
cell wall	polysaccharide-amino acid	some forms, various types	chitin and/or polysaccharides	cellulose and/or polysaccharides	absent
genetic recombination	rare: conjugation transduction transformation	common: fertilization and meiosis	common: fertilization and meiosis	common: fertilization and meiosis	common: fertilization and meiosis
nutrition	autotrophic (photo- or chemosynthetic) and/or heterotrophic	autotrophic (photosynthetic) and/or heterotrophic	heterotrophic: absorption	autotrophic photosynthetic	heterotrophic: digestion
motility (single cell)	bacterial flagella composed of flagellin	9+2 undulipodia, ameboid	nonmotile	9+2 undulipodia only in some gametes	9+2 undulipodia
cellularity	unicellular	various	various	multicellular	multicellular
cell division	mostly binary fission no centrioles, mitotic spindle or microtubules	varies	mitosis mitotic spindle centrioles absent	mitosis mitotic spindle centrioles absent	mitosis mitotic spindle centrioles absent
development	N/A	varies	spores	multicellular embryo enclosed in maternal tissues	blastula

MODES OF NUTRITION FOR LIFE ON EARTH[1]

Modes of nutrition vary as to energy source, electron source (hydrogen atom donor), and carbon source. All animals, fungi, many protoctists, and in fact most bacteria are chemoorganoheterotrophs—they break chemical bonds of ingested food for energy and obtain their carbon and hydrogen atoms from the same source. Green plants and their ancestors, the photosynthetic protoctists (algae), are photolithoautotrophs—sunlight is their primary energy source from which they "fix" inorganic carbon from atmospheric CO_2 into organic molecules via photosynthesis. Their sole source of hydrogen atoms for photosynthesis is from water (H_2O). The rest of the variation occurs within bacteria, which display an enormous range of metabolic pathways. Names are constructed by adding the suffix "-troph," for instance, photolithoautotroph.

ENERGY SOURCE (light vs. chemical compounds)	ELECTRON SOURCE (or hydrogen donor)	CARBON SOURCE	ORGANISMS	specific hydrogen atom or electron donors
PHOTO- (light)	LITHO- (inorganic compounds and C_1)	AUTO- (CO_2)	PROKARYOTES: Chlorobia / *Chromatium* / *Rhodospirillum* / Cyanobacteria PROTOCTISTA: algae GREEN PLANTS	H_2S, S / H_2S, S / H_2 / H_2O H_2O H_2O
		HETERO- $(CH_2O)_n$	NONE	
	ORGANO- (organic compounds)	AUTO-	NONE	
		HETERO-	PROKARYOTES: *Chromatium* / *Chloroflexa* / *Rhodospirillum* / *Rhodomicrobium* *Heliobacterium* / Halobacteria	small org. comp.[2] / small org. comp / small org. comp / C_2, C_3 compounds small org. comp
CHEMO- (chemical compounds)	LITHO- (inorganic compounds and C_1)	AUTO-	PROKARYOTES: "methanogens" / hydrogen oxidizers / methylotrophs / ammonia, nitrate oxidizers	H_2 / H_2 / CH_4, CHOH, etc. / NH_3, NO^-_2
		HETERO-	PROKARYOTES: manganese oxidizers / iron bacteria / sulfur bacteria / sulfide oxidizers (e.g., *Beggiatoa*) / sulfate reducers (e.g., *Desulfovibrio*)	Mn^{++} / Fe^{++} / S^{2-}, S small org. comp
	ORGANO- (organic compounds)	AUTO-	PROKARYOTES: Clostridia etc. (grown on CO_2 as sole C source)	H_2, $-CH_3$
		HETERO-	PROKARYOTES[3] (most) / PROTOCTISTA[4] (most) / FUNGI[4] / PLANTS[4] (achlorophyllous) / ANIMALS[4]	

[1]Modified from Table I, Modes of Nutrition for Life on Earth, *Illustrated Glossary of Protoctista*, edited by L. Margulis, H. McKhann and L. Olendzenski, Jones and Bartlett, Boston, 1993, and Table I-2, Modes of Nutrition known for Life on Earth, *Handbook of Protoctista*, edited by L. Margulis, J. Corliss, M. Melkonian and D. Chapman, Jones and Bartlett, Boston, 1990.

[2]Examples include acetate, propionate, pyruvate.

[3]Nitrate (NO_3^-), sulfate (SO_4^{2-}), oxygen (O_2), and even phosphate (PO_4^{3-}) may be terminal electron acceptors, leading to production of ammonia (NH_3^+), hydrogen sulfide (H_2S), water (H_2O), and phosphine (PH_3), respectively.

[4]Oxygen as terminal electron acceptor.

◆ APPENDIX XII ◆

PHYLOGENETIC CLASSIFICATION OF ORGANISMS

The classification of living organisms has always been a source of controversy in the biological community and of late has become even more contentious. The arrival of DNA typing has, in some cases, contradicted classification schemes based on physiological and morphological traits, especially in bacteria data. Thus, the field is in even more of a state of flux than usual. The outline listed below is not intended to be comprehensive and doubtless will disagree with one authority or another. It does, however, give a good general overview of the relationships between living organisms, and incorporates some of the more recent ideas on the subject. The general scheme follows Margulis and Schwartz, *Five Kingdoms*, 3rd edition. This includes the referral to all groupings at the phylum level as *phyla* rather than the historic *division* that has differentiated plant and fungal groupings at this hierarchical level. Arthropod orders are mainly from Raven and Johnson, *Biology*, and the *Peterson Field Guide to Insects* by Borror and White. Bird orders generally follow Gill's *Ornithology*, and mammalian orders follow the Smithsonian Institution's *Mammal Species of the World*. For an alternative, also widely accepted, bacterial classification system, please see *Bergey's Manual of Systematic Bacteriology*. The references listed at the end of the section provide some more specific and complete information. Many books and articles on the subject continue to be published. A vast amount of information (as well as misinformation) is also available on the World Wide Web.

Superkingdom Prokarya (Prokaryotae, Procaryotae, Prokarya)

KINGDOM BACTERIA (MONERA; Prokaryotic microorganisms and their descendants exclusive of nucleated organisms formed by symbiogenesis)

SUBKINGDOM ARCHAEA
  Phylum EURYARCHAEOTA (methanogens; methane-producing bacteria; halophils; salt-requiring bacteria)
  Phylum CRENARCHAEOTA (thermoacidophiles, heat/acid-loving bacteria)

SUBKINGDOM EUBACTERIA
Gracilicutes (Gram-negative bacteria)
  Phylum PROTEOBACTERIA (PURPLE BACTERIA; omnibacteria; respiring enteric bacteria, anoxygenic purple phototrophs, pseudomonads, nitrogen-fixing aerobic bacteria, chemoautotrophic bacteria, *Desulfovibrio*)
  Phylum SPIROCHAETAE (spirochetes; periplasmically flagellated swimmers)
  Phylum CYANOBACTERIA (oxygenic phototrophs; formerly blue-green algae)
  Phylum SAPROSPIRAE (fermenting gliders)
  Phylum CHLOROFLEXA (green nonsulfur phototrophs)
  Phylum CHLOROBIA (anoxygenic green sulfur phototrophs)

Tenericutes (wall-less eubacteria)
  Phylum APHRAGMABACTERIA (mycoplasmas, spiroplasmas)

Firmicutes (Gram-positive and protein-walled bacteria)
  Phylum ENDOSPORA (endospore-forming bacteria)
  Phylum PIRELLAE (stalked bacteria with proteinaceous walls)

Phylum ACTINOBACTERIA (actinomycetes; actinospore-forming filamentous bacteria)

Phylum DEINOCOCCI (aerobic radioresistant bacteria)

Phylum THERMOTOGAE (thermophilic fermenters)

Superkingdom Eukarya (Eukaryotae)

KINGDOM PROTOCTISTA (Protoctists, Protists, eukaryotic microorganisms and their descendants exclusive of plants, animals, and fungi)

Amitochondriates (Hypochondria, Archaeozoa; lack mitochondria)

Phylum ARCHAEPROTISTA (lack mitochondria)
    Class ARCHAMOEBAE
    Class METAMONADA
    Class PARABASALIA

Phylum MICROSPORA (micorsporans, microsporidian; amitochondriate parasites)

Amoeboids

Phylum RHIZOPODA (rhizopods; amebas)
    Class Naked and shelled amebae
    Class ACRASIOMYCOTA (ACRASEA, acrasids; cellular slime molds)

Phylum GRANULORETICULOSA (granuloreticulosans)
    Class RETICULOMYXIDS (reticulopods)
    Class FORAMINIFERA (foraminiferans, forams)

Phylum XENOPHYOPHORA (deep sea protists)

Phylum MYXOMYCOTA (myxomycotes; plasmodial slime molds; acellular slime molds)
    Class CERCOMONADS
    Class PROTOSTELIDS
    Class EUMYCETOZOA

Alveolates

Phylum DINOMASTIGOTA (DINOFLAGELLATA; dinoflagellates; dinomastigotes)

Phylum CILIOPHORA (ciliates)

Phylum APICOMPLEXA (apicomplexans; sporozoan parasites)

Swimming mastigotes

Phyum HAPTOMONADA (prymnesiophytes; coccolithophorids; phyto-flagellates)

Phylum CRYPTOMONADA(CRYPTOPHYTA; cryptophytes; cryptomonads)

Phylum DISCOMITOCHONDRIA(amoebomastigotes, kinetoplastids, euglenids, pseudocilates)
    Class AMOEBOMASTIGOTA (Schizopyrenida; amoebomastigotes; amoeboflagellates)
    Class KINETOPLASTIDS (flagellated parasites)
    Class EUGLENIDS (flagellate algae)
    Class PSEUDOCILIATES (flagellates)

Heterokonts (Stramenopiles)

Phylum CHRYSOMONADA (CHRYSOPHYTA; chrysophytes; golden-yellow algae)

Phylum XANTHOPHYTA (xanthophytes; yellow-green algae)
Phylum EUSTIGMATOPHYTA (eustigmatophytes; eustigs; eyespot algae)
Phylum BACILLARIOPHYTA (diatoms; phytoplankton)
Phylum PHAEOPHYTA (pheophytes; brown algae)
Phylum LABYRINTHULATA (slime nets)
Phylum PLASMODIOPHORA (plasmodial parasites; plant parasites)
Phylum OOMYCOTA (oomycotes; water molds)
Phylum HYPHOCHYTRIOMYCOTA (hyphochytrids; water molds)

Propagule-forming parasites
Phylum HAPLOSPORA (HAPLOSPORIDIA; haplosporidians; haplosporosome-forming parasites)
Phylum PARAMYXA (PARAMYXEA; nesting cell-forming parasites)
Phylum MYXOSPORA (MYXOZOA, MYXOSPORIDIA; myxosporidians; multicellular fish parasites)

Conjugating algae
Phylum RHODOPHYTA (rhodophytes; red algae)
Phylum GAMOPHYTA (CONJUGAPHYTA; gamophytes; conjugating green algae, desmids)

Radiolarates
Phylum ACTINOPODA (actinopods)
    Class HELIOZOA (heliozoans; sun animalcules)
    Class PHEODARIA (pheodarians; some "radiolarians")
    Class SPUMELLARIA (spumellarians; some "radiolarians")
    Class ACANTHARIA (acantharians)

Ancestral Phyla
Phylum CHLOROPHYTA (chlorophytes; motile green algae; plant ancestors)
Phylum CHYTRIDIOMYCOTA (chytrids; fungal ancestors)
    Class CHYTRIDIALES (chytrids, water molds)
    Class SPIZELLOMYCETALES (water molds)
    Class MONOBLEPHARIDALES (water molds)
    Class BLASTOCLADIALES (water molds)
Phylum ZOOMASTIGINA (zoomastigotes; aerobic motile single cells; animal ancestors)
    Class JAKOBIDS (mastigotes)
    Class BICOSOECIDS (loricate mastigotes)
    Class PROTEROMONADS (parasites)
    Class OPALINIDS (amphibian parasites)
    Class CHOANOMONADS (choanomastigotes; collared mastigotes)

KINGDOM FUNGI (mushrooms, fungus, molds and yeasts)
  Phylum ASCOMYCOTA (ascomycetes; molds, yeasts, mildews)
    DEUTEROMYCOTA (deuteromycetes; "fungi imperfecti")
    MYCOPHYCOPHYTA (lichens)
  Phylum BASIDIOMYCOTA (basidiomycetes; mushrooms, puffballs, smuts etc.)
  Phylum ZYGOMYCOTA (zygomycetes; zygospore-forming molds, mating molds)

KINGDOM PLANTAE (plants)
  Phylum Bryophyta (bryophytes; mosses, peat mosses)
  Phylum Hepatophyta (hepatophytes; liverworts)
  Phylum Anthocerophyta (anthocerophytes; hornworts)
  Phylum Coniferophyta (conifers)
  Phylum Cycadophyta (cycads)
  Phylum Ginkgophyta (ginkgo)
  Phylum Gnetophyta (gnetophytes)
  Phylum Lycophyta (lycophytes; lycopods; club mosses, ground pines)
  Phylum Psilophyta (psilophytes; whisk ferns)
  Phylum Filicinophyta (Pterophyta, Pterodatina; ferns)
  Phylum Sphenophyta (Equisetophyta, Arthrophyta; sphenophytes; horsetails)
  Phylum Anthophyta (anthocerophytes; angiosperms; flowering plants)
      Class Monocotyledons (monocots)
      Class Dicotyledons (dicots)

KINGDOM ANIMALIA (animals)

Subkingdom Parazoa (asymmetrical animals)
  Phylum Placozoa (placozoans)
  Phylum Porifera (sponges)
  Phylum Rhomboza (dicyemid or heterocyemids mesozoans)
  Phylum Orthonectida (orthonectid mesozoans)

Subkingdom Eumetazoa (symmetrical animals)
  Radiates
  Phylum Cnidaria (Coelenterata; cnidarians, coelenterates)
      Class Anthozoa (anthozoans; corals and sea anemones)
      Class Hydrozoa (hydrozoans; hydroids)
      Class Scyphozoa (scyphozoans; jellyfish)
      Class Cubozoa (cubozoans; box-jellies, sea-wasps)

Phylum Ctenophora (ctenophorans; comb jellies and sea walnuts)

Aperitoneal (acoelemate) worms
Phylum Gnathostomulida (gnathostomulids; jaw worms)

Phylum Platyhelminthes (flatworms)
  Class Cestoda (cestodans; tapeworms)
  Class Trematoda (trematodans; flukes)
  Class Monogenea (monogeneans; flukes)
  Class Turbellaria (turbellarians; free-living flatworms)

Phylum Gastrotricha (gastrotrichs)

Phylum Rotifera (rotifers, wheel animals, wheel animalcules)

Phylum Kinorhyncha (kinorhynchs)

Phylum Loricifera (loriciferans; loricifers)

Phylum Aсанthocephala (acanthocephalans; spiny-headed worms)

Phylum Nematoda (nematodes; eelworms and roundworms)

Phylum Nematomorpha (nematomorphs; Gordian worms; horsehair worms)

Phylum Priapulida (priapulids; cucumber worms)

Phylum Entoprocta (entoprocts)

Peritoneal (coelemate) worms
Phylum Rhynchocoela (Nemertinea, Nemertina; ribbon worms)

Phylum Sipuncula (sipunculans; peanut worms)

Phylum Echiura (echiurans; spoon worms)

Phylum Annelida (annelids; segmented worms)
    Class Hirudinea (leeches)
    Class Oligochaeta (earthworms)
    Class Polychaeta (Polychaetes; clamworms, plumed worms, scaleworms)

Phylum Pogonophora (Vestimentifera; giant tube or beard worms)

Phylum Mollusca (mollusks)
    Class Bivalvia (bivalves; clams, scallops, oysters, mussels)
    Class Cephalopoda (cephalopods; octopi, squids, nautilus)
    Class Gastropoda (gastropods; snails and slugs)
    Class Polyplacophora (polyplacophores; chitons)

Nonjointed-foot walkers
Phylum Tardigrada (tardigrades; water bears)

Phylum Pentastoma (pentastomes; tongue worms)

Phylum Onychophora (onychophores; velvet worms, walking worms)

Arthropods (Jointed-foot walkers)
Phylum Chelicerata (chelicerates; non-jawed arthropods)
    Class Arachnida (arachnids)
        Order Acari (mites)
        Order Araneae (spiders)
        Order Opiliones (daddy longlegs)
        Order Palpigrada (palpigrades)
        Order Pseudoscorpiones (false scorpions)
        Order Solifuga (Solipugida; solifugids)
        Order Scorpiones (scorpions)
    Class Merostomata (horseshoe crabs)
    Class Pycnogonida (sea spiders)

Phylum Mandibulata (Uniramia, Atelocerata; unbranched-legged arthropods)
   Class Chilopoda (centipedes)
   Class Diplopoda (millipedes)
   Class Insecta (insects)
      Order Anoplura (sucking lice)
      Order Coleoptera (beetles)
      Order Collembola (springtails)
      Order Dermaptera (earwigs)
      Order Dictyoptera (Blattoidea; cockroaches)
      Order Diplura (diplurans)
      Order Diptera (flies, midges, mosquitoes)
      Order Embioptera (webspinners)
      Order Ephemeroptera (mayflies)
      Order Hemiptera (true bugs; cicadas, hoppers, whiteflies, aphids, scale insects)
      Order Hymenoptera (sawflies, ants, wasps, bees)
      Order Isoptera (termites)
      Order Lepidoptera (butterflies, moths)
      Order Mecoptera (scorpionflies)
      Order Megaloptera (dobsonflies)
      Order Neuroptera (snakeflies, lacewings, antlions)
      Order Odonata (dragonflies, damselflies)
      Order Orthoptera (grasshoppers, katydids, crickets, mantids, walking-sticks)
      Order Plecoptera (stoneflies)
      Order Protura (bark lice)
      Order Psocoptera (booklice)
      Order Siphonaptera (fleas)
      Order Strepsiptera (twisted-winged parasites)
      Order Thysanoptera (thrips)
      Order Trichoptera (caddisflies)
      Order Zoraptera (zorapterans)
   Class Pauropoda (pauropods)
   Class Symphyla (symphylans)

Phylum Crustacea (crustaceans; lobsters, crabs, crayfish, shrimps)
   Class Branchiopoda; (gill-footed shrimps; brine shrimp, fairy shrimp, water fleas)
   Class Branchiura (fish lice)
   Class Cephalocarida (cephalocarids)
   Class Cirripedia (barnacles)
   Class Copepoda (copepods)
   Class Malacostraca (decapods; crabs, shrimps, lobsters, woodlice)
   Class Mystacocarida (mystracocarids)
   Class Ostracoda (ostracods)

Lophophorates
Phylum Bryozoa (Ectoprocta; ectoprocs; moss animals)

Phylum Phoronida (phoronids)

Phylum BRACHIOPODA (brachiopods; lamp shells)

Achordate deuterostomes
Phylum ECHINODERMATA (echinoderms)
　Subphylum PELMATOZOA
　　　Class CRINOIDEA (crinoids; sea lilies and feather stars)
　Subphylum Eleutherozoa
　　　Class ECHINOIDEA (echinoids; sea urchins and sand dollars)
　　　Class HOLOTHUROIDEA (holothuroids; sea cucumbers)
　　　Class STELLEROIDIA (stelleroids; asteroids; brittle stars, ophiuroids; sea stars)

Phylum CHAETOGNATHA (chaetognaths, arrow worms)

Phylum HEMICHORDATA (hemichordates; acorn worms)

Chordate deuterostomes
Phylum UROCHORDATA (urochordates)
　Class TUNICATA (Tunicates)

Phylum CEPHALOCHORDATA (cephalochordates; lancelets)

Phylum CRANIATA (VERTEBRATA; vertebrates)
　　Class CYCLOSTOMATA (AGNATHA; cyclostomes; lampreys, hagfish)
　　Class CHONDRICHTHYES (cartilaginous fishes; sharks, skates and rays)
　　Class OSTEICHTHYES (bony fishes)
　　　　Subclass ACTINOPTERYGII (ray-finned fishes)
　　　　Subclass SARCOPTERYGII (lobe-finned fishes)
　　　　Subclass CROSSOPTERYGII (coelecanths)

　　Superclass TETRAPODA
　　Class AMPHIBIA (amphibians)
　　　　Order ANURA (tailess amphibians; frogs and toads)
　　　　Order URODELA (CAUDATA; tailed amphibians; newts and salamanders)
　　　　Order APODA (GYMNOPHIONIA; caecilians)

　　Class REPTILIA (reptiles)
　　　　Order CHELONI (turtles and tortoises)
　　　　Order CROCODILIA (crocodilians; crocodiles, alligators and caimans)
　　　　Order SPHENODONTA (tuatara)
　　　　Order SQUAMATA (snakes)

　　Class AVES (birds; feathered reptiles)
　Subclass NEORNITHES (all living birds)
Superorder PALEOGNATHAE (flightless birds; ratites and tinamous)
　　　　Order CASUARIIFORMES (cassowaries)
　　　　Order DINORNITHIFORMES (moas and kiwis)
　　　　Order RHEIFORMES (rheas)

Order STRUTHIONEFORMES (ostriches)
Order TINAMIFORMES (tinamous)
Superorder NEOGNATHAE (modern birds)
  Order ANSERIFORMES (waterfowl; swans, ducks, geese)
  Order APODIFORMES (swifts, hummingbirds)
  Order CAPRIMULGIFORMES (nightjars)
  Order CHARADRIIFORMES (shorebirds, waders and seabirds: sandpipers, woodcocks, oystercatchers, plovers, gulls, terns, auks, puffins, murres, jaegers, skuas)
  Order CICONIIFORMES (herons, bitterns, storks, ibises, spoonbills,)
  Order COLIIFORMES (mousebirds)
  Order COLUMBIFORMES (pigeons, doves)
  Order CORACIIFORMES (kingfishers, hornbills, rollers, bee-eaters etc.)
  Order CUCULIFORMES (cuckoos)
  Order FALCONIFORMES (birds of prey; vultures, ospreys, hawks, eagles, falcons)
  Order GALLIFORMES (gamebirds; pheasants, quail, turkeys, curassows, moundbuilders)
  Order GAVIIFORMES (divers; loons)
  Order GRUIFORMES (Rails, coots, cranes, limpkins, bustards)
  Order MUSCOPHAGIFORMES (turacos, plantaineaters)
  Order PASSIFORMES (perching birds: flycatchers, thrushes, warblers, creepers, etc.)
  Order PELECANIFORMES (pelicans, tropicbirds, frigatebirds, boobies, gannets, cormorants, anhingas)
  Order PHOENICOPTERIFORMES (flamingos)
  Order PICIFORMES (woodpeckers)
  Order PODICIPEDIFORMES (grebes)
  Order PROCELLARIIFORMES (seabirds; albatrosses, shearwaters, petrels, fulmars)
  Order PSITTACIFORMES (parrots, macaws, cockatoos, lories)
  Order PTEROCLIDIFORMES (sandgrouse)
  Order SPHENISCIFORMES (penguins)
  Order STRIGIFORMES (owls)
  Order TROGONIFORMES (trogons, quetzals)

Class MAMMALIA (mammals)
Subclass PROTOTHERIA (egg-laying mammals)
  Order MONOTREMATA (monotremes; duck-billed platypus, spiny anteater)
Subclass THERIA (mammals bearing live young)
Infraclass METATHARIA (marsupials)
  Order DIPRODONTIA (diprododont marsupials; kangaroos)
  Order NOTORYCTEMORPHIA (marsupial moles)
  Order PERAMELEMORPHIA (PERAMELIDA; bandicoots)
  Order DIDELPHIMORPHIA (American marsupials; opossums)
  Order PAUCITUBERCULATA (opossum rats)
  Order DASYUROMORPHIA (marsupial carnivores)
  Order MICROBIOTHERIA (*Dromicops sp.*)
Infraclass EUTHERIA (placental mammals)
  Order ARTIODACTYLA (ungulates: even-toed hoofed mammals)

      Suborder SUINA (pigs, hippopotamus, etc.)
      Suborder TYLOPODA (camels, etc.)
      Suborder RUMINANTIA (deer, antelope, etc.)
Order CARNIVORA (carnivores)
Order CETACEA (whales and dolphins)
      Suborder ODONTOCETI (toothed whales and dolphins)
      Suborder MYSTICETI (baleen whales)
Order CHIROPTERA (bats)
Order DERMOPTERA (flying lemurs)
Order HYRACOIDEA (hyraxes)
Order HYSTRICOGNATHI (hystricognath rodents; porcupine, cane rats, mole rats, guinea pigs)
Order INSECTIVORA (insectivores; shrews, hedgehogs, moles)
Order LAGOMORPHA (lagomorphs; hares and rabbits)
Order MACROSCELIDEA (elephant shrews)
Order PERISSODACTYLA (ungulates; odd-toed hoofed mammals)
      Suborder HIPPOMORPHA (horses)
      Suborder CERATOMORPHA (rhinoceros and tapirs)
Order PHOLIDOTA (pangolins; scaly anteaters)
Order PINNIPEDIA (seals, sea lions, walruses)
Order PRIMATES (primates: humans, apes, chimpanzees, monkeys)
      Suborder ANTHROPOIDEA (monkeys, apes, man)
Order PROBOSIDEA (elephants)
Order RODENTIA (rodents)
Order SCANDENTIA (PROSIMII; tree shrews, lemurs)
Order SIRENIA (sirens; dugongs, manatees)
Order TUBULIDENTATA (aardvark; ant bear)
Order XENARTHRA (EDENTATA; sloths, armadillos, anteaters)

References

Barnes, R. S. K. (ed). *A Synoptic Classification of Lliving Organisms.* Oxford, U. K.: Blackwell Scientific, 1984.

Borror, D. J. and R. E. White. *Peterson Field Guides: Insects.* New York: Houghton Mifflin, 1970.

Gill, F. B. *Ornithology*, New York: W.H. Freeman & Co., 1989.

Krieg, N. R., and J. G. Holt, (eds). *Bergey's Manual of Systematic Bacteriology*, Vol. 1, 9th ed. Baltimore: Williams & Williams, 1994.

Margulis, L., J. Corliss, M. Melkonian, and D. Chapman, *Handbook of Protoctista*, Boston: Jones and Bartlett, 1990.

Margulis, L., H. McKhann, and L. Olendzenski (eds). *Illustrated Glossary of Protoctista.* Boston: Jones and Bartlett, 1993.

Margulis, L. and K. V. Schwartz. *Five Kingdoms,* 3rd ed. New York: W. H. Freeman, 1997.

Raven, P .H. and G. B. Johnson. *Biology*, St. Louis: Mosby-Year Book, Inc., 1992.

Wilson, D. E., and D. M. Reeder (eds). *Mammal Species of the World*, Washington, D.C.: Smithsonian Institution Press, 1993.

◆ APPENDIX XIII ◆

TIMELINE

The following timeline is intended as a series of snapshots taken throughout the history of biology. A summary of this length cannot be comprehensive. The exclusion of any particular person or event is not intended as a commentary on the importance of their individual contributions to the development of biology.

The Ionians

640?–546 B.C.   **Thales** was an Ionian philosopher who was one of the first to apply rational thought to the natural world.

500 B.C.   **Alcmaeon** was the first man to dissect animals for merely descriptive purposes rather than for divine messages. He discovered the tube connecting the ear to the throat, a piece of knowledge that was then lost until rediscovered by Eustachio in the 1500s.

460?–377? B.C.   **Hippocrates** was the first major name to be associated with modern biology and medicine. He removed the remnants of perceived divine intervention from the workings of the human body and founded the basis for modern medicine. Ironically, the "Hippocratic oath" was most likely composed at least six centuries after he lived.

The Greeks

384–322 B.C.   **Aristotle** was the first to categorize animals on the basis of similar characteristics. He is considered the founder of zoology.

380–287 B.C.   **Theophrastus** was a student of Aristotle's who initiated the categorization of plants and founded the science of botany.

~300 B.C.   **Herophilus** dissected the human body and defined the brain as the seat of intelligence.

~250 B.C.   **Erasistratus,** a pupil of Herophilus, connected convolutions in the brain with intelligence.

The Romans

~30   **Aulus Cornelius Celsus** collected and preserved a large body of Greek medical knowledge.

~60   **Dioscorides** collected and categorized hundreds of plant species. He paid special attention to their medical qualities and may be considered a founder of pharmacology.

23–79   **Gaius Plinius Secundus (Pliny)** wrote a thirty-seven volume encyclopedia collecting the knowledge of natural history, mostly from Aristotle. He also included a fair amount of superstition and was highly biased toward a human-centric view of the world.

130–200   **Galen** was the last real biologist of the ancient world. He performed many dissections, mostly limited to animals, and attempted to extend his knowledge to the workings of the human body.

The Dark Ages

980–1037   **abu-'Ali al-Husein ibn-Sïna (Avicenna)** was a Moslem Persian physician who wrote numerous books based on the medical theories of Hippocrates and Celcus.

1114–1187   **Gerard of Cremona**, an Italian scholar, translated the works of Hippocrates, Galen, and some of Aristotle into Latin.

1206–1280   **Albertus Magnus** was a German scholar who helped rediscover and propagate the teachings of Aristotle, reforming a foundation on which modern science could advance.

1225–1274   **Thomas Aquinas**, a pupil of Magnus, was an Italian rationalist who attempted to reconcile Aristotelian philosophy with Christian faith. His reasoning presumed a God-created mind that by definition could not arrive at a conclusion that was at odds with Christian teaching.

The Renaissance

1275–1326   **Mondino de'Luzzi** was an Italian anatomist who, against the tradition of the time, actually performed his own dissections, not leaving them to an underling. In his anatomy books, he rectified some, although by no means all, errors that had been perpetuated through lack of proper investigation.

1452–1519   **Leonardo da Vinci** was the most famous of the Italian artist-anatomists who dissected humans and other animals, and accurately illustrated the findings. He was the first to notice anatomical homology, laying groundwork for theories of evolution. Most of his work was not discovered until modern times.

1493–1541   **Theophrastus Bombastus von Hohenheim (Paracelcus)** was a Swiss physician who is considered a link to modern biology. Although his own theories were not much better than those of the ancients, his iconoclasm provided the impetus for novel thought and the eventual discarding of outdated Greek science.

1516–1565   **Konrad von Gesner** was a Swiss naturalist who accumulated a vast amount of knowledge, unfortunately without much regard for its veracity.

The Birth of Modern Biology

1543   **Nicolaus Copernicus**, a Polish astronomer, published his book describing the sun as the center of the universe. Although not directly relevant to biology, it marks the beginning of the "scientific revolution."

1543	**Andreas Vesalius**, a Belgian anatomist, published his revolutionary book *On the Structure of the Human Body*, the first accurate rendition of human anatomy ever. He insisted on doing his own dissections and was brave enough to correct errors when his eyes disagreed with old Greek doctrine.
1523–1562	**Gabriello Fallopio (Gabriel Fallopius)**, one of Vesalius' pupils, studied the reproductive system in particular and discovered the Fallopian tubes.
1500–1574	**Bartolommeo Eustachio (Eustachius)** was an opponent of Vesalius but also studied the human body. He rediscovered the Eustachian tube (see **Almaeon**), running from ear to throat.
1517–1590	**Ambroise Pare** was a French surgeon who rose from his beginnings as a barber-surgeon to reinstitute some of the gentler, more effective healing practices espoused by Hippocrates.
1537–1619	**Hieronymus Fabrizzi (Fabricius)** was an Italian anatomist who discovered valves in veins and went so far as to propose that they delayed (as opposed to stopped) backward flow.
1553–1617	**Prospero Alpini** was an Italian botanist who rediscovered sexuality in plants and was the first European to describe the coffee plant.
1564–1642	**Galileo Galilei** popularized the experimental method as an approach to science.
1577–1644	**Jan Baptista van Helmont**, a Flemish alchemist, performed the first chemical experiments on living organisms, in particular observing plant processes. He is considered the inventor of biochemistry.
1596–1650	**René Descartes,** a French philosopher described the physical human body as a complex mechanical device, with the exception of the mind and soul.
1608–1679	**Giovanni Alfonso Borelli,** an Italian physiologist, analyzed muscular action by treating muscles and bones as a system of levers.
1614–1672	**Franz de la Boe (Fransiscus Sylvius)** suggested the basic chemical nature of the workings of the body.
1620	**Francis Bacon**, an English philospher, published *Novum Organum*, the first formal description and theoretical backing of the scientific method for experimental science.
1628	**William Harvey**, an English student of Fabricius, studied the circulatory system in even greater detail and proposed unidirectional flow in his book *On the Motions of the Heart and Blood*.

1637–1680   **Jan Swammerdam**, a Dutch naturalist, was one of the first to use a primitive microscope to view and describe insects and blood.

1641–1712   **Nehemiah Grew** was an English botanist who studied plants, in particular their reproductive organs, under the new microscope lens. He described pollen grains.

1641–1673   **Regnier de Graaf** was a Dutch anatomist who first studied the reproductive organs of animals under the microscope. He described Graafian follicles in ovaries.

1628–1694   **Marcello Malpighi,** an Italian physiologist, was the first to detect capillaries, while studying frog lungs under the microscope.

1632–1723   **Anton van Leeuwenhoek**, a Dutch merchant to whom microscopy was an all-absorbing hobby, was the first to produce ground glass lenses free of interfering irregularities. He observed many phenomena, including spermatozoa, red blood cells moving through capillaries, protists (which he called animalcules) and even bacteria. Thus, he instituted the study of microbiology.

1635–1703   **Robert Hooke**, an English scientist, was the first to describe cells, which he observed through the microscope in a thin slice of cork. In 1665, he published *Micrographia*, a compilation of microscopic drawings.

1688   **Francesco Redi**, an Italian physician, performed experiments that disproved the theory of spontaneous generation. He showed that maggots developed on putrefying meat only when flies had access to it. His views were not universally accepted for some time.

1628–1705   **John Ray**, an English naturalist, was the first to make a major attempt at species classification. He published three volumes categorizing 18,600 plants and also a book on animals.

1660–1734   **Georg Ernst Stahl,** a German physician, published a book on medicine in 1707 in which he stated that living organisms were not governed by the same physical laws that applied to chemistry and physics. This was the theory of *vitalism*.

1668–1738   **Hermann Boerhaave**, a Dutch physician, published a book attempting to explain all biological functions in terms of chemistry and physics. This was the theory of *mechanism*.

1683–1757   **Rene Antoine Ferchault de Reaumur**, a French physicist, studied the process of digestion by using hawks, which regurgitate indigestible matter.

1707–1778   **Carl von Linne (Carolus Linnaeus)**, a Swedish naturalist, developed the system for classification of living organisms that was the direct ancestor of that which is used today.

1707–1788	**Georges Louis Leclerc, Compte de Buffon**, a French naturalist, was the first to point out vestigial structures and suggest a relationship to the, as yet unnamed concept of evolution.
1713–1781	**John Turberville Needham** apparently proved the theory of spontaneous generation by showing that microorganisms arose from boiled mutton broth. His efforts were obviously not enough to sterilize the broth.
1677–1761	**Stephen Hales**, an English botanist and chemist, published a book in 1727 describing experiments in plant physiology. He was the first to recognize that carbon dioxide provided nourishment to plants. He also decapitated a frog and was the first to observe automatic reflex action.
1765	**Lazzaro Spallanzani**, an Italian biologist, improved on Needham's experiment by boiling a nutritive solution for more than 30 minutes and showing that microorganisms then did not appear in a sealed flask. The skeptics continued to defer.
1731–1802	**Erasmus Darwin**, grandfather of Charles Darwin, was an English physician who wrote long poems about botany and zoology in which he accepted Linnaean classification. He also recognized the possibility of changes in species in response to the environment.
1733–1804	**Joseph Priestley**, an English chemist, discovered oxygen gas in 1774.
1733–1794	**Caspar Friedrich Wolff**, a German physiologist, published a book in 1759 describing his observations on developing plants, including the differentiation of embryological tissue. He later extended his observations to chick embryology.
1730–1799	**Jan Ingenhousz**, a Dutch physician, showed that the process by which plants consumed carbon dioxide and released oxygen occurred only in the presence of light.
1743–1794	**Antoine Laurent Lavoisier**, a French chemist, developed quantitative chemistry and used it to explain combustion. He demonstrated that respiration was a form of combustion.
1744–1829	**Jean Baptiste de Monet Chevalier de Lamarck**, a French naturalist, refined Linnaeus' classification system and introduced the concept of vertebrates and invertebrates. He is better known, however, for his suggestion that organs and structures grew or degenerated according to their use during life and that these characteristics were then inherited. The neck of the giraffe was his most infamous example. This theory is referred to as inheritance of acquired characteristics and, though defunct, opened the floodgates for a discussion of evolutionary theories.

1760   **Joseph Koelreuter,** a century before Mendel, was the first botanist to make and test plant hybrids. He noted the segregation of traits and the masking of alternative alleles but did not quantify his data.

1766–1834   **Thomas Robert Malthus,** an English economist, wrote a book entitled *An Essay on the Principle of Population*, which was to influence the evolutionary views of both **Darwin** and **Wallace** greatly.

1771–1802   **Marie François Xavier Bichat,** a French physician, showed that complex organs were made up of only a few varieties of tissues that were similar between organisms. He initiated the study of histology.

1747   **James Lind,** a Scottish physician, identified citrus fruit as providing a nutrient lacking in the diets of scurvy-ridden sailors.

1760s   **Albrecht von Haller,** a Swiss physiologist, first recognized nerves as irritable tissue and showed that sensory nerves all lead to the brain or spinal cord.

1773   **Otto Friedrich Müller,** a Danish microbiologist, was the first to see bacteria well enough to describe the shapes and forms of the various types.

1791   **Luigi Galvani,** an Italian anatomist, first discovered that the muscle of a dissected frog could be made to twitch under electrical stimulation.

1796   **Edward Jenner,** an English physician, inoculated a boy with the mild disease cowpox obtained from a milkmaid and showed that the boy was then immune to the deadly smallpox. He coined the word vaccine.

1796   **Franz Joseph Gall,** a German physician, began lecturing on the subject of the brain, in particular the concept of its functional division. This, unfortunately, was taken to extremes when his later followers developed the pseudoscience of phrenology—the ability to decipher emotional qualities by feeling bumps on the skull.

1785   **James Hutton,** a Scottish physician, suggested that the earth must be many millions of years old, much older than previously thought. This was to be a key point in evolutionary theory.

1807   **Jöns Jakob Berzelius,** a Swedish chemist, suggested that substances obtained from living (or once-living) organisms be termed organic. In 1836 he suggested the name catalysis for the breakdown of complex substances.

1800s   **William Smith,** an English surveyor turned geologist, was the first to identify rock strata by their fossil content.

1800s   **Georges Léopold Cuvier**, a French biologist, systematically studied fossils and founded the sciences of comparative anatomy and paleontology. He also extended Lamark's work on Linnaean classification and based his system on structure-function relationships rather than superficial similarities. His religious beliefs prevented him from contemplating evolutionary change.

1810   **Augustin Pyramus de Candolle**, a Swiss botanist, applied Cuvier's classification system to plants.

1816   **François Magendie**, a French physiologist, in experiments on dogs, showed that not only protein but complete protein was necessary for life.

1820   Achromatic microscopes were produced, eliminating the rings of color that had previously obscured small objects.

1827   **William Prout**, an English physician, was the first to recognize the three main groups of organic compounds now called carbohydrates, lipids, and proteins.

1827   **Karl Ernst von Baer**, a Russian biologist, discovered the mammalian egg within the ovary and went on to study the manner in which it developed. He founded the study of embryology.

1830s   **Johannes Evangelista Purkinje**, a Czech physiologist, first named protoplasm as the living substance within an egg. The Purkinje fibers in the heart as well as some specialized neurons in the brain are named for him.

1831   **Robert Brown**, a Scottish botanist, was the first to recognize a dense region near the center of each cell, now known as the nucleus.

1833   **Anselme Payen**, a French chemist, extracted a substance from sprouting barley that he termed diastase, and we now know as the enzyme amylase.

1838   **Matthias Jakob Schleiden**, a German botanist, proposed that all plants were composed of cells. The next year, **Theodor Schwann**, a German physiologist, extended the ideas of Schleiden to include animals. Together they are credited with the *cell theory* and initiating the science of cytology.

1840   **Emil Du Bois-Reymond**, a German physiologist, showed that nerve impulses were electrical in nature and that the passage of an impulse was accompanied by a change in the electrical state of the nerve.

1840s   **Freidrich Wohler**, a German chemist, inadvertently performed the first recognized synthesis of an organic compound out of the body. He produced urea by heating ammonium cyanate.

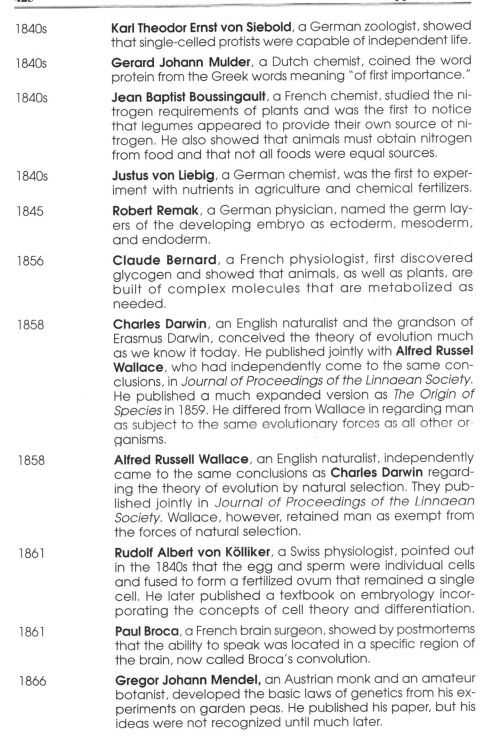

1840s	**Karl Theodor Ernst von Siebold**, a German zoologist, showed that single-celled protists were capable of independent life.
1840s	**Gerard Johann Mulder**, a Dutch chemist, coined the word protein from the Greek words meaning "of first importance."
1840s	**Jean Baptist Boussingault**, a French chemist, studied the nitrogen requirements of plants and was the first to notice that legumes appeared to provide their own source of nitrogen. He also showed that animals must obtain nitrogen from food and that not all foods were equal sources.
1840s	**Justus von Liebig**, a German chemist, was the first to experiment with nutrients in agriculture and chemical fertilizers.
1845	**Robert Remak**, a German physician, named the germ layers of the developing embryo as ectoderm, mesoderm, and endoderm.
1856	**Claude Bernard**, a French physiologist, first discovered glycogen and showed that animals, as well as plants, are built of complex molecules that are metabolized as needed.
1858	**Charles Darwin**, an English naturalist and the grandson of Erasmus Darwin, conceived the theory of evolution much as we know it today. He published jointly with **Alfred Russel Wallace**, who had independently come to the same conclusions, in *Journal of Proceedings of the Linnaean Society*. He published a much expanded version as *The Origin of Species* in 1859. He differed from Wallace in regarding man as subject to the same evolutionary forces as all other organisms.
1858	**Alfred Russell Wallace**, an English naturalist, independently came to the same conclusions as **Charles Darwin** regarding the theory of evolution by natural selection. They published jointly in *Journal of Proceedings of the Linnaean Society*. Wallace, however, retained man as exempt from the forces of natural selection.
1861	**Rudolf Albert von Kölliker**, a Swiss physiologist, pointed out in the 1840s that the egg and sperm were individual cells and fused to form a fertilized ovum that remained a single cell. He later published a textbook on embryology incorporating the concepts of cell theory and differentiation.
1861	**Paul Broca**, a French brain surgeon, showed by postmortems that the ability to speak was located in a specific region of the brain, now called Broca's convolution.
1866	**Gregor Johann Mendel**, an Austrian monk and an amateur botanist, developed the basic laws of genetics from his experiments on garden peas. He published his paper, but his ideas were not recognized until much later.

1850s	**Francis Galton**, a first cousin of **Charles Darwin**, was the first to stress the study of identical twins in deciphering nature vs. nurture. His research into fingerprints provided the British police with the first forensic fingerprinting system. He also did the world the disfavor of initiating the concept of eugenics.
1850s	**Pierre Eugène Marcelin Berthelot** synthesized many simple organic compounds from inorganic starting materials in the laboratory.
1850s	**H. B. D. Kettlewell** discovered industrial melanism among the peppered moths of England.
1856	An old, humanlike skull was unearthed in the Neanderthal valley of Germany's Rhineland.
1856	**Louis Pasteur**, a French chemist, was called in by the government to improve the efficiency of the beer and wine industry by preventing spoilage. He did this by inventing the process of pasteurization. He also performed the key experiment disproving spontaneous generation utilizing sterilized broth in flasks with S-shaped necks designed to trap any airborne particle.
1856	**W. H. Bates** first defined Batasian mimicry.
1860	**Ernst Heinrich Haeckel** popularized the overstated biogenetic law that ontogeny recapitulated phylogeny.
1878	**Fritz Mueller** first defined Mullarian mimicry.
1860s	**Ernst Heinrich Haeckel** proposed the overly extreme theory of recapitulation, where every step of evolutionary history is faithfully replayed in embryonic development.
1869	**Friedrich Miescher**, a Swiss biochemist, first discovered the existence of nucleic acids (which he called nuclein) in the cell nucleus.
1870	**Gustav Theodor Fritsch** and **Eduard Hitzig**, two German neurologists, exposed the brain of a living dog and were the first to map brain functions physically.
1872	**Ferdinand Julius Cohn**, a German botanist, made the first major attempt to classify bacteria.
1873	**Camillo Golgi**, an Italian cytologist, developed a cell stain of silver salts and revealed what we now call Golgi bodies. He also applied the technique to investigating neurons. He shared the Nobel prize in physiology or medicine with **Cajal** in 1906 for their work on the structure of the nervous sytem.
1876	**Robert Koch**, a German physician, was the first to culture bacteria on solid media, enabling the isolation of pure strains, including, in 1882, the cause of tuberculosis. His assistant, **Julius Richard Petri,** added a shallow dish with a cover, creating the Petri dish, which is used to this day in microorganismal research. Koch received the Nobel prize in physiology or medicine in 1905.

1876	**Wilhelm Kuhne**, a German physiologist, coined the word enzyme.
1879	**Hermann Fol,** a Swiss zoologist, witnessed the fertilization of a starfish egg by a spermatozoa.
1879	**Wilhelm Wundt**, a German physiologist, set up the first laboratory dedicated to experimental psychology.
1880	**Sydney Ringer**, an English physician, first developed an artificial solution, now known as Ringer's solution, in which organs, tissues, and cells could be kept alive outside the body.
1880s	**Sigmund Freud**, the Austrian physician-psychiatrist, first studied the ability of cocaine to deaden nerve endings before he turned his attentions to psychiatry. **Carl Koller** was the first to use it as a local anesthetic.
1880s	**Ivan Petrovich Pavlov**, a Russian physiologist, first introduced the concept of a conditioned reflex by inducing dogs to salivate at the sound of a bell. His work initiated the behaviorist school of psychology. American psychologists **John Broadus Watson** and **Burrhus Frederic Skinner** took this way of thinking to an extreme. Pavlov received the Nobel prize in physiology or medicine for his work on digestion.
1880s	**Albrech Kossel,** a German biochemist, first broke nucleic acids down into building blocks each containing phosphoric acid, ringed nitrogen-containing molecules, and a sugar moiety. He named them adenine, thymine, guanine, cytosine, and uracil. He also showed that there were two kinds of nucleic acids, now known to be RNA and DNA. He received the Nobel prize in physiology or medicine in 1910.
1882	**Walther Flemming**, a German cytologist, working with new synthetic stains, published his findings on the first observations of mitosis, which he named, and the coalescence of chromatin into what later came to be called chromosomes.
1884	**Max Rubner** showed that the energy released by foodstuffs in the body was exactly the same as if they had been consumed in a fire.
1884	**Ilya Ilitch Mechnikov**, a French-Russian biologist, showed that white blood cells flocked to the site of an infection. He shared the Nobel prize in physiology or medicine with Erlich in 1908.
1884	**Hans Christian Joachim Gram,** a Danish bacteriologist, developed the stain that bears his name. It distinguished between two main classes of bacterial cell walls, the characteristics of which were found to determine differential antibiotic sensitivity.

1885   **Oscar Hertwig** suggested that Miescher's nuclein (what is now called nucleic acid) was responsible for fertilization and the transmission of hereditary characteristics.

1887   **Eduard Joseph Louis Marie van Beneden** demonstrated that the number of chromosomes was constant in all the cells of an organism and characteristic of each species. He was the first to note the result of reductive division, or meiosis, in haploid reproductive cells.

1891   **Wilhelm von Waldeyer**, a German anatomist, first suggested neuron theory and showed that there were gaps between neurons, which are now called synapses. **Ramon y Cajal,** a Spanish neurologist, improved on **Golgi's** silver stain and confirmed the neuron theory, as well as examining the cellular structure of the central nervous system and the retina of the eye. Cajal shared the Nobel prize in physiology or medicine with **Golgi** in 1906 for their work on the structure of the nervous sytem.

1892   **Emil Adolf von Behring** and **Paul Ehrlich**, both assistants of **Koch**, first used inoculation of serum already containing antibodies to confer immunity to diphtheria. They founded the areas of study of both serology and immunology. von Behring received the Nobel prize for medicine or physiology in 1901. Erlich shared the prize with **Mechnikov** in 1908.

1892   **Dmitri Iosifovich Ivanovski,** a Russian botanist, showed the infectious nature of tobacco mosaic virus that was smaller than any known bacterium. This was discovered independently in 1895 by Dutch botanist **Martinus Willem Beijernick**.

1893   **Joseph von Merin** and **Oscar Minkowski**, two German physiologists, showed that excision of the pancreas from experimental animals induced severe diabetes.

1893   **August Weismann**, a German physician, published *The Germ-Plasm; a Theory of Heredity* in which he predicted the necessity for a reduction division during the maturation of germ cells.

1894   **Marie Eugène Francois Thomas Dubois**, a Dutch paleontologist, searched for a human missing link and published his findings of *Pithecanthropus erectus* fossils in Java.

1897   **Eduard Buchner**, a German chemist, showed that a cell-free extract of yeast would ferment sugar. It was a key experiment that shattered the vitalistic view in favor of a mechanistic one.

1898   **Friedrich August Johannes Löffler**, a German bacteriologist, demonstrated that hoof-and-mouth disease was caused by a virus.

1898   **Jules Jean Baptiste Vincent Bordet**, a Belgian bacteriologist, identified a complementary blood component needed to destroy bacteria tagged by antibodies. It was termed complement by **Ehrlich**. In 1901, Bordet showed the phenomenon of complement fixation, which was used by the German bacteriologist **August von Wassermann** to diagnose syphilis. Bordet received the Nobel prize in physiology or medicine in 1919.

1900   **Hugo de Vries**, a Dutch botanist, rediscovered Mendel's genetic theories and was the first to recognize the concept of mutation. He found and cited Mendel's paper when preparing to publish his work, as did his contemporaries **Karl Erich Correns** and **Erich Tschermak von Seysenegg**.

1900   **Karl Landsteiner**, an Austrian physician, first identified blood groups and by 1902 had devised the ABO system. In 1908, he also isolated the poliomyelitis virus. He received the Nobel prize in physiology or medicine in 1930.

1900s   **Charles Scott Sherrington**, an English physiologist, demonstrated the existence of the reflex arc and coined the term synapse. He founded the science of neurophysiology.

1901   **Jokichi Takamine**, a Japanese-American chemist, isolated the substance from the adrenal glands now known to be adrenaline. It was the first hormone to be isolated and to later have its structure determined.

1901   **Walter Reed**, an American surgeon, demonstrated a virus as the cause of yellow fever.

1902   **Walter S. Sutton**, an American cytologist, was among the first to associate chromosomes with Mendel's inheritance factors. He was inspired, in part, by the experiments of **Theodor Boveri**.

1902   **Ernest Henry Starling** and **William Maddock Bayliss**, two English physiologists, observed the action of secretin on the pancreas and two years later coined the term hormone.

1902   **Archibald Edward Garrod** first reported alcaptonuria, the first recognized inborn error in metabolism.

1902   **Gottleib Haberlandt** first proposed that all living plant cells are totipotent.

1903   **Willem Einthoven**, a Dutch physiologist, invented the EKG. He received the Nobel prize in physiology or medicine in 1924.

1905   **Alfred Binet,** a French psychologist, published his first IQ tests.

1905   **Arthur Harden**, an English biochemist, was the first to study intermediary metabolism in a yeast extract.

1905	**Nettie M. Stevens**, one of the few early twentieth century American scientists, performed the first work connecting sex determination to chromosomes.
1906	**Mikhail Semenovich Tsvett**, a Russian botanist, performed the first chromatography experiment.
1908	**G. H. Hardy and G. Weinberg** developed the mathematical equations for determining genetic allele frequencies in populations.
1909	**Wilhelm Ludwig Johannsen**, a Danish botanist, gave the name genes to the heritable characteristics thought to be located on chromosomes.
1909	**William Bateson** published *Mendel's Principles of Heredity*, promoting and expanding on Mendel's theories of genetics.
1909	**F. A. Janssens** first noted the existence of chiasmata in nuclei undergoing meiosis.
1911	**Francis Peyton Rous**, an American physician, discovered Rous Sarcoma virus, the first demonstration of an infectious agent that causes tumors, in this case in chickens.
1911	**Thomas Hunt Morgan**, an American geneticist who first introduced the fruit fly as a model organism for genetic study in 1907, first showed that genes are located on chromosomes in a linear sequence. He, along with his associates, **Sturtevant**, **Bridges**, **Muller,** and **Stern**, also pioneered concepts such as linkage and recombination. With Sturtevent, he drew the first genetic map. Morgan received the Nobel prize in physiology or medicine in 1933.
1912	**Casimir Funk**, a Polish-born biochemist, coined the word vitamins, originally vitamines, because he though that they were all amines.
1913	**Leonor Michaelis**, a German-American chemist, derived a mathematical equation describing enzyme-catalyzed reactions. It included what is still called the Michaelis constant.
1913	**Alfred H. Sturtevant**, a student of **Morgan**, published the first chromosome map of sex-linked factors in *Drosophilia*.
1915	**Frederick William Twort**, an English bacteriologist, first observed the action of what is now known as bacteriophages. In 1917, the Canadian bacteriologist **Félix Hubert d'Herelle,** made the same discovery independently and coined the term.
1916	**Edward Calvin Kendall,** an American biochemist, first isolated thyroxine. He and others such as the Polish-Swiss chemist, **Tadeus Reichstein** and **Philip Showalter Hench** went on to isolate the whole family of corticoids. The three shared the Nobel prize in physiology or medicine in 1950.

1918   **Otto Fritz Meyerhof**, a German biochemist, was the first to in-
            vestigate glycogen-lactic acid metabolism during anaero-
            bic muscle contraction. **Archibald Vivian Hill**, an English
            physiologist, came to the same conclusions contempora-
            neously. The biochemical details of this metabolic pathway
            were worked out by the Czech-American husband-andwife
            team of **Carl Ferdinand** and **Gerty Theresa Cori**. Meyerhof
            and Hill shared the Nobel prize in physiology or medicine in
            1922. The Coris shared the prize with **Houssay** in 1947.

1920s   **Hans Karl von Euler-Chelpin**, a German-Swedish chemist,
            worked out the structure of the first known coenzyme,
            whose existence had been postulated by **Harden** in 1904.

1921   **Otto Loewi,** a German physiologist, first demonstrated the
            chemical changes that accompany a nerve impulse.
            Shortly thereafter, English physiologist **Henry Hallet Dale**,
            identified acetylcholine. They shared the Nobel prize in
            physiology or medicine in 1936.

1921   **Frederick Grant Banting**, a Canadian physician, and his as-
            sistant **Charles Herbert Best** initiated the studies that led to
            the isolation of insulin.

1923   **Otto Heinrich Warburg** devised a system for measuring the
            rate of oxygen used in tissue slices. Today, it is still called the
            Warburg apparatus. He received the Nobel prize in physi-
            ology or medicine in 1931.

1923   **Theodor Svedberg**, a Swedish chemist, invented the ultra-
            centrifuge. The unit of sedimentation, the Svedberg unit, is
            named after him.

1924   **Bernardo Alberto Houssay**, an Argentinean physician, iden-
            tified the pituitary gland as being involved with sugar me-
            tabolism. The Chinese-American biochemist **Cho Hao Li**
            went on to isolate a number of different pituitary hormones,
            including growth hormone. Houssay shared the Nobel prize
            in physiology or medicine with the **Coris** in 1947.

1926   **James Batchellor Sumner**, an American biochemist, surrep-
            titiously obtained the first ever enzyme crystals, those of
            urease in the jack bean. He was further able to prove that
            they were protein.

1926   **Hermann Joseph Muller,** an American geneticist and stu-
            dent of **Morgan**, first used X-rays to increase the mutation
            rate of genes, generating more mutants for study. He also
            advocated a subtle form of eugenics that discouraged
            people with bad genes from reproducing so as to rid the
            human gene pool of its mutational load. Muller received
            the Nobel prize in physiology or medicine in 1946.

1928	**Alexander Fleming**, a Scottish bacteriologist, first noticed that some mold colonies, which had accidentally grown in his cultures, were inhibiting bacterial growth around themselves. He identified the mold as *Penicillium* and investigated some of its effects but never isolated the active compound. He shared the Nobel prize in physiology or medicine with **Chain** and **Florey** in 1945.
1929	**Adolph Friedrich Johannes Butenandt**, a German chemist, began the isolation of the various sex hormones.
1929	**Hans Berger**, a German psychiatrist, invented the EEG.
1930	**John Howard Northrop**, an American biochemist, crystallized a number of different digestive enzymes and confirmed that all enzymes are mainly protein.
1931	**Curt Stern** first demonstrated the physical crossing-over of chromosomes during Drosophila meiosis and correlated Mendelian traits with microscopically visible chromosomal abnormalities.
1930s	**William Clouser Boyd**, an American immunologist, collected large amounts of worldwide data on blood groups and showed that they generally followed a logical geographical division. He could, to some extent, trace prehistoric migrations and invasions.
1930s	**Fritz Albert Lipmann**, a German-American biochemist, explained the importance of phosphate groups in biochemical metabolism by showing that they could occur in low-energy and high-energy forms within molecules, thus providing convenient storage for chemical energy. He shared the Nobel prize in physiology or medicine with **Krebs** in 1953 for his discovery of coenzyme A.
1930s	The first electron microscope was built. **Vladimir Kosma Zworykin**, a Russian-American physicist, modified and refined it to become a practical tool in cytology.
1930s	**Phoebus Aaron Theodeor Levene**, a Russian-American chemist, further defined the building blocks of nucleic acid, termed them nucleotides, and showed how they connected, via a sugar-phosphate backbone, to form long chains. He also distinguished the differences in sugar type and base content between DNA and RNA. At the time, his work led many to the conclusion that nucleic acid did not contain enough variability to transmit genetic information.
1930s	**Calvin Bridges,** an associate of **Morgan,** first demonstrated the relationship between stained bands on giant chromosomes and the linear sequence of genes.
1935	**Paul Müller**, a Swiss chemist, first synthesized DDT. Ironically, he received the Nobel prize in physiology or medicine in 1948 before the hazards of this potent insecticide became known.

1935	**Rudolf Schoenheimer**, a German-American biochemist, was the first to make use of radioactive isotopes, in particular ^2H (heavy hydrogen) in biochemical research. In particular, he showed that stored fat was broken down for energy. He also used ^{15}N to tag amino acids.
1935	**Wendell Meredith Stanley**, an American biochemist, using tobacco mosaic virus, obtained the first purified crystalline virus particles.
1935	**Hans Spemann** and **Hilde Mangold** first discovered the process of induction in embryological development.
1937	**Arne Wilhelm Kaurin Tiselius,** a Swedish chemist, designed the first electrophoresis apparatus.
1937	**Albert Francis Blakeslee**, an American botanist, first used chemicals, in particular colchicine, to increase the mutation rate of chromosomes.
1939	**René Jules Dubos**, an American microbiologist, isolated the first antibiotic, tyrothricin, from a soil bacterium
1939	**Howard Walter Florey**, an Australian-English pathologist, together with **Ernst Boris Chain**, a German-English biochemist, isolated penicillin and produced it in quantity. They shared the Nobel prize in physiology or medicine with **Fleming** in 1945.
1940	**Hans Adolf Krebs**, a German-British biochemist, worked out the main steps in the portion of oxidative respiration now known as Kreb's cycle. He shared the Nobel prize in physiology or medicine with **Lipmann** in 1953. **Albert Szent-Gyorgyi von Nagyrapolt,** a Hungarian biochemist, was also active in deciphering the intermediate steps of various biochemical reactions, with particular reference to vitamin C and fumaric acid. He received the Nobel prize in physiology or medicine in 1937.
1940	**G. F. Gause** first defined the principle of competitive exclusion.
1940s	**Robert Tyron** first showed the existence of a genetic component of behavior by breeding rats for their ability to run a maze.
1940s	**Barbara McClintock** first discovered mobile genetic elements in maize. Her work was largely ignored until similar elements were discovered in bacteria and higher organisms in the 1960s. She was finally awarded the Nobel prize in physiology or medicine in 1983.
1940s	**David Ezra Green** and his associates isolated mitochondria and showed them to be the site of Kreb's cycle.
1940s	**Alexander Robertus Todd**, a Scottish chemist, was the first to synthesize various nucleotides.

1940s	**Harold Clayton Urey**, an American chemist, was a leading force in postulating a primordial atmosphere in which the elements for the origin of life were present.
1942	**Salvador Edward Luria,** an Italian-born American microbiologist, was the first to visualize bacteriophages using the electron microscope.
1943	**Selman Abraham Waksman**, a Russian-American bacteriologist, isolated streptomycin and also coined the word antibiotic.
1944	**Archer John Porter Martin** and **Richard Laurence Millington Synge**, a pair of English biochemists, worked out an elegant chromatography technique using filter paper. Paper chromatography is still used today.
1944	**Erwin Shrödinger**, a physicist, published *What Is Life?*, a book that influenced many scientists of the day, including **Watson** and **Crick**.
1944	**Oswald Theodore Avery**, an American bacteriologist, confirming phenomena observed by the British bacteriologist **Fred Griffith**, first showed that genetic information was contained in nucleic acid alone. In a paper published with **Colin Macleod** and **Maclyn McCartny**, he showed that an extract from one strain of bacteria, taken up by another strain, was able to produce changes in its physical structure and contained only nucleic acid.
1945	**George Wells Beadle** and his associate, **Edward Lawrie Tatum**, formulated the one gene-one enzyme hypothesis using genetic mutants in *Neurospora*. They shared the Nobel prize in physiology or medicine with **Lederberg** in 1969.
1945	**Salvador Edward Luria** and the American microbiologist **Alfred Day Hershey** independently showed that viruses had the capacity to mutate. They shared the Nobel prize in physiology or medicine with Delbrück in 1969.
1946	**Max Delbrück**, a German-born American microbiologist who formed the American phage group in 1945, and **Alfred Hershey** independently showed that viruses could exchange genetic material. They shared the Nobel prize in physiology or medicine with **Luria** in 1969.
1946	**Joshua Lederberg** along with **Edward Tatum** proved that genetic recombination occurred as a result of mating in bacteria. He shared the Nobel prize in physiology or medicine with **Beadle** and Tatum in 1958.
1948	**Erwin Chargaff**, an Austrian-American biochemist, recognized the strict pairing rule in nucleic acids of cytosine with guanine and adenine with either thymine or uracil.

1948   **George Snell,** an American geneticist, first mapped the
                MHC locus in mice. He shared the Nobel prize for physiol-
                ogy or medicine with **Jean Dausset** and **Baruj Benacerraf** in
                1980 for their discoveries concerning genetically deter-
                mined structures on the cell surface that regulate immuno-
                logical reactions.

1949   **Frank Macfarlane Burnet,** an Austrian physician, first sug-
                gested that antibodies are formed during the lifetime of an
                organism. An English biologist, **Peter Brian Medawar**, tested
                the hypothesis by inoculating mouse embryos with cells
                from another strain and showed that skin grafts from that
                strain to the adult mouse would not be rejected. They
                shared the Nobel prize for physiology and medicine in 1960
                for their discovery of acquired immunological tolerance.

1950   **Albert Claude**, a Belgian cytologist, discovered the endo-
                plasmic reticulum using the electron microscope. He
                shared the Nobel prize in physiology or medicine with
                George Palade and Christian de Duve in 1974.

The new age of molecular biology

1951   **Linus Pauling**, the American chemist who was to become
                famous for espousing the health benefits of vitamin C, first
                derived the helical form taken by amino acid chains. He
                received the Nobel Prize in chemistry in 1954 as well as the
                Nobel peace prize in 1962.

1950s   **Maurice High Frederick Wilkins**, a New Zealand-born British
                X-ray crystallographer, studied nucleic acids using X-ray dif-
                fraction, accumulating the data that would lead to the
                solving of the structure of DNA by **Watson** and **Crick**. He
                shared the Nobel prize for physiology or medicine with
                Watson and Crick in 1962.

1950s   **Ulf Von Euler,** a Swedish physiologist, identified the role of
                epinephrine and discovered prostoglandins. **Julius Axelrod,**
                an American biochemist, performed research on neural
                transmitters that eventually led to the development of
                Tylenol. They shared the 1970 Nobel prize for physiology or
                medicine with **Sir Bernard Katz**, a British biophysicist, for their
                independent work on the storage and release of humoral
                neurotransmitters from nerve terminals.

1952   **Robert William Riggs** and **Thomas J. King** were the first to
                transplant nuclei from one frog embryo to another, prov-
                ing that the nucleus is the site of genetic information in
                eukaryotes.

1952   **Rosalind Franklin**, a British X-ray crystallographer, studied
                nucleic acids using X-ray diffraction, accumulating the
                data that would lead (somewhat surreptitiously) to the solv-
                ing of the structure of DNA by **Watson** and **Crick**. She was
                not acknowledged in the 1962 Nobel prize for the discov-
                ery of the structure of DNA.

1952	**Rosalyn Sussman Yalow**, an American biophysicist, developed the radioimmune assay. She shared the Nobel prize for physiology or medicine with **Roger Guillemin** and **Andrew Schally** in 1977.

1952   **Alfred Hershey**, along with **Martha Chase**, performed a key experiment suggesting that DNA served as the genetic material. They labeled protein and DNA respectively with radioactive sulphur and phosphorus to follow the path of phage infection in their famous pulse-chase experiment. He received the Nobel prize in physiology or medicine in 1969.

1952   **Rita Levi-Montalcini** discovered nerve growth factor. She shared the Nobel prize for physiology or medicine with **Stanley Cohen** in 1986.

1953   **Fredrick Sanger**, an English biochemist, using partial digestion and paper chromatography, deduced the exact order of amino acids in the insulin molecule. He received his first Nobel prize for chemistry in 1958.

1953   **Francis Harry Compton Crick,** an English biochemist, and **James Dewey Watson**, an American biochemist and former "Quiz Kid" using the data obtained from both **Wilkins** and **Franklin** (in this case surreptitiously) and inspired by **Pauling's** elucidation of the helical nature of proteins, worked out the double helical structure of DNA. They shared the Nobel prize for physiology or medicine with Wilkins in 1962.

1953   **Stanley Lloyd Miller**, an American chemist and a student of Urey, first recreated the possible origins of life on earth in the laboratory. He circulated the gases presumed to be present in the primordial atmosphere past an electric discharge and was able to produce simple organic compounds including some small amino acids. Similar experiments, with slightly varying conditions, have since produced even more impressive results.

1954   **Vincent du Vigneaud**, an American biochemist, produced the first synthetic hormone, oxytocin, and showed it to be identical to the natural hormone.

1955   **Heinz Fraenkel-Conrat**, a German-American biochemist, was the first to dissect a virus into its nucleic acid and protein portions, and reconstitute them, restoring its infectivity. He determined that in some viruses, RNA is the carrier of genetic information.

1956   Cyanocobalamine (vitamin B_{12}), first isolated in 1948, was the first molecule to be solved using computer-aided analysis of X-ray diffraction patterns.

1956   **George Emil Palade**, a Romanian-born American physiologist, discovered and named ribosomes. He shared the Nobel prize for physiology or medicine with **de Duve** and **Albert Claude** in 1974.

1956	**Arthur Kornberg** isolated the first DNA polymerase from *E. coli.* He shared the 1959 Nobel prize for physiology or medicine with his former mentor, **Severo Ochoa,** for their discovery of the mechanisms of nucleic acid synthesis.
1957	The existence of transfer RNAs (tRNAs) was confirmed by biochemists.
1957	**Vernon M. Ingram**, a British biochemist, following the work of **Linus Pauling**, identified a one amino acid change in the hemoglobin molecule responsible for sickle cell anemia.
1957	**Melvin Calvin**, an American biochemist, along with **Benson**, had been using radioactive ^{14}C since 1948 to work out many of the steps of photosynthesis, which he essentially completed around this time. The light independent reactions are still called the Calvin-Benson cycle. He was awarded the Nobel prize for chemistry in 1961.
1957	**Alick Isaacs**, a British bacteriologist, headed a group that discovered interferons.
1958	**F. C. Steward** first showed that, under the appropriate circumstances, individual plant cells are totipotent.
1958	**Matthew Meselson**, a graduate student with **Pauling** and **Frank W. Stahl,** a postdoctoral associate with **Delbrück**, first traced the fate of individual DNA strands in and proved the mechanism of semiconservative replication.
1958	**Jacques Lucien Monod** and **François Jacob**, both French molecular biologists, first put forth the operon theory of gene regulation. In 1965, they shared the Nobel prize in physiology or medicine with **André Lwoff** for their combined work on the genetic control of enzyme and virus synthesis.
1958	**Francis Crick** first enumerated the central dogma of molecular genetics.
1959	**Max Ferdinand Perutz**, an Austrian-British biochemist, used computer analysis of X-ray diffraction patterns to produce the first three-dimensional picture of the amino acid arrangement in a protein, hemoglobin. His student **Sir John Cowdery Kendrew**, did the same for myoglobin in 1960, and they shared the Nobel prize for chemistry in 1962.
1960	**Earl Wilbur Sutherland, Jr.**, an American pharmacologist, worked out the second messenger mechanism of hormone action. He was the sole recipient of the Nobel prize for physiology or medicine in 1971.
1960s	**Konrad Lorenz**, an Austrian zoologist, performed experiments on animal imprinting and is considered the founder of modern ethology. **Karl von Frisch**, an Austrian zoologist, deciphered the language of bees. **Nikolaas Tinbergen,** a Dutch zoologist who coined the fight-or-flight response, applied the theory of behavior patterns to humans. They shared the 1973 Nobel prize for physiology or medicine.

1960s	**Werner Arber,** a Swiss microbiologist, and his colleagues worked out the restriction modification system in bacteria that led to the discovery of restriction enzymes. Arber shared the Nobel prize for physiology or medicine with Nathans and Smith in 1978.
1961	**Marshall Nirenberg** used synthetic RNA in a cell-free system to decipher the first word in the genetic code—UUU for phenylalanine. The full genetic code was eventually worked out by several investigators including **Heinrich Matthaei, Severo Ochoa,** and **Har Gobind Khorana.** Nirenberg shared the Nobel prize for physiology or medicine with **Holley** and Khorana in 1968.
1965	**Robert William Holley** and his coworkers determined the first complete tRNA sequence, for alanine. He shared the 1968 Nobel prize for physiology or medicine with Khorana and Nirenberg.
1965	**Robert Bruce Merrifield** , an American biochemist, and **David Phillips**, a Welsh biochemist, independently first synthesized proteins in the laboratory, insulin and lysozyme, respectively.
1967	**Robert MacArthur** and **E. O. Wilson** popularized the concept of carrying capacity.
1967	**John B. Gurden**, a British biologist, applied nuclear transplantation to a frog and produced the first vertebrate clone.
1969	**Gerald Maurice Edelman**, an American biochemist, first deduced the structure of gamma globulin. He shared the 1972 Nobel prize for physiology or medicine with **Rodney Porter,** a British biochemist, for their work on the chemical structure of antibodies.
1970s	**Cesar Milstein,** an Argentinian-born British immunologist, developed monoclonal antibodies. He shared the 1984 Nobel prize in physiology or medicine with **Niels Kai Jerne** and **Georges J. F. Kohler** for their previous work on how antibodies are produced in response to antigens.
1970s	**Susumu Tonegawa,** a Japanese immunologist and molecular biologist, discovered the genetic basis for generation of antibody diversity. **Leroy Hood** and **Philip Leder** also contributed to these studies. Tonegawa received the Nobel prize in physiology or medicine in 1987.
1970s	**Edwin G. Krebs** and **Edmond H. Fisher** made important discoveries concerning reversible protein phosphorylation as a biological regulatory mechanism. They shared the Nobel prize in physiology or medicine in 1992.
1970	**Har Gobind Khorana** and his associates synthesized the first artificial gene, the tRNA for alanine.

1970	**Howard Temin** and **Satoshi Mizutani** reported the discovery of reverse transcriptase in Rous Sarcoma virus. Soon after, **David Baltimore** discovered the same enzyme in Rauscher murine leukemia virus. Temin and Baltimore shared the 1975 Nobel prize in physiology or medicine with **Renato Dulbecco.**
1970	**Hamilton Othanel Smith** and **Daniel Nathans**, American microbiologists, discovered restriction enzymes, thus officially initiating the age of recombinant DNA. They shared the Nobel prize in physiology or medicine with Werner Arber in 1978.
1970s	**E. O. Wilson, Richard Alexander,** and **Robert Trivers**, American scientists, initiated the study of sociobiology.
1970s	**Bruce Ames** developed the Ames test, using bacteria to test for carcinogenic compounds that could be mutagenic to humans.
1970s	**Erwin Neher** and **Bert Sakmann**, German biophysicists, developed the patch-clamp technique for nerve research. They shared the Nobel prize in physiology or medicine in 1991.
1970s	**Martin Robdell** discovered the role of G-proteins in signal transduction in cells. He shared the 1994 Nobel prize in physiology or medicine with **Alfred Gilman.**
1973	**Stanley Cohen** and **Herbert W. Boye**r were the first to engineer recombinant DNA and propagate it in bacteria, creating clones.
1975	**E. M. Southern** developed the Southern blot, one of the most used tools in molecular biology.
1976	**J. Michael Bishop**, an American virologist, and **Harold Varmu**s, an American physiologist, discovered the cellular origin of retroviral oncogenes. They shared the Nobel prize in physiology or medicine in 1989.
1977	**Richard Roberts**, an English-born American molecular biologist, and **Phillip Sharp**, an American molecular biologist, independently demonstrated the existence of introns. They shared the 1993 Nobel prize in physiology or medicine.
1977	**A. M. Maxam** and **W. Gilbert** developed a method of sequencing DNA, as did **F. Sanger** and his colleagues. Gilbert and Sanger shared the 1980 Nobel prize in physiology or medicine with **Paul Berg**, for his studies of recombinant DNA.
1977	**Alfred G. Gilman** identified and purified the G-protein originally discovered by **Robdell**. They shared the 1994 Nobel prize in physiology or medicine.
1978	**Robert Weinberg**, an American scientist, and his colleagues, first discovered and began to characterize oncogenes.

1979	**James Lovelock** and **Lynn Margulis** first proposed the Gaia hypothesis.
1980	**Luis Alvarez** first suggested that the extinction of the dinosaurs could be accounted for by a catastrophe such as a meteorite or asteroid.
1980	**David Botstein**, an American molecular biologist, and his colleagues, discovered restriction fragment length polymorphisms (RFLPs) and began mapping the human genome.
1980s	**Tom Cech** and **Sydney Altman** demonstrated the existence of protein-free, self-splicing RNA molecules known as ribozymes. They shared the 1989 Nobel prize for chemistry.
1980s	**Michael Smith** and others developed oligonucleotide site-directed mutagenesis of DNA. Smith shared the 1993 Nobel prize in chemistry with **Kary Mullis.**
1984	**Alec Jeffreys**, a British molecular biologist, first introduced DNA fingerprinting.
1986	**Kary B. Mullis** invented the polymerase chain reaction (PCR), which was further developed by **Henry Erlich** and others at Cetus Corporation. Mullis shared the 1993 Nobel prize in chemistry with **Michael Smith.**
1990	**R. Michael Blaese** and **W. French Anderson**, American physicians, conducted the first successful human gene therapy.
1990	The Human Genome Project was officially launched.

References

Asimov, I. *A Short History of Biology*, The Natural History Press, NY, 1964.

Asimov, I. *Asimov's Chronology of Science & Discovery*, Harper Reference, NY, 1994.

Magnear, L. N. *A History of the Life Sciences*, 2nd ed., Marcel Dekker, Inc., NY, 1994.

◆ APPENDIX XIV ◆

NOBEL PRIZE WINNERS IN PHYSIOLOGY OR MEDICINE

(1901-1996)

1996—The prize was awarded jointly to
ROLF M. ZINKERNAGEL, experimental immunology institute at Zurich University, Switzerland
PETER C. DOHERTY, University of Tennessee, Memphis, U.S.A. for their work done at the John Curtin School of Medical Research in Canberra that helped clarify how the body's network of immune cells identifies invading viruses.

1995—The prize was awarded jointly to
EDWARD B. LEWIS
CHRISTIANE NÜSSLEIN-VOLHARD
ERIC F. WIESCHAUS for their discoveries concerning the genetic control of early embryonic development.

1994—The prize was awarded jointly to
ALFRED G. GILMAN, U.S.A., University of Texas Southwestern Medical Center, Dallas, TX, (1941-)
MARTIN RODBELL, U.S.A., National Institute of Environmental Health Sciences, Research Triangle Park, NC, (1925-) for their discovery of G-proteins and the role of these proteins in signal transduction in cells.

1993—The prize was awarded jointly to
RICHARD J. ROBERTS, England, New England Biolabs, Beverly, MA, U.S.A., (1943-)
PHILLIP A. SHARP, U.S.A., Center for Cancer Research, Massachusetts Institute of Technology, Cambridge, MA, (1944-) for their discoveries of split genes.

1992—The prize was awarded jointly to
EDMOND H. FISCHER, U.S.A., University of Washington, Seattle WA (in Shanghai, China), (1920-)
EDWIN G. KREBS, U.S.A., University of Washington, Seattle WA, (1918-) for their discoveries concerning reversible protein phosphorylation as a biological regulatory mechanism.

1991—The prize was awarded jointly to
ERWIN NEHER, Germany, Max-Planck-Institut für Biophysikalische Chemie, Göttingen, (1944-)
BERT SAKMANN, Germany, Max-Planck-Institut für Medizinische Forschung, Heidelberg, (1942-) for their discoveries concerning the function of single ion channels in cells.

1990—The prize was awarded jointly to
JOSEPH E. MURRAY, U.S.A., Brigham and Women's Hospital, Boston, MA, (1919-)
E. DONNALL THOMAS, U.S.A., Fred Hutchinson Cancer Research Center, Seattle, WA, (1920-) for their discoveries concerning Organ and Cell Transplantation in the Treatment of Human Disease.

1989—The prize was awarded jointly to
J. MICHAEL BISHOP, U.S.A., University of California School of Medicine, San Francisco, CA, (1936-)
HAROLD E. VARMUS, U.S.A., University of California School of Medicine, San Francisco, CA, (1939-) for their discovery of the cellular origin of retroviral oncogenes.

1988—The prize was awarded jointly to
SIR JAMES W. BLACK, Great Britain, King's College Hospital Medical School, University of London, London, Great Britain, (1924-)
GERTRUDE B. ELION, U.S.A., Wellcome Research Laboratories, Research Triangle Park, NC, (1918-)
GEORGE H. HITCHINGS, U.S.A., Wellcome Research Laboratories, Research Triangle Park, NC, (1905-) for their discoveries of important principles for drug treatment.

1987—SUSUMU TONEGAWA, Japan, Massachusetts Institute of Technology (MIT), Cambridge, MA, U.S.A., (1939-) for his discovery of the genetic principle for generation of antibody diversity.

1986—The prize was awarded jointly to
STANLEY COHEN, U.S.A., Vanderbilt University School of Medicine, Nashville, TN, (1922-)
RITA LEVI-MONTALCINI, Italy and U.S.A., Institute of Cell Biology of the C.N.R., Rome, Italy (in Turin, Italy), (1909-) for their discoveries of growth factors.

1985—The prize was awarded jointly to
MICHAEL S. BROWN, U.S.A., University of Texas Health Science Center at Dallas, Dallas, TX, (1941-)
JOSEPH L. GOLDSTEIN, U.S.A., University of Texas Health Science Center at Dallas, Dallas, TX, (1940-) for their discoveries concerning the regulation of cholesterol metabolism.

1984—The prize was awarded jointly to
NIELS K. JERNE, Denmark, Basel Institute for Immunology, Basel, Switzerland, (1911-1994)
GEORGES J.F. KÖHLER, Federal Republic of Germany, Basel Institute for Immunology, Basel, Switzerland, (1946-)
CÉSAR MILSTEIN, Great Britain and Argentina, MRC Laboratory of Molecular Biology, Cambridge (in Bahia Blanca, Argentina), (1927-) for theories concerning the specificity in development and control of the immune system and the discovery of the principle for production of monoclonal antibodies.

1983
BARBARA MC CLINTOCK, U.S.A., Cold Spring Harbor Laboratory, Cold Spring Harbor, NY, (1902-1992) for her discovery of mobile genetic elements.

1982—The prize was awarded jointly to
SUNE K. BERGSTRÖM, Sweden, The Karolinska Institute, Stockholm, (1916-)
BENGT I. SAMUELSSON, Sweden, The Karolinska Institute, Stockholm, (1934-)
SIR JOHN R. VANE, Great Britain, The Wellcome Research Laboratories, Beckenham, (1927-) for their discoveries concerning prostaglandins and related biologically active substances.

1981—The prize was awarded by one-half to
ROGER W. SPERRY, U.S.A., California Institute of Technology, Pasadena, CA, (1913-) for his discoveries concerning the functional specialization of the cerebral hemispheres
and the other half jointly to
DAVID H. HUBEL, U.S.A., Harvard Medical School, Boston, MA (in Canada) (1926-)
TORSTEN N. WIESEL, Sweden, Harvard Medical School, Boston, MA, U.S.A., (1924-) for their discoveries concerning information processing in the visual system.

1980—The prize was awarded jointly to
BARUJ BENACERRAF, U.S.A., Harvard Medical School, Boston, MA (Caracas, Venezuela), (1920-)
JEAN DAUSSET, France, Université de Paris, Laboratoire ImmunoHématologi, Paris, (1916-)
GEORGE D. SNELL, U.S.A., Jackson Laboratory, Bar Harbor, ME, (1903-) for their discoveries concerning genetically determined structures on the cell surface that regulate immunological reactions.

1979—The prize was awarded jointly to
ALAN M. CORMACK, U.S.A., Tufts University, Medford, MA (in Johannesburg, South Africa), (1924-)
SIR GODFREY N. HOUNSFIELD, Great Britain, Central Research Laboratories, EMI, London, (1919-) for the development of computer assisted tomography.

1978—The prize was awarded jointly to
WERNER ARBER, Switzerland, Biozentrum der Universität, Basel, (1929-)
DANIEL NATHANS, U.S.A., Johns Hopkins University School of Medicine, Baltimore, MD, (1928)
HAMILTON O. SMITH, U.S.A., Johns Hopkins University School of Medicine, Baltimore, MD, (1931-) for the discovery of restriction enzymes and their application to problems of molecular genetics.

1977—The prize was divided, one half being awarded jointly to
ROGER GUILLEMIN, U.S.A., The Salk Institute, San Diego, CA (in Dijon, France), (1924-)
ANDREW V. SCHALLY, U.S.A., Veterans Administration Hospital, New Orleans, LA (in Wilno, Poland), (1926-) for their discoveries concerning the peptide hormone production of the brain and the other half being awarded to
ROSALYN YALOW, U.S.A., Veterans Administration Hospital, Bronx, NY, (1921-) for the development of radioimmunoassays of peptide hormones.

1976—The prize was awarded jointly to
BARUCH S. BLUMBERG, U.S.A., The Institute for Cancer Research, Philadelphia, PA, (1925-)
D. CARLETON GAJDUSEK, U.S.A., National Institutes of Health, Bethesda, MD, (1923-) for their discoveries concerning new mechanisms for the origin and dissemination of infectious diseases.

1975—The prize was awarded jointly to
DAVID BALTIMORE, U.S.A., Massachusetts Institute of Technology (MIT), Cambridge, MA, (1938-)

RENATO DULBECCO, U.S.A., Imperial Cancer Research Fund Laboratory, London (in Catanzaro, Italy), (1914-)
HOWARD MARTIN TEMIN, U.S.A., University of Wisconsin, Madison, WI, (1934-1994) for their discoveries concerning the interaction between tumour viruses and the genetic material of the cell.

1974—The prize was awarded jointly to
ALBERT CLAUDE, Belgium, Université Catholique de Louvain, Louvain, (1899-1983)
CHRISTIAN DE DUVE, Belgium, The Rockefeller University, New York, NY, (1917-)
GEORGE E. PALADE, U.S.A., Yale University School of Medicine, New Haven, CT (in Iasi, Roumania), (1912-) for their discoveries concerning the structural and functional organization of the cell.

1973—The prize was awarded jointly to
KARL VON FRISCH, Federal Republic of Germany, Zoologisches Institut der Universität München, Munich (in Vienna, Austria), (1886-1982)
KONRAD LORENZ, Austria, Österreichische Akademie der Wissenschaften, Institut für vergleichende Verhaltensforschung, Altenberg, (1903-1989)
NIKOLAAS TINBERGEN, Great Britain, Department of Zoology, University Museum, Oxford (in the Hague, the Netherlands), (1907-1988) for their discoveries concerning organization and elicitation of individual and social behavior patterns.

1972—The prize was awarded jointly to
GERALD M. EDELMAN, U.S.A., Rockefeller University, New York, NY, (1929-)
RODNEY R. PORTER, Great Britain, University of Oxford, (1917-1985) for their discoveries concerning the chemical structure of antibodies.

1971
EARL W. JR. SUTHERLAND, U.S.A., Vanderbilt University, Nashville, TN, (1915-1974) for his discoveries concerning the mechanisms of the action of hormones.

1970—The prize was awarded jointly to
SIR BERNARD KATZ, Great Britain, University College, London, (1911-)
ULF VON EULER, Sweden, The Karolinska Institute, Stockholm, (1905-1983)
JULIUS AXELROD, U.S.A., National Institutes of Health, Bethesda, MD, (1912-) for their discoveries concerning the humoral transmittors in the nerve terminals and the mechanism for their storage, release, and inactivation.

1969—The prize was awarded jointly to
MAX DELBRÜCK, U.S.A., California Institute of Technology, Pasadena, CA (in Berlin, Germany), (1906-1981)
ALFRED D. HERSHEY, U.S.A., Carnegie Institution of Washington, Long Island, New York, NY, (1908-)
SALVADOR E. LURIA, U.S.A., Massachusetts Institute of Technology (MIT), Cambridge, MA (in Torino, Italy), (1912-1991) for their discoveries concerning the replication mechanism and the gentic structure of viruses.

1968—The prize was awarded jointly to
ROBERT W. HOLLEY, U.S.A., Cornell University, Ithaca, NY, (1922-)
HAR GOBIND KHORANA, U.S.A., University of Wisconsin, Madison, WI (in Raipur, India), (1922-)

MARSHALL W. NIRENBERG, U.S.A., National Institutes of Health, Bethesda, MD, (1927-) for their interpretation of the genetic code and its function in protein synthesis.

1967—The prize was awarded jointly to
RAGNAR GRANIT, Sweden, The Karolinska Institute, Stockholm, (in Helsinki, Finland), (1900-1991)
HALDAN KEFFER HARTLINE, U.S.A., The Rockefeller University, New York, NY, (1903-1983)
GEORGE WALD, U.S.A., Harvard University, Cambridge, MA, (1906-) for their discoveries concerning the primary physiological and chemical visual processes in the eye.

1966—The prize was divided equally between
PEYTON ROUS, U.S.A., Rockefeller University, New York, NY, (1879-1970) for his discovery of tumor-inducing viruses
CHARLES BRENTON HUGGINS, U.S.A., Ben May Laboratory for Cancer Research, University of Chicago, Chicago, IL, (1901-) for his discoveries concerning hormonal treatment of prostatic cancer.

1965—The prize was awarded jointly to
FRANÇOIS JACOB, France, Institut Pasteur, Paris, (1920-)
ANDRÉ LWOFF, France, Institut Pasteur, Paris, (1902-1994)
JACQUES MONOD, France, Institut Pasteur, Paris, (1910-1976) for their discoveries concerning genetic control of enzyme and virus synthesis.

1964—The prize was awarded jointly to
KONRAD BLOCH, U.S.A., Harvard University, Cambridge, MA (in Neisse, Germany), (1912-)
FEODOR LYNEN, Germany, Max-Planck-Institut für Zellchemie, Munich, (1911-1979) for their discoveries concerning the mechanism and regulation of cholesterol and fatty acid metabolism.

1963—The prize was awarded jointly to
SIR JOHN CAREW ECCLES, Australia, Australian National University, Canberra, (1903-)
SIR ALAN LLOYD HODGKIN, Great Britain, Cambridge University, Cambridge, (1914-)
SIR ANDREW FIELDING HUXLEY, Great Britain, London University, (1917-) for their discoveries concerning the ionic mechanisms involved in excitation and inhibition in the peripheral and central portions of the nerve cell membrane.

1962—The prize was awarded jointly to
FRANCIS HARRY COMPTON CRICK, Great Britain, Institute of Molecular Biology, Cambridge, (1916-)
JAMES DEWEY WATSON, U.S.A., Harvard University, Cambridge, MA, (1928-)
MAURICE HUGH FREDERICK WILKINS, Great Britain, University of London, (1916-) for their discoveries concerning the molecular structure of nuclear acids and its significance for information transfer in living material.

1961
GEORG VON BÉKÉSY, U.S.A., Harvard University, Cambridge, MA (in Budapest, Hungary), (1899-1972) for his discoveries of the physical mechanism of stimulation within the cochlea.

1960—The prize was awarded jointly to
SIR FRANK MACFARLANE BURNET, Australia, Walter and Eliza Hall Institute for Medical Research, Melbourne, (1899-1985)
SIR PETER BRIAN MEDAWAR, Great Britain, University College, London, (1915-1987) for discovery of acquired immunological tolerance.

1959—The prize was awarded jointly to
SEVERO OCHOA, U.S.A., New York University, College of Medicine, New York, NY (in Luarca, Spain), (1905-1993)
ARTHUR KORNBERG, U.S.A., Stanford University, Stanford, CA, (1918-) for their discovery of the mechanisms in the biological synthesis of ribonucleic acid and deoxiribonucleic acid.

1958—The prize was divided, one-half being awarded jointly to
GEORGE WELLS BEADLE, U.S.A., California Institute of Technology, Pasadena, CA, (1903-1989)
EDWARD LAWRIE TATUM, U.S.A., Rockefeller Institute for Medical Research, New York, NY, (1909-1975) for their discovery that genes act by regulating definite chemical events and the other half to
JOSHUA LEDERBERG, U.S.A., Wisconsin University, Madison, WI, (1925-) for his discoveries concerning genetic recombination and the organization of the genetic material of bacteria.

1957
DANIEL BOVET, Italy, Istituto Superiore di Sanità (Chief Institute of Public Health), Rome (in Neuchâtel, Switzerland), (1907-1992) for his discoveries relating to synthetic compounds that inhibit the action of certain body substances and especially their action on the vascular system and the skeletal muscles.

1956—The prize was awarded jointly to
ANDRÉ FRÉDÉRIC COURNAND, U.S.A., Cardio-Pulmonary Laboratory, Columbia University Division, Bellevue Hospital, New York, NY (in Paris, France), (1895-1988)
WERNER FORSSMANN, Germany, Mainz University and Bad Kreuznach, (1904-1979)
DICKINSON W. RICHARDS, U.S.A., Columbia University, New York, NY, (1895-1973) for their discoveries concerning heart catherization and pathological changes in the circulatory system.

1955
AXEL HUGO THEODOR THEORELL, Sweden, Nobel Medical Institute, Stockholm, (1903-1982) for his discoveries concerning the nature and mode of action of oxidation enzymes.

1954—The prize was awarded jointly to
JOHN FRANKLIN ENDERS, U.S.A., Harvard Medical School, Boston, MA, Research Division of Infectious Diseases, Children's Medical Center, Boston, MA, (1897-1985)

THOMAS HUCKLE WELLER, U.S.A., Research Division of Infectious Diseases, Children's Medical Center, Boston, MA, (1915-)
FREDERICK CHAPMAN ROBBINS, U.S.A., Western Reserve University, Cleveland, OH, (1916-) for their discovery of the ability of poliomyelitis viruses to grow in cultures of various types of tissue.

1953—The prize was divided equally between
SIR HANS ADOLF KREBS, Great Britain, Sheffield University (in Hildesheim, Germany), (1900-1981) for his discovery of the citric acid cycle.
FRITZ ALBERT LIPMANN, U.S.A., Harvard Medical School and Massachusetts General Hospital, Boston, MA (in Koenigsberg, then Germany), (1899-1986) for his discovery of coenzyme A and its importance for intermediary metabolism.

1952
SELMAN ABRAHAM WAKSMAN, U.S.A., Rutgers University, New Brunswick, NJ (in Priluka, Ukraine, Russia), (1888-1973) for his discovery of streptomycin, the first antibiotic effective against tuberculosis.

1951
MAX THEILER, Union of South Africa, Laboratories Division of Medicine and Public Health, Rockefeller Foundation, New York, NY, U.S.A., (1899-1972) for his discoveries concerning yellow fever and how to combat it.

1950—The prize was awarded jointly to
EDWARD CALVIN KENDALL, U.S.A., Mayo Clinic, Rochester, MN, (1886-1972)
TADEUS REICHSTEIN, Switzerland, Basel University (in Wloclawek, Poland) (1897-)
PHILIP SHOWALTER HENCH, U.S.A., Mayo Clinic, Rochester, MN, (1896-1965) for their discoveries relating to the hormones of the adrenal cortex, their structure, and biological effects.

1949—The prize was divided equally between
WALTER RUDOLF HESS, Switzerland, Zurich University, (1881-1973) for his discovery of the functional organization of the interbrain as a coordinator of the activities of the internal organs.
ANTONIO CAETANO DE ABREU FREIRE EGAS MONIZ, Portugal, University of Lisbon, Neurological Institute, Lisbon, (1874-1955) for his discovery of the therapeutic value of leucotomy in certain psychoses.

1948
PAUL HERMANN MÜLLER, Switzerland, Laboratorium der Farben-Fabriken J.R. Geigy A.G. (Laboratory of the J.R. Geigy Dye-Factory Co.), Basel, (1899-1965) for his discovery of the high efficiency of DDT as a contact poison against several arthropods.

1947—The prize was divided, one-half being awarded jointly to
CARL FERDINAND CORI, U.S.A., Washington University, St. Louis, MO (in Prague, then Austria), (1896-1984)
GERTY THERESA CORI, née RADNITZ, U.S.A., Washington University, St. Louis, MO (in Prague, then Austria), (1896-1957) for their discovery of the course of the catalytic conversion of glycogen the other half being awarded to
BERNARDO ALBERTO HOUSSAY, Argentina, Instituto de Biologia y Medicina Experimental (Institute for Biolog and Experimental Medicine), Buenos Aires, (1887-1971) for his discovery of the part played by the hormone of the anterior pituitary lobe in the metabolism of sugar.

1946
HERMANN JOSEPH MULLER, U.S.A., Indiana University, Bloomington, IN, (1890-1967) for the discovery of the production of mutations by means of X-ray irradiation.

1945—The prize was awarded jointly to
SIR ALEXANDER FLEMING, Great Britain, London University (in Lochfield, Scotland), (1881-1955)
SIR ERNST BORIS CHAIN, Great Britain, Oxford University (in Berlin, Germany), (1906-1979)
LORD HOWARD WALTER FLOREY, Great Britain, Oxford University (in Adelaide, Australia), (1898-1968) for the discovery of penicillin and its curative effect in various infectious diseases.

1944—The prize was awarded jointly to
JOSEPH ERLANGER, U.S.A., Washington University, St. Louis, MO, (1874-1965)
HERBERT SPENCER GASSER, U.S.A., Rockefeller Institute for Medical Research, New York, NY, (1888-1963) for their discoveries relating to the highly differentiated functions of single nerve fibres.

1943—The prize was divided equally between
HENRIK CARL PETER DAM, Denmark, Polytechnic Institute, Copenhagen, (1895-1976) for his discovery of vitamin K.
EDWARD ADELBERT DOISY, U.S.A., Saint Louis University, St. Louis, MO, (1893-1986) for his discovery of the chemical nature of vitamin K.

1942-1940
The prize money was allocated to the Main Fund (1/3) and to the Special Fund (2/3) of this prize section.

1939
GERHARD DOMAGK, Germany, Munster University, (1895-1964) for the discovery of the antibacterial effects of prontosil. (Caused by the authorities of his country to decline the award, he later received the diploma and the medal.)

1938
CORNEILLE JEAN FRANÇOIS HEYMANS, Belgium, Ghent University, (1892-1968) for the discovery of the role played by the sinus and aortic mechanisms in the regulation of respiration.

1937
ALBERT SZENT-GYÖRGYI VON NAGYRAPOLT, Hungary, Szeged University, (1893-1986) for his discoveries in connection with the biological combustion processes, with special reference to vitamin C and the catalysis of fumaric acid.

1936—The prize was awarded jointly to
SIR HENRY HALLETT DALE, Great Britain, National Institute for Medical Research, London, (1875-1968)
OTTO LOEWI, Austria, Graz University (in Frankfurt-on the-Main, Germany), (1873-1961) for their discoveries relating to chemical transmission of nerve impulses.

1935
HANS SPEMANN, Germany, University of Freiburg im Breisgau, (1869-1941) for his discovery of the organizer effect in embryonic development.

1934—The prize was awarded jointly to
GEORGE HOYT WHIPPLE, U.S.A., Rochester University, Rochester, NY, (1878-1976)
GEORGE RICHARDS MINOT, U.S.A., Harvard University, Cambridge, MA, (1885-1950)
WILLIAM PARRY MURPHY, U.S.A., Harvard University, Cambridge, MA and Peter Brent Brigham Hospital, Boston, MA, (1892-1987) for their discoveries concerning liver therapy in cases of anaemia.

1933
THOMAS HUNT MORGAN, U.S.A., California Institute of Technology, Pasadena, CA, (1866-1945) for his discoveries concerning the role played by the chromosome in heredity.

1932—The prize was awarded jointly to
SIR CHARLES SCOTT SHERRINGTON, Great Britain, Oxford University, (1857-1952)
LORD EDGAR DOUGLAS ADRIAN, Great Britain, Cambridge University, (1889-1977) for their discoveries regarding the functions of neurons.

1931
OTTO HEINRICH WARBURG, Germany, Kaiser-Wilhelm-Institut (now Max-Planck-Institut) für Biologie, Berlin-Dahlem, (1883-1970) for his discovery of the nature and mode of action of the respiratory enzyme.

1930
KARL LANDSTEINER, Austria, Rockefeller Institute for Medical Research, New York, NY, (1868-1943) for his discovery of human blood groups.

1929—The prize was divided equally between
CHRISTIAAN EIJKMAN, the Netherlands, Utrecht University, (1858-1930) for his discovery of the antineuritic vitamin
SIR FREDERICK GOWLAND HOPKINS, Great Britain, Cambridge University, (1861-1947) for his discovery of the growth-stimulating vitamins.

1928
CHARLES JULES HENRI NICOLLE, France, Institut Pasteur, Tunis, (1866-1936) for his work on typhus.

1927
JULIUS WAGNER-JAUREGG, Austria, Vienna University, (1857-1940) for his discovery of the therapeutic value of malaria inoculation in the treatment of dementia paralytica.

1926
JOHANNES ANDREAS GRIB FIBIGER, Denmark, Copenhagen University, (1867-1928) for his discovery of the Spiroptera carcinoma.

1925
The prize money for 1925 was allocated to the Special Fund of this prize section.

1924
WILLEM EINTHOVEN, the Netherlands, Leyden University (in Semarang, Java, then Dutch East Indies), (1860-1927) for his discovery of the mechanism of the electrocardiogram.

1923—The prize was divided equally between
SIR FREDERICK GRANT BANTING, Canada, Toronto University, (1891-1941)
JOHN JAMES RICHARD MACLEOD, Canada, Toronto University (in Cluny, Scotland), (1876-1935) for the discovery of insulin.

1922—The prize was divided equally between
SIR ARCHIBALD VIVIAN HILL, Great Britain, London University, (1886-1977) for his discovery relating to the production of heat in the muscle.
OTTO FRITZ MEYERHOF, Germany, Kiel University, (1884-1951) for his discovery of the fixed relationship between the consumption of oxygen and the metabolism of lactid acid in the muscle.

1921
The prize money for 1921 was allocated to the Special Fund of this prize section.

1920
SCHACK AUGUST STEENBERGER KROGH, Denmark, Copenhagen University, (1874-1949) for his discovery of the capillary motor regulating mechanism.

1919
JULES BORDET, Belgium, Brussels University, (1870-1961) for his discoveries relating to immunity.

1918-1915
The prize money for 1918-1915 was allocated to the Special Fund of this prize section.

1914
ROBERT BÁRÁNY, Austria, Vienna University, (1876-1936) for his work on the physiology and pathology of the vestibular apparatus.

1913
CHARLES ROBERT RICHET, France, Sorbonne University, Paris, (1850-1935) in recognition of his work on anaphylaxis.

1912
ALEXIS CARREL, France, Rockefeller Institute for Medical Research, New York, NY, (1873-1944) in recognition of his work on vascular suture and the transplantation of blood vessels and organs.

1911
ALLVAR GULLSTRAND, Sweden, Uppsala University, (1862-1930) for his work on the dioptrics of the eye.

1910
ALBRECHT KOSSEL, Germany, Heidelberg University, (1853-1927) in recognition of the contributions to our knowledge of cell chemistry made through his work on proteins, including the nucleic substances.

1909
EMIL THEODOR KOCHER, Switzerland, Berne University, (1841-1917) for his work on the physiology, pathology, and surgery of the thyroid gland.

1908—The prize was awarded jointly to
ILYA ILYICH MECHNIKOV, Russia, Institut Pasteur, Paris, France, (1845-1916)
PAUL EHRLICH, Germany, Goettingen University and Königliches Institut für experimentelle Therapie (Royal Institute for Experimental Therapy), Frankfurt-on-the-Main, (1854-1915) in recognition of their work on immunity.

1907
CHARLES LOUIS ALPHONSE LAVERAN, France, Institut Pasteur, Paris, (1845-1922) in recognition of his work on the role played by protozoa in causing diseases.

1906—The prize was awarded jointly to
CAMILLO GOLGI, Italy, Pavia University, (1843-1926)
SANTIAGO RAMON Y CAJAL, Spain, Madrid University, (1852-1934) in recognition of their work on the structure of the nervous system.

1905
ROBERT KOCH, Germany, Institut für Infektions-Krankheiten (Institute for Infectious Diseases), Berlin, (1843-1910) for his investigations and discoveries in relation to tuberculosis.

1904
IVAN PETROVICH PAVLOV, Russia, Military Medical Academy, St. Petersburg (now Leningrad), (1849-1936) in recognition of his work on the physiology of digestion, through which knowledge on vital aspects of the subject has been transformed and enlarged.

1903
NIELS RYBERG FINSEN, Denmark, Finsen Medical Light Institute, Copenhagen (in Thorshavn, Faroe Islands), (1860-1904) in recognition of his contribution to the treatment of diseases, especially lupus vulgaris, with concentrated light radiation, whereby he has opened a new avenue for medical science.

1902
SIR RONALD ROSS, Great Britain, University College, Liverpool (in Almora, India), (1857-1932) for his work on malaria, by which he showed how it enters the organism and thereby has laid the foundation for successful research on this disease and methods of combating it.

1901
EMIL ADOLF VON BEHRING, Germany, Marburg University, (1854-1917) for his work on serum therapy, especially its application against diphtheria, by which he has opened a new road in the domain of medical science and thereby placed in the hands of the physician a victorious weapon against illness and deaths.

Nobel Prize Winners 1901-1995 for Physiology or Medicine. ©1995, The Swedish Institute and the Nobel Foundation.

♦ APPENDIX XV ♦

NOBEL PRIZE WINNERS IN CHEMISTRY

(1901-1996)

1996—The prize was awarded jointly to
ROBERT CURL, Rice University, Houston, TX, U.S.A.
RICHARD SMALLEY, Rice University, Houston, TX, U.S.A.
HAROLD KROTO, University of Sussex, Great Britain for their 1985 discovery of fullereness, new forms of the element carbon in which the atoms are arranged in closed cells.

1995—The prize was awarded jointly to
PAUL CRUTZEN, Max-Planck-Institute for Chemistry, Mainz, Germany (Dutch citizen)
MARIO MOLINA, Department of Earth, Atmospheric, and Planetary Sciences and Department of Chemistry, MIT, Cambridge, MA, U.S.A.,
F. SHERWOOD ROWLAND, Department of Chemistry, University of California, Irvine, CA, for their work in atmospheric chemistry, particularly concerning the formation and decomposition of ozone.

1994
GEORGE A. OLAH, University of Southern California, CA, U.S.A. (in Budapest, Hungary), (1927-) for his contribution to carbocation chemistry.

1993—The prize was awarded for contributions to the developments of methods within DNA-based chemistry equally between
KARY B. MULLIS, U.S.A., La Jolla, CA, (1944-) for his invention of the polymerase chain reaction (PCR) method.
MICHAEL SMITH, Canada, University of British Columbia, Vancouver, Canada (in Blackpool, England), (1932-) for his fundamental contributions to the establishment of oligonucleotide-based, site-directed mutagenesis and its development for protein studies.

1992
RUDOLPH A. MARCUS, U.S.A., California Institute of Technology, Pasdena, CA (in Montreal, Canada), (1923-) for his contributions to the theory of electron transfer reactions in chemical systems.

1991
RICHARD R. ERNST, Switzerland, Eidgenössische Technische Hochschule Zürich, (1933-) for his contributions to the development of the methodology of high resolution nuclear magnetic resonance (NMR) spectroscopy.

1990
ELIAS JAMES COREY, U.S.A., Harvard University, Cambridge, MA, (1928-) for his development of the theory and methodology of organic synthesis.

1989—The prize was awarded jointly to
SIDNEY ALTMAN, U.S.A., and Canada, Yale University, New Haven, CT, (1939-)

THOMAS R. CECH, U.S.A., University of Colorado, Boulder, CO, (1947-) for their discovery of catalytic properties of RNA.

1988—The prize was awarded jointly to
JOHANN DEISENHOFER, Federal Republic of Germany, Howard Hughes Medical Institute and Department of Biochemistry, University of Texas Southwestern Medical Center at Dallas, TX, U.S.A., (1943-)
ROBERT HUBER, Federal Republic of Germany, Max-Planck-Inslilut für Biochemie, Martinsried, (1937-)
HARTMUT MICHEL, Federal Republic of Germany, Max-Planck-Institut für Biophysik, Frankfurt/Main, (1948-) for the determination of the three-dimensional structure of a photosynthetic reaction center.

1987—The prize was awarded jointly to
DONALD J. CRAM, U.S.A., University of California, Los Angeles, CA, (1919-)
JEAN-MARIE LEHN, France, Université Louis Pasteur, Strasbourg, and Collège de France, Paris, (1939-)
CHARLES J. PEDERSEN, U.S.A., Du Pont, Wilmington, DE (in Fusan, Korea, as a Norwegian citizen), (1904-1989) for their development and use of molecules with structure-specific interactions of high selectivity.

1986—The prize was awarded jointly to
DUDLEY R. HERSCHBACH, U.S.A., Harvard University, Cambridge, MA, (1932-)
YUAN T. LEE, U.S.A., University of California, Berkeley, CA (in Hsinchu, Taiwan), (1936-)
JOHN C. POLANYI , Canada, University of Toronto, Toronto, (1929-) for their contributions concerning the dynamics of chemical elementary processes.

1985—The prize was awarded jointly to
HERBERT A. HAUPTMAN, U.S.A., The Medical Foundation of Buffalo, Buffalo, NY, (1917-)
JEROME KARLE, U.S.A., U.S. Naval Research Laboratory, Washington, DC, (1918-) for their outstanding achievements in the development of direct methods for the determination of crystal structures.

1984
ROBERT BRUCE MERRIFIELD, U.S.A., Rockefeller University, New York, NY, (1921-) for his development of methodology for chemical synthesis on a solid matrix.

1983
HENRY TAUBE, U.S.A., Stanford University, Stanford, CA (in Saskatoon, Canada), (1915-) for his work on the mechanisms of electron transfer reactions, especially in metal complexes.

1982
SIR AARON KLUG, Great Britain, MRC Laboratory of Molecular Biology, Cambridge (in Lithuania), (1926-) for his development of crystallographic electron microscopy and his structural elucidation of biologically important nuclei acid-protein complexes.

1981—The prize was awarded jointly to
KENICHI FUKUI, Japan, Kyoto University, Kyoto, (1918-)
ROALD HOFFMANN, U.S.A., Cornell University, Ithaca, NY (in Zloczow,

Poland), (1937-) for their theories, developed independently, concerning the course of chemical reactions.

1980—The prize was divided, one-half being awarded to
PAUL BERG, U.S.A., Stanford University, Stanford, CA, (1926-) for his fundamental studies of the biochemistry of nucleic acids, with particular regard to recombinant DNA
and the other half jointly to
WALTER GILBERT, U.S.A., Biological Laboratories, Cambridge, MA, (1932-)
FREDERICK SANGER, U.S.A., Great Britain, MRC Laboratory of Molecular Biology, Cambridge, (1918-) for their contributions concerning the determination of base sequences in nucleic acids.

1979—The prize was divided equally between
HERBERT C. BROWN, U.S.A., Purdue University, West Lafayette, IN (in London, Great Britain), (1912-)
GEORG WITTIG, Federal Republic of Germany, University of Heidelberg, (1897-1987) for their development of the use of boron- and phosphorus-containing compounds, respectively, into important reagents in organic synthesis.

1978
PETER D. MITCHELL, Great Britain, Glynn Research Laboratories, Bodmin, (1920-1992) for his contribution to the understanding of biological energy transfer through the formulation of the chemiosmotic theory.

1977
ILYA PRIGOGINE, Belgium, Université Libre de Bruxelles, Brussells (University of Texas, U.S.A.) (in Moscow, Russia), (1917-) for his contributions to nonequilibrium thermodynamics, particularly the theory of dissipative structures.

1976
WILLIAM N. LIPSCOMB, U.S.A, Harvard University, Cambridge, MA, (1919-) for his studies on the structure of boranes illuminating problems of chemical bonding.

1975—The prize was divided equally between
SIR JOHN WARCUP CORNFORTH, Australia and Great Britain, University of Sussex, Brighton, (1917-) for his work on the stereochemistry of enzyme-catalyzed reactions.
VLADIMIR PRELOG, Switzerland, Eidgenössische Technische Hochschule, Zurich (in Sarajevo, Bosnia), (1906-) for his research into the stereochemistry of organic molecules and reactions.

1974
PAUL J. FLORY, U.S.A., Stanford University, Stanford, CA (1910-1985) for his fundamental achievements, both theoretical and experimental, in the physical chemistry of macromolecules.

1973—The prize was divided equally between
ERNST OTTO FISCHER, Federal Republic of Germany, Technical University of Munich, Munich, (1918-)
SIR GEOFFREY WILKINSON, Great Britain, Imperial College, London, (1921-) for their pioneering work, performed independently, on the chemistry of the organometallic, so-called sandwich compounds.

1972— The prize was divided, one-half being awarded to
CHRISTIAN B. ANFINSEN, U.S.A., National Institutes of Health, Bethesda, MD, (1916-) for his work on ribonuclease, especially concerning the connection between the amino acid sequence and the biologically active confirmation
and the other half jointly to
STANFORD MOORE, U.S.A., Rockefeller University, New York, NY. (1913-1982)
WILLIAM H. STEIN, U.S.A., Rockefeller University, New York, NY, (1911-1980) for their contribution to the understanding of the connection between chemical structure and catalytic activity of the active centre of the ribonuclease molecule.

1971
GERHARD HERZBERG, Canada, National Research Council of Canada, Ottava (in Hamburg, Germany), (1904-) for his contributions to the knowledge of electronic structure and geometry of molecules, particularly free radicals.

1970
LUIS F. LELOIR, Argentina, Institute for Biochemical Research, Buenos Aires, (1906-1987) for his discovery of sugar nucleotides and their role in the biosynthesis of carbohydrates.

1969—The prize was divided equally between
SIR DEREK H. R. BARTON, Great Britain, Imperial College of Science and Technology, London, (1918-)
ODD HASSEL, Norway, Kjemisk Institutt, Oslo University, Oslo, (1897-1981) for their contributions to the development of the concept of conformation and its application in chemistry.

1968
LARS ONSAGER, U.S.A., Yale University, New Haven, CT (in Olso, Norway), (1903-1976) for the discovery of the reciprocal relations bearing his name, which are fundamental for the thermodynamics of irreversible processes.

1967—The prize was divided, one-half being awarded to
MANFRED EIGEN, Federal Republic of Germany, Max-Planck-Institut für Physikalische Chemie, Goettingen, (1927-)
and the other half jointly to
RONALD GEORGE WREYFORD NORRISH, Great Britain, Institute of Physical Chemistry, Cambridge, (1897-1978)
LORD GEORGE PORTER, Great Britain, The Royal Institution, London, (1920-) for their studies of extremely fast chemical reactions, effected by disturbing the equlibrium by means of very short pulses of energy.

1966
ROBERT S. MULLIKEN, U.S.A., University of Chicago, Chicago, IL, (1896-1986) for his fundamental work concerning chemical bonds and the electronic structure of molecules by the molecular orbital method.

1965
ROBERT BURNS WOODWARD, U.S.A., Harvard University, Cambridge, MA, (1917-1979) for his outstanding achievements in the art of organic synthesis.

1964
DOROTHY CROWFOOT HODGKIN, Great Britain, Royal Society, Oxford University, (1910-) for her determinations by X-ray techniques of the structures of important biochemical substances.

1963—The prize was divided equally between
KARL ZIEGLER, Germany, Max-Planck-Institut für Kohlenforschung (Max-Planck-Institute for Carbon Research) Mülheim/Ruhr, (1898-1973)
GIULIO NATTA, Italy, Institute of Technology, Milan, (1903-1979) for their discoveries in the field of the chemistry and technology of high polymers.

1962—The prize was divided equally between
MAX FERDINAND PERUTZ, Great Britain, Laboratory of Molecular Biology, Cambridge (in Vienna, Austria), (1914-)
SIR JOHN COWDERY KENDREW, Great Britain, Laboratory of Molecular Biology, Cambridge, (1917-) for their studies of the structures of globular proteins.

1961
MELVIN CALVIN, U.S.A., University of California, Berkeley, CA, (1911-) for his research on carbon dioxide assimilation in plants.

1960
WILLARD FRANK LIBBY, U.S.A., University of California, Los Angeles, CA, (1908-1980) for his method to use carbon 14 for age determination in archaeology, geology, geophysics, and other branches of science.

1959
JAROSLAV HEYROVSKY, Czechoslovakia, Polarographic Institute of the Czechoslovak Academy of Science, Prague, (1890-1967) for his discovery and development of the polarographic methods of analysis.

1958
FREDERICK SANGER, Great Britain, Cambridge University, (1918-) for his work on the structure of proteins, especially that of insulin.

1957
LORD ALEXANDER R. TODD, Great Britain, Cambridge University, (1907-) for his work on nucleotides and nucleotide coenzymes.

1956—The prize was awarded jointly to
SIR CYRIL NORMAN HINSHELWOOD, Great Britain, Oxford University, (1897-1967)
NIKOLAY NIKOLAEVICH SEMENOV, USSR, Institute for Chemical Physics of the Academy of Sciences of the USSR, Moscow, (1896-1986) for their researches into the mechanism of chemical reactions.

1955
VINCENT DU VIGNEAUD, U.S.A., Cornell University, New York, NY, (1901-1978) for his work on biochemically important sulphur compounds, especially for the first synthesis of a polypeptide hormone.

1954
LINUS CARL PAULING, U.S.A., California Institute of Technology, Pasadena, CA, (1901-) for his research into the nature of the chemical bond and its application to the elucidation of the structure of complex substances.

1953
HERMANN STAUDINGER, Germany, University of Freiburg im Breisgau and Staatliches Institut für makromolekulare Chemie (State Research Institute for Macromolecular Chemistry), Freiburg in Br., (1881-1965) for his discoveries in the field of macromolecular chemistry.

1952—The prize was awarded jointly to
ARCHER JOHN PORTER MARTIN, Great Britain, National Institute for Medical Research, London, (1910-)
RICHARD LAURENCE MILLINGTON SYNGE, Great Britain, Rowett Research Institute, Bucksburn (Scotland), (1914-) for their invention of partition chromatography.

1951—The prize was awarded jointly to
EDWIN MATTISON MC MILLAN, U.S.A., University of California, Berkeley, CA, (1907-1991)
GLENN THEODORE SEABORG, U.S.A., University of California, Berkeley, CA, (1912-) for their discoveries in the chemistry of the transuranium elements.

1950—The prize was awarded jointly to
PAUL HERMANN DIELS, OTTO, Germany, Kiel University, (1876-1954)
KURT ALDER, Germany, Cologne University, (1902-1958) for their discovery and development of diene synthesis.

1949
WILLIAM FRANCIS GIAUQUE, U.S.A., University of California, Berkeley, CA, (1895-1982) for his contributions in the field of chemical thermodynamics, particularly concerning the behavior of substances at extremely low temperatures.

1948
ARNE WILHELM KAURIN TISELIUS, Sweden, Uppsala University, (1902-1971) for his research on electrophoresis and adsorption analysis, especially for his discoveries concerning the complex nature of serum proteins.

1947
SIR ROBERT ROBINSON, Great Britain, Oxford University, (1886-1975) for his investigations on plant products of biological importance, especially the alkaloids.

1946—The prize was divided, one-half being awarded to
JAMES BATCHELLER SUMNER, U.S.A., Cornell University, Ithaca, NY, (1887-1955) for his discovery that enzymes can be crystallized
the other half jointly to
JOHN HOWARD NORTHROP, U.S.A., Rockefeller Institute for Medical Research, Princeton, NJ, (1891-1987)
WENDELL MEREDITH STANLEY, U.S.A., Rockefeller Institute for Medical Research, Princeton, NJ, (1904-1971) for their preparation of enzymes and virus proteins in a pure form.

1945
ARTTURI ILMARI VIRTANEN, Finland, Helsinki University, (1895-1973) for his research and inventions in agricultural and nutrition chemistry, especially for his fodder preservation method.

1944
OTTO HAHN, Germany, Kaiser-Wilhelm-Institut, (now Max-Planck Institut) für Chemie, Berlin-Dahlem, (1879-1968) for his discovery of the fission of heavy nuclei.

1943
GEORGE DE HEVESY, Hungary, Stockholm University, Sweden, (1885-1936) for his work on the use of isotopes as tracers in the study of chemical processes.

1942-1940
The prize money was allocated to the Main Fund (1/3) and to the Special Fund (2/3) of this prize section.

1939—The prize was divided equally between
ADOLF FRIEDRICH JOHANN BUTENANDT, Germany, Berlin University and Kaiser-Wilhelm-Institut (now Max-Planck-Institut) für Biochemie, Berlin-Dahlem, (1903-) for his work on sex hormones. (Caused by the authorities of his country to decline the award but later received the diploma and the medal).
LEOPOLD RU ZI CKA, Switzerland, Eidgenössiche Technische Hochschule, (Federal Institute of Technology) Zurich, (in Vukovar, then Austria-Hungary), (1887-1976) for his work on polymethylenes and higher terpenes.

1938
RICHARD KUHN, Germany, Heidelberg University and Kaiser-Wilhelm-Institut (now Max-Planck-Institut) für medizinische Forschung, Heidelberg (in Vienna,Austria), (1900-1967) for his work on carotenoids and vitamins. (Caused by the authorities of his country to decline the award but later received the diploma and the medal.)

1937—The prize was divided equally between
SIR WALTER NORMAN HAWORTH, Great Britain, Birmingham University, (1889-1971) for his investigations on carbohydrates and vitamin C.
PAUL KARRER, Switzerland, Zurich University, (1889-1971) for his investigations on carotenoids, flavins, and vitamins A and B2.

1936
PETRUS (PETER) JOSEPHUS WILHELMUS DEBYE, the Netherlands, Berlin University, and Kaiser-Wilhelm- Institut (now Max-Planck-Institut) für Physik, Berlin- Dahlem, Germany, (1884-1966) for his contributions to our knowledge of molecular structure through his investigations on dipole moments and on the diffraction of X-rays and electrons in gases.

1935—The prize was awarded jointly to
FRÉDÉRIC JOLIOT, France, Institut du Radium, Paris, (1900-1958)
IRÈNE JOLIOT-CURIE, France, Institut du Radium, Paris, (1897-1956) in recognition of their synthesis of new radioactive elements.

1934
HAROLD CLAYTON UREY, U.S.A., Columbia University, New York, NY, (1893-1981) for his discovery of heavy hydrogen.

1933
The prize money was allocated to the Main Fund (1/3) and to the Special Fund (2/3) of this prize section.

1932
IRVING LANGMUIR, U.S.A., General Electric Co., Schenectady, NY, (1881-1957) for his discoveries and investigations in surface chemistry.

1931—The prize was awarded jointly to
CARL BOSCH, Germany, Heidelberg University and I.G. Farbenindustrie A.G., Heidelberg, (1874-1940)
FRIEDRICH BERGIUS, Germany, Heidelberg University and I.G. Farbenindustrie A.G. Mannheim-Rheinau, (1884-1949) in recognition of their contributions to the invention and development of chemical high pressure methods.

1930
HANS FISCHER, Germany, Technische Hochschule (Institute of Technology), Munich, (1881-1945) for his researches into the constitution of haemin and chlorophyll and especially for his synthesis of haemin.

1929—The prize was divided equally between
SIR ARTHUR HARDEN, Great Britain, London University, (1865-1940)
HANS KARL AUGUST SIMON VON EULER-CHELPIN, Sweden, Stockholm University (in Augsburg, Germany), (1873-1964) for their investigations on the fermentation of sugar and fermentative enzymes.

1928
ADOLF OTTO REINHOLD WINDAUS, Germany, Goettingen University, (1876-1959) for the services rendered through his research into the constitution of sterols and their connection with vitamins.

1927
HEINRICH OTTO WIELAND, Germany, Munich University, (1877-1957) for his investigations of the constitution of the bile acids and related substances.

1926
THE (THEODOR) SVEDBERG, Sweden, Uppsala University, (1884-1971) for his work on disperse systems.

1925
RICHARD ADOLF ZSIGMONDY, Germany, Goettingen University (in Vienna, Austria), (1865-1929) for his demonstration of the heterogenous nature of colloid solutions and for the methods he used, which have since become fundamental in modern colloid chemistry.

1924
The prize money for 1924 was allocated to the Special Fund of this prize section.

1923
FRITZ PREGL, Austria, Graz University, (1869-1930) for his invention of the method of microanalysis of organic substances.

1922
FRANCIS WILLIAM ASTON, Great Britain, Cambridge University, (1877-1945) for his discovery, by means of his mass spectrograph, of isotopes in a large number of nonradioactive elements and for his enunciation of the whole number rule.

1921
FREDERICK SODDY, Great Britain, Oxford University, (1877-1956) for his con-

tributions to our knowledge of the chemistry of radioactive substances and his investigations into the origin and nature of isotopes.

1920
WALTHER HERMANN NERNST, Germany, Berlin University, (1864-1941) in recognition of his work in thermochemistry.

1919
The prize money for 1919 was allocated to the Special Fund of this prize section.

1918
FRITZ HABER, Germany, Kaiser-Wilhelm-Institut (now Fritz-Haber-Institut) für physikalische Chemie und Electrochemie, Berlin-Dahlem, (1868-1934) for the synthesis of ammonia from its elements.

1917-1916
The prize money was allocated to the Special Fund of this prize section.

1915
RICHARD MARTIN WILLSTÄTTER, Germany, Munich University, (1872-1942) for his researches on plant pigments, especially chlorophyll.

1914
THEODORE WILLIAM RICHARDS, U.S.A., Harvard University, Cambridge, MA, (1868-1928) in recognition of his accurate determinations of the atomic weight of a large number of chemical elements.

1913
ALFRED WERNER, Switzerland, Zurich University, (in Mulhouse, Alsace, then Germany), (1866-1919) in recognition of his work on the linkage of atoms in molecules by which he has thrown new light on earlier investigations and opened up new fields of research especially in inorganic chemistry.

1912—The prize was divided equally between
VICTOR GRIGNARD, France, Nancy University, (1871-1935) for the discovery of the so-called Grignard reagent, which in recent years has greatly advanced the progress of organic chemistry.
PAUL SABATIER, France, Toulouse University, (1854-1941) for his method of hydrogenating organic compounds in the presence of finely disintegrated metals whereby the progress of organic chemistry has been greatly advanced in recent years.

1911
MARIE CURIE, France, Sorbonne University, Paris (in Warsaw, Poland), (1867-1934) in recognition of her services to the advancement of chemistry by the discovery of the elements radium and polonium, by the isolation of radium, and by the study of the nature and compounds of this remarkable element.

1910
OTTO WALLACH, Germany, Goettingen University, (1847-1931) in recognition of his services to organic chemistry and the chemical industry by his pioneer work in the field of alicyclic compounds.

1909
WILHELM OSTWALD, Germany, Leipzig University (in Riga, then Russia),

(1853-1932) in recognition of his work on catalysis and for his investigations into the fundamental principles governing chemical equilibria and rates of reaction.

1908
LORD ERNEST RUTHERFORD, Great Britain, Victoria University, Manchester (in Nelson, New Zealand), (1871-1937) for his investigations into the disintegration of the elements and the chemistry of radioactive substances.

1907
EDUARD BUCHNER, Germany, Landwirtschaftliche Hochschule, (Agricultural College), Berlin, (1860-1917) for his biochemical researches and his discovery of cell-free fermentation.

1906
HENRI MOISSAN, France, Sorbonne University, Paris, (1852-1907) in recognition of the great services rendered by him in his investigation and isolation of the element fluorine and for the adoption in the service of science of the electric furnace called after him.

1905
JOHANN FRIEDRICH WILHELM ADOLF VON BAEYER, Germany, Munich University, (1835-1917) in recognition of his services in the advancement of organic chemistry and the chemical industry through his work on organic dyes and hydroaromatic compounds.

1904
SIR WILLIAM RAMSAY, Great Britain, London University, (1852-1916) in recognition of his services in the discovery of the inert gaseous elements in air and his determination of their place in the periodic system.

1903
SVANTE AUGUST ARRHENIUS, Sweden, Stockholm University, (1859-1927) in recognition of the extraordinary services he has rendered to the advancement of chemistry by his electrolytic theory of dissociation.

1902
HERMANN EMIL FISCHER, Germany, Berlin University, (1852-1919) in recognition of the extraordinary services he has rendered by his work on sugar and purine syntheses.

1901
JACOBUS HENRICUS VAN'T HOFF, the Netherlands, Berlin University, Germany, (1852-1911) in recognition of the extraordinary services he has rendered by the discovery of the laws of chemical dynamics and osmotic pressure in solutions.

◆ APPENDIX XVI ◆

LIST OF ENDANGERED SPECIES
FROM THE U.S. FISH AND WILDLIFE SERVICE, DIVISION OF
ENDANGERED SPECIES
as of November 30, 1995

U.S. Species

VERTEBRATES

Mammals
E —Bat, gray (Myotis grisescens)
E —Bat, Hawaiian hoary (Lasiurus cinereus semotus)
E —Bat, Indiana (Myotis sodalis)
E —Bat, lesser (Sanborn's) long-nosed (Leptonycteris curasoae yerbabuenae)
E —Bat, little Mariana fruit (Pteropus tokudae)
E —Bat, Mariana fruit (Pteropus mariannus mariannus)
E —Bat, Mexican long-nosed (Leptonycteris nivalis)
E —Bat, Ozark big-eared (Plecotus townsendii ingens)
E —Bat, Virginia big-eared (Plecotus townsendii virginianus)
T —Bear, American black (Ursus americanus) (S/A)
T —Bear, grizzly (Ursus arctos)
T —Bear, Louisiana black (Ursus americanus luteolus)
    Caribou, woodland (Rangifer tarandus caribou)
E —Cougar, eastern (Felis concolor couguar)
E —Deer, Columbian white-tailed (Odocoileus virginianus leucurus)
E —Deer, key (Odocoileus virginianus clavium)
E —Ferret, black-footed (Mustela nigripes)
XN —Ferret, black-footed (Mustela nigripes)
E —Fox, San Joaquin kit (Vulpes macrotis mutica)
E —Jaguarundi (Felis yagouaroundi cacomitli)
E —Jaguarundi (Felis yagouaroundi tolteca)
E —Kangaroo rat, Fresno (Dipodomys nitratoides exilis)
E —Kangaroo rat, giant (Dipodomys ingens)
E —Kangaroo rat, Morro Bay (Dipodomys heermanni morroensis)
E —Kangaroo rat, Stephens' (Dipodomys stephensi (incl. D. cascus))
E —Kangaroo rat, Tipton (Dipodomys nitratoides nitratoides)
T —Lion, mountain (Felis concolor (all subsp. except coryi)) (S/A)
E —Manatee, West Indian (Florida) (Trichechus manatus)
E —Mountain beaver, Point Arena (Aplodontia rufa nigra)
E —Mouse, Alabama beach (Peromyscus polionotus ammobates)
E —Mouse, Anastasia Island beach (Peromyscus polionotus phasma)
E —Mouse, Choctawahatchee beach (Peromyscus polionotus allophrys)
E —Mouse, Key Largo cotton (Peromyscus gossypinus allapaticola)
E —Mouse, Pacific pocket (Perognathus longimembris pacificus)
E —Mouse, Perdido Key beach (Peromyscus polionotus trissyllepsis)
E —Mouse, salt marsh harvest (Reithrodontomys raviventris)
T —Mouse, southeastern beach (Peromyscus polionotus niveiventris)
E —Ocelot (Felis pardalis)
T —Otter, southern sea (Enhydra lutris nereis)

XN —Otter, southern sea (Enhydra lutris nereis)
E —Panther, Florida (Felis concolor coryi)
T —Prairie dog, Utah (Cynomys parvidens)
E —Pronghorn, Sonoran (Antilocapra americana sonoriensis)
E —Rabbit, Lower Keys (Sylvilagus palustris hefneri)
E —Rice rat (silver rice rat) (Oryzomys palustris natator)
T —Sea-lion, Steller (northern) (Eumetopias jubatus)
E —Seal, Caribbean monk (Monachus tropicalis)
T —Seal, guadalupe fur (Arctocephalus townsendi)
E —Seal, Hawaiian monk (Monachus schauinslandi)
T —Shrew, Dismal Swamp southeastern (Sorex longirostris fisheri)
E —Squirrel, Carolina northern flying (Glaucomys sabrinus coloratus)
E —Squirrel, Delmarva Peninsula fox (Sciurus niger cinereus)
XN —Squirrel, Delmarva Peninsula fox (Sciurus niger cinereus)
E —Squirrel, Mount Graham red (Tamiasciurus hudsonicus grahamensis)
E —Squirrel, Virginia northern flying (Glaucomys sabrinus fuscus)
E —Vole, Amargosa (Microtus californicus scirpensis)
E —Vole, Florida salt marsh (Microtus pennsylvanicus dukecampbelli)
E —Vole, Hualapai Mexican (Microtus mexicanus hualpaiensis)
E —Whale, blue (Balaenoptera musculus)
E —Whale, bowhead (Balaena mysticetus)
E —Whale, finback (Balaenoptera physalus)
E —Whale, humpback (Megaptera novaeangliae)
E —Whale, right (Balaena glacialis (incl. australis))
E —Whale, Sei (Balaenoptera borealis)
E —Whale, sperm (Physeter macrocephalus (catodon))
E —Wolf, gray (Canis lupus)
XN —Wolf, gray (Canis lupus)
XN —Wolf, gray (Canis lupus)
F —Wolf, red (Canis rufus)
XN —Wolf, red (Canis rufus)
E —Woodrat, Key Largo (Neotoma floridana smalli)

Birds

E —`Akepa, Hawaii (honeycreeper) (Loxops coccineus coccineus)
E —`Akepa, Maui (honeycreeper) (Loxops coccineus ochraceus)
E —`Akialoa, Kauai (honeycreeper) (Hemignathus procerus)
E —`Akiapola`au (honeycreeper) (Hemignathus munroi)
E —Blackbird, yellow-shouldered (Agelaius xanthomus)
E —Bobwhite, masked (quail) (Colinus virginianus ridgwayi)
E —Broadbill, Guam (Myiagra freycineti)
T —Caracara, Audubon's crested (Polyborus plancus audubonii)
E —Condor, California (Gymnogyps californianus)
E —Coot, Hawaiian (`alae-ke`oke`o) (Fulica americana alai)
E —Crane, Mississippi sandhill (Grus canadensis pulla)
E —Crane, whooping (Grus americana)
XN —Crane, whooping (Grus americana)
E —Creeper, Hawaii (Oreomystis mana)
E —Creeper, Molokai (kakawahie) (Paroreomyza flammea)
E —Creeper, Oahu (alauwahio) (Paroreomyza maculata)
E —Crow, Hawaiian (`alala) (Corvus hawaiiensis)
E —Crow, Mariana (Corvus kubaryi)

E —Curlew, Eskimo (Numenius borealis)
E —Duck, Hawaiian (koloa) (Anas wyvilliana)
E —Duck, Laysan (Anas laysanensis)
T —Eagle, bald (Haliaeetus leucocephalus)
T —Eider, spectacled (Somateria fischeri)
E —Falcon, American peregrine (Falco peregrinus anatum)
E —Falcon, northern aplomado (Falco femoralis septentrionalis)
E —Falcon, peregrine (Falco peregrinus) (S/A)
E —Finch, Laysan (honeycreeper) (Telespyza cantans)
E —Finch, Nihoa (honeycreeper) (Telespyza ultima)
E —Flycatcher, Southwestern willow (Empidonax traillii extimus)
T —Gnatcatcher, coastal California (Polioptila californica californica)
T —Goose, Aleutian Canada (Branta canadensis leucopareia)
E —Goose, Hawaiian (nene) (Nesochen sandvicensis)
E —Hawk, Hawaiian (io) (Buteo solitarius)
E —Hawk, Puerto Rican broad-winged (Buteo platypterus brunnescens)
E —Hawk, Puerto Rican sharp-shinned (Accipiter striatus venator)
E —Honeycreeper, crested (`akohekohe) (Palmeria dolei)
T —Jay, Florida scrub (Aphelocoma coerulescens coerulescens)
E —Kingfisher, Guam Micronesia (Halcyon cinnamomina cinnamomina)
E —Kite, Everglade snail (Rostrhamus sociabilis plumbeus)
E —Mallard, Mariana (Anas oustaleti)
E —Megapode, Micronesian (La Perouse's) (Megapodius laperouse)
E —Millerbird, Nihoa (old world warbler) (Acrocephalus familiaris kingi)
T —Monarch, Tinian (Monarcha takatsukasae)
E —Moorhen (gallinule), Hawaiian common (Gallinula chloropus sandvicensis)
E —Moorhen (gallinule), Mariana common (Gallinula chloropus guami)
T —Murrelet, marbled (Brachyramphus marmoratus marmoratus)
E —Nightjar, Puerto Rican (whip-poor-will) (Caprimulgus noctitherus)
E —Nukupu`u (honeycreeper) (Hemignathus lucidus)
E —`O`o, Kauai (`o`o `a`a) (honeyeater) (Moho braccatus)
E —`O`u (honeycreeper) (Psittirostra psittacea)
T —Owl, Mexican spotted (Strix occidentalis lucida)
T —Owl, northern spotted (Strix occidentalis caurina)
E —Palila (honeycreeper) (Loxioides bailleui)
E —Parrot, Puerto Rican (Amazona vittata)
E —Parrotbill, Maui (honeycreeper) (Pseudonestor xanthophrys)
E —Pelican, brown (Pelecanus occidentalis)
E —Petrel, Hawaiian dark-rumped (Pterodroma phaeopygia sandwichensis)
E —Pigeon, Puerto Rican plain (Columba inornata wetmorei)
E —Plover, piping (Charadrius melodus)
T —Plover, piping (Charadrius melodus)
T —Plover, western snowy (Charadrius alexandrinus nivosus)
E —Po`ouli (honeycreeper) (Melamprosops phaeosoma)
E —Prairie-chicken, Attwater's greater (Tympanuchus cupido attwateri)
E —Rail, California clapper (Rallus longirostris obsoletus)
E —Rail, Guam (Rallus owstoni)
XN —Rail, Guam (Rallus owstoni)
E —Rail, light-footed clapper (Rallus longirostris levipes)
E —Rail, Yuma clapper (Rallus longirostris yumanensis)
T —Shearwater, Newell's Townsend's (formerly Manx) (`a`o) (Puffinus auricularis
    newelli)

E —Shrike, San Clemente loggerhead (Lanius ludovicianus mearnsi)
E —Sparrow, Cape Sable seaside (Ammodramus maritimus mirabilis)
E —Sparrow, Florida grasshopper (Ammodramus savannarum floridanus)
T —Sparrow, San Clemente sage (Amphispiza belli clementeae)
E —Stilt, Hawaiian (ae`o) (Himantopus mexicanus knudseni)
E —Stork, wood (Mycteria americana)
E —Swiftlet, Mariana gray (vanikoro) (Aerodramus vanikorensis bartschi)
E —Tern, California least (Sterna antillarum browni)
E —Tern, least (Sterna antillarum)
E —Tern, roseate (Sterna dougallii dougallii)
T —Tern, roseate (Sterna dougallii dougallii)
E —Thrush, large Kauai (Myadestes myadestinus)
E —Thrush, Molokai (oloma`o) (Myadestes lanaiensis rutha)
E —Thrush, small Kauai (puaiohi) (Myadestes palmeri)
T —Towhee, Inyo California (brown) (Pipilo crissalis eremophilus)
E —Vireo, black-capped (Vireo atricapillus)
E —Vireo, least Bell's (Vireo bellii pusillus)
E —Warbler, nightingale reed (Acrocephalus luscinia)
E —Warbler, Bachman's (Vermivora bachmanii)
E —Warbler, golden-cheeked (Dendroica chrysoparia)
E —Warbler, Kirtland's (Dendroica kirtlandii)
E —White-eye, bridled (Zosterops conspicillatus conspicillatus)
E —Woodpecker, ivory-billed (Campephilus principalis)
E —Woodpecker, red-cockaded (Picoides borealis)

Reptiles
T —Alligator, American (Alligator mississippiensis) (S/A)
T — Anole, Culebra Island giant (Anolis roosevelti)
T —Boa, Mona (Epicrates monensis monensis)
E —Boa, Puerto Rican (Epicrates inornatus)
E —Boa, Virgin Islands tree (Epicrates monensis granti)
E —Crocodile, American (Crocodylus acutus)
E —Gecko, Monito (Sphaerodactylus micropithecus)
T —Iguana, Mona ground (Cyclura stejnegeri)
E —Lizard, blunt-nosed leopard (Gambelia silus)
T —Lizard, Coachella Valley fringe-toed (Uma inornata)
T —Lizard, Island night (Xantusia riversiana)
E —Lizard, St. Croix ground (Ameiva polops)
T —Rattlesnake, New Mexican ridge-nosed (Crotalus willardi obscurus)
T —Skink, bluetail (blue-tailed) mole (Eumeces egregius lividus)
T —Skink, sand (Neoseps reynoldsi)
T —Snake, Atlantic salt marsh (Nerodia clarkii taeniata)
T —Snake, Concho water (Nerodia paucimaculata)
T —Snake, eastern indigo (Drymarchon corais couperi)
T —Snake, giant garter (Thamnophis gigas)
E —Snake, San Francisco garter (Thamnophis sirtalis tetrataenia)
T —Tortoise, desert (Gopherus agassizii)
T —Tortoise, desert (Gopherus (Xerobates, Scaptochelys) agassizii) (S/A)
T —Tortoise, gopher (Gopherus polyphemus)
E —Turtle, Alabama redbelly (red-bellied) (Pseudemys alabamensis)
T —Turtle, flattened musk (Sternotherus depressus)
E —Turtle, green sea (Chelonia mydas)

T —Turtle, green sea (Chelonia mydas)
E —Turtle, hawksbill sea (Eretmochelys imbricata)
E —Turtle, Kemp's (Atlantic) ridley sea (Lepidochelys kempii)
E —Turtle, leatherback sea (Dermochelys coriacea)
T —Turtle, loggerhead sea (Caretta caretta)
T —Turtle, olive (Pacific) ridley sea (Lepidochelys olivacea)
E —Turtle, Plymouth redbelly (red-bellied) (Pseudemys rubriventris bangsi)
T —Turtle, ringed map (sawback) (Graptemys oculifera)
T —Turtle, yellow-blotched map (sawback) (Graptemys flavimaculata)

Amphibians
T —Coqui, golden (Eleutherodactylus jasperi)
T —Salamander, Cheat Mountain (Plethodon nettingi)
E —Salamander, desert slender (Batrachoseps aridus)
T —Salamander, Red Hills (Phaeognathus hubrichti)
T —Salamander, San Marcos (Eurycea nana)
E —Salamander, Santa Cruz long-toed (Ambystoma macrodactylum croceum)
E —Salamander, Shenandoah (Plethodon shenandoah)
E —Salamander, Texas blind (Typhlomolge rathbuni)
E —Toad, arroyo southwestern (Bufo microscaphus californicus)
E —Toad, Houston (Bufo houstonensis)
T —Toad, Puerto Rican crested (Peltophryne lemur)
E —Toad, Wyoming (Bufo hemiophrys baxteri)

Fishes
T —Catfish, Yaqui (Ictalurus pricei)
E —Cavefish, Alabama (Speoplatyrhinus poulsoni)
T —Cavefish, Ozark (Amblyopsis rosae)
E —Chub, bonytail (Gila elegans)
E —Chub, Borax Lake (Gila boraxobius)
T —Chub, Chihuahua (Gila nigrescens)
E —Chub, humpback (Gila cypha)
T —Chub, Hutton tui (Gila bicolor ssp.)
E —Chub, Mohave tui (Gila bicolor mohavensis)
E —Chub, Oregon (Oregonichthys (Hybopsis) crameri)
E —Chub, Owens tui (Gila bicolor snyderi)
E —Chub, Pahranagat roundtail (bonytail) (Gila robusta jordani)
T —Chub, slender (Erimystax (Hybopsis) cahni)
T —Chub, Sonora (Gila ditaenia)
T —Chub, spotfin (turquoise shiner) (Cyprinella (Hybopsis) monacha)
E —Chub, Virgin River (Gila robusta semidnuda)
E —Chub, Yaqui (Gila purpurea)
E —Cui-ui (Chasmistes cujus)
E —Dace, Ash Meadows speckled (Rhinichthys osculus nevadensis)
T —Dace, blackside (Phoxinus cumberlandensis)
E —Dace, Clover Valley speckled (Rhinichthys osculus oligoporus)
T —Dace, desert (Eremichthys acros)
T —Dace, Foskett speckled (Rhinichthys osculus ssp.)
E —Dace, Independence Valley speckled (Rhinichthys osculus lethoporus)
E —Dace, Kendall Warm Springs (Rhinichthys osculus thermalis)
E —Dace, Moapa (Moapa coriacea)
E —Darter, amber (Percina antesella)

T —Darter, bayou (Etheostoma rubrum)
E —Darter, bluemask (jewel) (Etheostoma (Doration) sp.)
E —Darter, boulder (Elk River) (Etheostoma wapiti)
T —Darter, Cherokee (Etheostoma (Ulocentra) sp.)
E —Darter, duskytail (Etheostoma (Catonotus) sp.)
E —Darter, Etowah (Etheostoma etowahae)
F —Darter, fountain (Etheostoma fonticola)
T —Darter, goldline (Percina aurolineata)
T —Darter, leopard (Percina pantherina)
E —Darter, Maryland (Etheostoma sellare)
T —Darter, Niangua (Etheostoma nianguae)
E —Darter, Okaloosa (Etheostoma okaloosae)
E —Darter, relict (Etheostoma (Catonotus) chienense)
T —Darter, slackwater (Etheostoma boschungi)
T —Darter, snail (Percina tanasi)
E —Darter, watercress (Etheostoma nuchale)
E —Gambusia, Big Bend (Gambusia gaigei)
E —Gambusia, Clear Creek (Gambusia heterochir)
E —Gambusia, Pecos (Gambusia nobilis)
E —Gambusia, San Marcos (Gambusia georgei)
E —Goby, tidewater (Eucyclogobius newberryi)
E —Logperch, Conasauga (Percina jenkinsi)
E —Logperch, Roanoke (Percina rex)
T —Madtom, Neosho (Noturus placidus)
E —Madtom, pygmy (Noturus stanauli)
E —Madtom, Scioto (Noturus trautmani)
E —Madtom, Smoky (Noturus baileyi)
T —Madtom, yellowfin (Noturus flavipinnis)
XN —Madtom, yellowfin (Noturus flavipinnis)
T —Minnow, loach (Rhinichthys (Tiaroga) cobitis)
E —Minnow, Rio Grande silvery (Hybognathus amarus)
E —Poolfish (killifish), Pahrump (Empetrichthys latos)
E —Pupfish, Ash Meadows Amargosa (Cyprinodon nevadensis mionectes)
E —Pupfish, Comanche Springs (Cyprinodon elegans)
E —Pupfish, desert (Cyprinodon macularius)
E —Pupfish, Devils Hole (Cyprinodon diabolis)
E —Pupfish, Leon Springs (Cyprinodon bovinus)
E —Pupfish, Owens (Cyprinodon radiosus)
E —Pupfish, Warm Springs (Cyprinodon nevadensis pectoralis)
T —Salmon, chinook (winter Sacramento R.) (Oncorhynchus tshawytscha)
T —Salmon, chinook (spring/summer Snake R.)(Oncorhynchus tshawytscha)
T —Salmon, chinook (fall Snake R.)(Oncorhynchus tshawytscha)
E —Salmon, sockeye (red, blueback) (Oncorhynchus nerka)
T —Sculpin, pygmy (Cottus pygmaeus)
T —Shiner, beautiful (Cyprinella (Notropis) formosa)
T —Shiner, blue (Cyprinella (Notropis) caerulea)
E —Shiner, Cahaba (Notropis cahabae)
E —Shiner, Cape Fear (Notropis mekistocholas)
E —Shiner, Palezone (Notropis sp.)
T —Shiner, Pecos bluntnose (Notropis simus pecosensis)
T —Silverside, Waccamaw (Menidia extensa)
T —Smelt, delta (Hypomesus transpacificus)

T —Spikedace (Meda fulgida)
T —Spinedace, Big Spring (Lepidomeda mollispinis pratensis)
T —Spinedace, Little Colorado (Lepidomeda vittata)
E —Spinedace, White River (Lepidomeda albivallis)
E —Springfish, Hiko White River (Crenichthys baileyi grandis)
T —Springfish, Railroad Valley (Crenichthys nevadae)
E —Springfish, White River (Crenichthys baileyi baileyi)
E —Squawfish, Colorado (Ptychocheilus lucius)
XN —Squawfish, Colorado (Ptychocheilus lucius)
E —Stickleback, unarmored threespine (Gasterosteus aculeatus williamsoni)
T —Sturgeon, Gulf (Acipenser oxyrhynchus desotoi)
E —Sturgeon, pallid (Scaphirhynchus albus)
E —Sturgeon, shortnose (Acipenser brevirostrum)
E —Sturgeon, white (Kootenai River pop.) (Acipenser transmontanus)
E —Sucker, June (Chasmistes liorus)
E —Sucker, Lost River (Deltistes luxatus)
E —Sucker, Modoc (Catostomus microps)
E —Sucker, razorback (Xyrauchen texanus)
E —Sucker, shortnose (Chasmistes brevirostris)
T —Sucker, Warner (Catostomus warnerensis)
E —Topminnow, Gila (incl. Yaqui) (Poeciliopsis occidentalis)
T —Trout, Apache (Arizona) (Oncorhynchus (Salmo) apache)
E —Trout, Gila (Oncorhynchus (Salmo) gilae)
T —Trout, greenback cutthroat (Oncorhynchus (Salmo) clarki stomias)
T —Trout, Lahontan cutthroat (Oncorhynchus (Salmo) clarki henshawi)
T —Trout, Little Kern golden (Oncorhynchus (Salmo) aguabonita whitei)
T —Trout, Paiute cutthroat (Oncorhynchus (Salmo) clarki seleniris)
E —Woundfin (Plagopterus argentissimus)
XN —Woundfin (Plagopterus argentissimus)

INVERTEBRATES

Clams
E —Acornshell, southern (Epioblasma othcaloogensis)
E —Clubshell, black (Curtus' mussel) (Pleurobema curtum)
E —Clubshell, ovate (Pleurobema perovatum)
E —Clubshell, southern (Pleurobema decisum)
E —Clubshell (Pleurobema clava)
E —Combshell, southern (penitent mussel) (Epioblasma penita)
E —Combshell, upland (Epioblasma metastriata)
E —Elktoe, Appalachian (Alasmidonta raveneliana)
E —Fanshell (Cyprogenia stegaria)
T —Fatmucket, Arkansas (Lampsilis powelli)
E —Heelsplitter, Carolina (Lasmigona decorata)
T —Heelsplitter, inflated (Potamilus inflatus)
E —Kidneyshell, triangular (Ptychobranchus greeni)
E —Lampmussel, Alabama (Lampsilis virescens)
T —Moccasinshell, Alabama (Medionidus acutissimus)
E —Moccasinshell, Coosa (Medionidus parvulus)
T —Mucket, orange-nacre (Lampsilis perovalis)
E —Mussel, dwarf wedge (Alasmidonta heterodon)
E —Mussel, ring pink (golf stick pearly) (Obovaria retusa)

E —Mussel, winged mapleleaf (Quadrula fragosa)
T —Pearlshell, Louisiana (Margaritifera hembeli)
E —Pearlymussel, Appalachian monkeyface (Quadrula sparsa)
E —Pearlymussel, birdwing (Conradilla caelata)
E —Pearlymussel, cracking (Hemistena lata)
E —Pearlymussel, Cumberland bean (Villosa trabalis)
E —Pearlymussel, Cumberland monkeyface (Quadrula intermedia)
E —Pearlymussel, Curtis' (Epioblasma (Dysnomia) florentina curtisi)
E —Pearlymussel, dromedary (Dromus dromas)
E —Pearlymussel, green-blossom (Epioblasma torulosa gubernaculum)
E —Pearlymussel, Higgins' eye (Lampsilis higginsi)
E —Pearlymussel, little-wing (Pegias fabula)
E —Pearlymussel, orange-foot pimple back (Plethobasus cooperianus)
E —Pearlymussel, pale lilliput (Toxolasma cylindrellus)
E —Pearlymussel, pink mucket (Lampsilis abrupta)
E —Pearlymussel, purple cat's paw (Epioblasma obliquata obliquata)
E —Pearlymussel, tubercled-blossom (Epioblasma torulosa torulosa)
E —Pearlymussel, turgid-blossom (Epioblasma turgidula)
E —Pearlymussel, white cat's paw (Epioblasma obliquata perobliqua (sulcata delicata))
E —Pearlymussel, white wartyback (Plethobasus cicatricosus)
E —Pearlymussel, yellow-blossom (Epioblasma florentina florentina)
E —Pigtoe, Cumberland (Cumberland pigtoe mussel) (Pleurobema gibberum)
E —Pigtoe, dark (Pleurobema furvum)
E —Pigtoe, fine-rayed (Fusconaia cuneolus)
E —Pigtoe, flat (Marshall's mussel) (Pleurobema marshalli)
E —Pigtoe, heavy (Judge Tait's mussel) (Pleurobema taitianum)
E —Pigtoe, rough (Pleurobema plenum)
E —Pigtoe, shiny (Fusconaia cor (edgariana))
E —Pigtoe, southern (Pleurobema georgianum)
E —Pocketbook, fat (Potamilus (Proptera) capax)
T —Pocketbook, fine lined (Lampsilis altilis)
E —Pocketbook, speckled (Lampsilis streckeri)
E —Riffleshell, northern (Epioblasma torulosa rangiana)
E —Riffleshell, tan (Epioblasma walkeri)
E —Rock-pocketbook, Ouachita (Wheeler's pearly mussel) (Arkansia wheeleri)
E —Spinymussel, James River (Virginia) (Pleurobema collina)
E —Spinymussel, Tar River (Elliptio steinstansana)
E —Stirrupshell (Quadrula stapes)

Snails
E —Ambersnail, Kanab (Oxyloma haydeni kanabensis)
E —Limpet, Banbury Springs (Lanx sp.)
E —Marstonia (snail), royalobese) (Pyrgulopsis (Marstonia) ogmoraphe)
E —Riversnail, Anthony's (Athearnia anthonyi)
T —Shagreen, Magazine Mountain (Mesodon magazinensis)
T —Snail, Bliss Rapids (Taylorconcha serpenticola)
T —Snail, Chittenango ovate amber (Succinea chittenangoensis)
T —Snail, flat-spired three-toothed (Triodopsis platysayoides)
E —Snail, Iowa Pleistocene (Discus macclintocki)
E —Snail, Morro shoulderband (banded dune) (Helminthoglypta walkeriana)
T —Snail, noonday (Mesodon clarki nantahala)

T —Snail, painted snake coiled forest (Anguispira picta)
E —Snail, Snake River physa (Physa natricina)
T —Snail, Stock Island tree (Orthalicus reses (not incl. nesodryas))
E —Snail, tulotoma (Alabama live-bearing) (Tulotoma magnifica)
E —Snail, Utah valvata (Valvata utahensis)
E —Snail, Virginia fringed mountain (Polygyriscus virginianus)
E —Snails, Oahu tree (Achatinella spp.)
E —Springsnail, Alamosa (Tryonia alamosae)
E —Springsnail, Bruneau Hot (Pyrgulopsis bruneauensis)
E —Springsnail, Idaho (Fontelicella idahoensis)
E —Springsnail, Socorro (Pyrgulopsis neomexicana)

Insects
E —Beetle, American burying (giant carrion) (Nicrophorus americanus)
E —Beetle, Coffin Cave mold (Batrisodes texanus)
T —Beetle, delta green ground (Elaphrus viridis)
E —Beetle, Hungerford's crawling water (Brychius hungerfordi)
E —Beetle, Kretschmarr Cave mold (Texamaurops reddelli)
T —Beetle, northeastern beach tiger (Cicindela dorsalis dorsalis)
T —Beetle, Puritan tiger (Cicindela puritana)
E —Beetle, Tooth Cave ground (Rhadine persephone)
T —Beetle, valley elderberry longhorn (Desmocerus californicus dimorphus)
T —Butterfly, bay checkerspot (Euphydryas editha bayensis)
E —Butterfly, El Segundo blue (Euphilotes battoides allyni)
E —Butterfly, Karner blue (Lycaeides melissa samuelis)
E —Butterfly, Lange's metalmark (Apodemia mormo langei)
E —Butterfly, lotis blue (Lycaeides argyrognomon lotis)
E —Butterfly, mission blue (Icaricia icarioides missionensis)
E —Butterfly, Mitchell's satyr (Neonympha mitchellii mitchellii)
E —Butterfly, Myrtle's silverspot (Speyeria zerene myrtleae)
T —Butterfly, Oregon silverspot (Speyeria zerene hippolyta)
E —Butterfly, Palos Verdes blue (Glaucopsyche lygdamus palosverdesensis)
E —Butterfly, Saint Francis' satyr (Neonympha mitchellii francisci)
E —Butterfly, San Bruno elfin (Callophrys mossii bayensis)
E —Butterfly, Schaus swallowtail (Heraclides (Papilio) aristodemus ponceanus)
E —Butterfly, Smith's blue (Euphilotes enoptes smithi)
E —Butterfly, Uncompahgre fritillary (Boloria acrocnema)
E —Dragonfly, Hine's emerald (Somatochlora hineana)
E —Fly, Delhi Sands flower-loving (Rhaphiomidas terminatus abdominalis)
T —Moth, Kern primrose sphinx (Euproserpinus euterpe)
T —Naucorid, Ash Meadows (Ambrysus amargosus)
T —Skipper, Pawnee montane (Hesperia leonardus (pawnee) montana)

Arachnids
E —Harvestman, Bee Creek Cave (Texella reddelli)
E —Harvestman, Bone Cave (Texella reyesi)
E —Pseudoscorpion, Tooth Cave (Microcreagris texana)
E —Spider, spruce-fir moss (Microhexura montivaga)
E —Spider, Tooth Cave (Leptoneta myopica)

Crustaceans
E —Amphipod, Hay's Spring (Stygobromus hayi)

E —Crayfish, cave (no common name) (Cambarus aculabrum)
E —Crayfish, cave (no common name) (Cambarus zophonastes)
E —Crayfish, Nashville (Orconectes shoupi)
E —Crayfish, Shasta (placid) (Pacifastacus fortis)
E —Fairy shrimp, Conservancy (Branchinecta conservatio)
E —Fairy shrimp, longhorn (Branchinecta longiantenna)
E   Fairy shrimp, riverside (Streptocephalus woottoni)
T —Fairy shrimp, vernal pool (Branchinecta lynchi)
E —Isopod, Lee County cave (Lirceus usdagalun)
T —Isopod, Madison Cave (Antrolana lira)
E —Isopod, Socorro (Thermosphaeroma (Exosphaeroma) thermophilus)
E —Shrimp, Alabama cave (Palaemonias alabamae)
E —Shrimp, California freshwater (Syncaris pacifica)
E —Shrimp, Kentucky cave (Palaemonias ganteri)
T —Shrimp, Squirrel Chimney Cave (Florida cave) (Palaemonetes cummingi)
E —Tadpole shrimp, vernal pool (Lepidurus packardi)

FLOWERING PLANTS

E —Large-fruited sand-verbena (Abronia macrocarpa)
E —Abutilon eremitopetalum (Plant, no common name)
E —Ko`oloa`ula (Abutilon menziesii)
E —Abutilon sandwicense (Plant, no common name)
E —Liliwai (Acaena exigua)
E —San Mateo thornmint (Acanthomintha obovata ssp. duttonii)
E —Round-leaved chaff-flower (Achyranthes splendens var. rotundata)
T —Northern wild monkshood (Aconitum noveboracense)
T —Sensitive joint-vetch (Aeschynomene virginica)
E —Sandplain gerardia (Agalinis acuta)
E —Arizona agave (Agave arizonica)
E —Mahoe (Alectryon macrococcus)
E —Alsinidendron obovatum (Plant, no common name)
E —Alsinidendron trinerve (Plant, no common name)
T —Seabeach amaranth (Amaranthus pumilus)
E —South Texas ambrosia (Ambrosia cheiranthifolia)
E —Crenulate lead-plant (Amorpha crenulata)
T —Little amphianthus (Amphianthus pusillus)
E —Large-flowered fiddleneck (Amsinckia grandiflora)
E —Kearney's blue-star (Amsonia kearneyana)
E —Tobusch fishhook cactus (Ancistrocactus tobuschii)
T —Price's potato-bean (Apios priceana)
E —McDonald's rock-cress (Arabis mcdonaldiana)
E —Rock cress (Arabis perstellata)
E —Shale barren rock-cress (Arabis serotina)
E —Dwarf bear-poppy (Arctomecon humilis)
E —Presidio (Raven's) manzanita (Arctostaphylos hookeri var. ravenii)
T —Morro manzanita (Arctostaphylos morroensis)
E —Cumberland sandwort (Arenaria cumberlandensis)
E —Marsh sandwort (Arenaria paludicola)
E —Sacramento prickly-poppy (Argemone pleiacantha ssp. pinnatisecta)
E —Ka`u silversword (Argyroxiphium kauense)
E —`Ahinahina (Haleakala silversword) (Argyroxiphium sandwicense ssp. macrocephalum)

E —ʻAhinahina (Mauna Kea silversword) (Argyroxiphium sandwicense ssp. sandwicense)
E —Aristida chaseae (Plant, no common name)
E —Pelos del diablo (Aristida portoricensis)
T —Mead's milkweed (Asclepias meadii)
T —Welsh's milkweed (Asclepias welshii)
E —Four-petal pawpaw (Asimina tetramera)
E —Cushenbury milk-vetch (Astragalus albens)
E —Applegate's milk-vetch (Astragalus applegatei)
E —Pyne's (Guthrie's) ground-plum (Astragalus bibullatus)
E —Sentry milk-vetch (Astragalus cremnophylax var. cremnophylax)
E —Mancos milk-vetch (Astragalus humillimus)
T —Heliotrope milk-vetch (Astragalus montii)
E —Osterhout milk-vetch (Astragalus osterhoutii)
T —Ash Meadows milk-vetch (Astragalus phoenix)
E —Jesup's milk-vetch (Astragalus robbinsii var. jesupi)
E —Star cactus (Astrophytum asterias)
E —Auerodendron pauciflorum (Plant, no common name)
E —Texas ayenia (Ayenia limitaris)
E —Palo de Ramón (Banara vanderbiltii)
E —Hairy rattleweed (Baptisia arachnifera)
E —Truckee barberry (Berberis sonnei)
T —Virginia round-leaf birch (Betula uber)
E —Cuneate bidens (Bidens cuneata)
E —Koʻokoʻolau (Bidens micrantha ssp. kalealaha)
E —Koʻokoʻolau (Bidens wiebkei)
E —Sonoma sunshine (Baker's stickyseed) (Blennosperma bakeri)
T —Decurrent false aster (Boltonia decurrens)
T —Florida bonamia (Bonamia grandiflora)
E —Bonamia menziesii (Plant, no common name)
E —ʻOlulu (Brighamia insignis)
E —Pua ʻala (Brighamia rockii)
E —Vahl's boxwood (Buxus vahlii)
E —Uhiuhi (Caesalpinia kavaiense)
E —Capá rosa (péndula cimarrona) (Callicarpa ampla)
E —Texas poppy-mallow (Callirhoe scabriuscula)
T —Tiburon mariposa lily (Calochortus tiburonensis)
E —Calyptranthes thomasiana (Plant, no common name)
T —Palma de manaca or manac palm (Calyptronoma rivalis)
T —San Benito evening-primrose (Camissonia benitensis)
E —Brooksville (Robins') bellflower (Campanula robinsiae)
E —ʻAwikiwiki (Canavalia molokaiensis)
E —Small-anthered bittercress (Cardamine micranthera)
T —Navajo sedge (Carex specuicola)
E —Tiburon paintbrush (Castilleja affinis ssp. neglecta)
E —San Clemente Island Indian paintbrush (Castilleja grisea)
E —California jewelflower (Caulanthus californicus)
E —Coyote ceanothus (Coyote Valley California-lilac) (Ceanothus ferrisae)
T —Spring-loving centaury (Centaurium namophilum)
E —ʻAwiwi (Centaurium sebaeoides)
E —Fragrant prickly-apple (Cereus eriophorus var. fragrans)
E —Chamaecrista glandulosa var. mirabilis (Cassia mirabilis) (Plant, no com-

mon name)
E —Deltoid spurge (Chamaesyce deltoidea ssp. deltoidea)
E —`Akoko (Chamaesyce deppeana (Euphorbia d.))
T —Garber's spurge (Chamaesyce garberi)
E —`Ewa Plains `akoko (Chamaesyce skottsbergii var. kalaeloana)
E —`Akoko (Chamaesyce celastroides var. kaenana)
E —Chamaesyce halemanui (Plant, no common name)
E —`Akoko (Chamaesyce kuwaleana)
E —Pygmy fringe-tree (Chionanthus pygmaeus)
E —Howell's spineflower (Chorizanthe howellii)
E —Ben Lomond spineflower (Chorizanthe pungens var. hartwegiana)
T —Monterey spineflower (Chorizanthe pungens var. pungens)
E —Robust spineflower (includes Scotts Valley spineflower) (Chorizanthe
   robusta)
E —Sonoma spineflower (Chorizanthe valida)
E —Florida golden aster (Chrysopsis floridana)
E —Fountain thistle (Cirsium fontinale var. fontinale)
E —Chorro Creek bog thistle (Cirsium fontinale obispoense)
T —Pitcher's thistle (Cirsium pitcheri)
T —Sacramento Mountains thistle (Cirsium vinaceum)
E —Presidio clarkia (Clarkia franciscana)
E —Pismo clarkia (Clarkia speciosa immaculata)
E —Morefield's leather-flower (Clematis morefieldii)
E —Alabama leather-flower (Clematis socialis)
E —`Oha wai (Clermontia lindseyana)
E —`Oha wai (Clermontia oblongifolia ssp. brevipes)
E —`Oha wai (Clermontia oblongifolia ssp. mauiensis)
E —`Oha wai (Clermontia peleana)
E   `Oha wai (Clermontia pyrularia)
T —Pigeon wings (Clitoria fragrans)
E —Kauila (Colubrina oppositifolia)
E —Short-leaved rosemary (Conradina brevifolia)
E —Etonia rosemary (Conradina etonia)
E —Apalachicola rosemary (Conradina glabra)
T —Cumberland rosemary (Conradina verticillata)
E —Salt marsh bird's-beak (Cordylanthus maritimus ssp. maritimus)
E —Palmate-bracted bird's-beak (Cordylanthus palmatus)
E —Pennell's bird's-beak (Cordylanthus tenuis ssp. capillaris)
E —Palo de nigua (cap jug™erilla) (Cornutia obovata)
T —Cochise pincushion cactus (Coryphantha (Escobaria) robbinsorum)
E —Nellie cory cactus (Coryphantha (Escobaria) minima)
T —Bunched cory cactus (Coryphantha ramillosa)
E —Pima pineapple cactus (Coryphantha scheeri var. robustispina)
T —Lee pincushion cactus (Coryphantha sneedii var. leei)
E —Sneed pincushion cactus (Coryphantha sneedii var. sneedii)
E —Cranichis ricartii (Plant, no common name)
E —Higuero de Sierra (Crescentia portoricensis)
E —Avon Park harebells (Crotalaria avonensis)
E —Terlingua Creek cats-eye (Cryptantha crassipes)
E —Okeechobee gourd (Cucurbita okeechobeensis ssp. okeechobeensis)
E —Haha (Cyanea asarifolia)
E —Haha (Cyanea copelandii ssp. copelandii)

E —Haha (Cyanea grimesiana ssp. obatae)
E —Haha (Cyanea hamatiflora ssp. carlsonii)
E —Haha (Cyanea lobata)
E —Cyanea macrostegia ssp. gibsonii (Plant, no common name)
E —Haha (Cyanea mannii)
E —Haha (Cyanea mceldowneyi)
E —Haha (Cyanea pinnatifida)
E —Haha (Cyanea procera)
E —Haha (Cyanea shipmannii)
E —Haha (Cyanea stictophylla)
E —Cyanea superba (Plant, no common name)
E —Haha (Cyanea truncata)
E —Cyanea undulata (Plant, no common name)
T —Jones cycladenia (Cycladenia humilis var. jonesii)
E —Ha`iwale (Cyrtandra crenata)
E —Ha`iwale (Cyrtandra giffardii)
T —Ha`iwale (Cyrtandra limahuliensis)
E —Ha`iwale (Cyrtandra munroi)
E —Ha`iwale (Cyrtandra polyantha)
E —Ha`iwale (Cyrtandra tintinnabula)
E —Leafy prairie-clover (Dalea (Petalostemum) foliosa)
E —Daphnopsis hellerana (Plant, no common name)
E —Beautiful pawpaw (Deeringothamnus pulchellus)
E —Rugel's pawpaw (Deeringothamnus rugelii)
E —Delissea rhytidosperma (Plant, no common name)
E —San Clemente Island larkspur (Delphinium variegatum ssp. kinkiense)
E —Garrett's mint (Dicerandra christmanii)
E —Longspurred mint (Dicerandra cornutissima)
E —Scrub mint (Dicerandra frutescens)
E —Lakela's mint (Dicerandra immaculata)
E —Slender-horned spineflower (Dodecahema leptoceras)
E —Na`ena`e (Dubautia herbstobatae)
E —Dubautia latifolia (Plant, no common name)
E —Dubautia pauciflorula (Plant, no common name)
E —Santa Clara Valley dudleya (Dudleya setchellii)
E —Santa Barbara Island liveforever (Dudleya traskiae)
E —Smooth coneflower (Echinacea laevigata)
E —Tennessee purple coneflower (Echinacea tennesseensis)
E —Nichol's Turk's head cactus (Echinocactus horizonthalonius var. nicholii)
T —Chisos Mountain hedgehog cactus (Echinocereus chisoensis var. chisoensis)
E —Kuenzler hedgehog cactus (Echinocereus fendleri var. kuenzleri)
E —Lloyd's hedgehog cactus (Echinocereus lloydii)
E —Black lace cactus (Echinocereus reichenbachii (melanocentrus) var. albertii)
E —Arizona hedgehog cactus (Echinocereus triglochidiatus var. arizonicus)
E —Davis' green pitaya (Echinocereus viridiflorus var. davisii)
T —Lloyd's Mariposa cactus (Echinomastus (Sclerocactus) mariposensis)
T —Ash Meadows sunray (Enceliopsis nudicaulis var. corrugata)
E —Kern mallow (Eremalche kernensis)
E —Santa Ana River woolly-star (Eriastrum densifolium ssp. sanctorum)
T —Hoover's woolly-star (Eriastrum hooveri)
E —Maguire daisy (Erigeron maguirei var. maguirei)

T —Parish's daisy (Erigeron parishii)
T —Zuni (rhizome) fleabane (Erigeron rhizomatus)
E —Indian Knob mountain balm (Eriodictyon altissimum)
T —Gypsum wild-buckwheat (Eriogonum gypsophilum)
T —Scrub buckwheat (Eriogonum longifolium var. gnaphalifolium)
E —Cushenbury buckwheat (Eriogonum ovalifolium var. vineum)
E —Steamboat buckwheat (Eriogonum ovalifolium var. williamsiac)
E —Clay-loving wild-buckwheat (Eriogonum pelinophilum)
E —San Mateo woolly sunflower (Eriophyllum latilobum)
E —San Diego button-celery (Eryngium aristulatum var. parishii)
E —Loch Lomond coyote-thistle (Eryngium constancei)
E —Snakeroot (Eryngium cuneifolium)
E —Contra Costa wallflower (Erysimum capitatum var. angustatum)
E —Menzies' wallflower (Erysimum menziesii)
E —Ben Lomond wallflower (Erysimum teretifolium)
E —Minnesota trout lily (Erythronium propullans)
E —Uvillo (Eugenia haematocarpa)
E —Nioi (Eugenia koolauensis)
E —Eugenia woodburyana (Plant, no common name)
T —Telephus spurge (Euphorbia telephioides)
T —Penland alpine fen mustard (Eutrema penlandii)
E —Heau (Exocarpos luteolus)
E —Mehamehame (Flueggea neowawraea)
E —Johnston's frankenia (Frankenia johnstonii)
E —Gahnia lanaiensis (Plant, no common name)
E —Small's milkpea (Galactia smallii)
E —Na`u or Hawaiian gardenia (Gardenia brighamii)
T —Geocarpon minimum (Plant, no common name)
E —Hawaiian red-flowered geranium (Geranium arboreum)
E —Nohoanu (Geranium multiflorum)
T —Gesneria pauciflora (Plant, no common name)
E —Spreading avens (Geum radiatum)
E —Monterey gilia (Gilia tenuiflora ssp. arenaria)
E —Beautiful goetzea or matabuey (Goetzea elegans)
E —Gouania hillebrandii (Plant, no common name)
E —Gouania meyenii (Plant, no common name)
E —Gouania vitifolia (Plant, no common name)
T —Ash Meadows gumplant (Grindelia fraxino-pratensis)
E —Haplostachys haplostachya (Plant, no common name)
E —Harper's beauty (Harperocallis flava)
T —Higo chumbo (Harrisia portoricensis)
E —Todsen's pennyroyal (Hedeoma todsenii)
E —`Awiwi (Hedyotis cookiana)
E —Kio`ele (Hedyotis coriacea)
E —Hedyotis degeneri (Plant, no common name)
E —Pilo (Hedyotis mannii)
E —Hedyotis parvula (Plant, no common name)
E —Roan Mountain bluet (Hedyotis purpurea var. montana)
E —Na Pali beach hedyotis (Hedyotis st.-johnii)
E —Schweinitz's sunflower (Helianthus schweinitzii)
T —Swamp pink (Helonias bullata)
T —Marin dwarf-flax (Hesperolinon congestum)

E —Hesperomannia arborescens (Plant, no common name)
E —Hesperomannia arbuscula (Plant, no common name)
E —Hesperomannia lydgatei (Plant, no common name)
T —Dwarf-flowered heartleaf (Hexastylis naniflora)
E —Kauai hau kuahiwi (Hibiscadelphus distans)
E —Koki`o ke`oke`o (Hibiscus arnottianus ssp. immaculatus)
E —Ma`o hau hele (Hibiscus brackenridgei)
E —Clay's hibiscus (Hibiscus clayi)
E —Slender rush-pea (Hoffmannseggia tenella)
T —Water howellia (Howellia aquatilis)
T —Mountain golden heather (Hudsonia montana)
T —Lakeside daisy (Hymenoxys herbacea)
E —Texas prairie dawn-flower (Texas bitterweed) (Hymenoxys texana)
E —Highlands scrub hypericum (Hypericum cumulicola)
E —Cook's holly (Ilex cookii)
E —Ilex sintenisii (Plant, no common name)
E —Peter's Mountain mallow (Iliamna corei)
E —Holy Ghost ipomopsis (Ipomopsis sancti-spiritus)
T —Dwarf lake iris (Iris lacustris)
E —Hilo ischaemum (Ischaemum byrone)
E —Aupaka (Isodendrion hosakae)
E —Wahine noho kula (Isodendrion pyrifolium)
T —Small whorled pogonia (Isotria medeoloides)
T —Ash Meadows ivesia (Ivesia kingii var. eremica)
E —Beach jacquemontia (Jacquemontia reclinata)
E —Cooley's water-willow (Justicia cooleyi)
E —Cooke's koki`o (Kokia cookei)
E —Koki`o (hau-hele`ula or Hawaii tree cotton) (Kokia drynarioides)
E —Kamakahala (Labordia lydgatei)
E —Burke's goldfields (Lasthenia burkei)
E —Beach layia (Layia carnosa)
E —San Joaquin wooly-threads (Lembertia congdonii)
E —Lepanthes eltoroensis (Plant, no common name)
E —Barneby ridge-cress (peppercress) (Lepidium barnebyanum)
E —Leptocereus grantianus (Plant, no common name)
T —Prairie bush-clover (Lespedeza leptostachya)
T —Dudley Bluffs bladderpod (Lesquerella congesta)
E —Missouri bladderpod (Lesquerella filiformis)
E —San Bernardino Mountains bladderpod (Lesquerella kingii ssp. bernardina)
T —Lyrate bladderpod (Lesquerella lyrata)
E —White bladderpod (Lesquerella pallida)
E —Kodachrome bladderpod (Lesquerella tumulosa)
T —Heller's blazingstar (Liatris helleri)
E —Scrub blazingstar (Liatris ohlingerae)
E —Western lily (Lilium occidental)
E —Butte County meadowfoam (Limnanthes floccosa ssp. californica)
E —Sebastopol meadowfoam (Limnanthes vinculans)
E —Pondberry (Lindera melissifolia)
E —Nehe (Lipochaeta fauriei)
E —Nehe (Lipochaeta kamolensis)
E —Nehe (Lipochaeta lobata var. leptophylla)
E —Nehe (Lipochaeta micrantha)

E —Nehe (Lipochaeta tenuifolia)
E —Lipochaeta venosa (Plant, no common name)
E —Nehe (Lipochaeta waimeaensis)
E —Lobelia niihauensis (Plant, no common name)
E —Lobelia oahuensis (Plant, no common name)
E —Bradshaw's desert-parsley (lomatium) (Lomatium bradshawii)
E —San Clemente Island broom (Lotus dendroideus ssp. traskiae)
E —Scrub lupine (Lupinus aridorum)
E —Clover lupine (Lupinus tidestromii)
E —Lyonia truncata var. proctorii (Plant, no common name)
E —Rough-leaved loosestrife (Lysimachia asperulaefolia)
E —Lysimachia filifolia (Plant, no common name)
E —Lysimachia lydgatei (Plant, no common name)
T —White birds-in-a-nest (Macbridea alba)
E —San Clemente Island bush-mallow (Malacothamnus clementinus)
E —Walker's manioc (Manihot walkerae)
E —Mariscus fauriei (Plant, no common name)
E —Mariscus pennatiformis (Plant, no common name)
T —Mohr's Barbara's buttons (Marshallia mohrii)
E —Alani (Melicope lydgatei)
E —Alani (Melicope mucronulata)
E —Alani (Melicope adscendens)
E —Alani (Melicope ballouii)
E —Alani (Melicope haupuensis)
E —Alani (Melicope knudsenii)
E —Alani (Melicope ovalis)
E —Alani (Melicope pallida)
E —Alani (Melicope quadrangularis)
E —Alani (Melicope reflexa)
T —Ash Meadows blazing-star (Mentzelia leucophylla)
E —Michigan monkey-flower (Mimulus glabratus var. michiganensis)
E —MacFarlane's four-o'clock (Mirabilis macfarlanei)
E —Mitracarpus maxwelliae (Plant, no common name)
E —Mitracarpus polycladus (Plant, no common name)
E —Munroidendron racemosum (Plant, no common name)
E —Myrcia paganii (Plant, no common name)
E —Neraudia angulata (Plant, no common name)
E —Neraudia sericea (Plant, no common name)
E —Amargosa niterwort (Nitrophila mohavensis)
E —Britton's beargrass (Nolina brittoniana)
E —`Aiea (Nothocestrum breviflorum)
E —`Aiea (Nothocestrum peltatum)
E —Kulu`i (Nototrichium humile)
E —Holei (Ochrosia kilaueaensis)
E —Eureka Valley evening-primrose (Oenothera avita ssp. eurekensis)
E —Antioch Dunes evening-primrose (Oenothera deltoides ssp. howellii)
E —Bakersfield cactus (Opuntia treleasei)
E —California Orcutt grass (Orcuttia californica)
E —Palo de rosa (Ottoschulzia rhodoxylon)
E —Canby's dropwort (Oxypolis canbyi)
E —Cushenbury oxytheca (Oxytheca parishii var. goodmaniana)
T —Fassett's locoweed (Oxytropis campestris var. chartacea)

E —Carter's panicgrass (Panicum fauriei var. carteri)
T —Papery whitlow-wort (Paronychia chartacea)
E —Furbish lousewort (Pedicularis furbishiae)
T —Siler pincushion cactus (Pediocactus sileri)
E —Peebles Navajo cactus (Pediocactus peeblesianus var. peeblesianus)
E —Brady pincushion cactus (Pediocactus bradyi)
E —Knowlton cactus (Pediocactus knowltonii)
E —San Rafael cactus (Pediocactus despainii)
E —Blowout penstemon (Penstemon haydenii)
E —Penland beardtongue (Penstemon penlandii)
E —White-rayed pentachaeta (Pentachaeta bellidiflora)
E —Wheeler's peperomia (Peperomia wheeleri)
T —Makou (Peucedanum sandwicense)
E —Clay phacelia (Phacelia argillacea)
E —North Park phacelia (Phacelia formosula)
E —Texas trailing phlox (Phlox nivalis ssp. texensis)
E —Phyllostegia glabra var. lanaiensis (Plant, no common name)
E —Phyllostegia mannii (Plant, no common name)
E —Phyllostegia mollis (Plant, no common name)
E —Phyllostegia waimeae (Plant, no common name)
T —Dudley Bluffs twinpod (Physaria obcordata)
E —Key tree-cactus (Pilosocereus robinii (Cereus r.))
T —Godfrey's butterwort (Pinguicula ionantha)
E —Ruth's golden aster (Pityopsis (Heterotheca Chrysopsis) ruthii)
E —Laukahi kuahiwi (Plantago hawaiensis)
E —Laukahi kuahiwi (Plantago princeps)
T —Eastern prairie fringed orchid (Platanthera leucophaea)
T —Western prairie fringed orchid (Platanthera praeclara)
E —Chupacallos (Chupagallo) (Pleodendron macranthum)
E —Mann's bluegrass (Poa mannii)
E —Hawaiian bluegrass (Poa sandvicensis)
E —Poa siphonoglossa (Plant, no common name)
E —San Diego mesa mint (Pogogyne abramsii)
E —Otay mesa mint (Pogogyne nudiuscula)
E —Lewton's polygala (Polygala lewtonii)
E —Tiny polygala (Polygala smallii)
E —Wireweed (Polygonella basiramia)
E —Sandlace (Polygonella myriophylla)
E —Po`e (Portulaca sclerocarpa)
E —Little Aguja pondweed (Potamogeton clystocarpus)
E —Robbins' cinquefoil (Potentilla robbinsiana)
T —Maguire primrose (Primula maguirei)
E —Loulu (Pritchardia affinis)
E —Loulu (Pritchardia munroi)
E —Scrub plum (Prunus geniculata)
E —Kaulu (Pteralyxia kauaiensis)
E —Harperella (Ptilimnium nodosum (fluviatile))
E —Arizona cliffrose (Purshia subintegra)
T —Hinckley's oak (Quercus hinckleyi)
E —Autumn buttercup (Ranunculus acriformis var. aestivalis)
E —Remya kauaiensis (Plant, no common name)
E —Maui remya (Remya mauiensis)

E —Remya montgomeryi (Plant, no common name)
E —Chapman rhododendron (Rhododendron chapmanii)
E —Michaux's sumac (Rhus michauxii)
T —Knieskern's beaked-rush (Rhynchospora knieskernii)
T —Miccosukee gooseberry (Ribes echinellum)
E —Rollandia crispa (Plant, no common name)
E —Gambel's watercress (Rorippa gambellii)
E —Bunched arrowhead (Sagittaria fasciculata)
T —Kral's water-plantain (Sagittaria secundifolia)
E —Sanicula mariversa (Plant, no common name)
E —Lanai sandalwood or `iliahi (Santalum freycinetianum var. lanaiense)
E —Green pitcher-plant (Sarracenia oreophila)
E —Alabama canebrake pitcher-plant (Sarracenia rubra ssp. alabamensis)
E —Mountain sweet pitcher-plant (Sarracenia rubra ssp. jonesii)
E —Dwarf naupaka (Scaevola coriacea)
E —Diamond Head schiedea (Schiedea adamantis)
E —Ma`oli`oli (Schiedea apokremnos)
E —Schiedea haleakalensis (Plant, no common name)
E —Schiedea kaalae (Plant, no common name)
E —Schiedea lydgatei (Plant, no common name)
E —Schiedea spergulina var. leiopoda (Plant, no common name)
T —Schiedea spergulina var. spergulina (Plant, no common name)
E —Shrubby reed-mustard (toad-flax cress) (Schoenocrambe suffrutescens)
T —Clay reed-mustard (Schoenocrambe argillacea)
E —Barneby reed-mustard (Schoenocrambe barnebyi)
T —Schoepfia arenaria (Plant, no common name)
E —American chaffseed (Schwalbea americana)
E —Northeastern (Barbed bristle) bulrush (Scirpus ancistrochaetus)
T   Mesa Verde cactus (Sclerocactus mesae-verdae)
T —Uinta Basin hookless cactus (Sclerocactus glaucus)
E —Wright fishhook cactus (Sclerocactus wrightiae)
T —Florida skullcap (Scutellaria floridana)
E —Large-flowered skullcap (Scutellaria montana)
T —Leedy's roseroot (Sedum integrifolium ssp. leedyi)
T —San Francisco Peaks groundsel (Senecio franciscanus)
E —Hayun lagu (Guam), Tronkon guafi (Rota) (Serianthes nelsonii)
E —`Ohai (Sesbania tomentosa)
T —Nelson's checker-mallow (Sidalcea nelsoniana)
E —Pedate checker-mallow (Sidalcea pedata)
E —Silene alexandri (Plant, no common name)
T —Silene hawaiiensis (Plant, no common name)
E —Silene lanceolata (Plant, no common name)
E —Silene perlmanii (Plant, no common name)
E —Fringed campion (Silene polypetala)
E —White irisette (Sisyrinchium dichotomum)
E —Erubia (Solanum drymophilum)
E —Popolo ku mai (Solanum incompletum)
E —`Aiakeakua, popolo (Solanum sandwicense)
T —White-haired goldenrod (Solidago albopilosa)
T —Houghton's goldenrod (Solidago houghtonii)
E —Short's goldenrod (Solidago shortii)
T —Blue Ridge goldenrod (Solidago spithamaea)

E —Spermolepis hawaiiensis (Plant, no common name)
E —Gentian pinkroot (Spigelia gentianoides)
T —Virginia spiraea (Spiraea virginiana)
T —Ute ladies'-tresses (Spiranthes diluvialis)
E —Navasota ladies'-tresses (Spiranthes parksii)
T —Cóbana negra (Stahlia monosperma)
E —Stenogyne angustifolia (Plant, no common name)
E —Stenogyne bifida (Plant, no common name)
E —Stenogyne campanulata (Plant, no common name)
E —Stenogyne kanehoana (Plant, no common name)
E —Malheur wire-lettuce (Stephanomeria malheurensis)
E —Metcalf Canyon jewelflower (Streptanthus albidus ssp. albidus)
E —Tiburon jewelflower (Streptanthus niger)
E —Palo de jazmín (Styrax portoricensis)
E —Texas snowbells (Styrax texana)
E —California seablite (Suaeda californica)
E —Eureka Dune grass (Swallenia alexandrae)
E —Palo colorado (Ternstroemia luquillensis)
E —Ternstroemia subsessilis (Plant, no common name)
E —Tetramolopium arenarium (Plant, no common name)
E —Pamakani (Tetramolopium capillare)
E —Tetramolopium filiforme (Plant, no common name)
E —Tetramolopium lepidotum ssp. lepidotum (Plant, no common name)
E —Tetramolopium remyi (Plant, no common name)
T —Tetramolopium rockii (Plant, no common name)
E —`Ohe`ohe (Tetraplasandra gymnocarpa)
E —Cooley's meadowrue (Thalictrum cooleyi)
E —Slender-petaled mustard (Thelypodium stenopetalum)
E —Ashy dogweed (Thymophylla tephroleuca)
T —Last Chance townsendia (Townsendia aprica)
E —Bariaco (guayabacón) (Trichilia triacantha)
E —Running buffalo clover (Trifolium stoloniferum)
E —Persistent trillium (Trillium persistens)
E —Relict trillium (Trillium reliquum)
E —Solano grass (Tuctoria mucronata)
E —Opuhe (Urera kaalae)
E —Vernonia proctorii (Plant, no common name)
E —Hawaiian vetch (Vicia menziesii)
E —Vigna o-wahuensis (Plant, no common name)
E —Pamakani (Viola chamissoniana chamissoniana)
E —Viola helenae (Plant, no common name)
E —Viola lanaiensis (Plant, no common name)
E —Wide-leaf warea (Warea amplexifolia)
E —Carter's mustard (Warea carteri)
E —Dwarf iliau (Wilkesia hobdyi)
E —Xylosma crenatum (Plant, no common name)
E —Tennessee yellow-eyed grass (Xyris tennesseensis)
E —A`e (Zanthoxylum hawaiiense)
E —St. Thomas prickly-ash (Zanthoxylum thomasianum)
E —Texas wild-rice (Zizania texana)
E —Florida ziziphus (Ziziphus celata)

NONFLOWERING PLANTS

Conifers & Cycads
E —Santa Cruz cypress (Cupressus abramsiana)
E —Florida torreya (Torreya taxifolia)

Ferns & Allies
E —Pendant kihi fern (Adenophorus periens)
E —Adiantum vivesii (Fern, no common name)
T —American hart's-tongue fern (Asplenium scolopendrium var. americanum)
E —Asplenium fragile var. insulare (Fern, no common name)
E —Pauoa (Ctenitis squamigera)
E —Elfin tree fern (Cyathea dryopteroides)
E —Asplenium-leaved diellia (Diellia erecta)
E —Diellia falcata (Fern, no common name)
E —Diellia pallida (Fern, no common name)
E —Diellia unisora (Fern, no common name)
E —Diplazium molokaiense (Fern, no common name)
E —Elaphoglossum serpens (Fern, no common name)
E —Wawae`iole (Huperzia mannii)
E —Louisiana quillwort (Isoetes louisianensis)
E —Black-spored quillwort (Isoetes melanospora)
E —Mat-forming quillwort (Isoetes tegetiformans)
E —Wawae`iole (Lycopodium nutans)
E —`Ihi`ihi (Marsilea villosa)
E —Aleutian shield-fern (Aleutian holly-fern) (Polystichum aleuticum)
E —Polystichum calderonense (Fern, no common name)
E —Pteris lidgatei (Fern, no common name)
E —Tectaria estremerana (Fern, no common name)
T —Alabama streak-sorus fern (Thelypteris pilosa var. alabamensis)
E —Thelypteris inabonensis (Fern, no common name)
E —Thelypteris verecunda (Fern, no common name)
E —Thelypteris yaucoensis (Fern, no common name)

Lichens
Florida perforate cladonia (Cladonia perforata)
Rock gnome lichen (Gymnoderma lineare)

Foreign Species

VERTEBRATES

Mammals

E —Anoa, lowland (Bubalus depressicornis (B. anoa depressicornis) 1 sp.
E —Anoa, mountain (Bubalus quarlesi (B. anoa quarlesi)) 1 sp.
E —Antelope, giant sable (Hippotragus niger variani) 1 sp.
E —Argali (Ovis ammon) 1 sp.
T —Argali (Ovis ammon) 1 sp.
E —Armadillo, giant (Priodontes maximus (giganteus)) 1 sp.
E —Armadillo, pink fairy (Chlamyphorus truncatus) 1 sp.
E —Ass, African wild (Equus asinus (africanus)) 1 sp.
E —Ass, Asian wild (kulan, onager) (Equus hemionus) 1 sp.
E —Avahi (Avahi (Lichanotus) laniger (entire genus)) 1 sp.
E —Aye-aye (Daubentonia madagascariensis) 1 sp.
E —Babirusa (Babyrousa babyrussa) 1 sp.
T —Baboon, gelada (Theropithecus gelada) 1 sp.
E —Bandicoot, barred (Perameles bougainville) 1 sp.
E —Bandicoot, desert (Perameles eremiana) 1 sp.
E —Bandicoot, lesser rabbit (Macrotis leucura) 1 sp.
E —Bandicoot, pig-footed (Chaeropus ecaudatus) 1 sp.
E —Bandicoot, rabbit (Macrotis lagotis) 1 sp.
E —Banteng (Bos javanicus (banteng)) 1 sp.
E —Bat, Bulmer's fruit (flying fox) (Aproteles bulmerae) 1 sp.
E —Bat, bumblebee (Craseonycteris thonglongyai) 1 sp.
E —Bat, Rodrigues fruit (flying fox) (Pteropus rodricensis) 1 sp.
E —Bat, Singapore roundleaf horseshoe (Hipposideros ridleyi) 1 sp.
E —Bear, Baluchistan (Ursus thibetanus gedrosianus) 1 sp.
E —Bear, brown (Ursus arctos arctos) 1 sp.
E —Bear, brown (Ursus arctos pruinosus) 1 sp.
E —Bear, Mexican grizzly (Ursus arctos (U. a. nelsoni)) 1 sp.
E —Beaver (Castor fiber birulai) 1 sp.
E —Bison, wood (Bison bison athabascae) 1 sp.
E —Bobcat (Felis rufus escuinapae) 1 sp.
E —Bontebok (antelope) (Damaliscus dorcas dorcas) 1 sp.
E —Camel, Bactrian (Camelus bactrianus (ferus)) 1 sp.
E —Cat, Andean (Felis jacobita) 1 sp.
E —Cat, black-footed (Felis nigripes) 1 sp.
E —Cat, flat-headed (Felis planiceps) 1 sp.
E —Cat, Iriomote (Felis (Mayailurus) iriomotensis) 1 sp.
E —Cat, leopard (Felis bengalensis bengalensis) 1 sp.
E —Cat, marbled (Felis marmorata) 1 sp.
E —Cat, Pakistan sand (Felis margarita scheffeli) 1 sp.
E —Cat, Temminck's (golden cat) (Felis temmincki) 1 sp.
E —Cat, tiger (Felis tigrinus) 1 sp.
E —Chamois, Apennine (Rupicapra rupicapra ornata) 1 sp.
E —Cheetah (Acinonyx jubatus) 1 sp.
E —Chimpanzee, pygmy (Pan paniscus) 1 sp.
E —Chimpanzee (Pan troglodytes) 1 sp.
T —Chimpanzee (Pan troglodytes) 1 sp.
E —Chinchilla (Chinchilla brevicaudata boliviana) 1 sp.

E —Civet, Malabar large-spotted (Viverra megaspila civettina) 1 sp.
E —Cochito (Gulf of California harbor porpoise) (Phocoena sinus) 1 sp.
E —Deer, Bactrian (Cervus elaphus bactrianus) 1 sp.
E —Deer, Barbary (Cervus elaphus barbarus) 1 sp.
E —Deer, Bawean (Axis (Cervus) porcinus kuhli) 1 sp.
E —Deer, Cedros Island mule (Odocoileus hemionus cedrosensis) 1 sp.
E —Deer, Corsican red (Cervus elaphus corsicanus) 1 sp.
E —Deer, Eld's brow-antlered (Cervus eldi) 1 sp.
E —Deer, Formosan sika (Cervus nippon taiouanus) 1 sp.
E —Deer, hog (Axis (Cervus) porcinus annamiticus) 1 sp.
E —Deer, marsh (Blastocerus dichotomus) 1 sp.
E —Deer, McNeill's (Cervus elaphus macneilii) 1 sp.
E —Deer, musk (Moschus spp. (all species)) —2 spp.
E —Deer, North China sika (Cervus nippon mandarinus) 1 sp.
E —Deer, pampas (Ozotoceros bezoarticus) 1 sp.
E —Deer, Persian fallow (Dama dama mesopotamica) 1 sp.
E —Deer, Philippine (Axis (Cervus) porcinus calamianensis) 1 sp.
E —Deer, Ryukyu sika (Cervus nippon keramae) 1 sp.
E —Deer, Shansi sika (Cervus nippon grassianus) 1 sp.
E —Deer, South China sika (Cervus nippon kopschi) 1 sp.
E —Deer, swamp (barasingha) (Cervus duvauceli) 1 sp.
E —Deer, Visayan (Cervus alfredi) 1 sp.
E —Deer, Yarkand (Cervus elaphus yarkandensis) 1 sp.
E —Dhole (Asiatic wild dog) (Cuon alpinus) 1 sp.
E —Dibbler (Antechinus apicalis) 1 sp.
E —Dog, African wild (Lycaon pictus) 1 sp.
E —Dolphin, Chinese river (whitefin) (Lipotes vexillifer) 1 sp.
E —Dolphin, Indus River (Platanista minor) 1 sp.
E —Drill (Papio leucophaeus) 1 sp.
E —Dugong (Dugong dugon) 1 sp.
E —Duiker, Jentink's (Cephalophus jentinki) 1 sp.
E —Eland, western giant (Taurotragus derbianus derbianus) 1 sp.
T —Elephant, African (Loxodonta africana) 1 sp.
E —Elephant, Asian (Elephas maximus) 1 sp.
E —Fox, northern swift (Vulpes velox hebes) 1 sp.
E —Fox, Simien (Canis (Simenia) simensis) 1 sp.
E —Gazelle, Arabian (Gazella gazella) 1 sp.
E —Gazelle, Clark's (Dibatag) (Ammodorcas clarkei) 1 sp.
E —Gazelle, Cuvier's (Gazella cuvieri) 1 sp.
E —Gazelle, Mhorr (Gazella dama mhorr) 1 sp.
E —Gazelle, Moroccan (Dorcas) (Gazella dorcas massaesyla) 1 sp.
E —Gazelle, Pelzeln's (Gazella dorcas pelzelni) 1 sp.
E —Gazelle, Rio de Oro Dama (Gazella dama lozanoi) 1 sp.
E —Gazelle, sand (Gazella subgutturosa marica) 1 sp.
E —Gazelle, Saudi Arabian (Gazella dorcas saudiya) 1 sp.
E —Gazelle, slender-horned (Rhim) (Gazella leptoceros) 1 sp.
E —Gibbons (Hylobates spp. (including Nomascus)) —9 spp.
E —Goat, wild (Chiltan markhor) (Capra aegagrus (falconeri) chiltanensis) 1 sp.
E —Goral (Nemorhaedus goral) 1 sp.
E —Gorilla (Gorilla gorilla) 1 sp.
E —Hare, hispid (Caprolagus hispidus) 1 sp.
E —Hartebeest, Swayne's (Alcelaphus buselaphus swaynei) 1 sp.

E —Hartebeest, Tora (Alcelaphus buselaphus tora) 1 sp.
E —Hog, pygmy (Sus salvanius) 1 sp.
E —Horse, Przewalski's (Equus przewalskii) 1 sp.
E —Huemul, north Andean (Hippocamelus antisensis) 1 sp.
E —Huemul, South Andean (Hippocamelus bisulcus) 1 sp.
E —Hutia, Cabrera's (Capromys angelcabrerai) 1 sp.
E —Hutia, dwarf (Capromys nana) 1 sp.
E —Hutia, large-eared (Capromys auritus) 1 sp.
E —Hutia, little earth (Capromys sanfelipensis) 1 sp.
E —Hyena, Barbary (Hyaena hyaena barbara) 1 sp.
E —Hyena, brown (Hyaena brunnea) 1 sp.
E —Ibex, Pyrenean (Capra pyrenaica pyrenaica) 1 sp.
E —Ibex, Walia (Capra walie) 1 sp.
E —Impala, black-faced (Aepyceros melampus petersi) 1 sp.
E —Indri (Indri indri (entire genus)) 1 sp.
E —Jaguar (Panthera onca) 1 sp.
E —Jaguarundi (Felis yagouaroundi fossata) 1 sp.
E —Jaguarundi (Felis yagouaroundi panamensis) 1 sp.
E —Kouprey (Bos sauveli) 1 sp.
E —Langur, capped (Presbytis pileata) 1 sp.
E —Langur, Douc (Pygathrix nemaeus) 1 sp.
E —Langur, entellus (Presbytis entellus) 1 sp.
E —Langur, Francois' (Presbytis francoisi) 1 sp.
E —Langur, golden (Presbytis geei) 1 sp.
T —Langur, long-tailed (Presbytis potenziani) 1 sp.
E —Langur, Pagi Island (Nasalis (Simias) concolor) 1 sp.
T —Langur, purple-faced (Presbytis senex) 1 sp.
T —Lechwe, red (Kobus leche) 1 sp.
E —Lemurs (All species of the family Lemuridae) —9 spp.
E —Leopard, clouded (Neofelis nebulosa) 1 sp.
E —Leopard, snow (Panthera uncia) 1 sp.
E —Leopard (Panthera pardus) 1 sp.
T —Leopard (Panthera pardus) 1 sp.
E —Linsang, spotted (Prionodon pardicolor) 1 sp.
E —Lion, Asiatic (Panthera leo persica) 1 sp.
T —Loris, lesser slow (Nycticebus pygmaeus) 1 sp.
E —Lynx, Spanish (Felis (Lynx) pardina) 1 sp.
T —Macaque, Formosan rock (Macaca cyclopis) 1 sp.
T —Macaque, Japanese (Macaca fuscata) 1 sp.
E —Macaque, lion-tailed (Macaca silenus) 1 sp.
T —Macaque, stump-tailed (Macaca arctoides) 1 sp.
T —Macaque, Toque (Macaca sinica) 1 sp.
E —Manatee, Amazonian (Trichechus inunguis) 1 sp.
T —Manatee, West African (Trichechus senegalensis) 1 sp.
E —Mandrill (Papio sphinx) 1 sp.
E —Mangabey, Tana River (Cercocebus galeritus) 1 sp.
E —Mangabey, white-collared (Cercocebus torquatus) 1 sp.
E —Margay (Felis wiedii) 1 sp.
E —Markhor, Kabal (Capra falconeri megaceros) 1 sp.
E —Markhor, straight-horned (Capra falconeri jerdoni) 1 sp.
E —Marmoset, buff-headed (Callithrix flaviceps) 1 sp.
E —Marmoset, buffy tufled-ear (Callithrix jacchus aurita) 1 sp.

E —Marmoset, cotton-top (Saguinus oedipus) 1 sp.
E —Marmoset, Goeldi's (Callimico goeldii) 1 sp.
E —Marmot, Vancouver Island (Marmota vancouverensis) 1 sp.
E —Marsupial, eastern jerboa (Antechinomys laniger) 1 sp.
E —Marsupial-mouse, large desert (Sminthopsis psammophila) 1 sp.
E —Marsupial-mouse, long-tailed (Sminthopsis longicaudata) 1 sp.
E — Marten, Formosan yellow-throated (Martes flavigula chrysospila) 1 sp.
E —Monkey (langur), Guizhou snub-nosed (Rhinopithecus (Pygathrix) brelichi) 1 sp.
E —Monkey (langur), Sichuan snub-nosed (Rhinopithecus (Pygathrix) roxellana) 1 sp.
E —Monkey (langur), Tonkin snub-nosed (Rhinopithecus (Pygathrix) avunculus) 1 sp.
E —Monkey (langur), Yunnan snub-nosed (Rhinopithecus (Pygathrix) bieti) 1 sp.
E —Monkey, black colobus (Colobus satanas) 1 sp.
T —Monkey, black howler (Alouatta pigra) 1 sp.
E —Monkey, Diana (Cercopithecus diana) 1 sp.
E —Monkey, howler (Alouatta palliata (villosa)) 1 sp.
E —Monkey, L'hoest's (Cercopithecus lhoesti) 1 sp.
E —Monkey, Preuss' red colobus (Colobus badius preussi) 1 sp.
E —Monkey, proboscis (Nasalis larvatus) 1 sp.
E —Monkey, red-backed squirrel (Saimiri oerstedii) 1 sp.
E —Monkey, red-bellied (Cercopithecus erythrogaster) 1 sp.
E —Monkey, red-eared nose-spotted (Cercopithecus erythrotis) 1 sp.
E —Monkey, spider (Ateles geoffroyi frontatus) 1 sp.
E —Monkey, spider (Ateles geoffroyl panamensis) 1 sp.
E —Monkey, Tana River red colobus (Colobus rufomitratus (badius) rufomitra-
    tus) 1 sp.
E —Monkey, woolly spider (Brachyteles arachnoides) 1 sp.
E —Monkey, yellow-tailed woolly (Lagothrix flavicauda) 1 sp.
E —Monkey, Zanzibar red colobus (Colobus kirki) 1 sp.
E —Mouse, Australian native (Notomys aquilo) 1 sp.
E —Mouse, Australian native (Zyzomys (Notomys) pedunculatus) 1 sp.
E —Mouse, Field's (Pseudomys fieldi) 1 sp.
E —Mouse, Gould's (Pseudomys gouldII) 1 sp.
E —Mouse, New Holland (Pseudomys novaehollandiae) 1 sp.
E —Mouse, Shark Bay (Pseudomys praeconis) 1 sp.
E —Mouse, Shortridge's (Pseudomys shortridgei) 1 sp.
E —Mouse, Smoky (Pseudomys fumeus) 1 sp.
E —Mouse, western (Pseudomys occidentalis) 1 sp.
E —Muntjac, Fea's (Muntiacus feae) 1 sp.
E —Native-cat, eastern (Dasyurus viverrinus) 1 sp.
E —Numbat (Myrmecobius fasciatus) 1 sp.
E —Orangutan (Pongo pygmaeus) 1 sp.
E —Oryx, Arabian (Oryx leucoryx) 1 sp.
E —Otter, Cameroon clawless (Aonyx (Paraonyx) congica microdon) 1 sp.
E —Otter, giant (Pteronura brasiliensis) 1 sp.
E —Otter, long-tailed (Lutra longicaudis (incl. platensis)) 1 sp.
E —Otter, marine (Lutra felina) 1 sp.
E —Otter, southern river (Lutra provocax) 1 sp.
E —Panda, giant (Ailuropoda melanoleuca) 1 sp.
E —Pangolin (scaly anteater) (Manis temmincki) 1 sp.
E —Planigale, little (Planigale ingrami subtilissima) 1 sp.
E —Planigale, southern (Planigale tenuirostris) 1 sp.
E —Porcupine, thin-spined (Chaetomys subspinosus) 1 sp.

E —Possum, Leadbeater's (Gymnobelideus leadbeateri) 1 sp.
E —Possum, mountain pygmy (Burramys parvus) 1 sp.
E —Possum, scaly-tailed (Wyulda squamicaudata) 1 sp.
E —Prairie dog, Mexican (Cynomys mexicanus) 1 sp.
E —Pronghorn, peninsular (Antilocapra americana peninsularis) 1 sp.
E —Pudu (Pudu pudu) 1 sp.
E —Puma, Costa Rican (Felis concolor costaricensis) 1 sp.
E —Quokka (Setonix brachyurus) 1 sp.
E —Rabbit, Ryukyu (Pentalagus furnessi) 1 sp.
E —Rabbit, volcano (Romerolagus diazi) 1 sp.
E —Rat, false water (Xeromys myoides) 1 sp.
E —Rat, stick-nest (Leporillus conditor) 1 sp.
E —Rat-kangaroo, brush-tailed (Bettongia penicillata) 1 sp.
E —Rat-kangaroo, Gaimard's (Bettongia gaimardi) 1 sp.
E —Rat-kangaroo, Lesuer's (Bettongia lesueur) 1 sp.
E —Rat-kangaroo, plain (Caloprymnus campestris) 1 sp.
E —Rat-kangaroo, Queensland (Bettongia tropica) 1 sp.
E —Rhinoceros, black (Diceros bicornis) 1 sp.
E —Rhinoceros, great Indian (Rhinoceros unicornis) 1 sp.
E —Rhinoceros, Javan (Rhinoceros sondaicus) 1 sp.
E —Rhinoceros, northern white (Ceratotherium simum cottoni) 1 sp.
E —Rhinoceros, Sumatran (Dicerorhinus (Didermoceros) sumatrensis) 1 sp.
E —Saiga, Mongolian (antelope) (Saiga tatarica mongolica) 1 sp.
E —Saki, southern beared (Chiropotes satanas satanas) 1 sp.
E —Saki, white-nosed (Chiropotes albinasus) 1 sp.
E —Seal, Mediterranean monk (Monachus monachus) 1 sp.
E —Seal, Saimaa (Phoca hispida saimensis) 1 sp.
E —Seledang (Gaur) (Bos gaurus) 1 sp.
E —Serow (Capricornis sumatraensis) 1 sp.
E —Serval, Barbary (Felis serval constantina) 1 sp.
E —Shapo (Ovis vignei vignei) 1 sp.
E —Shou (Cervus elaphus wallichi) 1 sp.
E —Siamang (Symphalangus syndactylus) 1 sp.
E —Sifakas (Propithecus spp.) —2 spp.
E —Sloth, Brazilian three-toed (Bradypus torquatus) 1 sp.
E —Solenodon, Cuban (Solenodon (Atopogale) cubanus) 1 sp.
E —Solenodon, Haitian (Solenodon paradoxus) 1 sp.
E —Stag, Barbary (Cervus elaphus barbarus) 1 sp.
E —Stag, Kashmir (Cervus elaphus hanglu) 1 sp.
E —Suni, Zanzibar (Neotragus (Nesotragus) moschatus moschatus) 1 sp.
E —Tahr, Arabian (Hemitragus jayakari) 1 sp.
E —Tamaraw (Bubalus mindorensis) 1 sp.
E —Tamarin (marmoset), golden-rumped (golden-headed) (Leontopithecus
    (Leontideus) spp.) 1 sp.
E —Tamarin, pied (Saguinus bicolor) 1 sp.
T —Tamarin, white-footed (Saguinus leucopus) 1 sp.
E —Tapir, Asian (Tapirus indicus) 1 sp.
E —Tapir, Brazilian (Tapirus terrestris) 1 sp.
E —Tapir, Central American (Tapirus bairdii) 1 sp.
E —Tapir, mountain (Tapirus pinchaque) 1 sp.
T —Tarsier, Philippine (Tarsius syrichta) 1 sp.
E —Tiger, Tasmanian (Thylacine) (Thylacinus cynocephalus) 1 sp.

E —Tiger (Panthera tigris) 1 sp.
E —Uakari (all species) (Cacajao spp.) —2 spp.
E —Urial (Ovis musimon (orientalis) ophion) 1 sp.
E —Vicuna (Vicugna vicugna) 1 sp.
E —Wallaby, banded hare (Lagostrophus fasciatus) 1 sp.
E —Wallaby, brindled nail-tailed (Onychogalea fraenata) 1 sp.
E —Wallaby, crescent nail-tailed (Onychogalea lunata) 1 sp.
E —Wallaby, Parma (Macropus parma) 1 sp.
E —Wallaby, western hare (Lagorchestes hirsutus) 1 sp.
E —Wallaby, yellow-footed rock (Petrogale xanthopus) 1 sp.
E —Whale, gray (Eschrichtius robustus) 1 sp.
E —Wolf, maned (Chrysocyon brachyurus) 1 sp.
E —Wombat, hairy-nosed (Barnard's and Queensland hairy-nosed) (Lasiorhinus krefftii) 1 sp.
E —Yak, wild (Bos grunniens mutus) 1 sp.
T —Zebra, Grevy's (Equus grevyi) 1 sp.
T —Zebra, Hartmann's mountain (Equus zebra hartmannae) 1 sp.
E —Zebra, mountain (Equus zebra zebra) 1 sp.

Birds
E —Albatross, Amsterdam (Diomedia amsterdamensis) 1 sp.
E —Albatross, short-tailed (Diomedea albatrus) 1 sp.
E —Alethe, Thyolo (Alethe choloensis) 1 sp.
E —Booby, Abbott's (Sula abbotti) 1 sp.
E —Bristlebird, western rufous (Dasyornis broadbenti littoralis) 1 sp.
E —Bristlebird, western (Dasyornis brachypterus longirostris) 1 sp.
E —Bulbul, Mauritius olivaceous (Hypsipetes borbonicus olivaceus) 1 sp.
E —Bullfinch, Sao Miguel (finch) (Pyrrhula pyrrhula murina) 1 sp.
T —Bush-shrike, Ulugura (Malaconotus alius) 1 sp.
E —Bushwren, New Zealand (Xenicus longipes) 1 sp.
E —Bustard, great Indian (Choriotis nigriceps) 1 sp.
E —Cahow (Bermuda petrel) (Pterodroma cahow) 1 sp.
E —Condor, Andean (Vultur gryphus) 1 sp.
E —Cotinga, banded (Cotinga maculata) 1 sp.
E —Cotinga, white-winged (Xipholena atropurpurea) 1 sp.
E —Crane, black-necked (Grus nigricollis) 1 sp.
E —Crane, Cuba sandhill (Grus canadensis nesiotes) 1 sp.
E —Crane, hooded (Grus monacha) 1 sp.
E —Crane, Japanese (Grus japonensis) 1 sp.
E —Crane, Siberian white (Grus leucogeranus) 1 sp.
E —Crane, white-naped (Grus vipio) 1 sp.
E —Crow, white-necked (Corvus leucognaphalus) 1 sp.
E —Cuckoo-shrike, Mauritius (Coquus (Coracina) typicus) 1 sp.
E —Cuckoo-shrike, Reunion (Coquus (Coracina) newtoni) 1 sp.
E —Curassow, razor-billed (Mitu (Crax) mitu mitu) 1 sp.
E —Curassow, red-billed (Crax blumenbachii) 1 sp.
E —Curassow, Trinidad white-headed (Pipile pipile pipile) 1 sp.
E —Dove, cloven-feathered (Drepanoptila holosericea) 1 sp.
E —Dove, Grenada gray-fronted (Leptotila rufaxilla wellsi) 1 sp.
E —Duck, pink-headed (Rhodonessa caryophyllacea) 1 sp.
E —Duck, white-winged wood (Cairina scutulata) 1 sp.
E —Eagle, Greenland white-tailed (Haliaeetus albicilla groenlandicus) 1 sp.

E —Eagle, harpy (Harpia harpyja) 1 sp.
E —Eagle, Madagascar sea (Haliaeetus vociferoides) 1 sp.
E —Eagle, Madagascar serpent (Eutriorchis astur) 1 sp.
E —Eagle, Philippine (monkey-eating) (Pithecophaga jefferyi) 1 sp.
E —Eagle, Spanish imperial (Aquila heliaca adalberti) 1 sp.
E —Egret, Chinese (Egretta eulophotes) 1 sp.
E —Falcon, Eurasian peregrine (Falco peregrinus peregrinus) 1 sp.
E —Flycatcher, Euler's (Empidonax euleri johnstonei) 1 sp.
E —Flycatcher, Seychelles paradise (Terpsiphone corvina) 1 sp.
E —Flycatcher, Tahiti (Pomarea nigra) 1 sp.
E —Fody, Mauritius (Foudia rubra) 1 sp.
E —Fody, Rodrigues (Foudia flavicans) 1 sp.
E —Fody, Seychelles (weaver-finch) (Foudia sechellarum) 1 sp.
E —Francolin, Djibouti (Francolinus ochropectus) 1 sp.
E —Frigatebird, Andrew's (Fregata andrewsi) 1 sp.
E —Goshawk, Christmas Island (Accipiter fasciatus natalis) 1 sp.
E —Grackle, slender-billed (Quisicalus (Cassidix) palustris) 1 sp.
E —Grasswren, Eyrean (flycatcher) (Amytornis goyderi) 1 sp.
E —Grebe, Alaotra (Tachybaptus rufoflavatus) 1 sp.
E —Grebe, Atitlan (Podilymbus gigas) 1 sp.
E —Greenshank, Nordmann's (Tringa guttifer) 1 sp.
E —Guan, horned (Oreophasis derbianus) 1 sp.
E —Guan, white-winged (Penelope albipennis) 1 sp.
T —Guineafowl, white-breasted (Agelastes meleagrides) 1 sp.
E —Gull, Audouin's (Larus audouinii) 1 sp.
E —Gull, relict (Larus relictus) 1 sp.
E —Hawk, Anjouan Island sparrow (Accipiter francesii pusillus) 1 sp.
E —Hawk, Galapagos (Buteo galapagoensis) 1 sp.
E —Hermit, hook-billed (hummingbird) (Glaucis (Ramphodon) dohrnii) 1 sp.
E —Honeyeater, helmeted (Meliphaga cassidix) 1 sp.
E —Hornbill, helmeted (Rhinoplax vigil) 1 sp.
E —Ibis, Japanese crested (Nipponia nippon) 1 sp.
E —Ibis, northern bald (Geronticus eremita) 1 sp.
E —Kagu (Rhynochetos jubatus) 1 sp.
E —Kakapo (owl-parrot) (Strigops habroptilus) 1 sp.
E —Kestrel, Mauritius (Falco punctatus) 1 sp.
E —Kestrel, Seychelles (Falco araea) 1 sp.
E —Kite, Cuba hook-billed (Chondrohierax uncinatus wilsonii) 1 sp.
E —Kite, Grenada hook-billed (Chondrohierax uncinatus mirus) 1 sp.
E —Kokako (wattlebird) (Callaeas cinerea) 1 sp.
E —Lark, Raso (Alauda razae) 1 sp.
E —Macaw, glaucous (Anodorhynchus glaucus) 1 sp.
E —Macaw, indigo (Anodorhynchus leari) 1 sp.
E —Macaw, little blue (Cyanopsitta spixii) 1 sp.
E —Magpie-robin, Seychelles (thrush) (Copsychus sechellarum) 1 sp.
E —Malimbe, Ibadan (Malimbus ibadanensis) 1 sp.
E —Malkoha, red-faced (cuckoo) (Phaenicophaeus pyrrhocephalus) 1 sp.
E —Megapode, Maleo (Macrocephalon maleo) 1 sp.
E —Nuthatch, Algerian (Sitta ledanti) 1 sp.
E —Ostrich, Arabian (Struthio camelus syriacus) 1 sp.
E —Ostrich, West African (Struthio camelus spatzi) 1 sp.
E —Owl, Anjouan scops (Otus rutilus capnodes) 1 sp.

E —Owl, giant scops (Otus gurneyi) 1 sp.
E —Owl, Madagascar red (Tyto soumagnei) 1 sp.
E —Owl, Seychelles (Otus insularis) 1 sp.
E —Owlet, Morden's (sokoke) (Otus ireneae) 1 sp.
E —Oystercatcher, Canarian black (Haematopus meadewaldoi) 1 sp.
E —Parakeet, Forbes' (Cyanoramphus auriceps forbesi) 1 sp.
E —Parakeet, golden-shouldered (hooded) (Psephotus chrysopterygius) 1 sp.
E —Parakeet, golden (Aratinga guarouba) 1 sp.
E —Parakeet, Mauritius (Psittacula echo) 1 sp.
E —Parakeet, Norfolk Island (Cyanoramphus novaezelandiae cookii) 1 sp.
E —Parakeet, ochre-marked (Pyrrhura cruentata) 1 sp.
E —Parakeet, orange-bellied (Neophema chrysogaster) 1 sp.
E —Parakeet, paradise (beautiful) (Psephotus pulcherrimus) 1 sp.
E —Parakeet, scarlet-chested (splendid) (Neophema splendida) 1 sp.
E —Parakeet, turquoise (Neophema pulchella) 1 sp.
E —Parrot, Australian (Geopsittacus occidentalis) 1 sp.
E —Parrot, Bahaman or Cuban (Amazona leucocephala) 1 sp.
E —Parrot, ground (Pezoporus wallicus) 1 sp.
E —Parrot, imperial (Amazona imperialis) 1 sp.
E —Parrot, red-browed (Amazona rhodocorytha) 1 sp.
E —Parrot, red-capped (Pionopsitta pileata) 1 sp.
E —Parrot, red-necked (Amazona arausiaca) 1 sp.
E —Parrot, red-spectacled (Amazona pretrei pretrei) 1 sp.
E —Parrot, red-tailed (Amazona brasiliensis) 1 sp.
E —Parrot, Seychelles lesser vasa (Coracopsis nigra barklyi) 1 sp.
E —Parrot, St. Lucia (Amazona versicolor) 1 sp.
E —Parrot, St. Vincent (Amazona guildingii) 1 sp.
E —Parrot, thick-billed (Rhynchopsitta pachyrhyncha) 1 sp.
E —Parrot, vinaceous-breasted (Amazona vinacea) 1 sp.
E —Penguin, Galapagos (Sphcniscus mendiculus) 1 sp.
E —Petrel, Madeira (freira) (Pterodroma madeira) 1 sp.
E —Petrel, Mascarene black (Pterodroma aterrima) 1 sp.
E —Pheasant, bar-tailed (Syrmaticus humaie) 1 sp.
E —Pheasant, Blyth's tragopan (Tragopan blythii) 1 sp.
E —Pheasant, brown eared (Crossoptilon mantchuricum) 1 sp.
E —Pheasant, Cabot's tragopan (Tragopan caboti) 1 sp.
E —Pheasant, cheer (Catreus wallichii) 1 sp.
E —Pheasant, Chinese monal (Lophophorus lhuysii) 1 sp.
E —Pheasant, Edward's (Lophura edwardsi) 1 sp.
E —Pheasant, Elliot's (Syrmaticus ellioti) 1 sp.
E —Pheasant, imperial (Lophura imperialis) 1 sp.
E —Pheasant, Mikado (Syrmaticus mikado) 1 sp.
E —Pheasant, Palawan peacock (Polyplectron emphanum) 1 sp.
E —Pheasant, Sclater's monal (Lophophorus sclateri) 1 sp.
E —Pheasant, Swinhoe's (Lophura swinhoii) 1 sp.
E —Pheasant, western tragopan (Tragopan melanocephalus) 1 sp.
E —Pheasant, white eared (Crossoptilon crossoptilon) 1 sp.
E —Pigeon, Azores wood (Columba palumbus azorica) 1 sp.
E —Pigeon, Chatham Island (Hemiphaga novaeseelandiae chathamensis) 1 sp.
E —Pigeon, Mindoro zone-tailed (Ducula mindorensis) 1 sp.
E —Pigeon, pink (Columba (Nesoenas) mayeri) 1 sp.
T —Pigeon, white-tailed laurel (Columba junoniae) 1 sp.

E —Piping-guan, black-fronted (Pipile jacutinga) 1 sp.
E —Pitta, Koch's (Pitta kochi) 1 sp.
E —Plover, New Zealand shore (Thinornis novaeseelandiae) 1 sp.
E —Pochard, Madagascar (Aythya innotata) 1 sp.
E —Quail, Merriam's Montezuma (Cyrtonyx montezumae merriami) 1 sp.
E —Quetzel, resplendent (Pharomachrus mocinno) 1 sp.
E —Rail, Aukland Island (Rallus pectoralis muelleri) 1 sp.
E —Rail, Lord Howe wood (Tricholimnas sylvestris) 1 sp.
E —Rhea, Darwin's (Pterocnemia pennata) 1 sp.
E —Robin, Chatham Island (Petroica traversi) 1 sp.
T —Robin, dappled mountain (Arcanator (Modulatrix) orostruthus) 1 sp.
E —Robin, scarlet-breasted (flycatcher) (Petroica multicolor multicolor) 1 sp.
E —Rockfowl, grey-necked (Picathartes oreas) 1 sp.
E —Rockfowl, white-necked (Picathartes gymnocephalus) 1 sp.
E —Roller, long-tailed ground (Uratelornis chimaera) 1 sp.
E —Scrub-bird, noisy (Atrichornis clamosus) 1 sp.
E —Shama, Cebu black (thrush) (Copsychus niger cebuensis) 1 sp.
E —Siskin, red (Carduelis (Spinus) cucullata) 1 sp.
E —Starling, Ponape mountain (Aplonis pelzelni) 1 sp.
E —Starling, Rothschild's (myna) (Leucopsar rothschildi) 1 sp.
E —Stork, oriental white (Ciconia ciconia boyciana) 1 sp.
E —Sunbird, Marungu (Nectarinia prigoginei) 1 sp.
E —Teal, Campbell Island flightless (Anas aucklandica nesiotis) 1 sp.
E —Thrasher, white-breasted (Ramphocinclus brachyurus) 1 sp.
E —Thrush, New Zealand (wattlebird) (Turnagra capensis) 1 sp.
E —Thrush, Taita (Turdus olivaceous helleri) 1 sp.
E —Tinamou, solitary (Tinamus solitarius) 1 sp.
E —Trembler, Martinique (thrasher) (Cinclocerthia ruficauda gutturalis) 1 sp.
E —Turaco, Bannerman's (Tauraco bannermani) 1 sp.
E —Turtle-dove, Seychelles (Streptopelia picturata rostrata) 1 sp.
T —Vanga, Pollen's (Xenopirostris polleni) 1 sp.
T —Vanga, Van Dam's (Xenopirostris damii) 1 sp.
E —Wanderer, plain (collared-hemipode) (Pedionomous torquatus) 1 sp.
E —Warbler (Old World), Aldabra (Nesillas aldabranus) 1 sp.
E —Warbler (Old World), Rodrigues (Bebrornis rodericanus) 1 sp.
E —Warbler (Old World), Seychelles (Bebrornis sechellensis) 1 sp.
E —Warbler (wood), Barbados yellow (Dendroica petechia petechia) 1 sp.
E —Warbler (wood), Semper's (Leucopeza semperi) 1 sp.
E —Wattle-eye, banded (Platysteira laticincta) 1 sp.
E —Weaver, Clarke's (Ploceus golandi) 1 sp.
E —Whipbird, western (Psophodes nigrogularis) 1 sp.
E —White-eye, Norfolk Island (Zosterops albogularis) 1 sp.
E —White-eye, Ponape greater (Rukia longirostra (sanfordi)) 1 sp.
E —White-eye, Seychelles (Zosterops modesta) 1 sp.
E —Woodpecker, imperial (Campephilus imperialis) 1 sp.
E —Woodpecker, Tristam's (Dryocopus javensis richardsi) 1 sp.
E —Wren, Guadeloupe house (Troglodytes aedon guadeloupensis) 1 sp.
E —Wren, St. Lucia house (Troglodytes aedon mesoleucus) 1 sp.

Reptiles
E —Alligator, Chinese (Alligator sinensis) 1 sp.
E —Boa, Jamaican (Epicrates subflavus) 1 sp.

E —Boa, Round Island (no common name) (Bolyeria multocarinata) 1 sp.
E —Boa, Round Island (no common name) (Casarea dussumieri) 1 sp.
E —Caiman, Apaporis River (Caiman crocodilus apaporiensis) 1 sp.
E —Caiman, black (Melanosuchus niger) 1 sp.
E —Caiman, broad-snouted (Caiman latirostris) 1 sp.
E —Caiman, Yacare (Caiman crocodilus yacare) 1 sp.
E —Chuckwalla, San Esteban Island (Sauromalus varius) 1 sp.
E —Crocodile, African dwarf (Osteolaemus tetraspis tetraspis) 1 sp.
E —Crocodile, African slender-snouted (Crocodylus cataphractus) 1 sp.
E —Crocodile, Ceylon mugger (Crocodylus palustris kimbula) 1 sp.
E —Crocodile, Congo dwarf (Osteolaemus tetraspis osborni) 1 sp.
E —Crocodile, Cuban (Crocodylus rhombifer) 1 sp.
E —Crocodile, Morelet's (Crocodylus moreletii) 1 sp.
E —Crocodile, mugger (Crocodylus palustris palustris) 1 sp.
T —Crocodile, Nile (Crocodylus niloticus) 1 sp.
E —Crocodile, Orinoco (Crocodylus intermedius) 1 sp.
E —Crocodile, Philippine (Crocodylus novaeguineae mindorensis) 1 sp.
E —Crocodile, saltwater (estuarine) (Crocodylus porosus) 1 sp.
E —Crocodile, Siamese (Crocodylus siamensis) 1 sp.
E —Gavial (gharial) (Gavialis gangeticus) 1 sp.
E —Gecko, day (Phelsuma edwardnewtoni) 1 sp.
E —Gecko, Round Island day (Phelsuma guentheri) 1 sp.
T —Gecko, Serpent Island (Cyrtodactylus serpensinsula) 1 sp.
T —Iguana, Acklins ground (Cyclura rileyi nuchalis) 1 sp.
T —Iguana, Allen's Cay (Cyclura cychlura inornata) 1 sp.
T —Iguana, Andros Island ground (Cyclura cychlura cychlura) 1 sp.
E —Iguana, Anegada ground (Cyclura pinguis) 1 sp.
E —Iguana, Barrington land (Conolophus pallidus) 1 sp.
T —Iguana, Cayman Brac ground (Cyclura nubila caymanensis) 1 sp.
T —Iguana, Cuban ground (Cyclura nubila nubila) 1 sp.
T —Iguana, Exuma Island (Cyclura cychlura figginsi) 1 sp.
E —Iguana, Fiji banded (Brachylophus fasciatus) 1 sp.
E —Iguana, Fiji crested (Brachylophus vitiensis) 1 sp.
E —Iguana, Grand Cayman ground (Cyclura nubila lewisi) 1 sp.
E —Iguana, Jamaican (Cyclura collei) 1 sp.
T —Iguana, Mayaguana (Cyclura carinata bartschi) 1 sp.
T —Iguana, Turks and Caicos (Cyclura carinata carinata) 1 sp.
E —Iguana, Watling Island ground (Cyclura rileyi rileyi) 1 sp.
T —Iguana, White Cay ground (Cyclura rileyi cristata) 1 sp.
E —Lizard, Hierro giant (Gallotia simonyi simonyi) 1 sp.
T —Lizard, Ibiza wall (Podarcis pityusensis) 1 sp.
E —Lizard, Maria Island ground (Cnemidophorus vanzoi) 1 sp.
E —Monitor, Bengal (Varanus bengalensis) 1 sp.
E —Monitor, desert (Varanus griseus) 1 sp.
E —Monitor, Komodo Island (Varanus komodoensis) 1 sp.
E —Monitor, yellow (Varanus flavescens) 1 sp.
E —Python, Indian (Python molurus molurus) 1 sp.
T —Rattlesnake, Aruba Island (Crotalus unicolor) 1 sp.
T —Skink, Round Island (Leiolopisma telfairi) 1 sp.
E —Snake, Maria Island (Liophus ornatus) 1 sp.
E —Tartaruga (Podocnemis expansa) 1 sp.
E —Terrapin, river (tuntong) (Batagur baska) 1 sp.

E —Tomistoma (Tomistoma schlegelii) 1 sp.
E —Tortoise, angulated (Geochelone yniphora) 1 sp.
E —Tortoise, Bolson (Gopherus flavomarginatus) 1 sp.
E —Tortoise, Galapagos (Geochelone elephantopus) 1 sp.
E —Tortoise, radiated (Geochelone (Testudo) radiata) 1 sp.
E —Tracaja (Podocnemis unifilis) 1 sp.
E —Tuatara (Sphenodon punctatus) 1 sp.
E —Turtle, aquatic box (Terrapene coahuila) 1 sp.
E —Turtle, black softshell (Trionyx nigricans) 1 sp.
E —Turtle, Brazilian (Hoge's) sideneck (Phrynops hogei) 1 sp.
E —Turtle, Burmese peacock (Morenia ocellata) 1 sp.
E —Turtle, Cat Island (Trachemys terrapen) 1 sp.
E —Turtle, Central American river (Dermatemys mawii) 1 sp.
E —Turtle, Cuatro Cienegas softshell (Trionyx ater) 1 sp.
E —Turtle, geometric (Psammobates (Geochelone) geometricus) 1 sp.
E —Turtle, Inagua Island (Trachemys stejnegeri malonei) 1 sp.
E —Turtle, Indian sawback (Kachuga tecta tecta) 1 sp.
E —Turtle, Indian softshell (Trionyx gangeticus) 1 sp.
E —Turtle, olive (Pacific) ridley sea (Lepidochelys olivacea) 1 sp.
E —Turtle, peacock softshell (Trionyx hurum) 1 sp.
E —Turtle, short-necked or western swamp (Pseudemydura umbrina) 1 sp.
E —Turtle, South American red-lined (Trachemys scripta callirostris) 1 sp.
E —Turtle, spotted pond (Geoclemys (Damonia) hamiltonii) 1 sp.
E —Turtle, three-keeled Asian (Melanochelys (Geoemyda, Nicoria) tricarinata) 1 sp.
E —Viper, Lar Valley (Vipera latifii) 1 sp.

Amphibians

T —Frog, Goliath (Conraua goliath) 1 sp.
E —Frog, Israel painted (Discoglossus nigriventer) 1 sp.
E —Frog, Panamanian golden (Atelopus varius zeteki) 1 sp.
E —Frog, Stephen Island (Leiopelma hamiltoni) 1 sp.
E —Salamander, Chinese giant (Andrias davidianus davidianus) 1 sp.
E —Salamander, Japanese giant (Andrias davidianus japonicus) 1 sp.
E —Toad, Cameroon (Bufo superciliaris) 1 sp.
E —Toad, Monte Verde (Bufo periglenes) 1 sp.
E —Toads, African viviparous (Nectophrynoides spp.) —2 spp.

Fishes

E —Ala Balik (trout) (Salmo platycephalus) 1 sp.
E —Ayumodoki (loach) (Hymenophysa (Botia) curta) 1 sp.
E —Blindcat, Mexican (catfish) (Prietella phreatophila) 1 sp.
E —Bonytongue, Asian (Scleropages formosus) 1 sp.
E —Catfish (no common name) (Pangasius sanitwongsei) 1 sp.
E —Catfish, giant (Pangasianodon gigas) 1 sp.
E —Cicek (minnow) (Acanthorutilus handlirschi) 1 sp.
E —Nekogigi (catfish) (Coreobagrus ichikawai) 1 sp.
E —Tango, Miyako (Tokyo bitterling) (Tanakia tango) 1 sp.
E —Temoleh, Ikan (minnow) (Probarbus jullieni) 1 sp.
E —Totoaba (seatrout or weakfish) (Cynoscion macdonaldi) 1 sp.

INVERTEBRATES

Clams
E —Pearlymussel, Nicklin's (Megalonaias nicklineana) 1 sp.
E —Pearlymussel, Tampico (Cyrtonaias tampicoensis tecomatensis) 1 sp.

Insects
E —Butterfly, Corsican swallowtail (Papilio hospiton) 1 sp.
E —Butterfly, Homerus swallowtail (Papilio homerus) 1 sp.
E —Butterfly, Luzon peacock swallowtail (Papilio chikae) 1 sp.
E —Butterfly, Queen Alexandra's birdwing (Troides (Ornithoptera) alexandrae) 1 sp.

Flowering Plants

E —Costa Rican jatropha (Jatropha costaricensis) 1 sp.

NONFLOWERING PLANTS

Conifers & Cycads
T —Pinabete or Guatemalan fir (Abies guatemalensis) 1 sp.
T —Alerce or Chilean false larch (Fitzroya cupressoides) 1 sp.

T ....Threatened
E ....Endangered
XN ....Extinct

References

Botkin, D. and E. Keller. *Environmental Science—Earth as a Living Planet.* New York: John Wiley & Sons, 1995.

Gill, F.B. *Ornithology.* New York: W. H. Freeman & Co., 1989.

Lehninger, A.L. *Biochemistry,* 2nd ed. New York: Worth Publishers, Inc., 1975.

Lehninger, A.L. and M.M. Cox. *Principles of Biochemistry,* 3rd ed. New York: Worth Publishers, 1993.

Margulis, L. and K.V. Schwartz. *Five Kingdoms. An Illustrated Guide to the Phyla of Life on Earth,* 2nd ed. New York: W.H. Freeman, 1988; 3rd ed. New York: W.H. Freeman, 1997.

Raven, P.H. and G.B. Johnson. *Biology,* 3rd ed. Missouri: Mosby-Year Book, 1992.

Stryer, L. *Biochemistry,* 4th ed. New York: W.H. Freeman, 1995.

Reader Input Request

The utmost care has been taken to ensure the accuracy of the definitions in this dictionary. However, in a volume this large, particularly a first edition, some errors will inevitably have been missed. In addition, the field of biology is expanding and changing at a phenomenal rate. By the time this book is on the shelves, some information will certainly have been superceded by more recent data.

If you believe you have found an error or out-of-date fact, I would be indebted if you would contact me so that the correct information can be included in the next edition.

Norah Rudin, Ph.D.

11780 San Pablo Ave.
Ste. 4C, Box 225
El Cerrito, CA 94530-1750

nrbiocom@uclink4.berkeley.edu

♦ Illustration Cross Reference Index ♦

FOR THESE ENTRIES	SEE ILLUSTRATION AT THESE ENTRIES
A	
acrosomal vesicle	fertilization
	spermatozoon
actin	muscle contraction
actomyosin	muscle contraction
adenosine, adenine	nucleotide
aerobic respiration	mitochondrion
agglutination	antigen-antibody interaction
alpha-actinin	sarcomere
amplification	PCR
ampulla	ear
anaphase	animal cells
	mitosis
anatropous	ovule
animal pole	egg
anther	flower
antibody	antigen-antibody interaction
anticodon	tRNA
antigen	antigen-antibody interaction
antigen-binding site	antibody
antipodal cells	embryo sac
antisense DNA	transcription
apical meristem	Dicotyledons
axillary bud	Dicotyledons
axon	neuron
B	
B-cell	cell-mediated immune response
	humoral immune response
base	nucleotide
base pair	DNA duplex
	DNA
biotic potential	growth curve
blade	Dicotyledons
blastula	differentiation
bloom	nitrogen cycle
blunt end	restriction enzyme
C	
campylotropous	ovule
capitulum	inflorescence
capping	splicing
carpel	flower
carrying capacity	growth curve
cell plate	mitosis
cell wall	cell
central nervous system	nervous system
cephalopod eye	eye

FOR THESE ENTRIES	SEE ILLUSTRATION AT THESE ENTRIES
chalaza	egg ovule
chromatography	gel filtration
cilium	axoneme
cochlea	ear
codon	tRNA
companion cell	mass flow
complement	humoral immune response non-specific immune response
complementary base pairing	DNA duplex
corymb	inflorescence
crossing-over	chromosome
cupula	ear
cuticle	leaf
cytidine, cytosine	nucleotide
cytokinesis	animal cells mitosis
cytotoxic T-cell	cell-mediated immune response
D	
deletion	chromosome
dendrite	neuron
denitrification	nitrogen cycle
deoxyribose	pentose
desmosome	intercellular junction
dichasial chyme	inflorescence
DNA	DNA duplex pentose phosphodiester bond semiconservative replication
DNA analysis	length polymorphisms PCR RFLP Southern blot
DNA polymerase	PCR
double helix	DNA
double-stranded DNA	DNA duplex
duplication	chromosome
E	
ear ossicles	ear
ectoderm	differentiation
electron transport chain	chloroplast mitochondrion
electrophoresis	length polymorphisms RFLP
embryo sac	ovule
endoderm	differentiation
endoplasmic reticulum	cell
epidermis	leaf

FOR THESE ENTRIES	SEE ILLUSTRATION AT THESE ENTRIES
epigynous	receptacle
equatorial division	animal cells
excision	plasmid
exon	splicing
	nuclease protection assay
F	
flagellum	axoneme
flower	receptacle
	inflorescence
food chain	food web
fossil fuels	carbon cycle
funicle	ovule
G	
gametophyte	life cycle
gap junction	intercellular junction
gastrula	differentiation
gene cloning	chromosome jumping
	chromosome walking
	homopolymeric tailing
	restriction enzyme
genetic engineering	chromosome jumping
	chromosome walking
	gene cloning
	homopolymeric tailing
	restriction enzyme
	site-directed mutagenesis
genetic map	chromosome jumping
	chromosome walking
genetic typing	length polymorphisms
	PCR
	RFLP
	Southern blot
glyoxisome	cell
Golgi complex	cell
guanosine, guanine	nucleotide
guard cell	leaf
H	
heavy chain	antibody
helix-loop-helix	DNA binding proteins
helix-turn-helix	DNA binding proteins
helper T-cell	cell-mediated immune response
	humoral immune response
heteroduplex	gene conversion
histamine	non-specific immune response
histone	chromosome
hnRNA	nuclease protection assay
homologous chromosomes	crossing over
homologous recombination	plasmid
	crossing over

FOR THESE ENTRIES	SEE ILLUSTRATION AT THESE ENTRIES
hydrogen bond	DNA duplex
hypervariable region	antibody
I	
immunoglobulin	antibody
induction	lytic
integration	plasmid
integuments	ovule
interferon	non-specific immune response
interleukin	non-specific immune response
interleukin-1	cell-mediated immune response
interleukin-2	cell-mediated immune response
internode	Dicotyledons
intrachromosomal	chromosome
interchromosomal	chromosome
intron	splicing
inversion	chromosome
inverted repeat	transposon
ion channel	sodium-potassium pump
L	
lateral bud	Dicotyledons
leaf	Dicotyledons
leguminous nodules	nitrogen cycle
length polymorphism	RFLP
leucine zipper	DNA binding proteins
ligation	gene cloning
	homopolymeric tailing
light chain	antibody
lysogenic	lytic
lysosome	cell
M	
macrophage	cell-mediated immune response
	non-specific immune response
macula	ear
megagametophyte	life cycle
	embryo sac
megasporangia	life cycle
megaspore	flower
	life cycle
meiosis	animal cells
membrane potential	action potential
mesoderm	differentiation
metaphase	animal cells
	mitosis
MHC proteins	cell-mediated immune response
microgametophyte	life cycle
micropyle	ovule

FOR THESE ENTRIES	SEE ILLUSTRATION AT THESE ENTRIES
microsporangia	life cycle
	flower
microspore	life cyclo
mitochondrion	cell
mitosis	cell cycle
mitotic cross-over	chromosome
monocyte	non-specific immune response
mother cell	flower
motor neuron	nervous system
mRNA	nuclease protection assay
muscle contraction	sarcomere
myelin sheath	neuron
myofilament	muscle contraction
myosin	muscle contraction
N	
nerve impulse	action potential
	sodium-potassium pump
	synaptic cleft
neural crest	differentiation
neuron	synaptic cleft
neurotransmitter	synaptic cleft
nitrification	nitrogen cycle
nitrogen-fixation	nitrogen cycle
node	Dicotyledons
node of Ranvier	neuron
notochord	differentiation
nucellus	flower
	ovule
nuclease	nuclease protection assay
nucleic acid	pentose
	phosphodiester bond
nucleolus	cell
nucleoside	nucleotide
nucleosome	chromosome
nucleus	cell
O	
ommatidium	compound eye
oogenesis	egg
oogonia	life cycle
organ of Corti	ear
orthotropous	ovule
ovaries	life cycle
oxidative phosphorylation	mitichondrion
P	
parasympathetic nervous system	nervous system
parenchyma	leaf

FOR THESE ENTRIES	SEE ILLUSTRATION AT THESE ENTRIES
pedicel	flower
perigynous	receptacle
peripheral nervous system	nervous system
peroxisome	cell
petal	flower
phloem	leaf
	mass flow
photophosphorylation	chloroplast
photosystem I	chloroplast
	Z scheme
photosystem II	chloroplast
	Z scheme
pith	Dicotyledons
plasma cell	humoral immune response
plasma membrane	cell
plasmid	gene cloning
polar nuclei	embryo sac
pollen sac	flower
poly(A) tail	splicing
polyadenylation	splicing
polymorphism	length polymorphisms
polypeptide	phosphodiester bond
postsynaptic membrane	synaptic cleft
presynaptic membrane	synaptic cleft
primary growth zone	Dicotyledons
primary mRNA transcript	splicing
primary oocyte	oogenesis
primary RNA transcript	nuclease protection assay
primary spermatocyte	spermatogenesis
prophage	lytic
prophase	animal cells
	mitosis
prostoglandin	non-specific immune response
protein synthesis	gene expression
	tRNA
protofilament	microtubule
pyramid of biomass	ecological pyramids
pyramid of energy	ecological pyramids
pyramid of numbers	ecological pyramids
pyrogen	non-specific immune response
R	
raceme	inflorescence
receptacle	flower
reciprocal translocation	chromosome
recombinant	gene cloning
	homopolymeric tailing
recombination	chromosome
reduction division	animal cells

FOR THESE ENTRIES	SEE ILLUSTRATION AT THESE ENTRIES
replication	semiconservative replication
resting potential	action potential
restriction endonuclease	gene cloning
	homopolymeric tailing
	Southern blot
restriction enzyme	palindrome
RFLP	length polymorphisms
	Southern blot
rhabdom	compound eye
ribose	pentose
ribosomes	cell
RNA	pentose
	phosphodiester bond
RNA polymerase	transcription
root	Dicotyledons
runoff	water cycle
S	
saltatory conduction	neuron
Schwann cell	neuron
secondary oocyte	oogenesis
semicircular canals	ear
sensory neuron	nervous system
sepal	flower
shot	Dicotyledons
sink	mass flow
solenoid	chromosome
source	mass flow
Southern blot	length polymorphisms
	RFLP
spermatid	spermatogenesis
spermatogenesis	spermatozooan
spermatogonia	spermatogenesis
	life cycle
spermatozooan	spermatogenesis
spike	inflorescence
splicing	nuclease protection assay
spore mother cells	life cycle
sporophyte	life cycle
stamen	flower
statocyst	ear
sticky end	restriction enzyme
stigma	flower
stoma	leaf
striated muscle	muscle contraction
stroma	chloroplast
style	flower
sugar-phosphate backbone	DNA duplex
	DNA
	phosphodiester bond

FOR THESE ENTRIES	SEE ILLUSTRATION AT THESE ENTRIES
suppressor T-cell	cell-mediated immune response
sympathetic nervous system	nervous system
synergid cells	embryo sac
T	
Taq polymerase	PCR
telophase	animal cells
	mitosis
terminal bud	Dicotyledons
testes	life cycle
thick filament	muscle contraction
	sarcomere
thin filament	muscle contraction
	sarcomere
thylakoid disc	chloroplast
thymidine, thymine	nucleotide
tight junction	intercellular junction
transcription	gene expression
translation	gene expression
	tRNA
transpiration	water cycle
tropomyosin	muscle contraction
troponin	muscle contraction
tubulin	microtubule
U	
umbel	inflorescence
undulipodium	axoneme
unequal cross-over	chromosome
uridine, uracil	nucleotide
V	
vacuole	cell
variable region	antibody
vascular cambium	Dicotyledons
vegetal pole	egg
vein	Dicotyledons
vertebrate eye	eye
vitelline membrane	fertilization
voluntary nervous system	nervous system
W	
water table	water cycle
X	
xylem	leaf
	mass flow
Z	
Z-line	sarcomere

FOR THESE ENTRIES	SEE ILLUSTRATION AT THESE ENTRIES
Z-scheme	chloroplast
zinc finger	DNA binding proteins
zygote	differentiation
	oogenesis

FOR THESE ENTRIES	SEE THESE APPENDICES
acid	pH scale
amino acid	genetic code
	amino acids
animals	five kingdoms
archaen	geologic time
autotrophic	modes of nutrition
base	pH scale
binomial nomenclature	classification
biochemical taxonomy	classification
buffer	pH scale
Cambrian	Phanerozoic time
carboniferous	Phanerozoic time
Cenozoic	geologic time
	Phanerozoic time
chemoautotrophism	modes of nutrition
chemoheterotrophism	modes of nutrition
chemosynthesis	modes of nutrition
chemotrophism	modes of nutrition
cladistics	classification
class	taxonomic hierarchy
Cretaceous	Phanerozoic time
Devonian	Phanerozoic time
division	taxonomic hierarchy
essential amino acids	amino acids
evolutionary taxonomy	classification
family	taxonomic hierarchy
fungi	five kingdoms
gene expression	genetic code
genetic code	genetic code
	amino acids
genus	taxonomic hierarchy
geologic time	geologic time
heterotrophic	modes of nutrition
indicator	pH scale
Jurassic	Phanerozoic time
kingdom	taxonomic hierarchy
lithotrophic	modes of nutrition
Mesozoic	geologic time
	Phanerozoic time
Miocene	Phanerozoic time
numerical taxonomy	classification
Oligocene	Phanerozoic time

FOR THESE ENTRIES	SEE THESE APPENDICES
order	taxonomic hierarchy
Ordovician	Phanerozoic time
organotrophic	modes of nutrition
Paleocene	Phanerozoic time
paleozoic	geologic time
	Phanerozoic time
Permian	Phanerozoic time
pH	pH scale
Phanerozoic	geologic time
	Phanerozoic time
phenetics	classification
photoautotrophism	modes of nutrition
photoheterotrophism	modes of nutrition
photosynthesis	modes of nutrition
phototrophism	modes of nutrition
phylogenetics	classification
phylum	taxonomic hierarchy
plants	five kingdoms
Precambrian	geologic time
prokaryotes	five kingdoms
protein synthesis	genetic code
proterozoic	geologic time
protists	five kingdoms
protoctist	five kingdoms
R-group	amino acids
rare species	endangered species
Silurian	Phanerozoic time
species	taxonomic hierarchy
taxonomy	taxonomic hierarchy
	classification
Tertiary	Phanerozoic time
threatened species	endangered species
translation	genetic code
Triassic	Phanerozoic time